Molecular Genetic Approaches in Conservation

Edited by

THOMAS B. SMITH

San Francisco State University
Department of Biology

and

ROBERT K. WAYNE

University of California, Los Angeles
Department of Biology

New York   Oxford
OXFORD UNIVERSITY PRESS
1996

Oxford University Press

Oxford   New York   Athens
Auckland   Bangkok   Bogota   Bombay
Buenos Aires   Calcutta   Cape Town   Dar es Salaam
Delhi   Florence   Hong Kong   Istanbul   Karachi
Kuala Lumpur   Madras   Madrid   Melbourne
Mexico City   Nairobi   Paris   Singapore
Taipei   Tokyo   Toronto

and associated companies in
Berlin   Ibadan

Library of Congress Cataloging-in-Publication Data
Molecular genetic approaches in conservation / edited by Thomas B. Smith.
Robert K. Wayne
    p.   cm.
Includes bibliographical references.
ISBN 0-19-509526-X
1. Conservation biology.   2. Molecular genetics.   3. Biological diversity
conservation.   I. Smith, Thomas B. (Thomas Bates)   II. Wayne, Robert K.
QH75.M65   1996
333.95'16'0157487328—dc20   96-24804

9 8 7 6 5 4 3 2 1

Printed in the United States of America
on acid-free paper

Preface

The role of molecular genetic approaches in conservation biology has expanded dramatically in the last decade. In addition to the use of these techniques in measuring genetic variation of dwindling populations, several other important uses have recently been recognized. The first relates to the realization that many threatened or endangered species are composed of populations whose persistence may depend on the maintenance of complex migration patterns. To maintain such "metapopulation structures," it is essential that patterns of genetic exchange among populations be understood and preserved. The second comes from the observation that species are often composed of populations that are genetically distinct and may require special recognition and separate conservation actions. In particular, the use of genetic techniques to define the "genetic units" for conservation has been one of the most significant applications of molecular techniques to managing threatened species. Finally, the genetic landscape of many species has been altered by genetic augmentation, such as by the intentional addition of hatchery-raised fish or by the invasion of non-native species that hybridize with native forms. Assessing the effects of such introductions on the genetic structure of native populations is difficult if not impossible without utilizing molecular genetic approaches.

The primary purpose of this book is to bring together in one volume the various molecular genetic approaches that may be useful in making informed conservation decisions. By focusing on the uses and relative cost-effectiveness of a wide range of molecular techniques, we hope to provide a kind of "tool box" for conservation biologists charged with implementing conservation research and management. Throughout the book, the descriptions and discussions are intended for researchers and managers who are interested in understanding or using molecular genetic methods but who may not have had formal training in molecular biology. Specifically, the volume is aimed toward academic researchers, and upper division college and graduate students in ecology, conservation, zoology, evolution, and botany, as well as managers in governmental and private organizations.

The first chapter entitled "An Overview of the Issues," summarizes the approaches and utility of applying molecular genetic approaches to conservation questions. The purpose of this section is to convince the reader that molecular genetic data is not only useful but may be of critical importance in making management decisions, and is best utilized when integrated with other information on ecology and demography.

The first part and main body of the book, entitled "Approaches," examines the uses of a wide range of molecular genetic techniques in conservation. Each chapter has a similar organization and begins with a brief description of the technique, its importance for conservation, a detailed technical description including estimates of relative costs, and finally case studies utilizing the technique. The authors of these chapters represent some of the pioneers who have applied molecular techniques to conservation issues. The last chapter in this section addresses future applications of molecular biology to conservation.

The second part, entitled "Analysis," critically describes approaches to analyzing molecular genetic data and identifies important caveats when making inferences from molecular genetic data.

Part III, entitled "Case Studies," consists of five chapters exploring the use of molecular data to answer specific conservation questions. In particular, this section highlights the importance of integrating results from several molecular genetic approaches to conservation programs.

The final part, entitled "Perspective," attempts to place molecular genetic techniques in the larger context of conservation genetics and to summarize some of the major themes of the volume.

The book was developed from a symposium held during the 75th Annual Meeting of the Pacific Division of the American Association for the Advancement of Science, held at San Francisco State University in June 1994. We especially thank Alan Leviton, Robert Bowman, and other members of the Pacific Division, AAAS organizing committee, for their tremendous assistance. Not all symposium papers are represented in this book, and some new ones have been added. We thank the many external reviewers who kindly made comments, criticisms, and suggestions about each manuscript, and the authors for utilizing most of the external reviewers' comments.

We are very grateful to the many institutions and individuals that contributed funds in support to the symposium and the book. These include: Thomas White, the San Francisco State University Foundation, Inc.; the University of California Systemwide Biotechnology, Research and Education program, Hoffmann-La Roche; Perkin-Elmer, Applied Biosystems Division; Genetic Resources Conservation Program, University of California; the Northern California Chapter, Society for Conservation Biology; and Department of Biology and the College of Science, San Francisco State University. We thank numerous individuals who were essential in making the symposium and subsequent book possible, including: Thomas White, Judah Rosenwald, Paul Fonteyn, John Hafernik, James Kelley, Mark Collins, Mike Vasey, Candida Kutz, Arlene Essex, Pat Mcquire, Karen Raby, Cristian Orrego, Frank Bayliss, Carole Conway, Staci Markos, Borja Milá, Kim O'Keefe, Derek Girman, Pamela Owings, Monique Fountain, Nancy Jo, Garen Grigorian, Hilde Spautz, Barbara Smith, Adrianna Smyth, Norm Gershenz, Vicki Case, Peggy Wade, Lisa Wayne, Lena Hileman, and Isabelle de Geofroy. Special thanks go to Klaus Koefli for assistance during the editorial process and for the challenging job of indexing the book.

San Francisco   T.B.S.
Los Angeles   R.K.W.

Contents

Contributors

Leila Agullana
Department of Genetics and Molecular Biology
John A. Burns School of Medicine
University of Hawaii at Manoa
Honolulu, Hawaii

Peter Arctander
Department of Population Biology
University of Copenhagen
Copenhagen, Denmark

C. Scott Baker
School of Biological Sciences
University of Auckland
Auckland, New Zealand

Eldredge Bermingham
Smithsonian Tropical Research Institute
Balboa, Republic of Panama

Michael W. Bruford
Conservation Genetics Group
Institute of Zoology
London, U.K.

Terry Burke
Department of Zoology
University of Leicester
Leicester, U.K.

Rebecca L. Cann
Department of Genetics and Molecular Biology
John A. Burns School of Medicine
University of Hawaii at Manoa
Honolulu, Hawaii

Diedre Carter
Roche Molecular Systems, Inc.
Alameda, California

David J. Cheesman
Conservation Genetics Group
Institute of Zoology
London, U.K.

Trevor Coote
Conservation Genetics Group
Institute of Zoology
London, U.K.

Scott V. Edwards
Department of Zoology and Burke Museum
University of Washington,
Seattle, Washington

Robert A. Feldman
Department of Genetics and Molecular Biology
John A. Burns School of Medicine
University of Hawaii at Manoa
Honolulu, Hawaii

Nicola Fildes
Roche Molecular Systems, Inc.
Alameda, California

Javier Francisco-Ortega
Department of Botany
University of Texas
Austin, Texas

Leonard A. Freed
Department of Zoology
University of Hawaii at Manoa
Honolulu, Hawaii

Peter Fritsch
Department of Botany
California Academy of Sciences
San Francisco, California

Ludovic Gielly
Laboratoire de Biologie des Populations
d'Altitude
Université Joseph Fourier
Grenoble, France

Nicholas Georgiadis
Department of Biology
Washington University
St. Louis, Missouri

Derek J. Girman
Department of Biology
University of California
Los Angeles, California

Harriet A.A. Green
Conservation Genetics Group
Institute of Zoology
London, U.K.

Susan A. Haines
Conservation Genetics Group
Institute of Zoology
London, U.K.

John Halley
NERC Centre for Population Biology
Imperial College
Ascot, Berks, U.K.

Olivier Hanotte
Department of Zoology
University of Leicester
Leicester, U.K.

Philip W. Hedrick
Department of Zoology
Arizona State University
Tempe, Arizona

A. Rus Hoelzel
National Cancer Institute
Frederick, Maryland

Robert K. Jansen
Department of Botany
University of Texas
Austin, Texas

Stephen A. Karl
Department of Biology
University of South Florida
Tampa, Florida

Pieter W. Kat
Department of Veterinary Pathology,
Microbiology and Immunology
University of California
Davis, California

Paul L. Leberg
Department of Biology
University of Southwestern Louisiana
Lafayette, Louisiana

Colin J. Limpus
Queensland Department of Environment
Queensland, Australia

Barbara L. Lundrigan
Museum of Zoology and Department of
Biology
University of Michigan
Ann Arbor, Michigan

Georgina Mace
Institute of Zoology
London, U.K.

Roberta J. Mason-Gamer
Harvard University Herbaria
Cambridge, Massachusetts

Phillip A. Morin
Department of Anthropology
University of California
Davis, California

Craig Moritz
Centre for Conservation Biology and
Department of Zoology
The University of Queensland
Queensland, Australia

Joseph E. Neigel
Department of Biology
University of Southwestern Louisiana
Lafayette, Louisiana

Richard A. Nichols
School of Biological Sciences
Queen Mary & Westfield College
London, U.K.

Jennifer L. Nielsen
UDSA Forest Service
Hopkins Marine Station
Stanford University
Pacific Grove, California

Colleen O'Ryan
Department of Chemical Pathology
University of Cape Town
Cape Town
Republic of South Africa

Stephen R. Palumbi
Department of Zoology and Kewalo
Marine Laboratory
University of Hawaii
Honolulu, Hawaii

Patricia G. Parker
Department of Zoology
Ohio State University
Columbus, Ohio

T. Peare
Department of Zoology
Ohio State University
Columbus, Ohio

Lisa Pope
Centre for Conservation Biology and
Department of Zoology
The University of Queensland
Queensland, Australia

Wayne K. Potts
Center for Mammalian Genetics and
Department of Pathology
University of Florida
Gainesville, Florida

Kornelia Rassmann
School of Biological Sciences
University College of North Wales
Bangor, Gwynedd, U.K.

Rebecca Reynolds
Roche Molecular Systems, Inc.
Alameda, California

Robert E. Ricklefs
Department of Biology
University of Pennsylvania
Philadelphia, Pennsylvania

Loren H. Rieseberg
Department of Biology
Indiana University
Bloomington, Indiana

Stanley Scher
Department of Environmental Science,
Policy, and Management
University of California
Berkeley, California

Gilles Seutin
Smithsonian Tropical Research Institute
Balboa, Republic of Panama

William B. Sherwin
Department of Biological Science
University of New South Wales
Kensington, New South Wales
Australia

Hans R. Siegismund
The Arboretum
Royal Veterinary and Agricultural
University
Horsholm, Denmark

Bo T. Simonsen
Department of Population Biology
University of Copenhagen
Copenhagen, Denmark

Thomas B. Smith
Department of Biology
San Francisco State University
San Francisco, California

Pierre Taberlet
Laboratoire de Biologie des Populations
d'Altitude
Université Joseph Fourier
Grenoble, France

Diethard Tautz
Zoologisches Institut der Universität
München
München, Germany

Andrea C. Taylor
Department of Biological Science
University of New South Wales
Kensington, New South Wales
Australia

Priscilla K. Tucker
Museum of Zoology and Department of
Biology
University of Michigan
Ann Arbor, Michigan

Iris van Pijlen
Department of Zoology
University of Leicester
Leicester, U.K.

T.A. Waite
School of Forestry

Michigan Technological University
Houghton, Michigan

Robert S. Wallace
Department of Botany
Iowa State University
Ames, Iowa

Robert K. Wayne
Department of Biology
University of California
Los Angeles, California

Thomas J. White
Roche Molecular Systems, Inc.
Alameda, California

Timothy R. Williams
Conservation Genetics Group
Institute of Zoology
London, U.K.; and
Durrell Institute of Conservation and
Ecology
University of Kent
Canterbury, Kent, U.K.

Jessica Worthington Wilmer
Centre for Conservation Biology and
Department of Zoology and Department
of Anatomy
The University of Queensland
Queensland, Australia

David S. Woodruff
Department of Biology
University of California, San Diego
La Jolla, California

Hans Zischler
Zoologisches Institut der Universität
München
München, Germany

Molecular Genetic Approaches in Conservation

1

An Overview of the Issues

GEORGINA M. MACE, THOMAS B. SMITH, MICHAEL W. BRUFORD, AND ROBERT K. WAYNE

Conservation planning operates at many levels, from whole ecosystems and communities down to individual organisms. At each of these levels, molecular genetic techniques may provide appropriate tools to evaluate processes and to develop management strategies. In this chapter, we present an overview of the questions that may be relevant at each of these hierarchical levels, discuss how molecular genetic analyses may be used to address these questions, and provide examples of the use of molecular techniques in conservation (see Table 1-1). We do not aim to present an exhaustive review but rather to focus on examples from our own research so as to introduce some of the approaches presented in later chapters.

There can be little doubt that extinction now threatens a large proportion of the world's species. Recent estimates suggest that impending rates of species extinctions are at least 4 to 5 times background rates seen in the fossil record (May et al., 1995; Lawton and May, 1995). The severity of the problem means that two kinds of conservation planning decisions are going to be especially important. First, the conservation action must be appropriate—threats need to be identified clearly and steps taken to alleviate them. With a good understanding of the underlying processes, there can be significant improvement (Caughley, 1994). Second, there is an increasing need to identify priorities for conservation action, that is, to single out taxa or sites where action needs to be taken immediately if the situation is not to deteriorate rapidly and irrevocably. In both cases, molecular genetic analyses can play a role.

During the 1980s, conservation genetics was an important and expanding scientific discipline with a strong theoretical framework. But in recent years the importance of demographic and especially environmental factors has rightly been emphasized (Lande, 1988; Caughley, 1994; Harcourt, 1995). However, the unfortunate result is that some now question whether genetic factors have any significance for conservation (e.g., Caro and Laurenson, 1994). This debate is likely to continue (May, 1995). Of course, over the short term, most species are threatened by extrinsic and environmental factors, but effective long-term conservation planning must incorporate genetic factors. Ultimately, the viability of species that survive short term demographic and environmental threats may

Table 1-1 Genetic Level and Questions, Appropriate Analyses and Examples of Their Use. The number of X's indicates the relative applicability of the technique

Levels/Questions	Morphology	Karyology	Allozymes	mtDNA	VNTR	RAPD	Single-copy DNA
Ecosystem/species/subspecies/metapopulation							
1. Phylogenetic distinction	XXX	XXX	XXX	XXX	X	X	X
2. Hybridization	X	XX	XX	XX	XX	?	XX
3. Phylogenetic history	X	X	XX	XXX	XX	?	?
Population							
4. Genetic variability and substructuring	X	X	XXX	XX	XXX	X	XX
Individual							
5. Breeding structure	X	X	XX	X	XXXX	XX	X
Examples	1–5	6–8	9–15	13, 16–19	20–24	25	26–30

Morphology: 1. Wayne et al. (1986a,b); 2. Cracraft and Plum (1988); 3. Wayne et al. (1991b); 4. Rinderer et al. (1991); 5. Vane-Wright et al. (1991).
Karyology: 6. R.K. Wayne et al. (1987); 7. Modi (1987); 8. Benischke and Kumamoto (1991).
Allozymes: 9. O'Brien et al. (1986); 10. O'Brien et al. (1987); 11. Wayne and O'Brien (1987); 12. Lacy and Foster (1988); 13. Brewer et al. (1990); 14. Daugherty et al. (1990); 15. O'Brien et al. (1990); 16. Wayne et al. (1991a,b) (see chapter 6).
mtDNA: 17. Lansman et al. (1983); 18. Avise and Nelson (1989); 19. Lehman et al. (1991).
VNTR: 20. Lehman et al. (1992); 21. Burke (1989); 22. Gilbert et al. (1990); 23. Rabenold et al (1990); 24. Schlotterer et al. (1991) (see chapter 17).
RAPD: 25. Williams et al. (1990) (see chapter 4).
Single-copy DNA: 26. Quinn et al. (1987); 27. Quinn and White (1987); 28. Karl and Avise (1993); 29. Palumbi and Baker (1994); 30. Tucker and Lundugan (1993) (see chapters 2, 3).

4

depend upon the genetic variability they possess, and genetic variation can interact with environmental factors in such a way that the two cannot be viewed as independent effects (Gilpin and Soule, 1986; Partridge and Bruford, 1994). Although it is relatively easy for species to lose genetic variation through bottlenecks in population size or inbreeding, it becomes vanishingly unlikely that a critical allele lost in this way will reevolve. We argue for the general importance of genetic factors on two grounds. First, the study of genetic variation provides an understanding of both the current and historical evolutionary processes that have generated biodiversity patterns, the preservation of which should be a important component of conservation plans (Smith et al., 1993). Second, from a precautionary point of view, future persistence of populations may depend upon the preservation of specific components of genetic diversity.

With rapid advances in molecular genetic techniques over the last ten years has come the realization that many can contribute to the management of threatened species and the conservation of biodiversity. However, despite a myriad of potential applications (Table 1-1), the usefulness of these techniques for making informed management decisions is generally under-appreciated. Molecular genetic techniques can be used to identify parents, offspring, and close relatives in captive and wild populations and hence improve the management and preservation of genetic diversity (chapters 5, 8, 15–19, 22, 25). They can be used to define the genetic units for conservation and the amount of gene flow between such units and, for forensic purposes, to match samples of uncertain provenance and species identity to populations from specific geographic regions (chapters 2–8, 10–12, 16–18, 20, 22, 22–24, 26–28). It is now possible to quantify the genetic variability of historical populations and their living descendants, and thus assess the rate at which genetic variation is being lost and restructured in populations in fragmented landscapes (chapters 8, 18). The quantity of material required for DNA analyses may be minute—single hairs, serum, or archaeological or museum samples of pelts and bone (chapters 8, 18). Consequently, genetic diversity can be measured using noninvasive methods, essential for studies of endangered species.

The relatively recent development of molecular genetic techniques and the perceived technical sophistication required to use them has meant that many have not yet been applied to the management and conservation of small populations. In this chapter, we consider a range of examples primarily from our own research, to illustrate the general applicability and utility of molecular techniques to conservation questions. The discussion is organized hierarchically, from questions of biodiversity at the ecosystem to individual levels (see Table 1-1).

ECOSYSTEM LEVEL

Important areas for conservation are often identified by the number of species they contain, that is, by some measure of species richness. Taxonomic experts may be sent to poorly documented regions to quantify the number of species found in a wide diversity of taxonomic groups. This approach provides a snapshot of species richness, but not of the historical processes that generate diversity, and it does not distinguish populations and regions with different evolutionary histories. However, molecular genetic techniques can provide an estimate of the number of distinct forms in an area as well as a measure of how different they are. Although many kinds of information should be used in evaluating different potential sites for conservation, areas that contain taxa representing long, distinct lineages are likely to preserve a greater degree of evolutionary heritage than ones with a similar number of species but containing a succession of closely allied forms (May, 1990; Vane-Wright et al., 1991). A number of different methods for quantifying evolutionary uniqueness have been proposed (Williams et al., 1993), but there is now an emerging consensus that selection should be based upon a phylogenetic tree constructed from molecular genetic data, preferably derived from multiple loci, and the aim should be to conserve a maximum diversity of evolutionary lineages (Faith, 1993, 1994; May, 1994). At this higher taxonomic level, encompassing many taxa, there is therefore a need to provide molecular genetic data to support decision making.

A related approach to selecting areas for conservation was recently proposed by Kremen (1992, 1994). She suggests selecting and analyzing divergence in a target taxon that is likely to be an indicator of biodiversity in other, unrelated taxa. Appropriate target taxa will be those that have undergone an endemic radiation in which specialized forms for a wide variety of unique habitats are likely to have evolved. Thus, a detailed study of the taxonomy and distribution of species within a target taxon may provide a conservative index of the overall biodiversity at a given locality. Moreover, the change in genetic biodiversity among localities could reflect the steepness of ecological gradients and the degree of evolutionary isolation of localities. Phylogenetic analysis of molecular and morphological data from each target taxon can reveal the complex historical forces that generated biodiversity in the region, identify those species which are evolutionarily distinct, and provide a means for evaluating the consequences of conservation actions directed at only part of the total area.

As a heuristic example, Figure 1-1 shows the distribution map and hypothetical molecular phylogeny of species from an indicator butterfly taxon, genus *Henotesia*, from five areas in Madagascar (Torres, Lees, Kremen, unpublished data). The taxonomy of each species based on morphological comparisons is indicated in the species labels. Species with the same upper case labels (e.g., "H") should generally be found in the same clade if morphology reflects evolutionary relationships as indicated by DNA sequence data. Some groupings based on morphology (e.g., "K") are not supported by the DNA sequence analysis. Of greater conservation relevance is the finding that although locality 1 has only a single species, it is the most phylogenetically distinct species in the entire group. This suggests that locality 1 may have had a long independent evolutionary history and contain other unique taxa. Locality 5 has the highest species number, but most of the species found there are from a single clade. Thus, a biodiversity analysis based on species number alone would rank locality 5 as having the highest biodiversity, whereas locality 1, with the most evolutionary distinct taxon, would have the lowest ranking. Conservation plans need to recognize phylogenetically unique areas, such as locality 1, as potentially important sources of evolutionary novelty.

Phylogenetic analyses using molecular data provide information about the

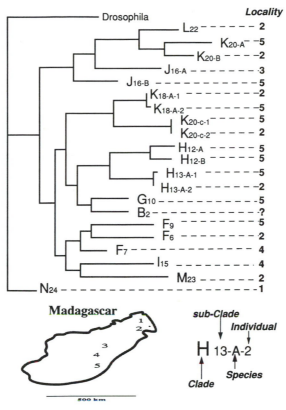

Figure 1-1 Hypothetical phylogeny of species from the butterfly genus *Henotesia* based on mitochondrial cytochrome *b* sequence (Torres, Lees, Kremen, unpublished data). The location of sampling localities is given in the lower left and the taxonomic scheme based on morphological analysis is given in the lower right.

historical processes that have generated biodiversity (chapter 9). An example of how this may be useful in distinguishing between processes and pattern is exemplified by the debate on how to preserve the high biodiversity characteristic of rainforests. Tropical rainforests are widely believed to exhibit the greatest species richness. However, the evolutionary mechanisms which have generated this tremendous diversity are controversial (Endler, 1982; Mayr and O'Hara, 1986; Patton and Smith, 1993; Patton et al., 1994). Early work stressed the importance of Pleistocene refugia in generating new species (Haffer, 1969) and although the hypothesis was not without critics (e.g., Endler, 1982) many conservation plans that focus on preservation of central core areas in rainforests are based on it. In fact, recent work in Central and South America provides little evidence that speciation in any group of organisms was actually tied to isolation in Pleistocene refugia (Flenley, 1993; Patton and Smith, 1993). For mammals, even the assertion that the greatest diversity is found in lowland forest is being challenged (Mares, 1992). Similar questions have been raised for African rainforests (Endler, 1982).

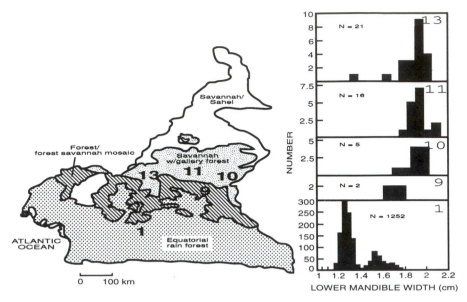

Figure 1-2 Frequency histograms for lower mandible width for male black-bellied seedcrackers (Estrildidae: *Pyrenestes ostrinus*), a seed-eating finch existing in ecotones and rainforests of Cameroon. The histograms show that the ecotone localities (9, 10, 11, and 13) have a large bill morph about 1.9 mm in width that is not found in the continuous rainforest. Locality 13 in the ecotone has all three bill morphs. In general, the greater diversity of bill morphs in ecotones reflects divergent selection for a wider diversity of seed types that exist in ecotones (see Smith, 1990, 1993). Ecotones may be important sources of evolutionary diversity.

The role that ecotones (between forest and savanna) play in generating genetic diversity may be significant. While there is little doubt that some centrally located regions show greater species richness, small isolated ecotone populations may exhibit greater interpopulation variation by being more subject to drift and differential directional selection (Figure 1-2; see Risser, 1995). For example, in tropical Africa many of the most important contact zones between divergent subspecies of passerine birds are located in savanna-rainforest ecotones (Figure 1-2; Chapin, 1932; Crowe and Crowe, 1982). Such peripheral ecotonal populations may provide an important source of new genetic variation which ultimately enriches more centrally located populations. Centrally located rainforest areas are presently the main focus of many of the conservation efforts, while ecotonal regions are rapidly being lost to overgrazing, burning, and wood harvesting with little or no effort toward their conservation. Molecular genetic analysis of ecotones can both document the level of genetic distinction of ecotone populations and better define the evolutionary processes that have generated rainforest biodiversity (see Smith et al., 1993).

Finally, an important application of molecular genetic approaches is to characterize overall biodiversity in microbial communities. It is estimated that only 20% of naturally occurring bacteria species have been isolated (L.G. Wayne et al., 1987). This is largely because of the difficulties in developing species-specific culture media. PCR-based techniques provide a novel and cost-effective means of measuring diversity in complex microbial communities and are likely to become the standard means of characterizing these communities in the future (Muyzer et al., 1993).

SPECIES LEVEL

Units for Conservation

The patterns of morphological divergence and geographical distributions of taxa may provide only partial information about their actual evolutionary relationships. Combined with molecular genetic data, however, we have a powerful tool for recognizing species and determining the appropriate evolutionary units for conservation. For example, the tuatara (genus *Sphenodon*) is a reptile which occurs on islands off the coast of New Zealand and represents the only extant genus of an entire reptilian order. Until recently, only one species, *S. punctatus*, was recognized. Molecular analyses have now revealed that in fact the species has a complex evolutionary history, and actually should be divided into three distinct groups (Daugherty et al., 1990). One taxon, which was described in the last century as *S. guntheri* and exists on one island only, was found to have three fixed allelic differences from *S. punctatus*, supporting its recognition as a distinct species. A genetically divergent population of *S. punctatus* was also recognized as a separate taxon and conservation plans are now being designed to preserve these distinct units.

A recent example concerns the rediscovery of Darwin's fox, *Dusicyon fulvipes* (Yahnke et al., 1996). Darwin's fox was known only from the temperate rainforests of Chiloé Island, Chile (Figure 1-3), and was classified as a subspecies of the mainland gray fox (*D. griseus*). The species is extremely rare—fewer than 500 individuals exist—but its conservation has been neglected because of the remoteness of the island habitat and its supposed close conspecific relationship to the mainland gray fox. However, about five years ago, Chilean researchers discovered a mainland population of Darwin's fox in Nahuelbuta National Park, 500 kilometers north of Chiloé Island, that was sympatric with the mainland gray fox, suggesting that the two were reproductively isolated. These researchers hypothesized that Darwin's fox was a distinct species and that the mainland population was a relict of a formerly more widely distributed species or the result of a historic transplantation of island foxes.

Analysis of DNA sequence data showed that the mainland population of Darwin's fox was closely related to, yet distinct from, the island population. This result suggested that the mainland population was not transplanted from the island but was instead a relict population. Moreover, phylogenetic analysis of mitochondrial DNA sequence data indicated that mainland and island Darwin's fox sequences form a monophyletic group ancestral to the two mainland species of foxes, *D. griseus* and *D. culpaeus* (Figure 1-3). Consequently, Darwin's fox is a phylogenetically distinct species and likely an earlier invader to the Chilean coast. Its distribution was probably more widespread before the destruction of mainland

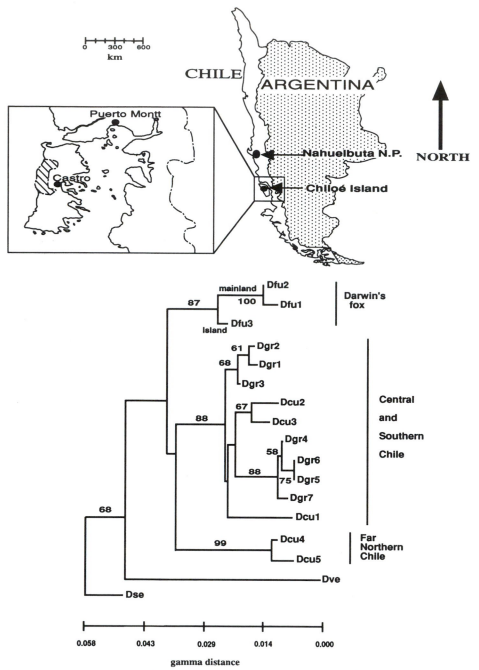

Figure 1-3 Above: Map of Chiloé Island which is located about 30 km directly west of the central Chilean coast. Location of the island temperate rainforest is indicated by hatching. Below: Neighbor-joining tree based on gamma distances between control region sequences from the three species of Chilean fox: the gray fox, *Dusicyon griseus* (Dgr 1–7); the culpeo fox, *D. culpaeus* (Dcu 1–5); and Darwin's fox, *D. fulvipes* (Dfu 1–3). The genetic distance scale is given below the phylogeny. Bootstrap support values over 50% are indicated (See Yahnke et al., 1996).

forests for timber and agriculture. However, there are rainforests on the mainland where Darwin's fox may still exist and might not have been discovered or has been misidentified as the gray fox (*D. griseus*). Darwin's fox is potentially a flagship species symbolizing the highly endangered and distinct temperate rainforests of Chiloé Island and mainland Chile. Increased protection for Darwin's fox might also result in better protection for the unique temperate rainforests of Chile.

Problems of Hybridization

Interbreeding among individuals from closely related species and from populations within the same species is common (Barton and Hewitt, 1989; O'Brien and Mayr, 1991), and may be of concern if it threatens the genetic integrity of an endemic rare or endangered species or population. Often hybridization may reflect habitat alterations or human introductions, such that nonnative forms become abundant and widespread and hybridization occurs between them and the native form. Examples include hybridization between coyotes and gray wolves in Minnesota and Ontario and in the American Southeast (Lehman et al., 1991; Wayne and Jenks, 1991), the displacement of native fish stocks by hatchery-raised fish, and the subspecific hybridization of the Florida puma and a nonnative subspecies (Waples and Teel, 1990; Roelke et al., 1993). Similarly, captive populations may often represent a blend of subspecies through unintentional interbreeding. For example, during a widespread survey of Asiatic lions in a zoo-based captive breeding program, many were found to contain genes only recorded from individuals of the African subspecies (O'Brien et al., 1987). This finding led to the removal of most animals from the program.

In conclusion, genetic screening of wild endangered populations or captive stocks suspected of hybridization may provide essential data for on-site conservation or captive breeding programs.

WITHIN SPECIES

Molecular genetic approaches are particularly useful for identifying biogeographic divisions that exist within species. Several species may show coincident phylogenetic divisions across their range, indicating the existence of biogeographic units having similar species composition but separate evolutionary histories. Examples of this are found in the sharp phylogenetic division existing between conspecific populations in the Gulf and Atlantic coasts of Florida (Avise, 1992). Divisions were found in all 18 taxa examined, including salt- and freshwater fish, birds, reptiles and invertebrates such as oysters and horseshoe crabs. The strong within-species genetic discontinuity between Gulf and Atlantic populations suggests that they are not only distinct biotic realms but should be conserved as separate units reflecting their unique and independent evolutionary heritage (Avise, 1992).

The genetic relationships among populations can often be used to deduce historical patterns of movement and the degree of gene flow among populations (see Avise, 1994). This information is especially useful for documenting the genetic

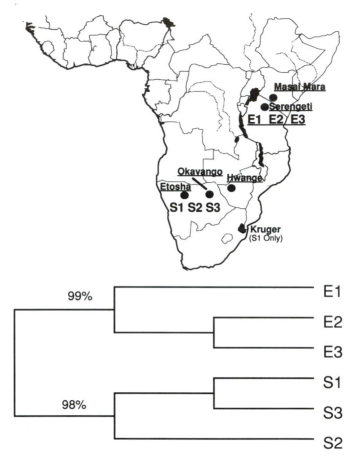

Figure 1-4 Sampling localities of east (Masai Mara/Serengeti) and southern (Etosha/Okavango/ Hwange/Kruger) African wild dogs (*Lycaon pictus*). The distribution of cytochome *b* sequences found in wild dogs is indicated on the map. E1–E3 are east African sequences and S1–S3 are southern African sequences. The phylogenetic relationship of these sequences and bootstrap support values are given in the lower half of the figure. Eastern and southern clades differ by about 1% in DNA sequence (see Girman et al., 1993).

isolation and distinction of vanishing populations, for tracing corridors of dispersal among populations and for identifying populations that might provide the source material for augmentation or reintroduction programs. Recent advances in the analysis of molecular data, such as those that separate population structure from population history (Templeton et al., 1995), are likely to lead to greater utility of molecular techniques in population-level conservation.

Augmentation of an endangered population may be necessary if demographic changes or inbreeding have significantly reduced the chances of population survival and this cannot be ameliorated by genetic management alone. Given these conditions, augmentation of genetically close individuals from a similar environment to that of the donor population may be desirable. For example, the endangered Florida panther is a population of about 50 that appears to be suffering from the effects of loss of genetic variability. Plans have been developed to augment the population genetically by introducing unrelated individuals from other parts of the species' range. Molecular genetic analysis of North American pumas has shown that the nearest population in Texas is genetically very similar and therefore provides appropriate stock for the augmentation (Roelke et al., 1993). An analogous case study concerns the reintroduction of Yellowstone wolves. Here, a mitochondrial DNA analysis has shown that all North American wolves are genetically very similar and high rates of gene flow have caused genetic uniformity over large geographic regions (Wayne et al., 1992). Hence, there need be no major concern about the reintroduction of wolves from British Columbia since these are likely to be quite similar to, and to have been in genetic contact with, the historical population of wolves living in Yellowstone.

Evolutionarily significant units for conservation can also be identified using molecular genetic techniques (Mortiz, 1994; see chapter 27). For example, the African hunting dog is a unique carnivore which has historically inhabited savannah-like habitats south of the Sahara desert (Figure 1-4). Wild dog populations have declined precipitously during the last 20 years and perhaps no more than several thousand now exist (Ginsberg and Macdonald, 1990). The east and west African populations are the most endangered, having recently suffered severe population declines. Molecular genetic analyses showed that eastern and southern populations form two genetically distinct clades that warrant separate preservation (Figure 1-4, Girman et al., 1993). Thus, plans to reintroduce southern African wild dogs to areas of eastern Africa where populations have recently gone extinct would mix populations with separate evolutionary histories. In addition to causing the potential loss of genes that are unique and adaptive in the east African population, introduced southern African dogs could put the native eastern population at risk with southern African diseases to which they have not developed immunity. Moreover, wild dogs in zoos are all of southern African origin, indicating that the captive breeding of the wild dog is not directed toward those populations most at risk.

INDIVIDUAL LEVEL

In small captive and wild populations there may be asymmetries in reproductive success among individuals and sexes so that the effective population size is reduced relative to the census number, and this can have important implications for conservation management. Molecular genetic techniques can be used to deduce parentage in wild and captive populations if all potential parents and their offspring are sampled (Burke, 1989). Even in populations for which little information is available, inferences can be made about the breeding structure from molecular genetic data. For example, Packer et al., (1991) were able to deduce that male coalitions of lions often consisted of individuals related as siblings or parent–offspring, and that the mating behavior of such coalitions was different from that in coalitions of unrelated males. Similarly, Lehman et al. (1992) deduced that wolf packs were predominantly composed of close relatives and were also able to deduce movement of close relatives among packs sharing common territorial boundaries. Finally, sex-specific reproductive success in a population may potentially be determined through the use of sex-linked minisatellite fragments or sex chromosome-specific probes, which in turn may lead to better estimates of effective population size (e.g., Rabenold et al., 1990; Longmire et al., 1991; Griffiths and Holland, 1990; see chapter 25).

An important aim of the genetic management of captive populations is to retain genetic variability, in part by reducing inbreeding. Selecting breeding pairs of low relatedness depends on the availability of complete information about parentage and about the relationships among the wild-caught founders of the captive population. This information is hardly ever available, and substructuring in small endangered populations can lead to inbreeding among wild-caught founders (Mace, 1986). Until recently, analysis of genetic variability in captive pedigreed populations was generally carried out through simulation models (MacLuer et al., 1986; Dobson et al., 1991; Hedrick and Miller, 1992), but hypervariable DNA analysis offers the opportunity to assess genetic variation directly and to refine and correct the wild or captive pedigree. Additionally, individuals can potentially be grouped into relatedness classes by using DNA fingerprinting and the more dissimilar individuals used as breeding pairs (Jeffreys et al., 1985a,b; Burke and Bruford, 1987; Packer et al., 1991; Wayne et al., 1991b; Lehman et al., 1992).

For example, in the captive breeding program for the critically endangered Mauritius pink pigeon (*Nesoenas mayeri*) there are good parentage records stretching back over several generations, but all captive birds are descended from just six founder birds and the relationships among these are not known. There is some evidence for inbreeding depression in captive pink pigeons (Jones et al., 1985), and therefore a need to investigate relationships further to improve genetic management. The relationship between fingerprint similarity (band sharing) and genetic relatedness in captive pink pigeons, including four of the six founders, was investigated (for methods see Burke and Bruford, 1987; Bruford et al., 1992) and band-sharing coefficients were initially calculated for a set of individuals selected at random from the captive population. These individuals shared a mean of 50% of their bands. Pairwise comparisons were then grouped into three classes; known first-degree relatives, known second-degree relatives, and individuals unrelated in the captive breeding program (Table 1-2). The band-sharing coefficients of all classes were significantly different, though there was limited overlap among them. DNA fingerprinting was useful for discriminating among individuals of different

Table 1-2 Mean Band-sharing Coefficients of the Mauritius Pink Pigeon for Relationship Classes and Probes (33.6 and 33.15)

| Probe | Overall | Mean Band-sharing Coefficient ($\pm$ SE) | | |
		1st Degree	2nd Degree	Unrelated
33.6	0.545 ± 0.018	0.607 ± 0.026	0.491 ± 0.023	0.468 ± 0.016
33.15	0.485 ± 0.018	0.582 ± 0.016	0.451 ± 0.023	0.333 ± 0.018
Combined	0.500 ± 0.014	0.583 ± 0.015	0.477 ± 0.017	0.388 ± 0.021

levels of relatedness within this captive population and practical management decisions could then be made about the breeding of individuals whose relationships were not known. Of course, some parentage information was required to first determine relatedness classes using this approach, but with the advent of more variable markers it should become easier to estimate relatedness reliably when breeding records are unavailable.

FORENSIC QUESTIONS

The forensic applications of molecular techniques are wide-ranging. The species and regions that samples have originated from may be deduced and samples from the same individual obtained from different places and at different times can be matched. Genetic typing can often be done on trace materials such as hair, horn, tusks, fecal material, ruminant boli, and mummified or museum-preserved remains (see chapter 18). One of the most intriguing recent cases discussed by Palumbi (chapter 2) concerns monitoring the legal market for whale meat in Japan. Whales taken under strict scientific collecting permits can be sold in Japan's domestic market, but concern has been raised that such legal sales allowed a route for illegal marketing. Palumbi and colleagues have shown that the whale meat for legal sale in Japan does involve several banned species, suggesting that the legal market is a vehicle for illegal trade and may be difficult to police.

GENE CONSERVATION

In small, isolated populations, allelic diversity and heterozygosity will be lost (chapters 20, 21). Molecular genetic techniques can be used to measure the loss of genetic variation and potentially variability affecting fitness (chapter 14). Genetic variation and effective population size may also be reduced through the intro- duction of genetically uniform hatchery-raised stock (Ryman, 1991; Waples and Teel, 1990).

One example of the use of molecular genetic techniques for measuring the loss of variability concerns the northern hairy-nosed wombat (*Lasiorhinus krefftii*; see chapter 27). Currently, this species exists as only a single colony of fewer than 80 individuals at Epping Forest National Park, Queensland, Australia. At the time of European settlement, the geographic range of the species spread several thousand kilometers from Queensland, through New South Wales and south to the border of Victoria (Figure 1-5). The combined effects of drought and cattle grazing are thought to have been in part responsible for the decline (Crossman et al., 1994). Because this remaining population is small and likely to be subdivided, random genetic drift will cause loss of genetic variability, particularly of rare alleles, over time. Also, the population has probably passed through several bottlenecks, as in 1985 when only 20–30 animals existed (Gordon et al., 1985). The current population is likely to retain only a small fraction of the variability formerly existing in the species.

Through the use of the polymerase chain reaction applied to DNA isolated

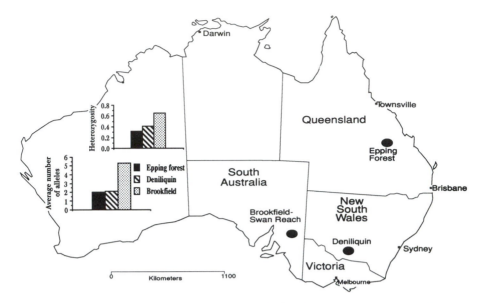

Figure 1-5 Locations of (1) the last remaining population of the northern hairy-nosed wombat (*Lasiorhinus krefftii*) in Epping forest, Queensland; (2) an extinct population at Deniliquin, New South Wales; and (3) two colonies of the abundant southern hairy nosed wombat (*L. latifrons.*) at Brookfield and Swan Reach, South Australia. The extinct colony at Deniliquin was based on samples from museum pelts. Heterozygosity and average number of alleles per locus for each of the three populations are indicated on the left side of the figure.

from museum samples, historical levels of genetic variability in the population from Deniliquin, New South Wales, were compared to the levels found in the current Epping Forest population, and to a related but abundant wombat species (Figure 1-5; Taylor et al., 1994). The molecular analysis showed that the northern hairy nosed wombat had only about 40% of the genetic variability expected in a widely distributed abundant species and that the loss of genetic variation was well advanced even in the historical population (Figure 1-5). Such information can be used to estimate effective population size over the period of decline in the Epping forest wombats and expected future losses in genetic variability. Moreover, if a captive breeding program is developed, microsatellite analysis may assist in choosing genetically dissimilar individuals to establish the breeding program (see above and chapter 27).

CONCLUSION

An integrated approach to conservation requires the use of ecological, demo-graphic, morphological and molecular genetic data to allow the preservation of significant amounts of evolutionary diversity. Data from the application of molecular genetic techniques may provide information with clear, tractable implications for conservation. We have illustrated in this chapter the importance of these techniques both for choosing appropriate units for conservation (either taxonomic or areal) and for making plans that will increase the prospects for long-term viability. However, there are limitations to the use of genetic techniques and problems in their application.

In the first place, there will be increasing demand for molecular genetic analyses, yet at present these are both technically demanding and relatively expensive. Therefore the development of cheaper, less demanding techniques and the availability of wider training opportunities are necessary if molecular approaches are to be more widely used. The laboratories where molecular techniques are already practiced have some difficult choices to make among an abundance of potential projects, especially because there is often not a positive relationship between the worthiness of a project and the possibility of raising funds to undertake it.

As will be clear in the case studies presented later in this book, productive application of molecular genetic techniques requires a carefully framed question, and quite often a fair amount of supporting information. For example, many studies that attempt to reconstruct pedigrees of families by DNA fingerprinting are unable to resolve the relationships fully, yet this might have been possible had observations been made on a breeding population to reduce the set of uncertainties for analysis. Also, especially in the case of studies with clear management implications, it is worth considering at the outset how alternative results from the molecular genetic analysis would be turned into practical actions. In some cases these alternatives are not clear; in others, supporting information would be required to reach a decision. For example, a study of differentiation between two isolated populations might be undertaken to help decide whether they should be treated as one or two units for conservation. However, this is a relative question and needs to be viewed with reference to levels of differentiation among other populations within the species. Moreover, ecological and demographic implications of treating populations as single or separate units need to be considered as well.

The importance of molecular genetic studies lies in the link that they provide between ecological, geographical, and evolutionary information, all of which are necessary for effective conservation planning. The studies in this book provide many examples of the way in which they have provided significant benefits for the conservation of biodiversity, and we anticipate that this trend will increase in the future.

REFERENCES

Avise, J.C. 1992. Molecular population structure and the biogeographic history of a regional fauna: a case history with lessons for conservation biology. Oikos 63: 62–76.

Avise, J.C. 1994. Molecular markers, natural history and evolution. New York: Chapman and Hall.

Avise, J.C. and W.S. Nelson. 1989. Molecular genetic relationships of extinct Dusky Seaside Sparrow. Science 243: 646–648.

Barton, N.H. and G.M. Hewitt. 1989. Adaptation, speciation, and hybrid zones. Nature 341: 497–503.

Benirschke, K. and T. Kumamoto. 1991. Mammalian cytogenetics and conservation of species. J. Hered. 82: 187–191.

Brewer, B., R.C. Lacy, M.L. Foster, and G. Alaks. 1990. Inbreeding depression in insular and central populations of *Permmosycus* mice. J. Hered. 81: 257–266.

Bruford, M.W., O. Hanotte, J.F.Y. Brookfield, and T. Burke. 1992. Single-locus and multilocus DNA fingerprinting. In: A.R. Hoelzel, ed. Molecular genetic analysis of populations: A practical approach, pp. 225–269. Oxford, U.K.: IRL Press.

Burke, T. 1989. DNA fingerprinting and other methods for the study of mating success. Trends Ecol. Evol. 4: 139–144.

Burke, T. and M.W. Bruford. 1987. DNA fingerprinting in birds. Nature 327: 149–152.

Caro, T.M. and M.K. Laurenson. 1994. Ecological and genetic factors in conservation: A cautionary tale. Science 263: 485–486.

Caughley, G. 1994. Directions in conservation biology. J. Animal Ecol. 63: 215–244.

Chapin, J.P. 1932. The birds of the Belgian Congo. Bull. Am. Mus. Nat. Hist. 3: 1–756.

Cracraft, J. and R.O. Prum. 1988. Pattern and process of diversification: speciation and historical congruence in some neotropical birds. Evolution 42: 603–620.

Crossman, D.G., C. Johnson, and A. Horsup. 1994. Trends in the population of the northern hairy-nosed wombat in Epping Forest National Park. Pacific Conserv. Biol. 1: 141–149.

Crowe, T.M. and A.A. Crowe. 1982. Patterns of distribution, diversity and endemism in Afrotropical birds. J. Zool. Soc., Lond. 198: 417–442.

Daugherty, C.H., A. Cree, J.M. Hay, and M.B. Thompson. 1990. Neglected taxonomy and continuing extinctions of tuatara (*Spenodon*). Nature 347: 177–179.

Dobson, A.P., G.M. Mace, J. Poole, and R.A. Brett. 1991. Conservation biology: the ecology and genetics of endangered species. In: R.J. Berry, T.J. Crawford, and G.M. Hewitt, eds. Genes in ecology, pp. 405–430. Oxford: Blackwell Science.

Endler, J.A. 1982. Pleistocene forest refuges: Fact or fancy. In: G.T. Prance, ed. Biological diversification in the tropics, pp. 641–657. New York. Columbia University Press.

Faith, D.P. 1993. Systematics and conservation: on predicting the feature diversity of subsets of taxa. Cladistics 8: 361–373.

Faith, D.P. 1994. Genetic diversity and taxonomic priorities for conservation. Biol. Conserv. 68: 69–74.

Flenley, J. 1993. The origins of diversity in tropical rain forests. Trends Ecol. Evol. 8: 119–120.

Gilbert, D.A., N. Lehman, S.J. O'Brien, and R.K. Wayne. 1990. Genetic fingerprinting reflects population differentiation in the Channel Island Fox. Nature 344: 764–767.

Gilpin, M.E. and M.E. Soule. 1986. Minimum viable population: the processes of population extinction. In: M.E. Soule, ed. Conservation biology: the science of scarcity and diversity, pp. 13–34. Sunderland, Mass: Sinaeur Associates.

Ginsberg, J.R. and D.W. Macdonald. 1990. Foxes, wolves, jackals and dogs. An action plan for the conservation of canids. Gland, Switzerland: International Union for Conservation of Nature and Natural Resources.

Girman, D.J., P.W. Kat, M.G.L. Mills, J.R. Ginsberg, M. Borner, V. Wilson, J.H. Fanshawe, C. Fitzgibbon, L.M. Lau, and R.K. Wayne. 1993. Molecular genetic and morphological analyses of the African wild dog (*Lycaon pictus*). J. Hered. 84: 450–459.

Gordon, G., T. Riney, J. Toop, B.C. Lawrie, and M.D. Godwin. 1985. Observations on the Queensland hairy-nosed wombat (*Lasiorhinus krefftii* (Owen)). Biol. Conserv. 33: 165–196.

Griffiths, R. and P.W.H. Holland. 1990. A novel avian w-chromosome DNA repeat sequence in the lesser black-backed gull (*larus-fuscus*). Chromosoma 99: 243–250.

Haffer, J. 1969. Speciation in Amazonian forest birds. Science, 165: 131–137.

Harcourt A.H. 1995. Population viability estimates—theory and practice for a wild gorilla population. Conserv. Biol. 9: 134–142.

Hedrick, P.W. and P.S. Miller. 1992. Conservation genetics—techniques and fundamentals. Ecol. Appl. 2: 30–46.

Jeffreys, A.J., Wilson, V., and Thein, S.L. 1985a. Hypervariable "minisatellite" regions in human DNA. Nature 314: 67–73.

Jeffreys, A.J., V. Wilson, and S.L. Thein. 1985b. Individual-specific "fingerprints" of human DNA. Nature 316: 76–79.

Jones, C.G., D.M. Todd, and Y. Mungroo. 1985. Mortality, morbidity and breeding success of the pink pigeon Columba (Nesoenas) mayeri. Proceedings of the Symposium on Disease and Management of Threatened Bird Populations. Ontario: ICBP.

Karl, S.A. and J.C. Avise. 1993. PCR-based assays of mendelian polymorphisms from anonymous single-copy nuclear DNA—techniques and applications for population genetics. Mol. Biol. Evol. 10: 342–361.

Kremen, C. 1992. Accessing the indicator properties of species assemblages for natural areas monitoring. Ecol. Appl. 2: 203–217.

Kremen, C. 1994. Biological inventory using target taxa: A case study of the butterflies of Madagascar. Ecol. Appl. 4: 407–422.

Lacy, R.C. and M.L. Foster 1988. Determination of pedigrees and taxa of primates by protein electrophoresis. Int. Zoo Yearbook. 27: 159–168.

Lande, R. 1988. Genetics and demography in biological conservation. Science 241: 1455–1460.

Lansman, R.A., J.C. Avise, C.F. Aquadro, J.F. Shipira, and S.W. Daniel. 1983. Extensive genetic variation in mitrochondrial DNAs among geographic populations of the deer mouse, Peromyscus maniculatus. Evolution 37: 1–16.

Lawton, J.H. and R.M. May. 1995. Extinction rates. Oxford, U.K.: Oxford University Press.

Lehman, N., A. Eisenhawer, K. Hansen, D.L. Mech, R.O. Peterson, P.J.P. Gogan, and R.K. Wayne. 1991. Introgression of coyote mitochondrial DNA into sympatric North American gray wolf populations. Evolution 45: 104–119.

Lehman, N., P. Clarkson, L.D. Mech, T.J. Meier, and R.K. Wayne. 1992. The use of DNA fingerprinting and mitochondrial DNA to study genetic relationships within and among wolf packs. Behav. Ecol. Sociobiol. 30: 83–94.

Longmire, J.L., R.E. Ambrose, N.C. Brown, T.J. Cade, T.L. Maechtle, W.S. Seegar, F.P. Ward, and C.M. White. 1991. Use of sex-linked minisatellite fragments to investigate genetic differentiation and migration of North American populations of the peregrine falcon (Falco peregrinus). In: T. Burke, G. Dolf, A.J. Jeffreys, and R. Wolff, eds. DNA fingerprinting: approaches and applications, pp. 217–229. Basel: Birkhäuser Verlag.

Mace, G.M. 1986. Genetic management of small populations. Int. Zoo Yearbook 24/25: 167–174.

MacLuer, J.W., J.L. VandeBerg, B. Read, and O.A. Ryder. 1986. Pedigree analysis by computer simulation. Zoo Biol. 5: 147–160.

Mares, M.A. 1992. Neotropical mammals and the myth of Amazonian biodiversity. Science 255: 976–979.

May, R. 1990. Taxonomy as destiny. Nature 347: 129–130.

May, R.M. 1994. Conceptual aspects of the quantification of the extent of biological diversity. Phil. Trans. R. S. London, Series B, Biol. Sci. 345: 13–20.

May, R. 1995. The cheetah controversy. Nature 374: 309–310.

May, R.M., Lawton, J.H. and Stork, N.E. 1995. Assessing extinction rates. In: J.H. Lawton and R.M. May, eds. Extinction rates. Oxford, U.K.: Oxford University Press.

Mayr, E. and R.J. O'Hara 1986. The biogeographical evidence supporting the Pleistocene forest refuge hypothesis. Evolution 40: 55–67.

Modi, W.S. 1987. Phylogenetic analysis of chromosomal banding patterns among the nearctic Arvicolidae (Mammalia, Rodentia). Syst. Zool. 36: 109–136.

Moritz, C. 1994. Defining evolutionarily significant units for conservation. Trends Ecol. Evol. 9: 373–375.

Muyzer, G., E. De Waal, and A.G. Utitterlinden. 1993. Profiling of complex microbial populations by denaturing gradient gel electrophoresis analysis of polymerase chain reaction-amplified genes coding for 16s rRNA. Appl. Environ. Microbiol. 59: 695–700.

O'Brien, S.J. and E. Mayr. 1991. Species hybridization and protection of endangered animals. Science 253: 251–252.

O'Brien, D., E. Wildt, and M. Bush. 1986. The African cheetah in genetic peril. Sci. Am. 254: 84–92.

O'Brien, S.J., P. Joslin, G.L. Smith, R. Wolfe, E. Heath, J. Otte-Joslin, P.P. Rawal, K.K. Bhattacherjee, and J.S. Martenson. 1987. Evidence for African origin of founders of the Asiatic lion species survival plan. Zoo Biol. 6: 99–116.

O'Brien, S.J., M.E. Rolke, N. Yuhki, K.W. Richards, W.E. Johnson, W.L. Franklin, A.E. Anderson, O.L. Bass, R.C. Belden, and J.S. Martenson. 1990. Genetic introgression within the Florida panther Felis concolor coryi. Natl. Geo. Res. 6: 485–494.

Packer, C., D.A. Gilbert, A.E. Pusey, and S.J. O'Brien. 1991. A molecular genetic analysis of kinship and cooperation in African lions. Nature 351: 562–565.

Palumbi, S.R. and C.S. Baker. 1994. Contrasting population structure from nuclear intron sequences and mtDNA of humpback whales. Mol. Biol. Evol. 11: 426–435.

Partridge, L. and M. Bruford. 1994. A crash course in survival. Nature, 372: 318–320.

Patton, J.L. and M.F. Smith. 1993. The diversification of South American murid rodents: evidence from mitochondrial DNA sequence data for the akodontine tribe. Biol. J. Linn. Soc. 50: 149–177.

Patton, J.L., M.N.F. Da Silva, and J. R. Malcolm. 1994. Gene genealogy and differentiation among arboreal spiny rats (rodentia: Echimyidae) of Amazon Basisn: A test of the riverine barrier hypothesis. Evolution 48: 1314–1323.

Quinn, T.W. and N.B. White. 1987. Identification of restriction fragment length polymorphisms in genomic DNA of the lesser snow geese (Anser caerulescens). Mol. Biol. Evol. 4: 126–143.

Quinn, T.W., J.S. Quinn, F. Cooke, and B.N. White. 1987. DNA marker analysis detects maternity and paternity in single broods of the lesser snow geese. Nature 362: 392–394.

Rabenold, P.P., K.N. Rabenold, W.H. Piper, J. Haydock, and S.W. Zack. 1990. Shared paternity revealed by genetic analysis in cooperatively breeding tropical wrens. (stripe-backed wren, Campylorhynchus nuchalis). Nature 348: 538–541.

Rinderer, T.E., J.A. Stelzer, B.P. Oldroyd, S.M. Buco, and W.L. Rubink. 1991. Hybridization between European and Africanized honey bees in the Neotropical Peninsula. Science 253: 309–311.

Risser, P.G. 1995. The status of the science of examining ecotones. Bioscience 45: 318–325.

Roelke, M.E., J.S. Martenson, and S.J. O'Brien. 1993. The consequences of demographic reduction and genetic depletion in the endangered Florida panther. Curr. Biol. 3: 340–350.

Ryman, N. 1991. Conservation genetics considerations in fishery management. J. Fish Biol. 39: 211–224.

Schlotterer, C., B. Amos, and D. Tautz. 1991. Conservation of polymorphic simple sequence loci in cetacean species. Nature 354: 63–65.

Smith, T.B. 1990. Natural selection on bill characters in the two bill morphs of the African finch Pyrenestes ostrinus. Evolution 44: 832–941.

Smith, T.B. 1993. Disruptive selection and the genetic basis of bill size polymorphism in the African finch *Pyrenestes*. Nature 363: 618–621.

Smith, T.B., M.W. Bruford, and R.K. Wayne. 1993. The preservation of process: the missing element of conservation programs. Biodiversity Lett. 1: 164–167.

Taylor, A.C., W.B. Sherwin, and R.K. Wayne. 1994. The use of simple sequence loci to measure genetic variation in bottlenecked species: the decline of the northern hairy-nosed wombat (*Lasiorhinus krefftii*). Mol. Ecol. 3: 277–290.

Templeton, A.R., E. Routman, and P.A. Christopher. 1995. Separating population structure from population history: A cladistic analysis of the geographical distribution of mtDNA Haplotypes in the Tiger Salamader, *Ambystoma tigrinum*. Genetics 140: 767–782.

Tucker, P.K. and B.L. Lundrigan. 1993. Rapid evolution of the sex determining locus in Old World mice and rats. Nature 364: 715–718.

Vane-Wright, R.I., C.J. Humphries, and P.H. Williams. 1991. What to protect—systematics and the agony of choice. Biol. Conserv. 55: 235–254.

Waples, R.S. and D.J. Teel. 1990. Conservation genetics of Pacific salmon I. Temporal changes in allele frequency. Science 241: 1455–1460.

Wayne, L.G., D.J. Brenner, R.R. Colwell, P.A.D. Grimont, O. Kandler, M.I. Krichevsky, L.H. Moore, W.E.C. Moore, R.G.E. Murry, E. Stackebrandt, M.P. Starr, and H.G. Truper. 1987. Report on the ad hoc committee on reconcilation of approaches to bacterial systematics. Int. J. Syst. Bacteriol. 37: 463–4.

Wayne, R.K. and S.M. Jenks. 1991. Mitochondrial DNA analysis supports extensive hybridization of the Endangered Red Wolf (*Canis rufus*). Nature 351: 565–568.

Wayne, R.K. and S.J. O'Brien. 1987. Allozyme divergence within the Canidae. Syst. Zool. 36: 339–355.

Wayne, R.K., L. Forman, A.K. Neuman, J.M. Simonson, and S.J. O'Brien. 1986a. Genetic monitors of zoo populations: Morphological and electrophoretic assays. Zoo Biol. 5: 215–232.

Wayne, R.K., W.S. Modi, and S.J. O'Brien. 1986b. Morphologic variability and asymmetry in the cheetah (*Acinonyx jubatus*), a genetically uniform species. Evolution 40: 78–85.

Wayne, R.K., W.G. Nash, and S.J. O'Brien. 1987. Chromosomal evolution of the Canidae: II. Divergence from the primitive carnivore karyotype. Cytogenet. Cell Genet. 44: 134–141.

Wayne, R.K., D. Gilbert, A. Eisenhawer, N. Lehman, K. Hansen, D. Girman, R.O. Peterson, L.D. Mech, P.J.P. Gogan, U.S. Seal, and R.J. Krumenaker. 1991a. Conservation genetics of the endangered Isle Royale gray wolf. Conserv. Biol. 5: 41–51.

Wayne, R.K., S. George, D. Gilbert, P. Collins, S. Kovach, D. Girman, and N. Lehman. 1991b. Genetic change in small, isolated populations: evolution of the Channel island fox, *Urocyon littoralis*. Evolution 45: 1849–1868.

Wayne, R.K., N. Lehman, M.W. Allard, and R.L. Honeycutt. 1992. Mitochondrial DNA variability of the gray wolf—genetic consequences of population decline and habitat fragmentation. Conserv. Biol. 6: 559–569.

Williams, J.G.K., A.R. Kubelik, K.J. Livak, J.A. Rafalski, and S.V. Tingey. 1990. DNA polymorphisms amplified by arbitrary primers are useful as genetic markers. Nucleic Acids Res. 18: 6531–6535.

Williams, P.H., R.I. Vane-Wright, and C.J. Humphries. 1993. Measuring biodiversity for choosing conservation areas. In: J. LaSalle and I.D. Gauld, eds. Hymenoptera and biodiversity, pp. 309–328. Wallingford: CAB International.

Yahnke, C.J., D. Smith, W. Johnson, E. Geffen, T. Fuller, B. Van Valkenburgh, F. Hertel, M.S. Roy, and R. K. Wayne. 1996. Darwin's fox: a distinct endangered species in a vanishing habitat. Conserv. Biol. 10: 1–11.

I

APPROACHES

2

Nuclear Genetic Analysis of Population Structure and Genetic Variation Using Intron Primers

STEPHEN R. PALUMBI AND C. SCOTT BAKER

A number of different approaches have been taken to develop sensitive assays of nuclear genetic variation. The first, protein allozyme analysis, revolutionized the field of population genetics, and provided what is still the largest database on genetic variation within and between species. Allozyme studies are based on the comparison of different alleles at protein-coding loci. In general, these alleles are distinguished by charge differences, and in some cases are known to show remarkably different biochemical properties (Koehn et al., 1980; Powers 1987). Directional and balancing selection are known (Koehn et al., 1980) or inferred (Karl and Avise, 1992) to be operating at some of these loci.

Analysis of allele frequency differences at allozyme loci is generally based on the assumption that every allele is equally distant phylogenetically. That is, if there are five alleles (*a,b,c,d,e*) at a locus, then allele *a* is assumed to be as distant from allele *b* as it is from allele *e*. Analyses of populations are based solely on the frequencies of alleles. Because of this, the relationships between populations that are inferred from allozyme data are based on the similarity of allele *frequencies*, not the similarity of the alleles themselves.

By contrast to allozymes, DNA sequence data allow both a phylogenetic and an allele-frequency approach to the analysis of population structure. Such approaches use the extra information contained in a phylogenetic analysis of the alleles themselves to help understand patterns of population variation. Failure to use the information in allele relationships can sometimes lead to erroneous conclusions about genetic structure of populations, especially when there are many rare or uncommon alleles in a population. When such alleles make up a significant fraction of the data, then phylogenetic approaches are especially important. mtDNA analysis has long included the phylogenetic component of genetic data in order to address questions of population structure. In general, this approach has been named "phylogeography" by Avise and coworkers (Avise et al., 1987; reviewed in Avise, 1994).

For nuclear genes, phylogenetic approaches to population structure have lagged behind those for mtDNA. However, several recently developed techniques allow access to nuclear DNA: microsatellite analysis, RAPDs, and anonymous single-copy polymorphisms (Hadrys et al., 1992; Karl and Avise, 1993; Queller et al., 1993). These techniques have a number of advantages, such as the ability to collect data from many individuals rapidly (microsatellites); need for minimal prior information about a species (RAPD); and random coverage of single-copy regions of the genome (anonymous RFLPs). However, they do not usually provide a powerful and parallel analysis of allele frequencies and phylogeny.

Development of a system for examining phylogeny and frequency of alleles at nuclear loci requires several technical and practical elements. First, the technique should be PCR-based to allow for the use of small tissue samples, preserved material or historic collections from museums. Second, PCR primers should be broadly useful taxonomically so that prior sequence information from a species need not be obtained before beginning a survey. Generally, this means that PCR primers should be in conserved regions of the genome. Third, the amplified DNA section should include a highly variable region likely to vary within most species. Fourth, the amplified product should be amenable to RFLP or sequence analysis.

DNA segments that fit the above criteria include introns in conserved nuclear genes. Introns occur in most nuclear genes, even those whose amino acid sequences are highly conserved among animal and plant phyla. Thus, PCR primers can be designed to anneal to the conserved regions of gene exons, and these primers can be of broad taxonomic utility. Intron positions are sometimes highly conserved among taxa (e.g., for the actin gene; Kowbel and Smith, 1989). Even when intron positions vary, it is often possible to choose two regions of a conserved gene which bracket a variety of intron positions for a variety of taxa (e.g., for the tubulin genes; see Palumbi, 1996). Because introns are noncoding sequences with only a few known functional constraints, their sequences tend to be highly variable within and between species. As a result, the product of the amplified DNA segment has highly conserved flanks with a central section of variable DNA. Sequence or RFLP analysis of these regions within and between species can be a powerful phylogenetic tool for the analysis of population structure. We have designed a number of exon priming, intron crossing (EPIC) primers that can be used in a wide variety of taxa (Palumbi, 1996).

TECHNICAL DESCRIPTION

Use of EPIC Primers

EPIC primers have been useful in obtaining amplifications of a number of different introns from a wide variety of taxa (see Figure 2-3). Although they do not always perform well on every animal phylum, they are of broad enough utility to be useful in preliminary amplifications from most taxa.

Use of EPIC primers is more time-consuming than use of mitochondrial primers because the amplification product represents a mixture of both alleles that are present in the nuclear genome of diploid animals and plants. Because of this, sequencing of the PCR products requires separation of the alleles, and this is best done by cloning the PCR product into a plasmid or phagemid vector (see Palumbi and Baker, 1994 for technical details, and Slade et al., 1993 for further discussion).

Palumbi and Baker (1994) discuss some of the pitfalls and dangers of using universal PCR primers. The two most severe are (1) amplification of "junk" DNA instead of the target intron, and (2) amplification and comparison of nonhomologous loci. The first problem is solved by comparison of DNA sequences of a clone to those that should flank the intron start and stop positions. If the amino acids coded for by the nucleotides read from the clone do not correspond to those predicted from the protein, then the clone is not a clone of the intron, and should be discarded.

The problem of nonhomologous loci is more difficult to solve. EPIC primers that work across taxa are usually designed in conserved regions of a protein coding sequence. Because such regions (by definition) occur in most taxa and in most copies of the target locus, multiple PCR products are common (Slade et al., 1993). Most nuclear genes occur in small gene families, or have closely related pseudogenes, so that this is likely to be a general result of using EPIC primers. Suppose two sequences are obtained from two different species: Does the nucleotide difference between them correspond to the evolutionary distance separating these species? Or does it correspond to the evolutionary distance separating two duplicates of this locus in the genome? The first would be true of the two sequences were truly homologous. The second would be true if they were the result of a gene duplication in the past.

Nonhomologous loci often have introns of different size (see Slade et al., 1993; Palumbi and Baker, 1994), and these can usually be separated on agarose gels. In the case of the whale introns described here, we selected the largest introns that were apparent after initial amplifications. Yet how can we confirm that a given set of sequences are from homologous loci? If the sequences obtained were from different loci, a phylogenetic analysis would reflect the duplication of the loci, not the divergence of the species being investigated. Thus, intron sequence data from a set of closely related species should be checked against expectations based on other molecular and morphological criteria. Concordance of an intron cladogram with other phylogenetic analyses suggests that intron sequences are from homologous loci.

A second approach is to test whether the distribution of sequence differences is as expected from Hardy–Weinberg equilibrium. Individuals of a population will each possess both copies of a duplicated gene, whereas copies of two alleles should be distributed randomly based on their gene frequencies. For example, in our humpback whale data, we have shown the existence of two intron sequence clades, and these clades are distributed in populations in accordance with Hardy–Weinberg expectations (Palumbi and Baker, 1994). Demonstration of the allelic nature of sequence variants and the homologous nature of sequenced sections of DNA is important in every new use of EPIC primers. The complex nature of the nuclear genome makes using these primers more difficult than using universal mtDNA primers, but the potential for examining many different independent loci makes these primers attractive choices in population genetic surveys.

APPLICATIONS IN CONSERVATION GENETICS

A Whale Studied Example

The baleen whales are wide-ranging species with populations in most of the world's oceans. Although there is only limited information available about genetic structure of these species, allozyme and mtDNA analyses have shown that populations in different oceans are genetically distinct (reviewed in Baker and Palumbi, 1996). Within oceans, population structure appears to be dominated by the social systems of baleen whales. In particular, the fidelity of baleen whales to particular migratory routes appears to play a strong role in population differentiation (Baker et al., 1990).

In the best documented case, mtDNA control region sequences from humpback whales (*Megaptera novaeangliae*) show a hierarchy of interoceanic and intraoceanic genetic differences (Baker et al., 1993). Within oceans, genetic differences closely match the different migratory corridors used by these animals as they move from summer feeding grounds near the poles to more equatorial wintering grounds. These genetic differences confirm that humpback whales following particular migration routes represent true populations with long-term stability. Gene flow analyses suggest that far fewer than one breeding female per generation moves from one migratory population to another.

However, the focus on mtDNA in these studies limits the generalizations that can be made. In particular, only female migration can be estimated because mtDNA is inherited only maternally. If males rove widely among migratory populations but females do not, then mtDNA differences may accumulate in the absence of significant nuclear DNA differentiation.

In order to assess possible gender-biased gene flow in humpback whales, we estimated genetic exchange among populations using variation in an actin intron (Palumbi and Baker, 1994). Actin primers were designed from published sequences available through Genbank. We focused on the first intron of actin because its position is highly conserved among different animal phyla. Initial amplifications using the universal primers generally give multiple bands because actin is a small multigene family. We chose the largest of these bands (about 1500 bp) to compare among different baleen whale species and between different humpback whale individuals.

Intron Variation between Species

Between blue (*Balaenoptera musculus*) and humpback whales, the 1477-bp actin intron that we compared showed an average of 2.8% sequence difference. There were 28 transition substitutions and 12 transversions. Insertions and deletions were rare. A sliding-window analysis of intron sequence divergence (Figure 2-1a) examines the percentage sequence difference in a 60-bp window along the length

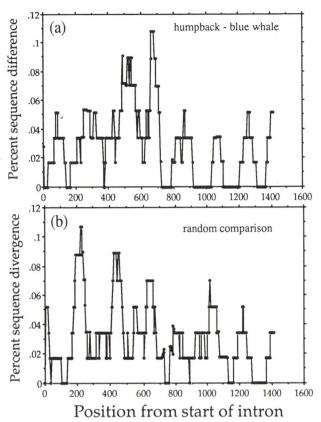

Figure 2-1 (a) Percentage sequence difference between actin introns from blue and humpback whales in a 60-bp window that is shifted sequentially along the length of the intron. Average sequence differences is 3%, but some windows have over 10% sequence difference, and some have zero. (b) Percentage sequence difference between the humpback actin intron and a sequence that has been subjected to 3% random substitution. As in the comparison of blue and humpback sequences, some 60-bp windows have over 10% variation, and some have zero. This comparison suggests that much of the variation in sequence divergence across the whale actin intron is random.

of the intron. Sequence divergence among these windows ranges from zero to about 11%, with a standard deviation of 2.5%. We can test whether there is spatial variation in sequence divergence rate along the intron by comparing the variance of sequence divergence from window to window with the variance obtained from a similar analysis of randomly mutated intron sequences. The humpback intron sequence was mutated at approximately 3% of its positions (chosen at random by Monte Carlo simulation, program available from S.R.P.). The distribution and magnitude of sequence differences along 60-bp windows (Figure 2-1b) is very similar to that obtained from the real comparison of humpback and blue whale sequences (Figure 2-1a). The range of sequence differences in the random comparison shown in Figure 2-1b is zero to 11%, with a standard deviation among windows of 2.5%. As a result, the variation in substitution rate that we observed across the actin intron probably reflects random stochastic processes, and there do not appear to be large blocks of sequence that are undergoing different substitution rates in this intron. As expected for noncoding sequences, nucleotide substitutions are not spatially clustered.

The rate of nucleotide substitution in introns can be estimated from these data using available information from the whale fossil record. The genus of humpback whales (*Megaptera*) is known from the early Pliocene–Late Miocene and has been dated at about 6 million years old. Fossil blue whales are unknown, but the genus of the blue whale (*Balaenoptera*) is thought to have arisen 15–20 million years ago. The molecular phylogeny of the baleen whales (Arnason et al., 1993; Baker et al. 1993) suggests that *Megaptera* arose from a *Balaenoptera* ancestor, and diverged from the lineage leading to blue whales in the middle Miocene. If we use the most conservative estimate of divergence dates between blue and humpback whales (6–20 million years), the rate of intron evolution falls between 0.14% and 0.46% per million years. This value is similar to that suggested by Schlotterer et al. (1991) (0.1% per million years) based on nucleotide substitutions in sequences flanking microsatellite repeats, and is slower than silent substitution rates reported for most other mammals (Wu and Li, 1985).

Intron Variation within Species: Humpback Whale Population Structure

In addition to variation in intron sequence between species, we also observed variation within species. We sequenced alleles of these actin introns by first cloning PCR products into T-tailed vectors (see Palumbi and Baker (1994) for details). This cloning step was required because the diploid nature of nuclear genes leads to heterogeneous intron sequences if different alleles amplified in the same PCR reaction are not separated.

Sequences of 10 introns of different alleles (hereafter called intron alleles, although the term alleles usually refers to the functional product of a locus) from 7 different humpback whales were examined phylogenetically (Figure 2-2). Alleles group into two clades that differ on average by 8–9 substitutions. Although this is a low percentage sequence difference ($\sim 0.5\%$), the length of the intron provides a large number of phylogenetically informative nucleotide substitutions.

The individual whales examined were either from the feeding aggregation that summers off the coast of central California or from the population that winters in Hawaii. These populations are known to show very different patterns of mitochondrial DNA variation: 59 out of 60 Hawaiian animals tested had haplotypes never seen in the California population (see Baker and Palumbi (1996) for review). Despite this distinction in mtDNA, the intron alleles did not cluster by geographic origin. Alleles from Californian and Hawaiian animals are interspersed in both clades (Figure 2-2). RFLP analyses of introns from additional animals confirms that Hawaii and California populations are not distinct at this locus (Palumbi and Baker, 1994). However, populations from the North Atlantic and Antarctic oceans differ from those in the north Pacific in the frequency of actin intron RFLPS, showing that over large spatial scales mtDNA and nuclear results are similar (Palumbi and Baker, 1994).

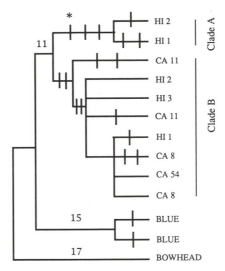

Figure 2-2   Allele phylogeny of actin intron sequences from humpback whales from Hawaii (HI), and California (CA). Sequences from blue and bowhead whales are used as outgroups. See Palumbi and Baker (1994) for full sequences and details of methods. Numbers above branches are interspecific branch lengths. Vertical bars represent polymorphic substitutions.

Discrepancies between Mitochondrial and Nuclear Results

A powerful reason for examining genetic patterns at different, independent loci is that multiple loci provide a more powerful view of the genetic structure of a species than if a single locus is examined. In the present case, however, two loci give very different results: mtDNA haplotypes differ between Californian and Hawaiian humpback whales, but nuclear actin introns do not. How can these results be reconciled, and what does this tell us about the population structure of humpbacks?

In principle there are several possible reasons for the discrepancy:

1. *Males move between whale populations but females do not.* Because mtDNA is maternally inherited, genetic subdivision at this locus indicates only that female movement among populations is low. However, males might have high migration rates, and might cause high gene flow at nuclear loci without increasing gene flow at mitochondrial loci. This explanation might also provide the mechanism whereby the winter-time "song of the humpback whale," a presumed mating display, can change in a similar fashion from year to year in Mexico (the presumed wintering grounds of the California whales) and in Hawaii (Payne and Guinee, 1983).

However appealing this explanation, the whales that have been seen both in Hawaii and in Mexico have not exclusively (or even mostly) been males. Furthermore, the single whale in our Hawaiian sample with a distinctly Californian mtDNA haplotype was a female (C. Scott Baker, unpublished observation). Males may move between populations *and breed* more often than do females, but more data are needed to demonstrate this.

2. *Population differences are recent, and nuclear loci have not yet diverged by genetic drift.* The mitochondrial locus is haploid and maternally inherited. As a result, the genetically effective population size at this locus in a given population is only 1/4 that of a diploid, biparentally inherited nuclear locus. Because of this difference, genetic drift is a more powerful force at mitochondrial loci, and populations that have been completely separated for a short time can show distinct mitochondrial haplotypes. Neigel and Avise (1986) showed that within about N generations (where N is the number of breeding females in a population), mitochondrial haplotypes will often be completely different in two populations (i.e., reciprocally monophyletic), whereas for nuclear loci this level of distinction takes about four times longer.

Thus, if the Hawaiian and Californian populations have been separated for longer than N generations but fewer than $4N$, mtDNA haplotypes would be reciprocally monophyletic (i.e., no alleles in common between populations) but nuclear alleles will still be shared. However, even though nuclear alleles might still be shared for $4N$ generations, the frequencies of these shared alleles are expected to diverge well before $4N$ generations. Thus, there may be only a narrow window in time during which the mitochondrial loci are reciprocally monophyletic and the nuclear loci are identical in frequency.

3. *Average nuclear loci are different between populations, but this actin locus is a random exception.* Although nuclear loci coalesce (i.e., become monophyletic) after about $4N$ generations, this is an average over many loci, and individual cases can vary from this average. Nei (1987) gives the standard deviation of fixation time of a substitution among loci as $2.14N$. Thus, some substitutions will become fixed in a population far sooner than others, and some substitutions will remain polymorphic for much longer than the $4N$-generation average that is so often quoted in the population genetic literature. This is one of the strongest reasons for using multiple loci to examine population structure of a species. Thus, it is possible that our results from a single actin intron do not reflect the average among all nuclear loci, but instead reflect the chance maintenance of actin intron polymorphisms at this particular locus.

4. *Most nuclear loci are different between populations, but this actin locus is under selection to maintain polymorphisms.* Directional selection is a powerful force that can change gene frequencies very rapidly within and between species (Nei, 1987). Although directional selection tends to decrease polymorphism, polymorphisms can be maintained in a population for long periods of time by several different types of selection. Balancing selection, frequency-dependent selection and over-dominant selection have been invoked recently to explain the persistence of polymorphisms over space (in oysters (Karl and Avise, 1992), or for long periods of time (in s-locus alleles of seed plants (Ioeger et al., 1990) or in MHC loci of mammals (Hughes and Nei, 1988)).

In the present case, actin intron variation may be maintained in Hawaiian and Californian populations by selection maintaining different actin protein alleles. Although this is a possibility that is difficult to disprove, actin is a very highly conserved protein, with over 95% identity of amino acids between phyla. Thus, it is unlikely that actin alleles differing in amino acid sequence are being maintained in whale populations by balancing selection. It is important to realize, however, that the dynamics of intron polymorphisms may in some cases be dominated by forces acting at the flanking protein-coding regions. Even though introns are largely neutral, they are linked to nonneutral protein coding genes.

Other Examples of Intron Sequences in Conservation Research

Lessa (1992) described a preliminary study of hemoglobin intron variation in pocket gophers. Three allelic variants in a small globin intron (126 bp) were detected in a small area in southern California and Arizona. Ruano et al. (1992) used sequences of noncoding region of the homeobox locus *Hox*2 to investigate genetic variation in hominoid species. Slade et al. (1993) screened a number of different EPIC primers in various vertebrate and invertebrate taxa (Figure 2-3). They showed that, in general, intron differences could be resolved with sequencing or with targeted digestion of PCR products using restriction enzymes known from sequencing to recognize polymorphic nucleotide positions. To date, few studies have taken full advantage of the phylogenetic information offered by intron sequence analysis. Yet such studies are likely to add uniquely to understanding of population biology of small populations. In particular, Roderick and Villablanca (1995) suggest that statistical analysis of colonization events and the effects of bottlenecks is made easier and more rigorous by a dual phylogenetic and allele-frequency approach.

A second use of intron sequences has been to confirm very low genetic variation in small populations of threatened species. The northern hairy-nosed wombat (*Lasiorhinus krefftii*) in Australia shows no mtDNA or intron sequence variation (Taylor et al., 1994). This low variation in normally variable sequences raised concerns about possible inbreeding effects due to small population size. Although microsatellite loci retained genetic variation in hairy-nosed wombats, this was only about 41% of that in populations of similar species or in historical collections of the same species. Thus, low sequence variation and low microsatellite variation appear to be correlated. The Hawaiian monk seal (*Monachus monachus*) also

Taxonomic breadth of EPIC primers				
Actin	X	X	X		X
Aldolase	X	X			
Arginine Kinase			X		
Creatine Kinase	X	X			X
Elongation factor	X	X	X	X	X
Histone	X				
Tubulin	X	X	X		X

Figure 2-3 Taxa for which EPIC primers have been tested. Primers are listed and further described in Slade et al. (1993), Lessa and Applebaum (1993), Palumbi and Baker (1994), and Palumbi (1996). Taxa depicted are mammals, birds and lower vertebrates, arthropods, cnidaria, and echinoderms. In the figure a X refers to successful amplification with the listed primers, although in some cases the amplified product may have been identified as a pseudogene or a intronless locus (Slade et al., 1993).

shows low heterogeneity in *d*-loop sequences (P. Armstrong, personal com-munication). Because these seals are the most endangered seals in the United States, and have been through two successive bottlenecks, variation at nuclear alleles might also be low. Sequences of introns from two tubulin loci confirm low genetic variation: no sequence differences have been found to date (P. Armstrong, personal communication). It is unclear whether this low genetic variation is due to exploitation by humans in the last century or is natural for this species.

CONCLUSIONS

The Need for Nuclear Loci in Conservation Genetics

Owing to its maternal mode of inheritance and absence of recombination (Wilson et al., 1985; Avise et al., 1987), mtDNA has been a favored genetic system for analysis of population structure. In general, mtDNA offers two important advantages over nuclear genetic markers. First, the phylogenetic relationship of mtDNA types reflects the history of maternal lineages within a population or species. Second, all else being equal, the effective population size of mtDNA genomes is one-fourth that of autosomal nuclear genes, leading to a higher rate of local differentiation by random drift (Neigel and Avise, 1986). The ability to detect local differentiation may also be enhanced by the rapid pace of mtDNA evolution, which is generally considered to be 5–10 times faster than that of nuclear DNA in most species of mammals (Wilson et al., 1985; but see Avise et al., 1992; Baker et al., 1993; Martin and Palumbi, 1993).

Despite its advantages, the analysis of mtDNA variation is widely recognized as suffering from weaknesses that limit its utility in some applications (Palumbi and Baker, 1994). First, the mitochondrial genome is only a single genetic locus. Lack of recombination links the 37 different genes of the animal mitochondrial genome into a single genetic entity transmitted in its entirety to progeny. Reliance on a single genetic locus diminishes power to detect significant spatial or temporal structure of populations. This is because genetic drift in populations involves random changes in gene frequencies and these changes will follow different trajectories at independent loci. Thus, an investigation of spatial or temporal genetic structure should ideally consider the variety of patterns seen in different loci (Slatkin and Maddison, 1990). In addition, the phylogenetic tree derived from a single locus may not accurately reflect the genetic history of a population, species or genus (Pamilo and Nei, 1988; Ball et al., 1990). It has been argued that the concordance or contrasts of phylogenetic patterns across several loci provides the most insight into historical processes (Ball et al., 1990 Karl and Avise, 1992).

A second weakness with exclusive reliance on mtDNA data is that such analyses allow only the reconstruction of maternal lineages (Wilson et al., 1985; Avise et al., 1987; Moritz et al., 1987). Although this can be a strength for describing the influence of maternal lineages on population structure (see above), mtDNA may not be useful for detecting paternal gene flow between populations.

How to Shop for a Genetic Locus

Although it's clear that our understanding of systematic relationships and population structure can benefit from complementary studies of nuclear and mtDNA variation, it is not always clear which nuclear markers are most suitable for what level of analysis. This volume reviews many of the different types of genetic approaches that have been developed. Although each technique has its own strengths and weaknesses, sequence analysis of single-copy nuclear DNA is likely to provide a level of resolution most comparable to those from mtDNA studies (Slade et al., 1993; Palumbi and Baker, 1994). A high-resolution approach to single-copy nuclear DNA variation can take advantage of current theoretical developments in phylogenetic reconstruction (Swofford and Olsen, 1990), the cladistic measurement of gene flow (Slatkin and Maddison, 1989, 1990) and the "coalescent" approach to population genetic models (e.g., Hudson 1990). This genealogical approach to allelic variation is not applicable to allozymes, multilocus DNA "fingerprinting," single-locus mini- and microsatellite systems. As a result, in cases where the genealogy of alleles can influence the interpretation of results, EPIC amplification and sequencing is an appropriate approach.

REFERENCES

Arnason, U., A. Gullberg, and B. Widegren. 1993. Cetacean mitochondrial DNA control region: sequences of all extant baleen whales and two sperm whale species. Mol. Biol. Evol. 10: 960–970.

Avise, J.C. 1994. Molecular markers, natural history, and evolution. New York: Chapman and Hall.

Avise, J.C., J. Arnold, R.M. Ball, E. Bermingham, T. Lamb, J.E. Neigel, C.A. Reeb, and N.C. Saunders. 1987. Intraspecific phylogeography: the mitochondrial DNA bridge between population genetics and systematics. Annu. Rev. Ecol. Syst. 18: 489–522.

Avise, J.C., B. Bowen, T. Lamb, A.B. Meylan, and E. Berminghom. 1992. Mitochondrial DNA evolution at a turtle's pace: evidence for low genetic variability and reduced microevolutionary rate in Testudines. Mol. Biol. Evol. 9: 457–473.

Baker, C.S. and S. R. Palumbi. 1996. Population structure, molecular systematics and forensic identification of whales and dolphins. In: J.C. Avise and J.L. Hamrick, eds. Vertebrate Conservation Genetics: Case Histories from Nature, pp. 10–49. New York: Chapman and Hall.

Baker, C.S., S.R. Palumbi, R.H. Lambertsen, M.T. Weinrich, J. Calambokidis, and S.J. O'Brien. 1990. The influence of seasonal migration on the distribution of mitochondrial DNA haplotypes in humpback whales. Nature 344: 238–240.

Baker, C.S., A. Perry, J.L. Bannister, M.T. Weinrich, R.B. Abernethy, J. Calambokidis, J., Lien, R.H. Lambertsen, J. Urban-Ramirez, O. Vasquez, P.J. Clapham, A. Alling, S.J. O'Brien, and S.R. Palumbi. 1993. Abundant mitochondrial DNA variation and world-wide population structure in humpback whales. Proc. Natl. Acad. Sci. U.S.A. 90: 8239–8243.

Ball, R.M., J.E. Neigel and J.C. Avise. 1990. Gene genealogies within the organismal pedigrees of random mating populations. Evolution 44: 360–370.

Hadrys, H., M. Balcik, and B. Schierwater. 1992. Application of random amplified polymorphic DNA (RAPD) in molecular ecology. Mol. Ecol. 1: 55–63.

Hudson, R. 1990. Gene genealogies and the coalescent process. In: D. Futuyma and J. Antonovics, eds. Oxford Surveys in Evolutionary Biology, vol. 7. Oxford, U.K.: Oxford University Press.

Hughes A. and M. Nei. 1988. Pattern of nucleotide substitution at major histocompatibility complex class I loci reveals over dominant selection. Nature 335: 167–170.

Ioerger, I.R., A.G. Clarke, and T.-H. Kao. 1990. Polymorphism at the self-incompatibility locus in Solanaceae predates speciation. Proc. Natl. Acad. Sci. U.S.A. 87: 9732–9735.

Karl, S.A. and J.C. Avise. 1992. Balancing selection at allozyme loci in oysters: Implications from nuclear RFLPs. Science 256: 100–102.

Karl, S.A. and J.C. Avise. 1993. PCR-based assays of Mendelian polymorphisms from anonymous single-copy nuclear DNA: techniques and applications for population genetics. Mol. Biol. Evol. 10: 342–361.

Koehn, R.K., R.I. Newell, and F. Immerman. 1980. Maintenance of an aminopeptidase allele frequency cline by natural selection. Proc. Natl. Acad. Sci. U.S.A. 77: 5385–5389.

Kowbel, D.J. and M.J. Smith. 1989. The genomic nucleotide sequences of two differentially expressed actin-coding genes from the sea star *Pisaster ochraceus*. Gene 77: 297–308.

Lessa, E. 1992. Rapid surveying of DNA sequence variation in natural populations. Mol. Biol. Evol. 9: 323–330.

Lessa, E. and G. Applebaum. 1993. Screening techniques for detecting allelic variation in DNA sequences, Mol. Ecol. 2: 119–129.

Martin, A. and S.R. Palumbi. 1993. Body size, metabolic rate, generation time, and the molecular clock. Proc. Natl. Acad. Sci. U.S.A. 90: 4087–4091.

Moritz, C., T. E. Dowling, and W.M. Brown. 1987. Evolution of animal mitochondrial DNA: relevance for population biology and systematics. Annu. Rev. Ecol. Syst. 18: 269–292.

Nei, M. 1987. Molecular evolutionary genetics. New York: Columbia University Press.

Neigel, J.E. and J.C. Avise. 1986. Phylogenetic relationships of mitochondrial DNA under various demographic models of speciation. In: E. Nevo and S. Karlin, eds. Evolutionary processes and theory, pp. 513–534. New York: Academic Press.

Palumbi, S.R. 1996. PCR and molecular systematics. In: D. Hillis, C. Moritz, and B.K. Mable, eds. Molecular systematics, 2nd ed., pp. 205–248. Sunderland, Mass.: Sinauer Press.

Palumbi S.R. and C.S. Baker. 1994. Contrasting population structure from nuclear intron sequences and mtDNA of humpback whales. Mol. Biol. Evol. 11: 426–435.

Pamilo, P. and M. Nei. 1988. Relationships between gene trees and species trees. Mol. Biol. Evol. 5: 568–583.

Payne, R. and L.N. Guinee. 1983. Humpback whale (*Megaptera novaeangliae*) songs as an indicator of "stocks." In: R. Payne, ed. Communication and behavior of whales. Boulder, Colo.: Westview Press.

Powers, D.A. 1987. A multidisciplinary approach to the study of genetic variation within species. In M.E. Feder, A. Bennet, W. Burggren, R.B. Huey, eds. New directions in physiological ecology, pp. 38–70. Cambridge, U.K.: University of Cambridge Press.

Queller, D.C. , J.E. Strassmann, and C.R. Hughes. 1993. Microsatellites and kinship. Trends Ecol. Evol. 8: 285–288.

Roderick, G. and F. Villablanca, 1995. Genetic and statistical analysis of colonization. In B.A. McPheron and G.J. Steck, eds. Economic fruit flies: A world assessment of their biology and management, pp. 281–290. St. Lucie. Press: Delray Beach, Florida.

Ruano, G., J. Rogers, A.C. Fergusonsmith, and K.K. Kidd. 1992. DNA sequence polymorphism within hominoid species exceeds the number of phylogenetically informative characters for a HOX2 locus. Mol. Biol. Evol. 9: 575–586.

Shlötterer, C., B. Amos, and D. Tautz. 1991. Conservation of polymorphic sequence loci in certain cetacean species. Nature 354: 63–65.

Slade, R.W., C. Moritz, A. Heideman, and P.T. Hale. 1993. Rapid assessment of single copy nuclear DNA variation in diverse species. Mol. Ecol. 2: 359–373.

Slatkin, M. and W.P. Maddison. 1989. A cladistic measure of gene flow inferred from the phylogenies of alleles. Genetics 123: 603–613.

Slatkin, M. and W.P. Maddison. 1990. Detecting isolation by distance using phylogenies of genes. Genetics 126: 249–260.

Swofford, D. and G. Olsen. 1990. Phylogeny reconstruction. In: D. Hillis and C. Moritz, eds. Molecular systematics, pp. 411–501. Sunderland, Mass.: Sinauer Press.

Taylor, A.C., W.B. Sherwin, and R.K. Wayne. 1994. Genetic variation of microsatellite loci in a bottlenecked species: the northern hairy-nosed wombat Lasiofhinus krefftii. Mol. Ecol. 3: 277–290.

Wilson, A.C., R.L. Cann, S.M. Carr, M. George, U.B. Gyllenstein, K.M. Helm-Bychowski, R.G. Higuchi, S.R. Palumbi, E.M. Pragger, R.D. Sage, and M. Stoneking. 1985. Mitochondrial DNA and two perspectives on evolutionary genetics. Biol. J. Linn. Soc. 26: 375–400.

Wu, C.-I. and W.-H. Li. 1985. Evidence for higher rates of nucleotide substitution in rodents than in man. Proc. Natl. Acad. Sci. U.S.A. 82: 1741–1745.

3

Application of Anonymous Nuclear Loci to Conservation Biology

STEPHEN A. KARL

A primary goal of many conservation programs is to determine the degree of genetic variation that exists in populations. By assessing genetic variation, conservation managers are able to estimate the degree of inbreeding in the population, determine the effective population size, identify individuals, identify populations, determine limits to species distinction, and monitor interspecific hybridization. There are currently many ways to assess genetic variation in natural populations, several of which are covered in this text. Historically, behavioral and morphological variation served as surrogates of genetic variation. More recently, techniques such as starch gel electrophoresis of soluble proteins (isozymes), mitochondrial and nuclear DNA restriction fragment analyses, DNA fingerprinting and mini- and microsatellite analyses have supplemented and greatly enhanced this endeavor. This chapter outlines the methodology and application of one of these approaches: the elucidation of DNA sequence variation in arbitrarily chosen, anonymous, single-copy nuclear DNA (ascnDNA) loci.

An ascnDNA approach has four main advantages for the determination of genetic variation: (1) it assays loci randomly distributed through the genome; (2) there is a high ratio of information output to the effort input; (3) the characters being assayed are generally not under direct selection pressure; and (4) these markers are widely applicable to many situations in conservation biology.

In principle, an almost unlimited number of ascnDNA regions can be screened. This allows a thorough representation of the overall genetic variation in a population under consideration. Each ascnDNA locus is arbitrarily chosen from the single-copy fraction of the entire nuclear genome and is as likely to represent a coding as a noncoding piece of DNA. The arbitrary nature of these DNA regions means that the data collected are from a diverse group of loci, some under selection, some neutral but linked to selected loci, and some strictly neutral. Assaying such a diverse group of markers helps to eliminate interpretation bias that can result from examining only a single class of loci. The type of genetic variation most easily revealed by this method is restriction endonuclease enzyme cut-site differences. The alleles at each locus are defined by the presence or absence of specific sites. The identity and frequency of each allele provides information on the kind, magnitude, and geographic arrangement of genetic variation occurring in individuals and populations. There is no reason to believe that, even when a locus itself is under selection, the presence or absence of a specific endonuclease restriction site is of any functional importance. In this case, the alleles per se most likely are neutral. This can be particularly useful since many studies and analyses assume that the markers being assayed are not under selection. The alleles also are co-dominantly inherited, which allows each individual to be unambiguously assigned a genotype. For loci with three or more alleles, the evolutionary relationship of the alleles can be determined. Taken together the particular traits of ascnDNA loci permit a fine-scale resolution of genetic relationships.

The detection of variation in ascnDNA involves the development of specific oligonucleotide primers flanking the ascnDNA locus. These primers are used for *in vitro* enzymatic amplification of the locus by the polymerase chain reaction (PCR) (Saiki et al., 1985, 1988; Mullis and Faloona, 1987). The inclusion of this enzymatic amplification step not only permits genetic analyses to be done nondestructively but eliminates the need for large amounts of radioactive nucleotides and long exposure times that are necessary with autoradiography. The efficiency of the ascnDNA technique allows the screening of a small number of individuals (about 12) for a large number of loci (about 40) or a large number of individuals (hundreds) at a small number of loci in only a few days.

A typical ascnDNA study results in the collection of haplotype or genotype data at several loci for each individual. This type of multilocus genotype data is useful for determining levels of heterozygosity (Hillis et al., 1991; Zhang et al., 1993), assessing parentage and kinship (Duvall et al., 1976; Quinn et al., 1989), elucidating population subdivision (Karl et al., 1992; Scribner et al., 1994), detecting hybridization (Pella and Milner, 1987; Hauser et al., 1991; Karl et al., 1995), and forensics analysis (Harvey, 1990). An ascnDNA study is most strongly recommended in circumstances where isozyme electrophoresis has failed to reveal sufficient variation, when existing data (such as isozyme and mitochondrial data) provide conflicting results, when it is imperative that genetic assays be done nondestructively, or when the biparentally inherited nuclear data are compared with maternally inherited mitochondrial DNA (mtDNA) to allow the elucidation of gender-specific or other ecological attributes.

TECHNICAL DESCRIPTION

Since the detailed strategy for the production and screening of ascnDNA loci is published elsewhere (Karl and Avise, 1993), only a brief discussion will be included here. In addition, the laboratory manual *Current Protocols in Molecular Biology* (Ausubel et al., 1993) is an excellent reference and can be consulted for details on most of the standard laboratory procedures.

The first step in this procedure is to create a recombinant nuclear DNA library. Nuclear DNA is isolated from a single individual and digested with a restriction endonuclease that cuts at many sites. The DNA fragments from 500 to 5000 bp are gel isolated and ligated into a phagemid vector (such as Stratagene's pBSSK +). Recombinant phagemids are recovered by transforming *Escherichia coli* cells (such as Stratagene's SURE cells) and plating them on the appropriate antibiotic selection media.

Each resulting colony contains a different fragment of nuclear DNA from the individual cloned. The size of the cloned fragment is determined by gel electrophoresis and the number of copies of that particular fragment in a haploid genome is estimated by a dot blot analysis (Figure 3-1; Hames and Higgins, 1985).

Next, sequences of the cloned fragments are determined by standard procedures (such as United States Biochemical's Sequenase Kit). Even though it is necessary only to determine the sequence of the first 100 to 200 nucleotides from each end of each cloned DNA fragment, sequence information from the entire cloned fragment is very useful. Not only can the accuracy of the subsequent restriction digestion fragment profiles be evaluated by comparison with the sequence information, but primer design is facilitated by a greater choice of

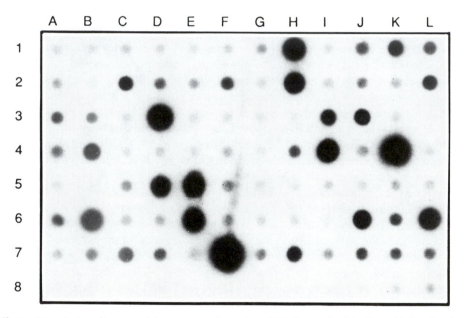

Figure 3-1   Autoradiogram of the green turtle nuclear DNA library dot blot (from Karl and Avise, 1993). Plasmid DNA extracted from each clone was fixed to a nylon membrane. The membrane was hybridized overnight with radioactively labeled total cell DNA and then washed under stringent conditions (Hames and Higgins, 1985). The intensity of the radioactive signal on the resulting autoradiogram is an approximate indication of the genomic copy number of the cloned fragment. For example, the cloned fragments that are in sectors 1H, 4K, and 7F were scored as highly repetitive. The clones in sectors 1K and 4B were scored as middle repetitive. The dot in sector 1A was a negative control containing phagemid DNA only. Dot sectors containing 150 ng of total cell DNA were also included as positive controls for hybridization efficiency (data not shown). Aliquots of DNA from the clones that produce little or no radioactive signal (such as 1B through 1F) were reapplied to a new filter and rehybridized. This second hybridization helps to eliminate false negatives (i.e., repetitive clones that appeared, for technical reasons, to be single-copy on the first blot). Repetitive DNA fragments such as mini- and microsatellite loci also can be identified by hybridizing the blot with the appropriate end-labeled probe (($GT)_{10}$, $(ATT)_6$, $(GATA)_5$, etc.).

priming sites as well as the ability to check the entire sequence for "false" priming sites.

Primers suitable for PCR amplification are designed from the sequence. Computer programs for this purpose (such as OLIGO 4.0; Rychlik and Rhoads, 1989; Rychlik, 1992) are available and extremely helpful. Suitable primers are from 18 to 25 nucleotides long and do not contain regions of significant secondary structure or interprimer base pairing. Each sequence is also analyzed for its ability to produce a protein by searching for putative open reading frames (ORFs). If an ORF is identified, the primers are designed so that the 3' end of the primer is located at the more highly conserved second nucleotide position of a putative amino acid codon. Once primer sequences are selected, they can be commercially synthesized in sufficient amounts for at least 2000 amplifications at the cost of about US$50–100 for each primer.

The optimal amplification conditions for each primer pair must be determined empirically. Many different amplification conditions can be used; however, testing each in turn and in combination would be prohibitive logistically and financially. Furthermore, previous experience has indicated that the only factors that significantly affect amplification success are the annealing temperature, $MgCl_2$ concentration, and the addition of bovine serum albumin (Karl and Avise, 1993). Test PCR reactions are performed on both cloned and total cell DNA using 50, 55, and 60°C as annealing temperatures. If none of these conditions results in the amplification of a single DNA fragment of the expected size, other reaction conditions can be tried. However, primers that fail to give good results under these conditions are most likely not of use (Karl and Avise, 1993). The time and expense of optimization must be weighed against that of making primers to a new locus.

Finally, polymorphisms are identified. To detect restriction endonuclease recognition sites that are variable in the population, nuclear DNA samples from 10–12 individuals most likely to be distantly related (such as from sites spanning the range of a species) are amplified with the ascnDNA primers. The amplified DNA is used directly in the restriction digestions without further purification. With the aid of 96-well microliter plates and multichannel pipettes, 12 individuals can be screened with 40 restriction endonucleases at a single locus in one day. Restriction site polymorphisms are indicated by the pattern of restriction fragment gains and losses seen in ethidium bromide-stained agarose gels (Figure 3-2). Once polymorphic enzyme sites are identified at a particular locus, population samples can be genotyped easily.

The data generated from ascnDNA loci have attributes of both isozyme electrophoretic and mtDNA assays. Isozyme data are unordered, co-dominant, diploid, multistate characters from several biparentally inherited loci. mtDNA data are phylogenetically ordered, haploid, multistate characters from a single maternally inherited locus. ascnDNA data are phylogenetically ordered, co-dominant, diploid, multistate characters from several biparentally inherited loci. ascnDNA data provide many of the benefits of both isozyme and mtDNA surveys. By assaying variation at the level of the DNA and by producing phylogenetically ordered characters, ascnDNA markers allow a more fine-scale resolution than isozymes. In addition, the multilocus aspect of ascnDNA data circumvents the potential bias

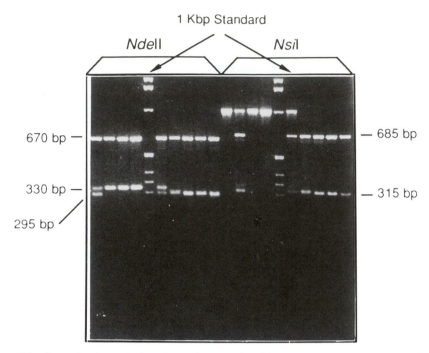

Figure 3-2 Example of an ethidium bromide-stained 2.5% agarose gel for the oyster ascnDNA locus CV-32. Lanes 1 to 3 and 6 to 10 are NdeII digestions of PCR-amplified DNA from six different individual oysters. Lanes 4 and 14 are amplified, digested DNA from the clone pdCV-32. Lanes 5 and 15 are 500 ng of 1 kb size standard (Gibco-BRL, Inc.). Lanes 11 to 14 and 16 to 20 are NsiI digestions of the same samples. All individuals for each enzyme are in the same order on the gel with the exception that the order of individuals in lanes 11 and 12 is reversed relative to 1 and 2. There was a single polymorphic NsiI recognition site. Individuals in lanes 11 and 13 were homozygous for the absence of this site, lanes 17 to 20 were homozygous for the presence of this site, and lanes 12 and 16 were heterozygous. The NdeII digestions illustrate the size polymorphism that also was present at this locus. The total size of all the restriction fragment in lanes 2 to 4 is 35 bp larger than that of lanes 7 to 10. Although it is possible that this difference is due to an additional restriction site producing an unobserved 35 bp fragment, there are two lines of evidence that indicate that this is not the case. First, a 35 bp size difference was consistently observed in these individuals with several different restriction endonucleases. Second, this size polymorphism can be directly detected in undigested amplified DNA.

introduced by assaying only mtDNA, which is a single-locus (Avise and Ball, 1990; Ball et al., 1990). One limitation to ascnDNA data, however, is that, given the relatively small size of the PCR-amplified DNA fragment (typically about 1 kb and the presumed slower rate of evolution in single-copy nuclear DNA relative to mitochondrial DNA (Brown et al., 1979; but see Vawter and Brown, 1986), normally few linked restriction site polymorphisms will be found. In fact, ascnDNA studies on green turtles, oysters, hydrothermal vent clams, and tube worms never revealed more than four polymorphic restriction sites at any single locus. This was true even though up to 40 restriction enzymes were routinely surveyed.

Fortunately, recent advances in PCR methodology may permit the efficient amplification of large DNA fragments (Barnes, 1994). Larger pieces of DNA will contain more restriction recognition sites and are therefore more likely to reveal polymorphisms. In addition, PCR-based sequencing protocols and the application of techniques such as single-strand conformational polymorphisms (SSCP) (Orita et al., 1989a,b; chapter 11) and denaturing gradient gel electrophoresis (Fischer and Lerman, 1983; Sheffield et al., 1990) promote efficient and effective ways of obtaining DNA sequence information from a large number of individuals (Lessa, 1992; Lessa and Applebaum, 1993). The collection of sequence information from ascnDNA loci should substantially increase the number of haplotypes observed since any base substitution at the locus will be revealed rather than only those substitutions that create or destroy restriction endonuclease recognition sites. In considering a haplotype analysis of ascnDNA, it should be noted that assumptions of complete linkage between polymorphic sites within a DNA fragment may occasionally be violated. This will create difficulty in inferring haplotypes from sequence (or restriction) data collected from diploid individuals. However, statistical methods are available that can detect recombination if it has occurred (Stephens, 1985; Hartl and Sawyer, 1991).

Overall, there is approximately a six-month development time necessary for any new ascnDNA survey. During this time there is some technical expertise required for the construction and screening of libraries and the sequencing and designing of primers. Once this initial phase is completed, however, the effort and expense of screening population samples are minimal and highly competitive with other molecular techniques (including isozyme electrophoresis). The true strength of the ascnDNA approach is that the data generated are from several unlinked loci each with co-dominant, ordered alleles, thereby providing data which are very useful for the elucidation and evaluation of many genetic attributes of populations.

APPLICATIONS IN CONSERVATION GENETICS

Given the recent introduction of the ascnDNA technique (Karl and Avise, 1993), few applications of this approach to conservation issues have been published. However, the studies available do provide examples of the general utility of this method. In this section, I will briefly discuss the application of ascnDNA markers to the determination of individual and population heterozygosity levels, parentage and kinship identification, the determination of hybrid status, the identification of population genetic subdivision, and forensics.

Genetic Variation

Estimating genetic variation within both individuals and populations is central to many conservation studies. An important factor in determining a population's viability is its effective population size. When the effective size of a population is reduced, inbreeding is increased and genetic variation is lost. It is this loss of genetic variation which reduces a species' capacity to adapt to changing environments. However, there has recently been considerable debate about the importance of maintaining heterozygosity in species under consideration for conservation

(Gilpin and Wills, 1991; Hughes, 1991; Miller and Hedrick, 1991; Vrijenhoek and Leberg, 1991). Nonetheless, the resolution of this debate will rely on the comparison of levels of heterozygosity in natural populations with the population's "health" or stability. ascnDNA markers are ideal for estimating genomic heterozygosity. In this application they are similar to isozymes in allowing the direct assessment of genetic variation at multiple, independent loci. ascnDNA markers are co-dominant alleles from a single locus; therefore, standard population parameters (allele frequency, proportion of loci polymorphic [P], average heterozygosity [H], etc.) can be estimated easily and directly. ascnDNA loci may be superior to isozymes since they are able to detect variation when isozymes do not exist (Karl et al., 1992; unpublished data). Furthermore, unlike isozymes, ascnDNAs most likely represent more than a single class of nuclear loci (Hedrick et al., 1986), providing an average view of genetic diversity across the entire genome.

Other nuclear DNA techniques such as random amplified polymorphic DNA (RAPD) (Chapter 4; Gibbs et al., 1994) and mini- and microsatellites (Chapters 15–17; Gilbert et al., 1990; Rassmann et al., 1994; Taylor et al., 1994) have also been used to estimate genetic variation within populations. However, the inability to determine the exact number of loci screened or to assign specific genotypes to individuals creates difficulty in quantifying individual and population heterozygosity levels except in the most extreme cases (Lynch and Milligan, 1994). ascnDNA as well as single-locus mini- and microsatellite loci may then be the markers of choice for species with little genetic variation due to evolutionarily small effective population sizes. These single-locus markers are generally highly variable and allow the unambiguous assignment of genotype to all individuals at several independent loci and can provide an accurate estimate of genome-wide heterozygosity.

Parentage and Kinship

Many conservation projects focus on populations that consist of only a few individuals (often extended families). It is frequently necessary and desirable to determine the level of genetic relatedness among all members of these populations. Knowing the familial relationships of all individuals, conservation managers are better able to design breeding programs which can equalize the genetic contribution of parents to subsequent generations. Single-locus and multilocus DNA fingerprints have been used to determining genetic relationships among closely related individuals (Burke and Bruford, 1987; McRae and Kovacs, 1994; Signer et al., 1994). Multilocus estimates can be difficult to interpret because the number of loci and the allelism at each locus are often unknown. Single-locus fingerprinting may be the most useful method for determining genetic relatedness among closely related individuals (Duvall et al., 1976; Quinn et al., 1989; Tautz, 1989; Weber and May, 1989). The assignment of genotypes at several loci permits the determination of parentage based on the joint probability of putative offspring–parent genotypes (Chakraborty et al., 1974; Weir, 1990). ascnDNA markers offer many of the advantages of single-locus fingerprinting, but are generally less variable. In situations where genetic variation is abundant, ascnDNA markers may be superior to single-locus fingerprints since they are likely to possess a more manageable number of alleles per locus. Both methods allow the identification of individual genotypes at several independent loci. An ideal study, therefore, might include data from both single-locus fingerprint and ascnDNA markers. By combining data from both types of loci, a large range of genetic variation can be assayed and analyzed. This approach is capable of providing data from various hierarchical levels: from siblings through populations and species.

Population Subdivision

By far the most common and familiar use of genetic data has been in the determination or documentation of population subdivision. Knowing the extent of population subdivision is critical to the development of stock-specific management practices that prevent the extirpation of isolated populations. Although nearly all of the genetic markers discussed in this text can be used for the estimation of population subdivision (Scribner et al., 1994), each will have its own strengths and weaknesses. Essentially the process of identifying subdivided populations is an extension of that for parentage and kinship discussed above. ascnDNA markers were originally developed in order to detect population subdivision (Karl and Avise, 1992; Karl et al., 1992) and therefore are well suited to this task. The next section of this chapter details two case studies that are particularly illustrative of the use of ascnDNA markers for population genetics. One of the primary strengths of ascnDNA markers is that they provide a ready source of genetic variation that can be analyzed using well-characterized and accepted population genetic algorithms (Wright, 1951; Slatkin, 1987; Slatkin and Barton, 1989; May and Krueger, 1990) as well as those used in mixed stock assessment (Pella and Milner, 1987).

Hybridization

Increasingly, the identification of hybrids and the study of mechanisms that promote hybridization are becoming important issues in conservation biology (Lehman et al, 1991; Wayne and Jenks, 1991; Avise, 1994). The effects of hybridization on rare species are both biological and legal. Biologically, hybridization affects rare species by causing the loss of locally adapted genotypes through genetic swamping. Hybrids also can have markedly reduced fitness due to the forcing together of incompatible genomes. Legally, hybridization can be used as an argument to remove a threatened or endangered species from consideration under the Endangered Species Act. Loss of protection for otherwise deserving organisms most likely will end in extinction.

Biparentally inherited markers are very helpful if not essential for identifying hybrids in natural populations (Pella and Milner, 1987; Karl et al., 1995). Isozyme markers can be difficult to use since closely related species often do not possess species-specific alleles (Campton, 1990; Hauser et al., 1991). In more distantly related taxa, isozymes can show allelic repression due to post-transcriptional processing (Whitt et al., 1972; Avise and Duvall, 1977). Both of these attributes will confound the accurate identification of natural hybrids. DNA markers circumvent these complications by assaying the more numerous polymorphisms in the DNA and because they are independent of gene expression. Multilocus DNA techniques (such as RAPD and DNA fingerprinting) have been used to document hybridization in natural populations (Gottelli et al., 1994; Roy et al., 1994) but the application of these techniques to studies of hybridization suffer from the same difficulties discussed under Genetic Variation and Parentage and Kinship, above. The inability to identify specific alleles at each locus greatly reduces the utility of these techniques (Lynch and Milligan, 1994). ascnDNA markers also may be superior to single-locus micro- and minisatellite markers because a presumed slower rate of evolution at ascnDNA loci reduces the probability of two individuals sharing alleles identical by kind but not by descent.

Forensics

The above discussions attest to the utility of ascnDNA markers in the identification of individuals, populations, species, and higher-level taxonomic categories. Although considerably more research and applications are needed before ascnDNA markers can meet the rigorous criteria for forensic use (Harvey, 1990), they undoubtedly still can be useful. ascnDNA shares many of the attributes of isozyme electrophoresis and therefore can easily be included with other techniques characterized as "...well established and generally accepted by the scientific community as valid mechanisms for evaluation of genetic information" (Harvey, 1990). Even though ascnDNA loci are generally not highly variable, they nonetheless provide specific, uncomplicated, reliable, population data on genotype frequencies (von Beroldingen et al., 1989) which can be used for forensic applications. Furthermore, since wildlife forensics involves a greater role for species identification than other forensic applications, techniques such as ascnDNA marker analysis are very useful because they offer specific and clear taxonomic information.

CASE STUDIES

The Green Turtle

One of the first applications of ascnDNA was the evaluation of male-mediated gene flow in the green turtle, *Chelonia mydas* (Karl et al., 1992). Previous protein electrophoretic surveys (Smith et al., 1977; Bonhomme et al., 1987) produced inconclusive results concerning population subdivision. A mtDNA survey revealed fixed or nearly fixed haplotype differences between several nesting beaches worldwide (Bowen et al., 1992). This was attributed to highly restricted gene flow and dispersal between nesting beaches due to female philopatry. To address questions of male-mediated gene flow, Karl et al. (1992) analyzed biparentally inherited ascnDNA markers. Table 3-1 summarizes some of the results from the ascnDNA loci used in the study. Substantial differences in intralocus variation were detected (from zero to 3.17% nucleotides polymorphic). Five of the seven loci surveyed were polymorphic with no locus having more than three alleles. Interlocality allele frequency variances (F_{st} values) among loci were estimated

Table 3-1 Summary of Anonymous Single-copy Nuclear DNA Variation. Nine individuals (18 haplotypes) from three geographically remote regions were surveyed with up to 40 enzymes to detect polymorphic restriction sites

Primer	Size	Total No. of Sites Observed	Total No. of bps Surveyed[a]	Percentage Sequence Surveyed	Percentage bps Polymorphic[b]	No. Sites Polymorphic	Percentage Sites Polymorphic
Green turtles[c]							
CM-01	1380	26	115	8.3%	0.0%	0	0.0%
CM-12	1195	29	140	11.7%	1.4%	2	6.9%
CM-14	930	19	79	8.5%	2.5%	2	10.5%
CM-28	1400	33	157	11.2%	0.0%	0	0.0%
CM-39	1350	23	106	7.9%	0.9%	1	4.3%
CM-45	1000	14	64	6.4%	1.6%	1	7.1%
CM-67	1160	22	97	8.4%	3.1%	3	13.6%
Overall	8850	166	758	9.0%[d]	1.4%[d]	9	5.4%[d]
American oyster[c]							
CV-07	1500	16	78	5.6%	2.5%	2	12.5%
CV-19	1500	24	121	8.4%	3.3%	4	16.7%
CV-32	1000	24	110	11.0%	0.9%	1	4.2%
CV-36	1050	13	60	6.5%	0.0%	0	0.0%
CV-195	770	5	22	3.5%	13.6%	3	60.0%
CV-233	830	7	34	5.9%	2.9%	1	14.3%
Overall	6650	89	425	6.8%[d]	3.9%[d]	11	12.4%[d]

[a] Calculated as the number of observed restriction sites for an enzyme multiplied by the number of base pairs in the recognition sequence (Nei, 1987), summed across all enzymes.

[b] Calculated as number of polymorphic sites divided by the total number of nucleotides surveyed, with the assumption that each polymorphic site represents a single nucleotide difference.

[c] From Karl and Avise (1993).

[d] Numbers averaged across all primer pairs.

(0.05–0.21) and indicated significant population genetic substructure. More interesting, however, were comparisons between the maternally inherited mtDNA and the biparentally inherited ascnDNA in the magnitude of gene flow between specific nesting beaches. Several pairs of populations exhibited fixed differences in mtDNA haplotypes but were not significantly different in allele frequencies at any of the ascnDNA loci. Although this may, in part, be due to a faster rate of mutation and smaller effective population size for mtDNA relative to nuclear DNA, the authors concluded that significant interpopulation gene flow most likely was occurring (Karl et al., 1992). The exact nature of this gene flow is unknown. The disparity between the maternal and biparentally inherited loci strongly indicates that gene flow is male-mediated. In this study, ascnDNA markers were central to the elucidation of gender-specific gene flow.

The American Oyster

This technique has also been applied to the analysis of geographic population genetic substructuring in the American oyster, *Crassostrea virginica* (Karl and Avise, 1992). Previous isozymes and mtDNA studies of this species resulted in conflicting conclusions. An isozyme electrophoretic study indicated little or no geographic population subdivision from Nova Scotia to Texas (Buroker, 1983). This led the author to conclude that there was extensive gene flow throughout the range. A subsequent mtDNA analysis (Reeb and Avise, 1990) revealed highly significant population subdivision with over 2% sequence divergence between Atlantic coast and Gulf of Mexico coast haplotypes. The authors concluded that these two subpopulations were genetically isolated for over a million years. Given the highly discordant nature of these results, Karl and Avise (1992) undertook an analysis of ascnDNA markers. The results of this survey, based on four polymorphic ascnDNA loci, indicated population subdivision with a sharp genetic discontinuity in ascnDNA allele frequencies at approximately the same geographic location as for the mtDNA. The authors concluded that the close concordance between mtDNA and ascnDNA data indicated that the discrepancy with the isozyme results is most likely due to "nonrepresentative evolutionary genetic properties of the allozyme systems" (Karl and Avise, 1992).

Table 3-1 summarizes some of the features of the ascnDNA loci used in this study. As was the case for green turtles, there was a large difference in the level of genetic variation observed among the loci. In addition, a larger number of polymorphic loci (five of six surveyed) and polymorphic enzyme sites (5.4% of the sites in turtles compared to 12.4% in oysters) also were observed. Furthermore, all the oyster ascnDNA loci, except one (CV-36), contained size polymorphisms, a trait that was virtually absent in turtle ascnDNA loci. Whether these differences are a reflection of a slower rate of evolution in turtles (Avise et al., 1992), larger effective population size for oysters, or both, is unknown. It may be noteworthy that the degree of isozyme variation in sea turtles also was less than that observed in oysters. In this, as in the marine turtle study, ascnDNA data was essential in identifying a possible resolution to the conflict created by two other, independent data sets.

Marine Turtle Hybrids

A final case study utilizing ascnDNA markers is the analysis of hybridization in marine turtles (Karl et al., 1995). Anecdotal observations of hybridization in marine turtle species have been difficult to establish on the basis of only morphological evidence. Karl et al. (1995) employed molecular genetic assays to document naturally occurring viable hybrids between species representing four of the five genera of cheloniid sea turtles. Specimens came from a variety of sources—some living hatchlings, some blood samples taken from adults, and some decomposing dead hatchlings. Regardless of the quality of the material, PCR amplifications generally worked well and genotype assignments were unambiguous. mtDNA analysis permitted identification of the maternal parental species in each cross. Data from five ascnDNA loci confirmed that these turtles were indeed hybrids and represent products from the mating of a ♂ loggerhead turtle (*Caretta caretta*) × ♀ Kemp's ridley turtle (*Lepidochelys kempii*), a ♂ loggerhead turtle × ♀ hawksbill turtle (*Eretmochelys imbricata*), a ♂ green turtle (*Chelonia mydas*) × ♀ loggerhead turtle, and a ♂ hawksbill turtle × ♀ green turtle. These species represent evolutionary lineages thought to have separated 10–75 million years ago, and may be some of the oldest vertebrate lineages capable of producing natural, viable hybrids. The authors concluded that human intervention in the life cycles of these animals, either through habitat alteration, captive rearing, or attempts to establish new breeding sites, may have increased the opportunities for the formation of hybrids. Again, positive, conclusive identification of the hybrid status of these turtles would have been almost impossible without employing ascnDNA markers.

CONCLUSIONS

Although ascnDNA markers are not applicable in all situations, the type of data collected has many of the advantages and few of the disadvantages of both isozymes and mtDNA and can generally be applied whenever either one is appropriate. ascnDNA markers have four main strengths. First, they are a random sample from the nuclear genome and do not represent a specific class of loci. Therefore, ascnDNA loci are more likely to be an accurate representation of the entire genome. Second, the type of variation most often surveyed is unlikely to be under direct natural selection, which is a necessary assumption in many applications. Third, ascnDNA alleles are phylogenetically ordered, co-dominant, diploid, multistate characters from several biparentally inherited loci and therefore can produce a large amount of data for a reasonable amount of effort. Finally, ascnDNA markers are amenable to well-known analytical tools and are applicable to many situations commonly encountered in conservation biology. ascnDNA data can be used in parentage analysis, estimation of gene flow, diversity estimation, the identification of hybridization, mixed stock assessment, and forensics; all of which find application in conservation biology.

REFERENCES

Ausubel, F.M., R. Brent, R.E. Kingston, D.D. Moore, J.G. Seidman, J.A. Smith, K. Struhl, P. Wang-Iverson, and S.G. Bonitz. 1993. Current protocols in molecular biology. New York: Greene Publishing Associates and Wiley-Interscience.

Avise, J.C. 1994. Molecular markers, natural history and evolution. New York: Chapman and Hall.

Avise, J.C. and R.M. Ball. 1990. Principles of genealogical concordance in species concepts and biological taxonomy. In: D. Futuyma and J. Antonovics eds. Oxford surveys in evolutionary biology, vol. 7, pp. 45–69. New York: Oxford University Press.

Avise, J.C. and S.W. Duvall. 1977. Allelic expression and genetic distance in hybrid macaque monkeys. J. Hered. 68: 23–30.

Avise, J.C., B.W. Bowen, T. Lamb, A.B. Meylan, and E. Bermingham. 1992. Mitochondrial DNA evolution at a turtle's pace: evidence for low genetic variability and reduced microevolutionary rate in the Testudines. Mol. Biol. Evol. 9: 457–473.

Ball, R.M., J.E. Neigel, and J.C. Avise. 1990. Gene genealogies within the organismal pedigrees of random-mating populations. Evolution 44: 360–370.

Barnes, W.M. 1994. PCR amplification of up to 35-Kb DNA with high fidelity and high yield from λ bacteriophage templates. Proc. Natl. Acad. Sci. U.S.A. 91: 2216–2220.

Bonhomme, F., S. Salvidio, A. LeBeau, and G. Pasteur. 1987. Comparaison génétique des tortues vertes (*Chelonia mydas*) des Oceans Atlantique, Indien et Pacifique: Une illustration apparente de la théorie mullerienne classique de la structure génétique des populations? Genetica 74: 89–94.

Bowen, B.W., A.B. Meylan, J.P. Ross, C.J. Limpus, G.H. Balazs, and J.C. Avise. 1992. Global population structure and natural history of the green turtle (*Chelonia mydas*) in terms of matriarchal phylogeny. Evolution 46: 865–881.

Brown, W.M., M. George, Jr., and A.C. Wilson. 1979. Rapid evolution of animal mitochondrial DNA. Proc. Natl. Acad. Sci. U.S.A. 76: 1967–1971.

Burke, T. and M.W. Bruford. 1987. DNA fingerprinting in birds. Nature 327: 149–152.

Buroker, N.E. 1983. Population genetics of the American oyster *Crassostrea virginica* along the Atlantic coast and the Gulf of Mexico. Marine Biol. 75: 99–112.

Campton, D.E. 1990. Application of biochemical and molecular genetic markers to analysis of hybridization. D.H. Whitmore, ed. Electrophoretic and isoelectric focusing techniques in fisheries management, pp. 241–263. Boca Raton, Fla.: CRC Press.

Chakraborty, R., M. Shaw, and W.J. Schull. 1974. Exclusion of paternity: the current state of the art. Am. J. Hum. Genet. 26: 477–488.

Duvall, S.W., I.S. Bernstein, and T.P. Gordon. 1976. Paternity and status in a rhesus monkey group. J. Reprod. Fertil. 47: 25–31.

Fischer, S.G. and L.S. Lerman. 1983. DNA fragments differing by single base-pair substitutions are separated in denaturing gradient gels: correspondence with melting theory. Proc. Natl. Acad. Sci. U.S.A. 80: 1579–1583.

Gibbs, H.L., K.A. Prior, and P.J Weatherhead. 1994. Genetic analysis of populations of threatened snake species using RAPD markers. Mol. Ecol. 3: 329–337.

Gilbert, D.A., N. Lehman, S.J. O'Brien, and R.K. Wayne. 1990. Genetic fingerprinting reflects population differentiation in the California Channel Island fox. Nature 344: 764–766.

Gilpin, M.E. and C. Wills. 1991. MHC and captive breeding: a rebuttal. Conserv. Biol. 5: 554–555.

Gottelli, D., C. Sillero-Zubiri, G.D. Applebaum, M.S. Roy, D.J. Girman, J. Garcia-Moreno, E.A. Ostrander, and R.K. Wayne. 1994. Molecular genetics of the most endangered canid: the Ethiopian wolf *Canis simensis*. Mol. Ecol. 3: 301–312.

Hames, B.D. and S.J. Higgins. 1985. Nucleic acid hybridisation: a practical approach. Washington, D.C.: IRL Press.

Hartl, D.L. and S.A. Sawyer. 1991. Inference of selection and recombination from nucleotide sequence data. J. Evol. Biol. 4: 519–532.

Harvey, W.D. 1990. Electrophoretic techniques in forensics and law enforcement. In: D.H. Whitmore, ed. Electrophoretic and isoelectric focusing techniques in fisheries management, pp. 313–321. Boca Raton, Fla.: CRC Press.

Hauser, L., A.R. Beaumont, G.T.H. Marshall, and R.J. Wyatt. 1991. Effects of sea trout stocking on the population genetics of landlocked brown trout, *Salmo trutta* L., in the Conwy River system, North Wales, U.K. J. Fish Biol. 39A: 109–116.

Hedrick, P.W., P.F. Brussard, F.W. Allendorf, J.A. Beardmore, and S. Orzack. 1986. Protein variation, fitness and captive propagation. Zoo Biol. 5: 91–99.

Hillis, D.M., M.T. Dixon, and A.L. Jones. 1991. Minimal genetic variation in a morphologically diverse species (Florida tree snail, *Liguus fasciatus*). J. Hered. 82: 282–286.

Hughes, A.L. 1991. MHC polymorphism and the design of captive breeding programs. Conserv. Biol. 5: 249–251.

Karl, S.A. and J.C. Avise. 1992. Balancing selection at allozyme loci in oysters: implications from nuclear RFLPs. Science 256: 100–102.

Karl, S.A. and J.C. Avise. 1993. PCR-based assays of Mendelian polymorphisms from anonymous single-copy nuclear DNA: techniques and applications for population genetics. Mol. Biol. Evol. 10: 342–361.

Karl, S.A., B.W. Bowen, and J.C. Avise. 1992. Global population genetic structure and male-mediated gene flow in the green turtle (*Chelonia mydas*): RFLP analysis of anonymous nuclear loci. Genetics 131: 163–173.

Karl, S.A., B.W. Bowen, and J.C. Avise. 1995. Hybridization among the ancient mariners: characterization of marine turtle hybrids with molecular genetic assays. J. Hered. 86: 262–268.

Lehman, N., A. Eisenhawer, K. Hansen, D.L. Mech, R.0. Peterson, J.P. Gogan, and R.K. Wayne. 1991. Introgression of coyote mitochondrial DNA into sympatric North American gray wolf populations. Evolution 45: 104–119.

Lessa, E.P. 1992. Rapid surveying of DNA sequence variation in natural populations. Mol. Biol. Evol. 9: 323–330.

Lessa, E.P. and G. Applebaum. 1993. Screening techniques for detecting allelic variation in DNA sequences. Mol. Ecol. 2: 119–129.

Lynch, M. and B.G. Milligan. 1994. Analysis of population genetic structure with RAPD markers. Mol. Ecol. 3: 91–99.

McRae, S.B. and K.M. Kovacs. 1994. Paternity exclusion by DNA fingerprinting, and mate guarding in the hooded seal *Cystophora cristata*. Mol. Ecol. 3: 101–107.

May, B. and C.C. Krueger. 1990. Use of allozyme data for population analysis. In: D.H. Whitmore, ed. Electrophoretic and isoelectric focusing techniques in fisheries management, pp. 157–171. Boca Raton, Fla.: CRC Press.

Miller, P.S. and P.W. Hedrick. 1991. MHC polymorphism and the design of captive breeding programs: simple solutions are not the answer. Conserv. Biol. 5: 556–558.

Mullis, K.B. and F.A. Faloona. 1987. Specific synthesis of DNA *in vitro* via a polymerase-catalyzed chain reaction. Methods Enzymol. 155: 335–350.

Nei, M. 1987. Molecular evolutionary genetics. New York: Columbia University Press.

Orita, M., H. Iwahana, H. Kanazawa, K. Hayashi, and T. Sekiyaa. 1989. Detection of polymorphisms of human DNA by gel electrophoresis as single-strand conformation polymorphisms. Proc. Natl. Acad. Sci. U.S.A. 86: 2766–2770.

Orita, M., Y. Suzuki, T. Sekiyaa, and K. Hayashi. 1989b. Rapid and sensitive detection of point mutations and DNA polymorphisms using the polymerase chain reaction. Genomics 5: 874–879.

Pella, J.J. and G.B. Milner. 1987. Use of genetic markers in stock composition analysis. In: N. Ryman and F. Utter, eds. Population genetics and fishery management, pp. 247–276. Seattle, Wash.: University of Washington Press.

Quinn, T.W., J.C. Davies, F. Cooke, and B.N. White. 1989. Genetic analysis of offspring of a female–female pair in the lesser snow goose (*Chen c. caerulescens*). The Auk 106: 177–184.

Rassmann, K., W. Arnold, and D. Tautz. 1994. Low genetic variability in a natural alpine marmot population (*Marmota marmota*, Sciuridae) revealed by DNA fingerprinting. Mol. Ecol. 3: 347–355.

Reeb, C.A. and J.C. Avise. 1990. A genetic discontinuity in a continuously distributed species: mitochondrial DNA in the American oyster, *Crassostrea virginica*. Genetics 124: 397–407.

Roy, M.S., E. Geffen, D. Smith, E.A. Ostrander, and R.K. Wayne. 1994. Patterns of differentiation and hybridization in North American wolflike canids, revealed by analysis of microsatellite loci. Mol. Biol. Evol. 11: 553–570.

Rychlik, W. 1992. OLIGO v. 4.0. National Biosciences, Inc., Plymouth, Minn.

Rychlik, W. and R.E. Rhoads. 1989. A computer program for choosing optimal amplification oligonucleotides for filter hybridizations, sequencing and *in vitro* amplification of DNA. Nucleic Acids Res. 17: 8543–8551.

Saiki, R., S. Scharf, F. Faloona, K.B. Mullis, G.T. Horn, and H.A. Erlich. 1985. Enzymatic amplification of β-globin geonomic sequences and restriction site analysis for diagnosis of sickle cell anemia. Science 230: 1350–1354.

Saiki, R.K., D.H. Gelfand, S. Stoffel, S.J. Scharf, R. Higuchi, G.T. Horn, K.B. Mullis, and H.A. Erlich. 1988. Primer-directed enzymatic amplification of DNA with a thermostable DNA polymerase. Science 239: 487–494.

Scribner, K.T., J.W. Arntzen, and T. Burke. 1994. Comparative analysis of intra- and inter-population genetic diversity in *Bufo bufo*, using allozyme, single-locus microsatellite, minisatellite, and multilocus minisatellite data. Mol. Biol. Evol. 11: 737–748.

Sheffield, V.C., D.R. Cox, and R.M. Myers. 1990. Identifying DNA polymorphisms by denaturing gradient gel electrophoresis. In: M.A. Innis, D.H. Gelfand, J.J. Sninsky, and T.J. White, eds. PCR protocols: A guide to methods and applications, pp. 206–218. New York: Academic Press.

Signer, E.N., C.R. Schmidt, and A.J. Jeffreys. 1994. DNA variability and parentage testing in captive Waldrapp ibises. Mol. Ecol. 3: 291–300.

Slatkin, M. 1987. Gene flow and the geographic structure of natural populations. Science 236: 787–792.

Slatkin, M. and N.H. Barton. 1989. A comparison of three indirect methods for estimating average levels of gene flow. Evolution 43: 1349–1368.

Smith, M.H., H.O. Hillestad, M.N. Manlove, D.O. Straney, and J.M. Dean. 1977. Management implication of genetic variability in loggerhead and green sea turtles. In: XIII International Congress of Game Biologists, Atlanta, Georgia, March 11–15, 1977, pp. 302–312. The Wildlife Society, Wildlife Management Institute.

Stephens, J.C. 1985. Statistical methods of DNA sequence analysis: detection of intragenic recombination or gene conversion. Mol. Biol. Evol. 2: 539–556.

Tautz, D. 1989. Hypervariability of simple sequences as a general source for polymorphic DNA markers. Nucleic Acids Res. 17: 6436–6471.

Taylor, A.C., W.B. Sherwin, and R.K. Wayne. 1994. Genetic variation of microsatellite loci in a bottlenecked species: the northern hairy-nosed wombat *Lasiorhinus krefftii*. Mol. Ecol. 3: 277–290.

Vawter, L. and W.M. Brown. 1986. Nuclear and mitochondrial DNA comparisons reveal extreme rate variation in the molecular clock. Science 234: 194–196.

von Beroldingen, C.H., E.T. Blake, R. Higuchi, G.F. Sensabaugh, and H. Erlich. 1989. Applications of PCR to the analysis of biological evidence. In: H. Erlich, ed. PCR technology: Principles and applications for DNA amplification, pp. 209–223. New York: Stockton Press.

Vrijenhoek, R.C. and P.L. Leberg. 1991. Let's not throw the baby out with the bathwater: a comment on management for MCH diversity in captive populations. Conserv. Biol. 5: 252–254.

Wayne, R.K. and S.M. Jenks. 1991. Mitochondrial DNA analysis implying extensive hybridization of the endangered red wolf *Canis rufus*. Nature 351: 565–568.

Weber, J.L. and P.E. May. 1989. Abundant class of human DNA polymorphisms which can be typed using the polymerase chain reaction. Am. J. Hum. Genet. 44: 388–396.

Weir, B. 1990. Genetic data analysis: methods for discrete population genetic data. Sunderland, Mass.: Sinauer Associates.

Whitt, G.S., P.L. Cho, and W.F. Childers. 1972. Preferential inhibition of allelic isozyme synthesis in an interspecific sunfish hybrid. J. Exp. Zool. 179: 271–282.

Wright, S. 1951. The genetical structure of populations. Ann. Eugen. 15: 323–354.

Zhang, Q., M.A.S. Maroof, and A. Kleinhofs. 1993. Comparative diversity analysis of RFLPs and isozymes within and among populations of *Hordeum vulgare* ssp. *spontaneum.* Genetics 134: 909–916.

4

The Use of Random Amplified Polymorphic DNA (RAPD) in Conservation Genetics

PETER FRITSCH AND LOREN H. RIESEBERG

A primary goal of conservation genetics is to estimate the level and distribution of genetic variation in endangered species. Genetic data are often critical to understanding the structure, evolutionary relationships, demographics, and taxonomy of populations, races, and species; thus, they often play a significant role in the formulation of appropriate management strategies directed toward the conservation of taxa (Beardmore, 1983; Allendorf and Leary, 1986; Lacy, 1988). Currently, the most popular and cost-effective means by which conservation geneticists infer variation at the molecular level is isozyme electrophoresis. Isozymes possess several positive attributes for use in genetic analysis: only small amounts of tissue are required, alleles exhibit simple Mendelian inheritance and codominant expression in most cases, and comparisons of homologous loci across populations and related species are straightforward (Schaal et al., 1991). Nevertheless, as applied to studies of genetic variation, isozymes have well-known limitations, the most serious of which are highly biased genomic sampling (only genes encoding soluble enzymes are surveyed) and the small number of loci (maximally about 40) available for study.

A more recent development in the detection of genetic polymorphisms is random amplified polymorphic DNA (RAPD; Welsh and McClelland, 1990; Williams et al., 1990). In contrast to isozymes, RAPD provides a more arbitrary sample of the genome and can generate essentially unlimited numbers of loci for use in genetic analysis. RAPD also has an advantage over another technique widely employed in genetic analysis—restriction fragment length polymorphism (RFLP)—in that it requires much less complex and labor-intensive procedures and a far smaller amount of genomic DNA (about as much as isozymes; Whitkus et al., 1994). RAPD is relatively new but has already proven useful in a wide variety of evolutionary and applied studies, including the construction of genetic maps, the production of genetic markers linked to specific phenotypic traits, parentage determination, clone and cultivar identification, and population dynamics (for reviews see Hadrys et al., 1993a; Williams et al., 1993; Whitkus et al., 1994). It has been most widely used in studies of seed plants, but is beginning to be employed with higher frequency in other organisms (e.g., Owen and Uyeda, 1991; Caswell-Chen et al., 1992; Smith et al., 1992; Woodward et al., 1992; Hadrys et al., 1993a,b; Puterka et al., 1993).

The advantages of RAPD in genetic analysis make it particularly suited to the analysis of rare and endangered organisms, where the availability of material and inability to detect polymorphic isozyme loci are often factors that hinder detailed analysis. Nevertheless, few studies using RAPD in rare and endangered taxa have been published to date, although several RAPD studies with conservation implications have been performed on plant strains and cultivars in economically important groups (e.g., Chalmers et al., 1992; Dawson et al., 1993; Russell et al., 1993; C. Orozco-Castillo et al., 1994). There may be multiple reasons for this, including the relatively recent advent of RAPD, its perceived expense, and problems in describing genetic variation resulting from one of its limiting properties: dominant expression. This chapter (1) describes the RAPD method, (2) outlines its various applications in conservation genetics, (3) reviews two case studies of its use, each exemplifying a different application of the method, and (4) notes the limitations and most appropriate uses of RAPD in conservation biology.

TECHNICAL DESCRIPTION

Procedure

The RAPD procedure (Williams et al., 1990) has not been substantially modified since it was originally described. Oligonucleotide primers (single-stranded DNA, usually 10-mers) of arbitrary sequence are deployed in a polymerase chain reaction (for basic PCR protocols, see Erlich, 1989; Innis et al., 1990; White, 1993) using genomic DNA as the template. Typically one primer is used and only 10-25 ng of DNA is necessary per reaction. If the primer binds to sites on different strands that are within about 3 kb of each other, the region between the ends of the priming sites will be amplified. Because this can occur at any number of locations within the genome for any given primer, more than one DNA fragment may result from a single reaction. Products of the reaction are visualized by agarose gel electrophoresis after staining with ethidium bromide (Figure 4-1).

10-mer primers employed for RAPD have a G-C content ranging from 50% to 70%; if they possess palindromes, the palindromic sequence is neither greater than six bases long nor complementary at the 3′ end; these constraints result in the most efficient amplification as detected by ethidium bromide staining (Williams et al., 1993). Most of these include three or all four bases in their sequence. A list of 10-mer primers thought to be useful in a wide range of plant groups is available (Fritsch et al., 1993). Recently, superior results in the form of increased numbers of bands per primer have been obtained with 10-mer primers composed of two bases. These primers may be amplifying regions with simple sequence repeats (D. Rhoades personal communication). 10-mer primer sets are commercially available from Operon Technologies, Inc., and from the Nucleic Acid-Protein Service Unit

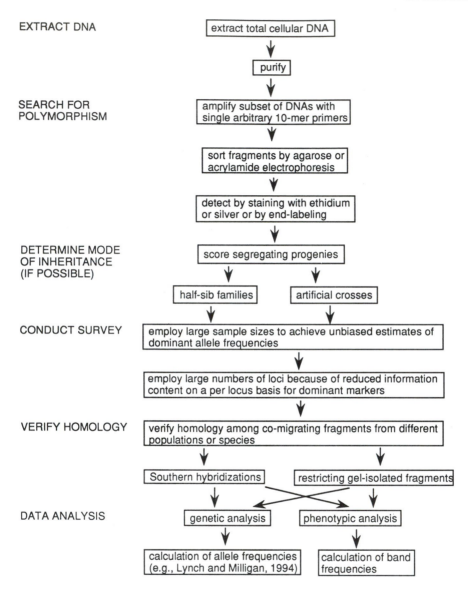

EXTRACT DNA

SEARCH FOR
POLYMORPHISM

DETERMINE MODE
OF INHERITANCE
(IF POSSIBLE)

CONDUCT SURVEY

VERIFY HOMOLOGY

DATA ANALYSIS

extract total cellular DNA

purify

amplify subset of DNAs with
single arbitrary 10-mer primers

sort fragments by agarose or
acrylamide electrophoresis

detect by staining with ethidium
or silver or by end-labeling

score segregating progenies

half-sib families

artificial crosses

employ large sample sizes to achieve unbiased estimates of
dominant allele frequencies

employ large numbers of loci because of reduced information
content on a per locus basis for dominant markers

verify homology among co-migrating fragments from different
populations or species

Southern hybridizations

restricting gel-isolated fragments

genetic analysis

phenotypic analysis

calculation of allele frequencies
(e.g., Lynch and Milligan, 1994)

calculation of band
frequencies

Figure 4-1   Flowchart depicting the rigorous analysis of RAPD data in natural populations.

at the University of British Columbia. A two-base 10-mer primer set is available
from UBC.

The above procedure is the most widely employed method for generating
RAPDs. However, variations of this procedure have been described and employed.
The AP-PCR method (arbitrarily primed PCR; Welsh and McClelland, 1990) is
very similar to RAPD except that it uses longer primers (20-mers) and PCR
products are radioactively labeled and resolved on polyacrylamide gels. DAF
(DNA Amplification Fingerprinting; Caetano-Anolles et al., 1991) employs shorter
primers (5- to 8-mers), and PCR products are visualized with silver staining, also

on polyacrylamide gels. These modifications result in increased detection and resolution of reaction products but are, at least initially, more expensive, and AP-PCR requires the use of radioactive materials, which RAPD avoids. UP-PCR (universally primed PCR; Bulat and Mironenko, 1990, 1992, cited in Bulat et al., 1994) appears to differ from RAPD only in the use of longer primers.

Another modification of potential significance is the use of truncated versions of *Taq* polymerase in the RAPD assay. These polymerases, marketed under various names (e.g., Stoffel Fragment, Perkin-Elmer; DeltaTaq, USB), have missing amino acids in their *N*-terminal region. They have higher thermostability than other versions of *Taq*, and no 3′ to 5′ or 5′ to 3′ exonuclease activity. Preliminary reports suggest more complex and thus informative banding patterns in RAPD.

Data Interpretation

RAPD variation can be genetically expressed in several ways, but the vast majority of polymorphisms exhibit a pattern in which bands are either present or absent and are inherited in a dominant fashion. These polymorphisms show 3:1 band-present: band-absent ratios in F_2 segregation analyses of diploid organisms. Most loci generated by RAPD are assumed to be dominant for band presence in the absence of segregation analysis. However, about 5% of RAPD loci tested show codominant expression resulting from differences in fragment length among the alleles (e.g., Williams et al., 1990). Codominance can be confirmed either by eluting one of the RAPD fragments of interest and using it as a labeled probe in a Southern hybridization (e.g., Rieseberg et al., 1993), or by excising the putative codominant fragments from the gel, treating them with restriction enzymes, and analyzing resulting band profiles for congruence (e.g., Fritsch and Rieseberg, 1992). Two other types of polymorphisms, band brightness differences and heteroduplex DNA bands, have been documented (Hunt and Page, 1992). Band brightness polymorphisms exhibit dominant expression for brightness differences, whereas heteroduplex DNA bands sometimes appear in individuals heterozygous for codominantly expressed loci. The heteroduplexes form as a PCR artifact and migrate to a position that depends on electrophoresis conditions.

RAPD bands of different molecular weight are interpreted as separate loci unless proven otherwise. Loci are usually scored for band presence or absence, and are assumed to possess only two alleles: the allele which results in amplification (the "band" allele) and the allele that does not result in amplification (the "null" allele). The null allele fails to amplify when either a primer site has been lost or an insertion increases the distance between primer sites to the point where amplification is no longer possible within the constraints of the PCR cycling regime. If desired, band homology assessments can be ascertained in the same way as was described above for codominance: by either restriction site analysis of excised fragments or Southern hybridization with labeled RAPD probes (Figure 4-1).

Expense

In the context of conservation genetics, it is probably safe to say that the cost of RAPD is intermediate between that of starch gel enzyme electrophoresis and two other popular methods of detecting genetic variation, restriction fragment length polymorphism (RFLP) and dispersed repetitive DNA (drDNA). When the cost of RAPD was compared to that of RFLP for the purposes of plant breeding (Ragot and Hoisington, 1993), RAPD was found to be more time- and cost-efficient for studies involving small sample sizes. This is of particular relevance in conservation studies, which often assess the genetic status of small numbers of individuals. Whitkus et al. (1994), in a review of the use of RAPD, RFLP, and drDNA markers in systematics and evolution, emphasize that the RAPD assay does not require labor-intensive and costly procedures for initial setup and maintenance, unlike RFLP and drDNA. RAPD costs more than isozymes principally because of the high cost of thermocyclers and commercially prepared *Taq* polymerase needed for DNA amplification. Fortunately, relatively simple protocols now exist for in-house *Taq* production (e.g., Engelke et al., 1990; Pluthero, 1992) of a quality adequate for most RAPD applications. If the need for commercial *Taq* is bypassed, then the cost of RAPD is much more comparable to that of isozymes.

However, the economy of RAPD depends to a large extent on the goals of the researcher. The analytical limitations of the technique (see below) caution against using RAPD simply to save money or because it is the latest method; the advantages of each technique relative to the question at hand and the amount of money available should be carefully considered before any data are collected. The most appropriate uses of RAPD, taking into account both total expense and analytical power relative to other methods, will be summarized at the end of the chapter.

APPLICATIONS IN CONSERVATION GENETICS

The RAPD assay can be used not only in conventional studies of sampling for genetic variation within and among populations of rare species, but also in cases where an understanding of the species' history is required for judicious decisions regarding long-term viability. In this section, the application of RAPD to issues relevant to conservation genetics is considered, including taxon identification and circumscription, hybridization and introgression, reproductive biology and the distribution of genetic variation within taxa. Because there have been relatively few studies in conservation genetics employing RAPD to date, examples will often be taken from studies that have applied RAPD to other issues, but exemplify the potential for use in conservation.

Taxon Identification and Circumscription

RAPD can be used to create genetic profiles for use in distinguishing among organisms at various hierarchical levels. Diagnostic markers can potentially be obtained at the individual level, or, in the case of extremely long-lived organisms, even within a clone that has undergone somatic mutation. Diagnostic markers are relevant to conservation biology particularly in cases where rare organisms are not easily distinguished from more common relatives by morphological analysis, where morphology is regionally ambiguous, or where it is desired to know the wild affiliation of ex situ organisms. Adams et al. (1993) developed species-specific RAPD markers in two species of *Juniperus*, one from west-central Asia and the other from Africa, and determined the placement of populations of uncertain affiliation. Similarly, Kambhampati et al. (1992) used RAPD to differentiate closely related mosquito species and identify unknown specimens. Diagnostic RAPD markers have been able to distinguish plant cultivars (Hu and Quiros, 1991) and populations (Chalmers et al., 1992), fungal mycelial clones (Smith et al., 1992), and strains of organisms ranging from bacteria (McMillan and Muldrow, 1992; Welsh et al., 1992) to mice (Welsh et al., 1991) When using RAPD for these applications, efforts should be made to obtain an adequate sample to ensure the diagnostic nature of the marker used, particularly if polymorphisms at the level of interest are high.

The RAPD method can also be used to assess the circumscription and taxonomic status of organisms, insight into which is often critical in influencing conservation decisions. Estimates of genetic distance can be calculated from RAPD marker frequencies in the organisms of interest; then ordination methods such as principal-components analysis (as implemented, for example, by the Computer program NTSYS-pc (Rohlf, 1993)), or clustering algorithms such as neighbor-joining (Saitou and Nei, 1987; as implemented, for example, by the computer program PHYLIP by J. Felsenstein (Felsenstein, 1989)) can graphically depict similarities based on these distance measures. We prefer principal-components analysis over cluster analysis in this case so as to avoid implying phylogeny. Studies have generated distances for these purposes with algorithms such as Gower's (1971) or Jaccard's (Jaccard, 1912) coefficients of similarity. However, neither of these measures is specifically intended for use with RAPD data. They may be problematic for use in estimating distances based on RAPD because of the dominant nature of allelic expression (see below). A recent method for estimating nucleotide divergence has been specifically designed for RAPD loci that corrects for dominance by assuming Hardy–Weinberg equilibrium; these divergence values can be used as distance measures for subsequent analysis (Clark and Lanigan, 1993).

An example of the use of RAPD data for taxonomic assessment with conservation implications is a RAPD study of the autumn buttercup, *Ranunculus acriformis* var. *aestivalis* (Van Buren et al., 1994). This taxon has been treated as a variety of *R. acris* as well as *R. acriforms*; morphological characters were not useful in resolving the disagreement. Genetic distances based on RAPD data clearly showed that the taxon under study was genetically more similar to the varieties of *R. acriformis* than to *R. acris*. Furthermore, the genetic similarity data were used in conjunction with information on the number of unique RAPD alleles and differences in ecology to justify the elevation of the taxon to species status.

Hybridization and Introgression

Hybridization and introgression between rare and common species has significant consequences for the preservation of genetic diversity. Rare species may be endangered either by outbreeding depression (Price and Waser, 1979) or by genetic assimilation (Cade, 1983). The detection of genetic mixing, therefore, plays an important role in conservation biology. RAPD has proven effective in the detection of hybridization and introgression in several groups of plants (Arnold et al., 1991; Crawford et al., 1993; Hanson 1993; Orozco-Castillo et al., 1994; Rieseberg and Gerber, 1995). Detection of introgression and hybridization with RAPD follows the same logic as for other molecular methods: species-specific markers are observed for complete additivity in putative F_1 hybrids and various degrees of additivity in putative introgressants. Currently, there are few studies documenting hybrids of rare organisms using RAPD markers. Rieseberg and Gerber (1995) employed RAPDs as a follow-up to an isozyme study of rare and widespread island plants in the genus *Cercocarpus* (Rieseberg, 1991; Rieseberg et al., 1989). They used RAPD to provide additional loci with which to document hybrid generations, because isozyme surveys provided only a single variable locus for this purpose. Another study documenting hybridization in rare plants was conducted by Crawford et al. (1993), which will be reviewed under Case Studies.

Reproductive Biology

In many instances, knowledge concerning the mating system, breeding behavior, and paternity of endangered organisms is desirable or essential for effective management and conservation (Karron, 1991; Templeton, 1991). For example, it may be useful to know the rate at which an endangered plant population propagates naturally via outcrossing whose individuals can be experimentally selfed, or to determine whether the same father is siring all progeny in uncontrolled animal breeding programs. Algorithms have been developed for dominant and mixed (dominant and codominant) inheritance systems to estimate outcrossing rates (e.g., a modification of Ritland and Jain, 1981) and to assign paternity (Lewis and Snow, 1992; Milligan and McMurry, 1993). The RAPD assay has been used both for analysis of breeding types and to determine paternity. Underwater cross-pollination has been documented in submersed plants not amenable to experimental manipulations using RAPD (Philbrick, 1993). Outcrossing rates have been estimated with RAPD in plant populations containing both male and potentially-selfing hermaphroditic plants (Fritsch and Rieseberg, 1992). Hadrys et al. (1993b) used RAPD markers as evidence for the relative success of the guarding male in siring offspring in different species of dragonflies.

Genetic Structure of Populations and Species

Assessing the level and distribution of genetic variation within and among populations of rare taxa is a primary goal in conservation genetics. RAPD may have its most significant impact in conservation biology as a means to provide robust data for this purpose, because the number of loci available for sampling is virtually unlimited. This makes RAPD particularly valuable when isozyme profiles are completely monomorphic.

However, there are two weaknesses in the standard use of the RAPD assay for describing the genetic structure of populations relative to isozymes and RFLPs: two alleles are assumed in the case of dominant loci, and heterozygotes are not usually detectable with RAPD using conventional procedures. Therefore, the mean number of alleles per locus (A) and observed heterozygosity (H_o), standard measures in population genetics, are not relevant to RAPD data, and other useful statistics, such as percentage polymorphic loci (P) and expected heterozygosity (H_e), cannot be calculated without invoking additional assumptions. The inability to detect heterozygotes does not mean that estimates of genetic diversity cannot be calculated, but it does mean that the statistical power of such calculations will be reduced relative to codominant systems. RAPD will tend to underestimate the amount of genetic diversity relative to codominant systems, because some individuals with monomorphic phenotypes (band allele always present) may be heterozygous (null allele present but not expressed). It will also result in more uncertain estimates of genetic structure because, for each locus, band frequencies above 50% may result in an underestimate of genetic diversity, whereas band frequencies below 50% may result in an overestimate.

Despite these shortcomings, RAPD has been shown to be highly useful in studies of genetic variation within species. It is a robust method for testing hypotheses of clonal population structure, and can also be used to assess the distribution of genetic variation within and among populations as long as certain precautions are taken. One of the best uses of RAPD in population studies to date has been for documenting the clonal nature of the basidiomycete fungus *Armillaria bulbosa* (Smith et al., 1992). Samples distributed among 15 hectares were found to have identical RAPD and RFLP profiles. Furthermore, segregation studies using haploid monospores revealed instances of heterozygous loci, which were present in all samples representing the putative clone; the probability of retaining heterozygosity from a mating of sibling monosporous isolates was shown to be significantly small. Therefore, it was concluded that the samples constituted a single immense clone. This study highlights the utility of RAPD not only for confirming clonal structure but also for studying haploid genomes. Dominance complications pose no difficulty in haploids, and this has made RAPD a popular method in the study of not only fungi but also bacteria (Welsh et al., 1992), conifers (megagametophytes; Bucci and Menozzi, 1993), and bees (drones; Hunt and Page, 1992). An application of RAPD to test the hypothesis of clonal structure (Swensen et al., 1995) will be reviewed under Case Studies.

Studies estimating genetic diversity have addressed the dominance problem in one of two ways: (1) Hardy–Weinberg equilibrium has been assumed, or (2) band phenotype has been scored (as presence/absence characters). The assumption of Hardy–Weinberg equilibrium allows one to calculate allele frequencies, because the frequency of the null–null genotype for each locus is known (band absent). Under Hardy–Weinberg, the frequency of the null allele is the square root of this frequency, and the frequency of the band allele is therefore one minus the null allele frequency. Once allele frequencies have been calculated, it is possible to estimate H_e and other descriptors. Lynch and Milligan (1994) have recently provided the calculations necessary for relatively unbiased measures of population structure using this approach (gene diversity within and among populations, population subdivision, degree of inbreeding, and individual relatedness; a computer program which calculates these statistics is currently being developed by M. Jordan, M. Lynch personal communication). They emphasize, however, that for the measures to be valid the marker alleles for most dominant loci should be in low frequency (e.g., < 0.94 for $N = 50$). Furthermore, two to ten times as many individuals (ideally at least 100) must be sampled per locus for dominant markers as opposed to codominant markers.

These criteria appear to limit the use of this method in conservation genetics. The first criterion is probably difficult to satisfy in organisms that show any substantial variation in allele frequencies among populations, even considering the essentially unlimited availability of loci with RAPD. The second criterion is likely to prove a formidable barrier in the study of rare organisms, where population sizes are often small. Furthermore, the assumption of Hardy–Weinberg equilibrium in rare populations is tenuous, since inbreeding is often the rule under these conditions. It would be wise to assess the degree of heterozygosity with isozyme loci, if possible, before invoking Hardy–Weinberg.

Several methods of analysis have been employed for estimating genetic diversity based on band phenotype (e.g., Shannon's information measure (Lewontin, 1971); Nei's gene diversity statistics (Nei, 1973); phi statistics from AMOVA (Excoffier and Smouse, 1994). All of these methods, like the previous method, will provide less accurate estimates than codominant systems. Furthermore, unlike the previous method, none of these has been designed specifically for use with RAPD markers, so at this point their utility is not clear. However, an advantage of these methods is that the Hardy–Weinberg assumption is not needed. Of the three, AMOVA shows the most promise, since measures of diversity can be tested for significance (Excoffier et al., 1992). This method has recently been applied to RAPD data from natural populations of buffalograss (Huff et al., 1993). A computer program for the AMOVA method is available (Excoffier, 1992), and an adaptation of the system for RAPD is underway (L. Excoffier, personal communication).

As has been pointed out by Dawson et al. (1993), the use of RAPD markers to estimate gene diversity is probably valid in cases where populations are predominantly inbred, because most individuals will likely be homozygous, thus minimizing error from heterozygote ambiguity. Therefore, phenotypic scoring methods will probably be most appropriate for RAPD data when independent evidence of inbreeding can be obtained. This has important implications for conservation genetics, where it is often the case that populations are inbred. Only a handful of studies have employed RAPD to assess genetic diversity in rare populations, and almost all have been in plants (Brauner et al., 1992, *Lactoris*; Hanson et al., 1992, *Pogogyne*; Milligan, 1992, *Aquilegia*; Friar et al., 1994, *Argyroxiphium*; Gibbs et al., 1994, snake species; Gustafsson and Gustafsson, 1994, *Vicia*; Van Buren et al., 1994, *Ranunculus*; Swensen et al., 1995, *Malacothamnus*). Brauner et al. (1992) found variation in populations of *Lactoris fernandeziana* after an isozyme study revealed no polymorphisms. Genetic variation in *Vicia pisiformis*, a rare European species, using both RFLP and RAPD, was found to be virtually absent; 82 RAPD fragments were scored. Hanson et al. (1992) studied a highly endangered group in the mint family. A high amount of genetic variation was found to occur among populations, leading to the conclusion that seven out of the eight existing populations must be preserved to ensure the maintenance of 95% of the genetic diversity of the species. In *Argyroxiphium sandwicense* (Friar et al., 1996), an endangered Hawaiian endemic, RAPD confirmed records that only one or two maternal plants served as the original seed source for an outplanted population, and suggested strategies for future management plans of this population.

CASE STUDIES

Malacothamnus

Swensen et al. (1995) employed RAPD, isozymes, and nuclear ribosomal DNA RFLP data to determine the genetic structure of an endangered plant endemic to Santa Cruz Island off the California coast, *Malacothamnus fasciculatus* var. *nesioticus*. This variety is one of five varieties constituting *M. fasciculatus*; it currently consists of two populations (18–19 individuals or ramets each) spaced several miles apart on Santa Cruz island (Figure 4-2). Because *Malacothamnus* commonly spreads vegetatively via rhizomes, it was suspected that each population consists of a single clone. The study was undertaken to confirm the clonal nature of populations, to estimate the level and distribution of any genetic variation detected within and among the two populations, and to compare genetic diversity and similarity to the other varieties of *M. fasciculatus*, which are native to other nearby islands and the mainland. This information would aid in the artificial establishment of new populations, if deemed necessary.

Twelve isozyme and well over 147 RAPD loci were monomorphic for all individuals within both populations, thus providing support for the idea of clonal structure (Figure 4-3). However, because fixed heterozygosity was not detected with isozymes, genetic drift could not be ruled out as an explanation for genetic uniformity. Also, one population possessed three RFLP patterns resulting from small length differences within nuclear ribosomal DNA. This could have been a result of limited intercrossing, or somatic mutation. The two RAPD genotypes corresponding to each population shared only 15 of 27 markers scored among the five *M. fasciculatus* varieties. Nei's mean genetic identity between *M. f. nesioticus* populations was 0.56, and the mean genetic identity between *M. f. nesioticus* and the other *M. fasciculatus* populations was 0.70. Principal-components analysis of the data showed that the study variety was genetically distinct from all other varieties within the species, and that the two populations were genetically distinct from each other (Figure 4-4).

This study highlights the utility of RAPD for assessing the clonal structure and genetic distinctiveness of populations. Although the clonal nature of the two populations is likely, the detection of slight variation in nrDNA suggests that some sexual reproduction, followed by clonal reproduction, may have taken place in the past. In any case, the data provide justification for the continued protection of both populations of *M. fasciculatus* var. *nesioticus*, and also provide baseline data for subsequent breeding experiments.

× Margyracaena

Crawford et al. (1993) used RAPD markers to test the hybrid origin of the rare flowering plant × *Margyracaena skottsbergii* endemic to Masatierra Island, one

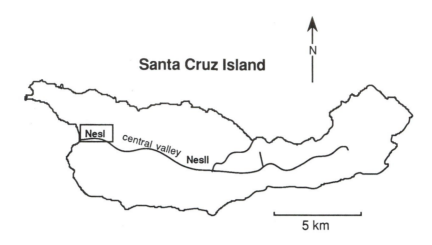

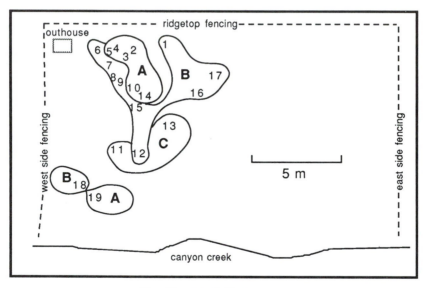

Nes(I) population

Figure 4-2 Map of Santa Cruz Island showing distribution of the two remaining populations of *Malacothamnus fasciculatus* var. *nesioticus*: NES(I) and NES(II). The inset shows the clonal population structure of the NES(I) population. Bushes sampled are indicated by numbers; bold letters indicate clonal individuals as determined by rDNA analysis (from Swensen et al., 1995).

of the Juan Fernandez Islands off the coast of Chile. This plant has been proposed as an intergeneric hybrid between the Masatierran endemic *Margyricarpus digynus* and the introduced noxious weed *Acaena argentea*. A native species of *Acaena*, *A. ovalifolia* was purported to have been involved in the ancestry of the putative hybrid on the basis of morphological features. × *Margyracaena* was found

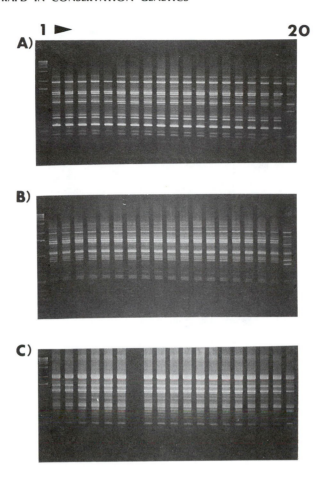

Figure 4-3 Photograph of RAPD amplification products in *Malacothamnus fasciculatus* var. *nesioticus*. DNA size standards are in lane 1 (1-kb ladder; Gibco BRL). Lanes 2–19 are amplification profiles of 18 different bushes from the NES(II) population, all of which appear to represent a single clone. Lane 20 is the amplification profile of one individual from the NES(I) population, which differs considerably in RAPD phenotype from the NES(II) population. (A) Primer UBC 149; (B) primer UBC 313; (C) primer UBC 222.

in two localities; in both instances it was growing with *A. argentea*. The plants of the putative hybrid were sterile, as were two out of three herbarium specimens collected in 1921. Despite intensive searching by the authors and associates, no wild individuals of *Margyricarpus* were found, even though it was common in certain canyons as of 1922. Apparently only a single plant of this species now exists; it is cultivated in a garden on Masatierra. As allozymes proved inconclusive regarding the parentage of × *Margyracaena*, RAPDs were surveyed to test for hybrid origin. DNA samples were obtained from the single plant of *Margyricarpus*, the two putative hybrids, and several plants each of the two *Acaena* species. It was important to sample both species for comparative purposes; if both of these

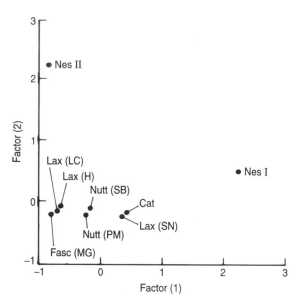

Figure 4-4   Two-dimensional principal-component analysis based on frequency data from 27 RAPD loci for populations of *Malacothamnus fasciculatus*. Variety designations as in Table 4-1 (from Swensen et al., 1995).

species showed identical profiles, no specific conclusions regarding hybrid origin could be made. The results revealed 68 bands (47% of the total) that were restricted to either putative parent and were shared by the putative hybrid; no bands were shared between *A. ovalifolia* and the putative hybrid. Thus, the F_1 hybrid status and hypothesized parentage of × *Margyracaena* was confirmed.

This study highlights two positive attributes of the RAPD assay: the potential for generating polymorphism when isozymes prove to be inadequate, and the ability to generate a large amount of data with extremely limited material. One of the putative parental species was known from only a single living specimen, yet enough material was available using nondestructive sampling to provide ample material for data collection. This study also has important conservation implications, in that a rare species is surviving naturally only through the existence of another species. Rieseberg (1991) has pointed out that in such cases it is imperative to protect the remaining germplasm, even though it is now mixed in the form of a hybrid that under normal circumstances would not be considered for protection. In this case, there is a chance to recover individuals similar to the original species if one of the hybrids flowers, because it could then be backcrossed to the cultivated individual (which apparently cannot self) to produce progeny with various percentages of the *Margyricarpus* genotype. If this situation arises, RAPD would likely be able to detect which progeny possess the highest percentage of *Margyricarpus* in their genomes for use in further breeding.

LIMITATIONS AND MOST APPROPRIATE APPLICATIONS

Limitations

The major advantages of RAPD are that the portion of the genome sampled is relatively unbiased, essentially unlimited numbers of loci can be generated, only a small amount of tissue is needed for analysis, tissue can be stored indefinitely before analysis, and the same primers can be deployed with success on any organism with a minimal or moderate amount of preliminary work. There are, however, some drawbacks to balance these advantages, some of which have already been discussed. Foremost among these is the problem of dominant allelic expression, and the associated problem of having to assume two alleles per locus. As a result, RAPD provides less genetic information on a per locus basis than codominant loci when applied to questions of population genetic structure, paternity, outcrossing rates, or hybridization. Improvements in the way RAPD data are handled are being actively pursued, but it is unlikely that the dominance problem will be overcome entirely. In the meantime, it is highly desirable in many situations (in assessing the distribution of genetic variation, for example) to gain insight into the structure of the population from an independent data set. An alternative approach is to treat RAPD fragments with restriction enzymes and search for variation; in this way, heterozygotes can be detected. This approach is feasible, but takes extra time.

Second, unambiguous locus homology assessment is not possible without extra work. Acrylamide gels can help alleviate this problem because they provide enhanced band resolution, but they add to the expense of the technique.

Third, the technique has, especially in the past, been hampered by reproducibility problems. The method is highly sensitive to the concentration of magnesium chloride, cleanliness of DNA, and the primer:DNA ratio used in the reaction mixture (Ellsworth et al., 1993). In this regard, proper laboratory technique and strict standardization of all reagents and protocols is well rewarded. The use of electronic pipetters is highly recommended for the procedure in order to increase accuracy. DNA should be measured with methods that are specific for it, such as DNA-intercalating dyes, rather than for nucleic acids in general, as there is often an unpredictable amount of RNA contamination in DNA isolates. These DNAs should subsequently be diluted to standard amounts, and the dilutions should be used as a working tube, the original being stored as a stock. DNA purification with systems such as Elu-quik (Schleicher and Schell) often improves amplification. Some workers find that a "hot start" (D'Aquila et al., 1991), in which *Taq* is added at 75°C as the last step in the preparation of the reaction tubes, improves amplification. DNA contamination by airbornes should be avoided as much as practically possible. Finally, primers must be properly maintained to avoid degradation. Each primer, upon arrival, should be distributed into aliquots, dried, and stored below freezing until use. Implementation of these procedures should result in good-quality, reproducible results.

Fourth, non-Mendelian inheritance has been observed to some extent in segregation studies. This may indicate RAPD artifact that may give misleading results in some applications. Heun and Helentjaris (1993) postulated that this could be a result of competition among priming sites (Williams et al., 1993), within individuals with differing genetic backgrounds. However, specific tests by Heun and Helentjaris (1993) found that primer competition did not greatly affect segregation ratios; they concluded that the most likely reason for the observation of non-Mendelian inheritance is the inclusion of "minor" and faint bands that may be highly sensitive to reaction conditions. Uniparental inheritance and meiotic drive could also be factors accounting for distorted segregation ratios. Their experiments in hybrid corn showed that over 90% of RAPD loci exhibited Mendelian inheritance.

Fifth, unlike other methods, the nature of the sampling regions and differences in band intensity in RAPD is poorly understood. There is evidence that RAPD primers are amplifying regions throughout the genome and that these regions belong to repeat categories ranging from low to high (Williams et al., 1993), which probably accounts for some of the intensity differences. Primer mismatches also affect the intensity of bands produced (Williams et al., 1993), and may result in segregating band-brightness polymorphisms.

Finally, nonparental RAPD bands have been reported in primates (Riedy et al., 1992). It is not known whether these bands were a result of PCR artifact or genomic mutation, or whether this problem is exclusive to primates. The general prevalence of nonparental artifactual bands would greatly reduce the utility of RAPD in molecular genetics, especially in paternity studies. Other workers have detected no nonparental bands during segregation analysis (e.g., Heun and Helentjaris, 1993). However, in light of the results of Riedy et al. (1992) it might be best to score only fragments that are found in two or more individuals and are thought or proven to be homologous. This procedure would also minimize the scoring of bands that are the result of random contamination.

Most Appropriate Applications

One of the consistent patterns emerging from genetic studies of endangered populations and species is that different types of molecular markers often reveal different aspects of the evolutionary history of these lineages (e.g., Avise, 1994; Rieseberg and Swensen, 1996). This finding is not unexpected given the variation in mode of inheritance, linkage relationships, and rates of evolutionary divergence observed for different classes of molecules (Avise, 1989; Doyle, 1992). Thus, we feel that the most appropriate and powerful application of RAPDs will be in cases where they are used in conjunction with other types of molecular markers (e.g., Swensen et al., 1995).

There are certain situations, however, where RAPDs are particularly useful relative to other kinds of markers. For example, RAPDs are often highly informative in cases where isozymes either fail to provide polymorphisms, show a level of polymorphism inadequate for the purpose at hand, or cannot provide robust conclusions because the number of loci is limited. The first two scenarios will most often be the case in the study of small populations; the latter will often be the case in studies of hybridization, paternity analysis, outcrossing estimation, and determination of clonal structure. Isozymes are less expensive, even if commercially prepared *Taq* can be avoided; however, RAPD is probably more cost-effective on a per-data-point basis when large numbers of loci are needed, because the number of loci available far exceeds that for isozymes. RAPD also can be used to estimate genetic diversity within and among populations if the appropriate degree of caution is applied.

In conclusion, under carefully controlled reaction conditions, and at the appropriate taxonomic level, RAPD can be a powerful tool for use in studies of the population genetics, systematics, and ecology of rare taxa. However, it is also clear that studies of endangered species should employ more than one class of molecular marker (which in some cases may be RAPDs), and the markers chosen should be both cost-effective and appropriate for the question to be addressed.

REFERENCES

Adams, R.P., T. Demeke, and H.A. Abulfatih. 1993. RAPD DNA fingerprints and terpenoids: Clues to past migrations of *Juniperus* in Arabia and east Africa. Theor. Appl. Genet. 87: 22–26.

Allendorf, F.W. and R.F. Leary. 1986. Heterozygosity and fitness in natural populations of animals. In: M.E. Soulé, ed. Conservation biology, pp. 57–76. Sunderland, Mass.: Sinauer Associates.

Arnold, M.L., C.M. Buckner, and J.J. Robinson. 1991. Pollen mediated introgression and hybrid speciation in Louisiana irises. Proc. Natl. Acad. Sci. USA 88: 1398–1402.

Avise, J.C. 1989. Gene trees and organismal histories: a phylogenetic approach to population biology. Evolution 43: 1192–1208.

Avise, J.C. 1994. Molecular markers, natural history, and evolution. New York: Chapman and Hall.

Beardmore, J.A. 1983. Extinction, survival and genetic variation. In: C.M. Schoenwald-Cox, S.M. Chambers, B. MacBryde, and L. Thomas, eds. Genetics and conservation, pp. 125–151. Menlo Park, Calif.: Benjamin-Cummings.

Brauner, S., D.J. Crawford, and T.F. Stuessy. 1992. Ribosomal DNA and RAPD variation in the rare plant family Lactoridaceae. Am. J. Bot. 79: 1436–1439.

Bucci, G.J. and P. Menozzi. 1993. Segregation analysis of random amplified polymorphic DNA (RAPD) markers in *Picea abies* Karst. Mol. Ecol. 2: 227-232.

Bulat, S.A. and N.V. Mironenko. 1990. Species identity of the phytopathogenic fungi *Pyrenophora teres* Drechsler and *P. graminea* Ito and Kuribayashi. Mikologia i Phytopathologia (Leningrad) 24: 435.

Bulat, S.A., and N.V. Mironenko. 1992. Polymorphism of yeast-like fungus *Aureobasidium pyllulans* (De Bary) revealed by universally primed polymerase chain reaction: Species divergence state. Genetica (Moscow) 28: 4–19.

Bulat, S.A., N.V. Mironenko, M.N. Lapteva, and P.P. Strelchenko. 1994. Polymerase chain reaction with universal primers (UP PCR) and its applications to plant genome analysis. In: R.P. Adams, J.S. Miller, E.M. Golenberg, and J.E. Adams, eds. Conservation of plant genes II: Utilization of ancient and modern DNA, pp. 113–120. St. Louis, Mo.: Missouri Botanical Garden.

Cade, T.J. 1983. Hybridization and gene exchange among birds in relation to conservation. In: C.M. Schoenwald-Cox, S.M. Chambers, B. MacBryde, and W.L. Thomas, eds. Genetics and conservation: A reference for managing wild animal and plant populations, pp. 288–310. Menlo Park, Calif.: Benjamin-Cummings.

Caetano-Anolles, G., G.J. Bassam, and P.M. Gresshof. 1991. High resolution DNA amplification fingerprinting using very short arbitrary oligonucleotide primers. Biotechnology 9: 553–556.

Caswell-Chen, E.P., V.M. Williamson, and F.F. Wu. 1992. Random amplified polymorphic DNA analysis of *Heterodera cruciferae* and *H. schachtii* populations. J. Nematol. 24: 343–351.

Chalmers, K.J., R. Waugh, J.I. Sprent, A.J. Simons, and W. Powell. 1992. Detection of genetic variation between and within populations of *Gliricidia sepiun* and *G. maculata* using RAPD markers. Heredity 69: 465–472.

Clark, A.G. and C.M.S. Lanigan. 1993. Prospects for estimating nucleotide divergence with RAPDS. Mol. Biol. Evol. 10: 1096–1111.

Crawford, D.J., S. Brauner, M.B. Cosner, and T.F. Stuessy. 1993. Use of RAPD markers to document the origin of the intergeneric hybrid × *Margyracaena skottsbergii* (Rosaceae) on the Juan Fernandez Islands. Am. J. Bot. 80: 89–92.

D'Aquila, R.T., L.J. Bechtel, J.A. Videler, J.J. Eron, P. Gorczyca, and J.C. Kaplan. 1991. Maximizing sensitivity and specificity of PCR by pre-amplification heating. Nuclear Acids Res. 19: 3749.

Dawson, I.K., K.J. Chalmers, R. Waugh, and W. Powell. 1993. Detection and analysis of genetic variation in *Hordeum spontaneum* populations from Israel using RAPD markers. Mol. Ecol. 2: 151–159.

Doyle, J.J. 1992. Gene trees and species trees: molecular systematics as one-character taxonomy. Syst. Bot. 17: 144–163.

Ellsworth, D.L., K.D. Rittenhouse, and R.L. Honeycutt. 1993. Artifactual variation in randomly amplified polymorphic DNA banding patterns. BioTechniques 14: 214–217.

Engelke, D.R., A. Krikos, M.E. Bruck, and D. Ginsburg. 1990. Purification of *Thermus aquaticus* DNA polymerase expressed in *Escherichia coli*. Anal. Biochem. 191: 396–400.

Erlich, H.A., ed. 1989. PCR technology: Principles and applications for DNA amplification. New York: Stockton Press.

Excoffier, L. 1992. Winamova 1.04. Analysis of Molecular Variance for Windows. Available by anonymous ftp from acasun1.unige.ch (in directory pub/anova).

Excoffier, L. and P.E. Smouse. 1994. Using allele frequencies and geographic subdivision to reconstruct gene trees within a species: Molecular variance parsimony. Genetics 136: 343–359.

Excoffier, L., P.E. Smouse, and J.M. Quattro. 1992. Analysis of molecular variance inferred from metric distances among DNA haplotypes: Application to human mitochondrial DNA restriction data. Genetics 131: 479–491.

Felsenstein, J. 1989. PHYLIP—Phylogeny Inference Package (Version 3.2). Cladistics 5: 164–166.

Friar, E.A., R. Robichaux, and D.W. Mount. 1996. Molecular genetic variation following a population crash in the endangered Mauna Kea Silversword, *Argyroxiphium sandwicense* ssp. *sandwicense* (Asteraceae). Molecular Ecology, in press.

Fritsch, P. and L.H. Rieseberg. 1992. Outcrossing rates are high in androdiocious populations of the flowering plant *Datisca glomerata*. Nature 359: 633–636.

Fritsch, P., M.A. Hanson, C.D. Spore, P.E. Pack, and L.H. Rieseberg. 1993. Constancy of RAPD primer amplification strength among distantly related taxa of flowering plants. Plant Mol. Biol. Rep. 11: 10–20.

Gibbs, H.L., K.A. Prior, and P. J. Weatherhead, 1994. Genetic analysis of populations of threatened snake species using RAPD markers. Mol. Ecol. 3: 329–338.

Gower, J.C. 1971. A general coefficient of similarity and some of its properties. Biometrics 27: 857–874.

Gustafsson, L. and P. Gustafsson. 1994. Low genetic variation in Swedish populations of the rare species *Vicia pisiformis* (Fabaceae) revealed with rflp (rDNA) and RAPD. Plant Syst. Evol. 189: 133–148.

Hadrys, H., M. Balick, and B. Schierwater. 1993a. Applications of random amplified polymorhic DNA (RAPD) in molecular ecology. Mol. Ecol. 1: 55–63.

Hadrys, H., B. Schierwater, S.L. Dellaporta, R. DeSalle, and L.W. Buss. 1993b. Determination of paternity in dragonflies by random amplified polymorphic DNA fingerprinting. Mol. Ecol. 2: 79–87.

Hanson, M.H. 1993. Dispersed unidirectional introgression from *Yucca schidigera* into *Y. baccata* (Agavaceae). Ph.D. dissertation, Claremont Graduate School, Claremont, Calif.

Hanson, M.H., P. Fritsch, and L.H. Rieseberg. 1992. Level and distribution of genetic diversity of endangered *Pogogyne* (Lamiaceae): Genetic evidence warrants recognition of a new species from Baja California. Am. J. Bot. 80(supplement): 152–153 [Abstract].

Heun, M. and T. Helentjaris. 1993. Inheritance of RAPDs in F_1 hybrids of corn. Theor. Appl. Genet. 85: 961–968.

Hu, J. and C. F. Quiros. 1991. Identification of broccoli and cauliflower cultivars with RAPD markers. Plant Cell Rep. 10: 505–511.

Huff, D.R., R. Peakall, and P.E. Smouse. 1993. RAPD variation within and among natural populations of outcrossing buffalograss (*Buchloë dactyloides* (Nutt.) Engelm.). Theor. Appl. Genet. 86: 927–934.

Hunt, G.J. and R.E. Page Jr. 1992. Patterns of inheritance with RAPD molecular markers reveal novel types of polymorphism in the honey bee. Theor. Appl. Genet. 85: 15–20.

Innis, M.A., D.H. Gelfand, J.J. Sninsky, and T.J. White, eds. 1990. PCR protocols: A guide to methods and applications. San Diego: Academic Press.

Jaccard, P. 1912. The distribution of the flora of the alpine zone. New Phytol. 11: 37–50.

Kambhampati, S., W.C. Black IV, and K.S. Rai. 1992. Random amplified polymorphic DNA of mosquito species and populations (Diptera: Culicidae): Techniques, statistical analysis, and applications. J. Med. Entomol. 29: 939–945.

Karron, J.D. 1991. Patterns of genetic variation and breeding systems in rare plant species. In: D.A. Falk, and K.E. Holsinger, eds. Genetics and conservation of rare plants, pp. 87–98. New York: Oxford University Press.

Lacy, R.C. 1988. A report on population genetics in conservation. Conserv. Biol. 2: 245–247.

Lewis, P.O. and A.A. Snow. 1992. Deterministic paternity exclusion using RAPD markers. Mol. Ecol. 1: 155–160.

Lewontin, R.C. 1972. The apportionment of human diversity. Evol. Biol. 6: 381–398.

Lynch, M. and B.G. Milligan. 1994. Analysis of population genetic structure with RAPD markers. Mol. Ecol. 3: 91–100.

McMillan, D.E. and L.L. Muldrow. 1992. Typing of toxic strains of *Clostridium difficile* using DNA fingerprints generated with arbitrary polymerase chain reaction primers. FEMS Microbiol. Lett. 92: 5–10.

Milligan, B.G. 1992. Genetic variation among west Texas populations of *Aquilegia*. In: R. Sivinski and K. Lightfoot, eds. Southwestern rare and endangered plants: Proceedings of the Southwestern Rare and Endangered Plant Conference, pp. 289–295. Santa Fe, N.M.: New Mexico Forestry and Resources Conservation Division.

Milligan, B.G. and C.K. McMurry. 1993. Dominant vs. codominant genetic markers in the estimation of male mating success. Mol. Ecol. 2: 275–283.

Nei, M. 1973. Analysis of gene diversity in subdivided populations. Proc. Natl. Acad. Sci. U.S.A. 70: 3321–3323.

Orozco-Castillo, C., K.J. Chalmers, R. Waugh, and W. Powell. 1994. Detection of genetic diversity and selective gene introgression in coffee using RAPD markers. Theor. Appl. Genet 87: 934–940.

Owen, J.L. and C.M. Uyeda. 1991. Single primer amplification of avian genomic DNA detects polymorphic loci. Animal Biotechnol. 2: 107–122.

Philbrick, C.T. 1993. Underwater cross-pollination in *Callitriche hermaphroditica* (Callitrichaceae): Evidence from random amplified polymorphic DNA markers. Am. J. Bot. 80: 391–394.

Pluthero, F.G. 1992. Rapid purification of high-activity *Taq* polymerase. Toronto: Abracax.

Price, M.V. and N.M. Waser. 1979. Pollen dispersal and optimal outcrossing in *Delphinium nelsoni*. Nature 277: 294–297.

Puterka, G.J., W.C. Black IV, W.M. Steiner, and R. L. Burton. 1993. Genetic variation and phylogenetic relationships among worldwide collections of the Russian wheat aphid, *Diuraphis noxia* (Mordvilko), inferred from allozyme and RAPD-PCR markers. Heredity 70: 604–618.

Ragot, M. and D.A. Hoisington. 1993. Molecular markers for plant breeding: Comparisons of RFLP and RAPD genotyping costs. Theor. Appl. Genet. 86: 975–984.

Riedy, M.F., W.J. Hamilton III, and C.F. Aquadro. 1992. Excess of non-parental bands in offspring from known primate pedigrees assayed using RAPD PCR. Nucleic Acids Res. 20: 918.

Rieseberg, L.H. 1991. Hybridization in rare plants: Insights from case studies in *Cercocarpus* and *Helianthus*. In: D.A. Falk and K.E. Holsinger, eds. Genetics and conservation of rare plants, pp 171–181. New York: Oxford University Press.

Rieseberg, L.H. and D. Gerber. 1995. Hybridization in the Catalina mahogany: RAPD evidence. Conserv. Biol. 9: 199–203.

Rieseberg, L.H. and S.M. Swensen. 1996. Conservation genetics of endangered island plants. In: J.C. Avise and J. Hamrick, eds. Conservation genetics: Case histories from nature, pp. 305–334. New York: Chapman and Hall.

Rieseberg, L.H., S. Zona, L. Aberbom, and T.D. Martin. 1989. Hybridization in the island endemic, Catalina mahogany. Conserv. Biol. 3: 52–58.

Rieseberg, L.H., H. Choi, R. Chan, and C.D. Spore. 1993. Genomic map of a diploid hybrid species. Heredity 70: 285–293.

Ritland, K. and S.K. Jain. 1981. A model for the estimation of outcrossing rate and gene frequencies using *n* independent loci. Heredity 47: 35–52.

Rohlf, F.J. 1993. NTSYS-pc. Numerical Taxonomy and Multivariate Analysis System, Version 1.80. Applied Biostatistics, Inc., Setaudet, N.Y.

Russell, J.R., F. Hosein, E. Johnson, R. Waugh, and W. Powell. 1993. Genetic differentiation of cocoa (*Theobroma cacao* L.) populations revealed by RAPD analysis. Mol. Ecol. 2: 89–97.

Saitou, N. and M. Nei 1987. The neighbor-joining method: A new method for reconstructing phylogenetic trees. Mol. Biol. Evol. 4: 406–425.

Schaal, B.A., W.J. Leverich, and S.H. Rogstad. 1991. A comparison of methods for assessing genetic variation in plant conservation biology. In: D.A. Falk and K.E. Holsinger, eds. Genetics and conservation of rare plants, pp. 123–134. New York: Oxford University Press.

Smith, M.L., J.N. Bruhn, and J.B. Anderson. 1992. The fungus *Armillaria bulbosa* is among the largest and oldest living organisms. Nature 356: 428–431.

Swensen, S.M., G.J. Allan, M. Howe, W.J. Elisens, S.A. Junak, and L.H. Rieseberg. 1995. Genetic analysis of the endangered island endemic *Malacothamnus fasciculatus* (Nutt.) Greene var. *nesioticus* (Rob.) Kearn. (Malvaceae). Cons. Biol. 9: 404–415.

Templeton, A.R. 1991. Off-site breeding of animals and implications for plant conservation strategies. In: D.A. Falk and K.E. Holsinger, eds. Genetics and conservation of rare plants, pp. 182–194. New York: Oxford University Press.

Van Buren, R., K.T. Harper, W.R. Andersen, D.J. Stanton, S. Seyoum, and J.L. England. 1994. Evaluating the relationship of autumn buttercup (*Ranunculus acriformis* var. *aestivalis*) to some close congeners using random amplified polymorphic DNA. Am. J. Bot. 81: 514–519.

Welsh, J. and M. McClelland. 1990. Fingerprinting genomes using PCR with arbitrary primers. Nucleic Acids Res. 18: 7213–7218.

Welsh, J., C. Petersen, and M. McClelland. 1991. Polymorphisms generated by arbitrarily primed PCR in the mouse: Application to strain identification and genetic mapping. Nucleic Acids Res. 19: 303–306.

Welsh, J., C. Pretzman, D. Postic, I. Saint Girons, G. Baranton, and M. McClelland. 1992. Genomic fingerprinting by arbitrarily primed polymerase chain reaction resolves *Borrelia burgsorferi* into three distinct groups. Int. J. Syst. Bacteriol. 42: 370–377.

White, B.A., ed. 1993. PCR protocols: Current methods and applications. Totowa, N.J.: Humana Press.

Whitkus, R., J. Doebley, and J.F. Wendel. 1994. Nuclear DNA markers in systematics and evolution. In: L. Phillips, and I.K. Vasil, eds. DNA-based markers in plants, pp. 116–141. Amsterdam, The Netherlands: Kluwer.

Williams, J.G.K., A.R. Kubelik, K.J. Livak, J.A. Rafalski, and S.V. Tingey. 1990. DNA polymorphisms amplified by arbitrary primers are useful as genetic markers. Nucleic Acids Res. 18: 6531–6535.

Williams, J.G.K., M.K. Hanafey, J.A. Rafalski, and S.V. Tingey. 1993. Genetic analysis using random amplified polymorphic DNA markers. Methods Enzymol. 218: 704–740.

Woodward, S.R., J. Sudweeks, and C. Teuscher. 1992. Random sequence oligonucleotide primers detect polymorphic DNA products which segregate in inbred strains of mice. Mammal. Genome 3: 73–78.

5

The Utility of Paternally Inherited Nuclear Genes in Conservation Genetics

PRISCILLA K. TUCKER AND BARBARA L. LUNDRIGAN

The mammalian Y chromosome is of special interest as a potential tool in the conservation genetics of mammalian species because of its unique mode of inheritance. With the exception of a small region that recombines with the X chromosome during meiosis, the Y chromosome is the only chromosome in the mammalian nuclear genome that does not recombine; thus, most Y chromosome-linked loci are clonally inherited from father to son.

Clonally inherited molecules are especially useful for inferring phylogenies or pedigrees, and for detecting and characterizing population phenomena, such as bottlenecks or founder events, because they are transmitted from one generation to the next without the shuffling effects of recombination. Mitochondrial genes, which are clonally inherited through maternal lineages in most vertebrates, have proven very useful in both intra- and interspecific studies (Wilson et al., 1985; Moritz et al., 1987; Avise et al., 1987). However, variation in the mitochondrial genome, and gene phylogenies and pedigrees established using mitochondrial DNA (mtDNA), represent only the maternal side of an organism's history. Y chromosome-linked loci provide an obvious complement to mtDNA because they represent the paternal side of an organism's history. In effect, phylogenies based on Y-linked loci supplement phylogenies constructed from maternally inherited or Mendelian inherited loci, and thus increase the explanatory power of phylogenetic analyses. Data from Y-linked loci may be particularly useful for resolving instances where phylogenies constructed from maternally inherited mtDNA do not coincide with phylogenies constructed from Mendelian inherited loci. Such cases, for example, could result from differential introgression of mitochondrial versus Mendelian inherited genes across hybrid zones (e.g., Ferris et al., 1983; Powell, 1983; Carr et al., 1986; Tegelstrom, 1987; Carr and Hughes, 1993).

Y-linked loci also contain historical information on male dispersal patterns which can be compared to historical information on female dispersal patterns provided by mitochondrial variation. Such comparisons are important in mammals, where social organization is often complex, the dominant mode of reproductive behavior is polygyny, and sex differences in dispersal are common. Finally, Y-linked loci can be used to determine an individual's sex. This may be particularly useful in instances where sex is difficult to determine (e.g., cetaceans; Palsboll et al., 1992).

To effectively use Y chromosome-linked loci in population studies, one needs to examine a locus that is evolving at an appropriate rate to permit the characterization of population phenomena such as bottlenecks, founder events, and dispersal. To use Y chromosome-linked loci for reconstructing phylogenies, one needs to examine a locus that is evolving rapidly enough to generate phylogenetic signal, but not so rapidly that convergent or reversal events are confounding factors. Thus, the utility of the Y chromosome for population and phylogenetic studies ultimately depends on the evolutionary tempo of loci found in the nonrecombining portion of the Y chromosome.

Much of the current knowledge on the evolution of Y chromosome-linked loci has come from comparative studies using Southern blot hybridization. In these studies, DNA sequences unique to the Y chromosome of humans or laboratory mice were used as probes to genomic DNA from other species. Most of the human Y chromosome sequences examined in this manner are either not located on the Y chromosome in other mammalian taxa, or, if located on the Y chromosome, have related copies elsewhere in the genome (Cooke et al., 1982; Kunkel and Smith, 1982; Page et al., 1984; Koenig et al., 1985; Erickson, 1987). However, a few sequences found exclusively on the Y chromosome in humans (i.e., Y-specific sequences) are also Y-specific in the great apes, family Pongidae (Wolfe et al., 1984a,b; Burk et al., 1985; Erickson, 1987; Guttenbach et al., 1992). Similarly, most laboratory mouse Y-specific sequences are Y-specific only among close relatives of the laboratory mouse (i.e., species belonging to the subgenus *Mus* of the genus *Mus*) (Nishioka and Lamothe, 1986; Platt and Dewey, 1987; Eicher et al., 1989). To date, only a few loci isolated from the nonrecombining portion of the Y chromosome are known to be Y-specific across a broad range of mammalian taxa. These are all functional loci and include *Zfy* (Page et al., 1987), *Sry* (Gubbay et al., 1990, Sinclair et al., 1990) *Ubely* (Mitchell et al., 1991), *Smcy* (Aguinik et al., 1994), and members of the YRRM gene family (Ma et al., 1993). These data suggest that there may be many Y-linked sequences appropriate for studies of closely related mammalian taxa, but only a few Y-linked sequences suitable for use in higher level systematic studies of mammals.

A few studies have investigated within-species variation of Y chromosome-linked loci using Southern blot analysis of restriction fragment length polymorphisms (RFLPs). Both multiple-copy Y-enriched (i.e., having more copies on the Y chromosome than on the X chromosome or autosomes) or Y-specific (i.e., found only on the Y chromosome) sequences and Y-specific single-copy sequences (Casanova, 1985; Lucotte and Ngo, 1985; Lucotte, et al., 1989, 1993; Oakey and Tyler-Smith, 1990; Lucotte and David, 1992; Tucker et al., 1992; Nagamine et al., 1992; Ritte et al., 1993; Spurdle et al., 1994) have been shown to be polymorphic in humans and/or house mice. However, while multiple-copy sequences are useful for identifiying variants within a population, they are problematic for inferring relationships because restriction fragments of repeat sequences cannot be mapped to specific sites. Likewise, RFLPs of single-copy Y-specific sequences are of limited use for inferring relationships within species because there is minimal intraspecific polymorphisms of these sequences.

Here we use a comparative sequencing approach to explore the utility of a Y chromosome-linked gene, *Sry*, the sex-determining locus responsible for testis determination (Sinclair et al., 1990; Berta et al., 1990; Jager et al., 1990; Gubbay et al., 1990; Koopman et al., 1991), for reconstructing historical relationships within and between species.

TECHNICAL DESCRIPTION

Sry is thought to contain a single exon consisting of a central HMG (high-mobility group) DNA-binding domain 79 amino acids in length, and flanking sequences. Comparative studies of *Sry* from over 20 mammalian species have shown that the DNA sequence of the HMG domain is similar across species of marsupial and placental mammals (Sinclair et al., 1990; Gubbay et al., 1990; Foster et al., 1992; Whitfield et al., 1993; Bianchi et al., 1993; Tucker and Lundrigan, 1993; Lundrigan and Tucker, 1994; Payen and Cotinot, 1994). However, there is little or no sequence similarity in flanking regions between these two infraclasses, or among orders within each infraclass. For example, the N-terminal region in primates is 58 amino acids long, but only 2 amino acids long in Old World mice and rats (family Muridae, subfamily Murinae). The C-terminal region in primates varies in length from 70 amino acids in humans to 94 amino acids in marmosets (Whitfield et al., 1993). The C-terminal region in Old World mice and rats varies in length from 92 amino acids in *Hylomyscus alleni* to 313 amino acids in *Mus musculus* (Tucker and Lundrigan, 1993). Significant length variation in the C-terminal region is evident even when comparing sibling species of house mice, *Mus musculus* and *Mus domesticus*: the C-terminal region of *Sry* in laboratory mice carrying a *Mus musculus* Y chromosome is 313 amino acids in length (Hacker et al., 1995); this same region in laboratory mice carrying a *Mus domesticus* Y chromosome ranges in length from 153 to 155 amino acids (Coward et al., 1994). The discrepancy in size between *Mus musculus* and *Mus domesticus Sry* sequences is due to a C-to-T transitional mutation at nucleotide position 9906 (numbers for nucleotide positions throughout the paper refer to the base pair position of *Sry* in GenBank entry X67204) resulting in the replacement of the amino acid glutamine in *M. musculus* with a stop codon in *Mus domesticus*. Variation in length of the protein in laboratory strains carrying different *Mus domesticus* Y chromosomes is due to variation in the number of CAG repeats in the interval spanning nucleotide positions 8812–8847 (Coward et al., 1994). Most of the size variation in the C-terminal region in other Old World mice and rats is due to the evolutionary expansion or contraction of CAG and CAC trinucleotide repeats (Tucker and Lundrigan, 1993). The lack of conservation in the N-terminal and C-terminal regions flanking the HMG domain of *Sry* makes comparative sequencing studies across broad taxonomic groups difficult to perform using PCR technology. We have been unable to amplify *Sry* in species outside the subfamily Murinae (Old World mice and rats) using PCR primer pairs designed from conserved murine *Sry* sequence found in the 5′ untranslated, N-terminal, and C-terminal regions. Our samples included species from the subfamilies Sigmodontinae (New World mice and rats), Arvicolinae (voles, lemmings, and muskrats),

Gerbillinae (gerbils, jirds, and sand rats), and Nesomyinae (Malagasy mice and rats). It is possible, however, to use primers designed from the conserved HMG domain to amplify a portion of this region from other murid rodents (Bianchi et al., 1993) and from representatives of other orders of mammals (Palsboll et al., 1992; Griffiths and Tiwari, 1993; Payen and Cotinot, 1993). Our comparative sequence data from this region indicate that this segment is too short and not sufficiently variable to be useful in investigating historical relationships within or between species of murine rodents (Tucker and Lundrigan, 1993; Lundrigan and Tucker, 1994). However, primers designed from the conserved HMG domain are useful in determining the sex of individuals from species representing a variety of orders of mammals. In these experiments, presence of the HMG domain is evidence of maleness, provided known male and female representatives of the same species and a second (non-Y-specific) gene are included as controls (for examples, see Palsboll et al., 1992; Griffiths and Tiwari, 1993).

We have been successful in amplifying and sequencing the entire *Sry* gene from murine taxa whose *Sry* sequence varies in overall (uncorrected) sequence divergence by as much as 11.7% (Tucker and Lundrigan, 1993). Detailed amplification and sequencing protocols have been published (Tucker and Lundrigan, 1993; Lundrigan and Tucker, 1994). Some important considerations for doing this work are presented below.

There are no "universal" primers for *Sry*. Rather, we have found that considerable experimentation with annealing temperature, magnesium chloride concentration, and primer combination is necessary to amplify *Sry* for each species examined. Our PCR amplification experiments always include male and female DNA from each taxon studied (Figure 5-1). The inclusion of female DNA in our

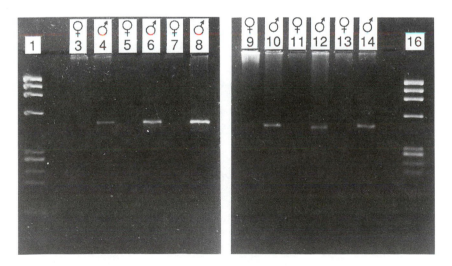

Figure 5-1 Ethidium bromide-stained agarose gels demonstrating male-specific amplification of a 471 bp portion of *Sry* from the following murine taxa: *Mus musculus* (lanes 3 and 4), *Hylomyscus alleni* (lanes 5 and 6), *Aethomys chrysophilus* (lanes 7 and 8), *Hybomys univitatus* (lanes 9 and 10), *Praomys fumatus* (lanes 11 and 12), and *Rattus everetti* (lanes 13 and 14). The size standard is *Hae*III-digested phiX174 RF DNA (lanes 1 and 16).

experiments is a necessary experimental control; amplification of the target sequence in males only is taken as evidence that the amplified product is unique to the Y chromosome. Such evidence is necessary, especially considering that *Sry* belongs to a gene family whose other members, the *Sox* (*Sry*-like HMG b*ox*) genes, are found on autosomes and the X chromosome (Gubbay et al., 1990; Stevanovic et al., 1993).

CASE STUDIES

We have generated comparative sequence data for *Sry* from 13 species of Old World mice and rats (Tucker and Lundrigan, 1993; Lundrigan and Tucker, 1994) (Table 5-1). Our sample consists of a 515 bp region (8218–8732) including 86 bp of 5' untranslated DNA, the N-terminal region (6 bp), the HMG box (237 bp), and a portion of the C-terminal region (186 bp). The remainder of the C-terminal region includes an imperfect trinucleotide repeat motif of variable length. Although

Table 5-1   Species and Collecting Localities of Specimens Used in the Study

Taxon[a]	Collecting Locality
Mus musculus musculus[b]	Japan: Kyushu
	Denmark: Viborg, Skive
	Czechoslovakia: Slovakia; Sladeckovce
castaneus	Thailand: Chonburi Province; Chonburi
Mus domesticus $(2n = 26)$	Switzerland: Grisons Canton; Zalende, Poschiavo
$(2n = 24)$	Italy: Sondrio; Tirano
$(2n = 40)$	Morocco: Tafilalt Oasis; Erfoud
$(2n = 40)$	United States: Maryland; Queen Anne's County, Centreville
$(2n = 22)^c$	Italy: Molise; 6.1 km W by road of Bonefro
$(2n = 40)^d$	Italy: Lazio; 11.4 km WNW by road of Cassino train station
Mus spicilegus	Austria: Burgenland Province; 6 km ENE Halbturn
	Yugoslavia: Debeljaca
Mus macedonicus	Yugoslavia: Gradsko
Mus spretus	Morocco: Azrou
	Spain: 8 km E Puerto Real
Mus cookii	Thailand: Tak province; Loei
Mus cervicolor popaeus	Thailand: Saraburi Province
cervicolor	Thailand: Chonburi Province
Mus caroli	Thailand: Chonburi Province
Mus pahari	Thailand: Tak Province
Mastomys hildebrantii	Kenya: Eastern Province; Machakos District
Hylomyscus alleni	Gabon: Estuaire Province; 1 km SE of Cape Esterias
Stochomys longicaudatus	Gabon; Estuaire Province; 1 km SE of Cape Esterias
Rattus exulans	Philippines: Negros Oriental Province; 3 km N, 17 km W of Dumaguete

[a] Sample size is 1 per locality unless designated otherwise.

[b] *Mus musculus musculus* from Japan is a hybrid of *Mus musculus castaneus* and *Mus musculus musculus* (Yonekawa et al., 1986, 1988; Bonhomme et al., 1989); its Y chromosome is of *M. musculus* origin (Tucker et al., 1992).

[c] Analysis of repeat based on a sample size of 13 males.

[d] Analysis of repeat based on a sample size of 12 males.

Table 5-2 Pairwise Percent Sequence Divergence (substitutions per 100 sites), Calculated by the Jukes and Cantor (1969) Method, for a 515 bp Region of Sry[a]

	Mmu(1)	Mmu(2)	Mdo	Mspi	Mma	Mspr	Mco	Mce	Mca	Mpa	Mhi	Hal	Slo	Rex
Mmu(1)	—	0.39	0.78	1.36	1.55	1.75	4.08	4.27	3.69	5.83	6.99	8.16	10.29	11.65
Mmu(2)		—	0.39	0.97	1.17	1.36	3.69	3.88	3.30	5.63	6.80	7.96	10.10	11.46
Mdo			—	0.58	0.78	0.97	3.30	3.50	2.91	5.24	6.41	7.57	10.10	11.07
Mspi				—	0.19	0.78	3.88	4.08	3.50	5.83	6.99	7.57	10.29	11.46
Mma					—	0.97	4.08	3.27	3.69	6.02	7.18	7.77	10.49	11.26
Mspr						—	4.27	4.47	3.88	6.21	7.38	7.96	10.29	11.84
Mco							—	2.52	2.33	5.44	6.21	8.16	10.68	11.65
Mce								—	2.14	5.24	6.02	7.96	10.49	11.07
Mca									—	4.66	5.44	7.38	9.51	10.49
Mpa										—	5.83	6.99	9.90	11.07
Mhi											—	4.65	7.96	9.51
Hal												—	8.54	9.90
Slo													—	11.84
Rex														—

[a] Mmu(1) = *Mus musculus musculus* from Japan; Mmu(2) = *Mus musculus castaneus* and European *Mus musculus musculus*; Mdo = *Mus domesticus*; Mspi = *Mus spicilegus*; Mma = *Mus macedonicus*; Mspr = *Mus spretus*; Mco = *Mus cookii*; Mce = *Mus cervicolor cervicolor*; Mca = *Mus caroli*; Mpa = *Mus pahari*; Mhi = *Mastomys hildebrantii*; Hal = *Hylomyscus alleni*; Slo = *Stochomys longicaudatus*; Rex = *Rattus exulans*.

we have sequenced this region in four species, it cannot be unambiguously aligned across taxa (Tucker and Lundrigan, 1993).

Percentage sequence divergence for the 515 bp fragment ranges from 0.19% (between *Mus macedonicus* and *Mus spicilegus*) to 11.84% (between *Rattus exulans* and *Stochomys longicaudatus*, and between *Rattus exulans* and *Mus spretus*) (Table 5-2). Overall, nucleotide substitutions have occurred more frequently in the C-terminal region of *Sry* than in the 5′ untranslated, N-terminal, and HMG box regions. A phylogenetic analysis based on sequence variation in *Sry* is in almost perfect concordance with trees constructed using mtDNA and allozyme data from the same group of taxa (Lundrigan and Tucker, 1994). However, the low level of substitution among closely related taxa, such as the Palearctic species of the genus *Mus*, renders the use of *Sry*, by itself, minimally useful for discerning relationships among those species (Figure 5-2).

Our investigation of population level variation included five species. We sequenced the 515 bp fragment from individual *Mus musculus* collected from four geographically distinct localities, and sequenced the same 515 bp region and the first two uninterrupted stretches of the trinucleotide repeat CAG (8733–8765 and 8811–8846) from individual *Mus domesticus* collected from six geographically distinct localities. The latter sample included within-population sampling for the repeat region at two localities ($n = 13$ and $n = 12$; Table 5-1). In addition, we sampled from two geographically distinct populations each of *Mus spretus*, *Mus spicilegus*, and *Mus cervicolor*.

Percentage sequence divergence within species ranges from zero in *Mus domesticus*, *M. spretus*, and *M. spicilegus*, to 0.39%, in *Mus musculus*. The variation (two base pair substitutions) found in *Mus musculus* distinguishes Asian *Mus musculus musculus* from Asian *Mus musculus castaneus* and European *Mus musculus musculus*. No variation was found between the geographically

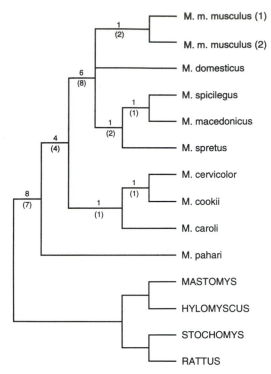

Figure 5-2 Results of a parsimony analysis based on 515 bp of *Sry*, performed using the branch-and-bound option in PAUP (version 3.1.1), with unordered, uniformly weighted characters. Outgroup taxa are in capital letters. The number of unique and unreversed synapomorphies is given above the branch, and the support index (Bremer, 1988), in parentheses, below the branch for each nonterminal clade. The single most-parsomonius tree has a length of 161 steps., a CI excluding uniformative characters of 0.694, and an RI of 0.825.

distinct populations of *Mus musculus castaneus* and European *Mus musculus musculus*. *M. cervicolor* had a single base pair difference distinguishing subspecies.

The complete lack of variation in this portion of *Sry* across the six geographically distinct populations of *Mus domesticus* is a somewhat surprising finding, especially considering that variation in allozymes and mitochondrial DNA occurs throughout the range of *Mus domesticus* (Britton-Davidian, 1990; Prager et al., 1993). In contrast to the 515 bp region, the first two uninterrupted regions of the trinucleotide repeat were polymorphic in length across the six geographically distinct populations of *Mus domesticus*. Repeat length varied from 9 to 11 CAG repeats in the first repeat region, and from 11 to 20 in the second repeat region. However, no length polymorphism in either repeat region was observed within the two nongeographic samples (Miller et al., 1995). Variation in length of the CAG repeats, in contrast to the 515 bp fragment, probably reflects the general instability of trinucleotide repeats relative to unique-sequence DNA (Richards and Sutherland, 1992, 1994).

In a recent study of a 1063 bp noncoding region flanking *Sry*, Nachman et al.

(1994) report the presence of two polymorphic base substitutions and two polymorphic insertion/deletion sites in a sample of 20 *Mus domesticus* collected from 11 geographically distinct localities. The slightly higher level of inter-population variation in this region, in contrast to that for the 515 bp fragment of *Sry*, may be due to actual differences in rate of evolution between these two regions and/or to sampling error.

The low level of polymorphism in the unique portion of *Sry* limits its use for investigating population substructure or historical relationships among populations. However, we were able to detect a probable founder event based on variation in the 515 bp portion of the *Sry* sequence (Tucker et al., 1992). Standard laboratory strains of inbred mice have either a *Mus musculus* or *Mus domesticus* Y chromosome. Based on the pattern of variation observed in *Sry* among populations of *Mus musculus*, we determined that the Y chromosomes of inbred strains carrying a *Mus musculus* Y chromosome are of Asian origin. This conclusion was corroborated by RFLP analysis using a multiple-copy Y-linked sequence as a hybridization probe (Tucker et al., 1992).

The variation found in the number of CAG repeats cannot, by itself, be used for investigating historical relationships because one cannot easily reconstruct the evolutionary history of the repeat. Stretches of an uninterrupted trinucleotide repeat cannot be unambiguously aligned between taxa, and the expansion or contraction of repeats does not necessarily occur by the addition or deletion of single repeat units (Richards and Sutherland, 1992, 1994).

There are two aspects of Y chromosome transmission that suggest low levels of variation might be characteristic of Y-linked loci. First, because the Y chromosome is clonally inherited through paternal lineages, the effective population of Y-linked loci is small relative to autosomal and X-linked loci. When the male-to-female breeding sex ratio is 1, Y-linked loci are only one-fourth as numerous as autosomal loci and only one-third as numerous as X-linked loci. In addition, there is evidence that house mice are polygynous (Singleton and Hay, 1983), reducing even further the effective population size of Y-linked loci. Second, lack of variation in this portion of *Sry* might result from genetic hitch-hiking (Maynard Smith and Haigh, 1974; Kaplan et al., 1989), a process whereby a selectively favored gene mutation sweeps through all populations of a species, effectively fixing alleles at other linked loci. Alternatively, lack of variation could result from background selection against linked deleterious mutations elsewhere on the Y chromosome (Charlesworth et al., 1993). Clearly, more empirical data are needed to determine whether the low level of variation in *Sry* within the species we surveyed is a result of either of these two processes. We are currently examining within and between-species sequence variation at an autosomal locus in the same individuals to assess whether the pattern of sequence evolution in *Sry* is different from that for autosomal loci.

LIMITATIONS AND APPLICATIONS

The attractiveness of mammalian Y chromosome-linked loci in conservation genetics results from its unique mode of clonal inheritance through paternal lineages. Y chromosome-linked loci are complementary to maternally inherited mitochondrial genes. Used in concert, they theoretically could provide a nearly complete record of male and female dispersal patterns, bottlenecks, and founder events in populations of mammals.

Our work on the evolution of Y chromosome-linked loci, in particular *Sry*, has direct implications for the usefulness of this gene, and possibly for the usefulness of other Y-linked loci in population studies. First, unlike many mitochondrial loci, universal primers are not available to amplify the entire *Sry* gene from a broad range of taxa. Although primers designed from the conserved HMG domain could be used to amplify a portion of the HMG domain across orders of mammals, and to determine the sex of individual animals, we have found that the HMG domain is both too short and not variable enough to provide much information for either interspecific or population studies. Therefore, if one wants to conduct a comparative sequence analysis of *Sry* in taxa other than primates, murine rodents, and bovids, for which the entire gene has been sequenced, one would first need to isolate *Sry* from a genomic library constructed from the study group of interest.

Second, and perhaps more importantly, our population study of sequence variation in *Sry* indicates that this gene is not highly polymorphic. If lack of variation at *Sry* is a consequence of the clonal nature of inheritance of the Y chromosome through paternal lineages, then other Y-linked loci may also not be variable enough to be useful in population studies. Although hypervariable multiple-copy Y-specific loci have been isolated and used in population studies, they are problematic for inferring relationships because they cannot be mapped to specific sites. Several other single-copy Y-specific genes have been isolated. However, it is not yet known whether they are variable enough for examination of intraspecific historical relationships. More research, including the further characterization of the evolution of these single-copy Y-specific genes, and the isolation and characterization of other mammalian Y-linked genes, is needed to determine the extent to which Y chromosome-linked genes will be useful in conservation biology.

Finally, while this chapter documents the utility of the *Sry* gene as a male-specific marker in conservation genetic studies of mammalian species, it should be noted that *Sry* is not conserved on the Y chromosome in other vertebrate groups. However, it is possible that male-specific markers can be isolated from the Y chromosome of other organisms with XX/XY sex determining systems (in which males are the heterogametic sex), to be similarly used in conservation genetic studies.

ACKNOWLEDGMENTS

We gratefully acknowledge M. Potter (NCI contract N01-CB-71085) and E. Eicher, for providing frozen organs or DNA from various strains of wild mice, and R. Sage (University of Missouri), B. Patterson, and L. Heaney (Field Museum of Natural History), R. Baker (The Museum, Texas Tech University), and M. Nachman for providing frozen organs from wild mice collected in the field. We thank R. Adkins for reviewing an earlier draft of this manuscript. This publication was supported by National Science Foundation grants, BSR-9009806 and DEB-9209950, by the University of Michigan Office of the Vice-President for Research, and by the Michigan Memorial Phoenix Project (all to P.K.T.).

REFERENCES

Aguinik, A.I., M.J. Mitchell, J.L. Lerner, D.R. Woods, and C.E. Bishop. 1994. A mouse Y chromosome gene encoded by a region essential for spermatogenesis and expression of male-specific minor histocompatibility antigens. Hum. Mol. Gen. 3: 873–878.

Avise, J.C., J. Arnold, R.M. Ball, F. Bermingham, T. Lamb, J.E. Neigel, C.A. Reeb, and N.C. Saunders. 1987. Intraspecific phylogeography: the mitochondrial DNA bridge between population genetics and systematics. Ann. Rev. Ecol. Syst. 18: 489–522.

Berta, P., J.R. Hawkins, A.H. Sinclair, A. Taylor, B.L. Griffiths, P.N. Goodfellow, and M. Fellous. 1990. Genetic evidence equating *SRY* and the testis determining factor. Nature 348: 448–450.

Bianchi, N.O., M.S. Bianchi, G. Bailliet, and A. de la Chapelle. 1993. Characterization and sequencing of the sex determining region Y gene (*Sry*) in *Akodon* (Cricetidae) species with sex reversed females. Chromosoma 102: 389–395.

Bonhomme, F., N. Miyashita, P. Boursot, J. Catalan, and K. Moriwaki. 1989. Genetical variation and polyphyletic origin in Japanese *Mus musculus*. Heredity 63: 299–308.

Bremer, K. 1988. The limits of amino-acid sequence data in angiosperm phylogenetic reconstruction. Evolution 42: 795–803.

Britton-Davidian, J. 1990. Genic differentiation in *M. m. domesticus* populations from Europe, the Middle East, and North Africa: geographic patterns and colonization events. Biol. J. Linn. Soc. 41: 27–45.

Burk, R.D., P. Ma, and K.D. Smith. 1985. Characterization and evolution of a single-copy sequence from the human Y chromosome. Mol. Cell. Biol. 5: 576–581.

Carr, S.M. and G.A. Hughes. 1993. Direction of introgressive hybridization between species of North American deer (*Odocoileus*) as inferred from mitochondrial-cytochrome-*b* sequences. J. Mammal. 74: 331–342.

Carr, S.M., S.W. Ballinger, J.N. Derr, L.H. Blankenship, and J.W. Bickham. 1986. Mitochondrial DNA analysis of hybridization between sympatric white-tailed deer and mule deer in west Texas. Proc. Natl. Acad. Sci. 83: 9576–9580.

Casanova, M., P. Leroy, C. Boucekkine, J. Weissenbach, C. Bishop, M. Fellous, M. Purrello, G. Fiori, and M. Siniscalco. 1985. A human Y-linked DNA polymorphism and its potential for estimating genetic and evolutionary distance. Science 230: 1403–1406.

Charlesworth, B., M.T. Morgan, and D. Charlesworth. 1993. The effect of deleterious mutations on neutral molecular variation. Genetics 134: 1289–1303.

Cooke, H.J., J. Schmidtke, and J.R. Gosden. 1982. Characterisation of a human Y chromosome repeated sequence and related sequences in higher primates. Chromosoma 87: 491–502.

Coward, P., K. Nagai, D. Chen, H.D. Thomas, C.M. Nagamine, C.M., and Y.-F.C. Lau. 1994. Polymorphism of a CAG trinucleotide repeat within *Sry* correlates with B6.Y[Dom] sex reversal. Nature Genetics 6: 245–250.

Eicher, E.M., K.W. Hutchinson, S.J. Phillips, P.K. Tucker, and B.K. Lee. 1989. A repeated segment on the mouse Y chromosome is composed of retroviral-related, Y-enriched, and Y-specific sequences. Genetics 122: 181–192.

Erickson, R.P. 1987. Evolution of four human Y chromosomal unique sequences. J. Mol. Evol. 25: 300–307.

Ferris, S.D., R.D. Sage, C.-M. Huang, J.T. Nielsen, U. Ritte, and A. C. Wilson. 1983. Flow of mitochondrial DNA across a species boundary. Proc. Natl. Acad. Sci. 80: 2290–2294.

Foster, J.W., F.E. Brennan, G.K. Hampikian, P.N. Goodfellow, A.H. Sinclair, R. Lovell-Badge, L. Selwood, M.B. Renfree, D.W. Cooper, and J.A. Marshall Graves. 1992. Evolution of sex determination and the Y chromosome: *SRY*-related sequences in marsupials. Nature 359: 531–533.

Griffiths, R. and B. Tiwari. 1993. Primers for the differential amplification of the sex-determining region Y gene in a range of mammal species. Mol. Ecol. 2: 405–406.

Gubbay, J., J. Collignon, P. Koopman, B. Capel, A. Economou, A. Munsterberg, N. Vivian, P. Goodfellow, and R. Lovell-Badge. 1990. A gene mapping to the sex-determining region of the mouse Y chromosome is a member of a novel family of embryonically expressed genes. Nature 346: 245–250.

Guttenbach, M., U. Muller, and M. Schmid. 1992. A human moderately repeated Y-specific DNA sequence is evolutionarily conserved in the Y chromosome of the great apes. Genomics 13: 363–367.

Hacker, A., B. Capel, P. Goodfellow, and R. Lovell-Badge. 1995. Expression of *Sry*, the mouse sex determining gene. Development 121: 1603–1614.

Jager, R.J., M. Anvret, K. Hall, and G. Scherer. 1990. A human XY female with a frame shift mutation in the candidate testis-determining gene *SRY*. Nature 348: 452–454.

Jukes, T.H. and C.R. Cantor. 1969. Evolution of protein molecules. In: H.N. Monro, ed. Mammalian protein metabolism III, pp. 21–120. New York: Academic Press.

Kaplan, N.L., R.R. Hudson, and C.H. Langley. 1989. The "hitchhiking effect" revisited. Genetics 123: 887–899.

Koenig, M., J.P. Moisan, R. Heilig, and J.L. Mandel. 1985. Homologies between X and Y chromosomes detected by DNA probes: localisation and evolution. Nucleic Acids Res. 13: 5485–5501.

Koopman, P., J. Gubbay, N. Vivian, P. Goodfellow, and R. Lovell-Badge. 1991. Male development of chromosomally female mice transgenic for *Sry*. Nature 351: 117–121.

Kunkel, L.M. and K.D. Smith. 1982. Evolution of human Y-chromosome DNA. Chromosoma 86: 209–228.

Lucotte, G. and F. David. 1992. Y-chromosome-specific haplotypes of Jews detected by probes 49f and 49a. Hum. Biol. 64: 757–761.

Lucotte, G. and K.Y. Ngo. 1985. p49, a highly polymorphic probe, that detects TaqI RFLPs on the human Y chromosome. Nucleic Acids Res. 13: 8285.

Lucotte, G., P. Smets, and J. Ruffie. 1993. Y-chromosome-specific haplotype diversity in Ashkenazic and Sephardic jews. Hum. Biol. 65: 835–840.

Lucotte, G., P. Guerin, L. Halle, F. Loirat, and S. Hazout. 1989. Y chromosome DNA polymorphisms in two African populations. Am. J. Hum. Genet. 45: 16–20.

Lundrigan, B.L. and P.K. Tucker. 1994. Tracing paternal ancestry in mice, using the Y-linked sex-determining locus, *Sry*. Mol. Biol. Evol. 11: 483–492.

Ma, K., J.D. Inglis, A. Sharkey, W.A. Bickmore, R.E. Hill, E.J. Prosser, R.M. Speed, E.J. Thomson, M. Jobling, K. Taylor, J. Wolfe, H.J. Cooke, T.B. Hargreave, and A.C. Chandley. 1993. A Y chromosome gene family with RNA-binding protein homology: candidates for the azoospermia factor AZF controlling human spermatogenesis. Cell 75: 1287–1295.

Maynard Smith, J. and J. Haigh. 1974. The hitch-hiking effect of a favorable gene. Genet. Res. 23: 23–35.

Miller, K. E., B.L. Lundrigan, and P.K. Tucker. 1995. Length variation of CAG repeats in *Sry* in populations of *Mus domesticus*. Mammal. Genome 6: 206–208.

Mitchell, M.M., D.R. Woods, P.K. Tucker, J.S. Opp, and C.E. Bishop. 1991. A candidate spermatogenic gene from the mouse Y chromosome may be a ubiquitin activating enzyme. Nature 354: 483–486.

Moritz, C., T.E. Dowling, and W.M. Brown. 1987. Evolution of mitochondrial DNA: relevance for population biology and systematics. Ann. Rev. Ecol. Syst. 18: 269–292.

Nachman, M.W. and C.A. Aquadro. 1994. Polymorphism and divergence at the 5′ flanking region of the sex-determining locus, *Sry*, in mice. Mol. Biol. Evol. 11: 539–547.

Nagamine, C.M., Y. Nishioka, K. Moriwaki, P. Boursot, F. Bonhomme, and Y.-F. C. Lau. 1992. The musculus-type Y chromosome of the laboratory mouse is of Asian origin. Mammal. Genome 3: 84–91.

Nishioka, Y., and E. Lamothe. 1986. Isolation and characterization of a mouse Y chromosomal repetitive sequence. Genetics 113: 417–432.

Oakey, R. and C. Tyler-Smith. 1990. Y chromosome DNA haplotyping suggests that most European and Asian men are descended from one or two males. Genomics 7: 325–330.

Page, D.C., M.E. Harper, J. Love, and D. Botstein. 1984. Occurrence of a transposition from the X-chromosome long arm to the Y-chromosome short arm during human evolution. Nature 311: 119–123.

Page, D.C., R. Mosher, E.M. Simpson, E.M.C. Fisher, G. Mardon, J. Pollack, B. McGillivray, A. de la Chapelle, and L. G. Brown. 1987. The sex determining region of the human Y chromosome encodes a finger protein. Cell 51: 1091–1104.

Palsboll, P.J., A. Vader, I. Bakke, and M.R. El-Gewely. 1992. Determination of gender in cetaceans by the polymerase chain reaction. Can. J. Zool. 70: 2166–2170.

Payen, E.J. and C.Y. Cotinot. 1993. Comparative HMG-box sequences of the *SRY* gene between sheep, cattle, and goats. Nucleic Acids Res. 21: 2772.

Payen, E.J. and C.Y. Cotinot. 1994. Sequence evolution of SRY gene within Bovidae family. Mammal. Genome 5: 723–725.

Platt, T.H.K. and M.J. Dewey. 1987. Multiple forms of male-specific simple repetitive sequences in the genus *Mus*. J. Mol. Evol. 25: 201–206.

Powell, J.R. 1983. Interspecific cytoplasmic gene flow in the absence of nuclear gene flow: evidence from *Drosophila*. Proc. Natl. Acad. Sci. U.S.A. 80: 492–495.

Prager, E.M., R.D. Sage, U. Gyllensten, W.K. Thomas, R. Hubner, C.S. Jones, L. Noble, J.M. Searle, and A.C. Wilson. 1993. Mitochondrial DNA sequence diversity and the colonization of Scandinavia by house mice from East Holstein. Biol. J. Linn. Soc. 50: 85–122.

Richards, R.I. and G.R. Sutherland. 1992. Heritable unstable DNA sequences. Nature Genetics 1: 7–9.

Richards, R.I. and G.R. Sutherland. 1994. Simple repeat DNA is not replicated simply. Nature Genetics 6: 114–116.

Ritte, U., E. Neufeld, M. Broit, D. Shavit, and U. Motro. 1993. The differences among Jewish communities, maternal and paternal contributions. J. Mol. Evol. 37: 435–440.

Sinclair, A.H., P. Berta, M.S. Palmer, J.R. Hawkins, B.L. Griffiths, M.J. Smith, J.W. Foster, A.-M. Frischauf, R. Lovell-Badge, and P.N. Goodfellow. 1990. A gene from the human sex-determining region encodes a protein with homology to a conserved DNA-binding motif. Nature 346: 240–244.

Singleton, G.R. and D.A. Hay. 1983. The effect of social organization on reproductive success and gene flow in colonies of wild house mice, *Mus musculus*. Behav. Ecol. Sociobiol. 12: 49–56.

Spurdle, A.B., M.F. Hammer, and T. Jenkins. 1994. The Y Alu polymorphism in southern African populations and its relationship to other Y-specific polymorphisms. Am. J. Hum. Genet. 54: 319–330.

Stevanovic, M., R. Lovell-Badge, J. Collignon, and P. N. Goodfellow. 1993. *SOX3* is an X-linked gene related to *SRY*. Hum. Mol. Genet. 2: 2013–2018.

Tegelstrom, H. 1987. Transfer of mitochondrial DNA from the northern red-backed vole (*Clethrionomys rutilus*) to the bank vole (*C. glareolus*). J. Mol. Evol. 24: 218–227.

Tucker, P.K. and B.L. Lundrigan. 1993. Rapid evolution of the sex determining locus in Old World mice and rats. Nature 364: 715–717.

Tucker, P.K., B.K. Lee, B.L. Lundrigan, and E.M. Eicher. 1992. Geographic origin of the Y chromosomes in "old" inbred strains of mice. Mammal Genome 3: 254–261.

Whitfield, L.S., R. Lovell-Badge, and P.N. Goodfellow. 1993. Rapid sequence evolution of the mammalian sex-determining gene *SRY*. Nature 364: 713–715.

Wilson, A.C., R.L. Cann, S.M. Carr, M. George, U.B. Gyllensten, K.M. Helm-Bychowski, R.G. Higuchi, S.R. Palumbi, E.M. Prager, R.D. Sage, and M. Stoneking, 1985. Mitochondrial DNA and two perspectives on evolutionary genetics. Biol. J. Linn. Soc. 26: 375–400.

Wolfe, J., R.P. Erickson, P.W.J. Rigby, and P.N. Goodfellow. 1984a. Regional localization of 3 Y-derived sequences on the human X and Y chromosomes. Ann. Hum. Genet. 48: 253–259.

Wolfe, J., R.P. Erickson, P.W.J. Rigby, and P.N. Goodfellow. 1984b. Cosmid clones derived from both euchromatic and heterochromatic regions of the human Y chromosome. EMBO 3: 1997–2003.

Yonekawa, H., O. Gotoh, Y. Tagashira, Y. Matsushima, L.-I. Shi, W.S. Cho, N. Miyashita, and K. Moriwaki. 1986. A hybrid origin of Japanese mice "*Mus musculus molossinus.*" Curr, Top. Microbiol. Immunol. 127: 62–67.

Yonekawa, H., K. Moriwaki, O. Gotoh, N. Miyashita, Y. Matsushima, S. Liming, W.S. Cho, Z. Xiao-Lan, and Y. Tagashira. 1988. Hybrid origin of Japanese mice "*Mus musculus molossinus*": evidence from restriction analysis of mitochondrial DNA. Mol. Biol. Evol. 5: 63–78.

6

Applications of Allozyme Electrophoresis in Conservation Biology

PAUL L. LEBERG

The influence of allozyme electrophoresis on our understanding of the genetic structure of natural populations is difficult to overstate. Since it was first applied to population genetics in the mid 1960s, allozyme electrophoresis has been widely used in systematics, evolution, ecology, and conservation biology. A survey of recent papers indicates that allozyme electrophoresis is still the predominant tool used to address questions in conservation genetics (Figure 6-1). As other molecular approaches become more readily available, the relative importance of allozyme electrophoresis will decrease. However, because of its relatively low cost and ease of application, this technique will continue to be used in many studies.

Allozyme electrophoresis refers to the separation of allelic variants of an enzyme or other protein locus in an electrophoretic field. Allozymes represent different allelic variants of a single locus. All the various forms of an enzyme, including products of several loci, are referred to as isozymes. Although both isozyme and allozyme variation are of interest to evolutionary biologists, the allelic nature of the latter is usually the focus of conservation geneticists.

Differential migration of allozyme variants is usually due to differences in their electrical charges. The shape or size of a protein can also affect its migration through some electrophoretic media. Differences in structure or charge between presumptive allelic variants are often assumed to be due to their amino acid composition, thereby reflecting differences in the DNA sequences encoding the protein.

TECHNICAL DESCRIPTION

Procedure

Detailed descriptions of allozyme electrophoresis have been published in Harris and Hopkinson (1976), Richardson et al. (1986), and Murphy et al. (1990), so my

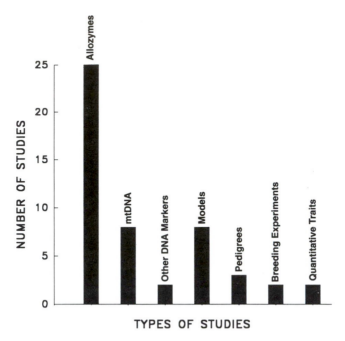

Figure 6-1   Survey of approaches used in 43 papers published in *Conservation Biology* during 1991–1994 that dealt with conservation genetics of natural or captive populations. Papers that combined several approaches were included in each of the appropriate categories.

discussion of methodological issues is brief. Proteins used in allozyme electrophoresis can by obtained from almost any tissue, such as muscle, kidney, or liver. In many cases, nondestructive sampling of body fluids such as blood is preferable to sacrificing the organism for internal tissues. Other tissues that can be collected from living organisms include fin clips and body mucus from fish (Robbins et al., 1987), and pulp of growing feathers from birds (Marsden and May 1984). Needle biopsies can be used to collect muscle tissue (Jordan et al., 1990). Unfortunately, isozyme expression is often tissue-specific; therefore, studies limited to nondestructive sampling can only assay a portion of the loci that can be examined when the complete organism is available. Samples should be stored at $-70°C$ if possible, although many laboratories use standard freezers with satisfactory results.

Samples are extracted from firm tissues by homogenization. Although the tissue homogenate can be used in electrophoresis, better resolution may be obtained by centrifuging the sample to remove any solids. The serum and hematocrit portions of blood are often analyzed separately after centrifugation.

Small portions of samples are placed in a gel for exposure to an electrophoretic gradient. Spaces between the samples allow for the differentiation of individual samples following completion of electrophoresis. Before and during electrophoresis, samples should be kept as cool as possible to avoid degradation.

Several gel media have been used in allozyme electrophoresis, the most common of which are hydrolyzed starch, cellulose acetate, agarose, and polyacrylamide (Harris and Hopkinson, 1976). Each medium has its advantages and disadvantages but the vast majority of allozyme studies on natural populations have used starch gels. Techniques associated with this medium have been recently reviewed by Murphy et al. (1990). Cellulose acetate gels are a viable alternative to starch, especially in cases where only small amounts of tissue are available. See Richardson et al. (1986) for procedures specific to cellulose acetate electrophoresis.

The electrophoretic field is typically established by placing the gel in contact with a pair of trays containing a buffer solution. Positive and negative wire leads connect the two trays to a power pack. The voltage applied to the system depends on the buffer solution. The choice of buffer solution may also affect the resolution of specific allozymes, so it is often necessary to evaluate several buffers to determine optimal electrophoretic conditions for each protein.

Following electrophoresis, gels are removed from contact with the power source. Prior to staining, starch gels are sliced into 5 to 7 slices that are approximately 1 mm thick. Most stains include a substrate specific to the allozyme of interest, a buffer solution, cofactors, and linking enzymes that produce a detectable reaction. Recipes for stains are provided in many sources, including Harris and Hopkinson (1976), Richardson et al. (1986), and Murphy et al. (1990). The stain solution is poured over the gel or gel slice as soon as possible after the electrophoresis has been completed. Sometimes agar is added to the stain solution to concentrate the chemicals over a portion of the gel. Depending on the specific stain recipe, the gel may be placed in an incubator to speed up the reaction of the enzyme with the substrate.

As the reaction products become visible on the gel, banding patterns are recorded. These banding patterns are often easily interpreted in terms of Mendelian genetics; however, in some cases the patterns can be very complex (Harris and Hopkinson, 1976; Richardson et al., 1986). Following the recording of banding patterns, gels may be photographed and stored for future reference.

Equipment and Expense

Major items of equipment needed for allozyme electrophoresis include power supplies, buffer trays, pipetters, a refrigerator-freezer, balance, incubator, and camera. Access to a small centrifuge and an ultralow freezer are beneficial but not absolutely necessary. With the exception of the ultralow freezer, most of this equipment can be purchased for a total investment of $4000–$6000. Because much of the equipment can be found in most laboratories, equipment costs are usually lower than this amount and are minor compared to the cost of supplies. A well-stocked laboratory will have a large number of different chemicals used in gel preparation and staining. Stocking such a laboratory from scratch would involve an initial investment of several thousand dollars. Costs per sample vary widely depending on the electrophoretic medium and the number of allozyme systems being assayed. One rough estimate of chemical costs for a typical study examining 50 loci using starch gel electrophoresis was approximately $10 per individual (P.J. Johns, personal communication). The expense of this technique will compare favorably with other molecular techniques when examination of a large number of loci and individuals is desired.

APPLICATIONS IN CONSERVATION GENETICS

Allozyme electrophoresis has been used to address many questions in genetics, evolution, and conservation biology (Richardson et al., 1986; Murphy et al., 1990). I will review those areas of investigation that are most pertinent to conservation biology and discuss them in light of recent papers in the journal *Conservation Biology* (Figure 6-2). Because this journal is one of the principal publications dealing with conservation of rare and endangered species, its content provides insight into the types of genetic studies most applicable to conservation biology.

Estimation of Genetic Variation

Allozyme electrophoresis is most often used in conservation biology to estimate genetic diversity in natural and captive populations (Figure 6-2). Common measurements of allozyme diversity include multiple locus heterozygosity, proportion of polymorphic loci, and average number of alleles per polymorphic locus. Low allozyme diversity is often thought to reflect the effects of past or current population bottlenecks on genetic diversity (Bonnell and Selander, 1974; Packer et al., 1991; Wayne et al., 1991; Hartl and Pucek, 1994). For example, low allozyme diversity in the cheetah (*Acinonyx jubatus*) was one piece of evidence O'Brien et al. (1983) used to conclude that the species had experienced a past constriction in population size. Allozyme diversity has also been used to assess the effects of reintroductions (Stuwe and Scribner 1989) and captive breeding programs (Templeton et al., 1987; Borlase et al., 1993, Leary et al., 1993) on genetic variation.

Associations of Genetic Diversity with Fitness and Population Viability

One motivation for estimating genetic diversity within populations is the belief that genetic diversity is positively associated with the viability of the population. Much of the evidence used to support this hypothesis is based on studies using allozyme markers. Many studies have explored correlations between single- or multiple-locus allozyme heterozygosity and individual fitness (Allendorf and Leary, 1986). Heterozygosity–fitness correlations may be due to the loci examined, closely linked loci, or inbreeding depression (Leary et al., 1987; Leberg et al., 1990). Regardless of the mechanism responsible for heterozygosity–fitness correlations, these relationships are often cited in discussions of the effects of genetic diversity on population viability. For example, Gilpin and Soulé (1986) use the existence of these correlations to suggest that populations that have lost genetic diversity will be more prone to extinction because of reduced population growth rates.

Failure to detect allozyme variation in a population has led some authors to reject a population as a source of individuals for reintroduction programs (Vrijenhoek et al., 1985), because they assumed that genetically depauperate populations would have reduced viability. Captive populations may be managed to maintain high levels of allozyme variation in captive populations through selective breeding (Wayne et al., 1986); however, this approach has been criticized because it can increase the loss of overall genetic diversity (Hedrick et al., 1986;

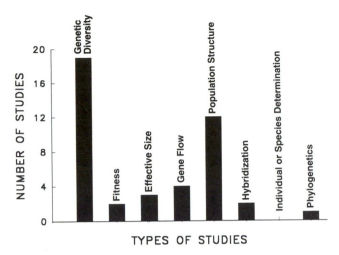

Figure 6-2 Survey of applications of allozyme electrophoresis used in 25 papers published in *Conservation Biology* during 1991–1994. Papers that used several approaches were included in each of the appropriate categories.

Haig et al., 1990). The existence of heterozygosity–fitness correlations has also been used to justify preservation of small populations of plants because selection favoring heterozygotes may have aided in the retention of genetic variation (Lesica and Allendorf, 1992).

Although genetic variation appears to affect many different population processes (Ayala, 1968; Leberg, 1991a), there have been few studies examining the effects of allozyme diversity on the viability of populations (Quattro and Vrijenhoek, 1989; Leberg, 1993). Compared to estimates of genetic diversity, assessments of relationships between allozyme variation and the fitness of individuals or the viability of populations are rare in the conservation literature (Figure 6-2). Although the assumption that genetic diversity is important for a population's survival appears to be widely accepted in conservation biology, there is little experimental support for this hypothesis.

Effective Population Size

Allele and genotype frequencies are often used by conservation geneticists to estimate the effective size of a population (Figure 6-2). Estimates of effective size are important because census population size can be a poor predictor of loss of allozyme diversity (Briscoe et al., 1992). Estimates of effective size can be made from changes in allozyme heterozygosity (Briscoe et al., 1992) and allele frequencies (Waples, 1989) over time. Using the latter approach, Waples and Teal (1990) estimated that the effective population sizes for hatchery populations of Pacific salmon (*Oncorhynchus* spp.) were often much smaller than census population sizes. Linkage disequilibrium at allozyme loci, generated by stochastic processes, can also be used to estimate effective population size (Hill, 1981). By measuring linkage disequilibrium, Bartley et al. (1992) were able to estimate the effective population sizes of several fishes.

Gene Flow

Gene flow between populations of an endangered species is often of interest because it reflects the degree to which the populations are isolated. If gene flow is rare or absent, it may be advisable to manage populations as distinct units. Gene flow among populations has been estimated using several indirect approaches based on differences in allele frequencies among populations (Slatkin, 1985a,b). These indirect estimators of gene flow are often applied to allozyme data when addressing conservation issues (Figure 6-2; Dole and Sun, 1992; Stangel et al., 1992b; Hickey et al., 1991).

Breeding Systems

Comparing genotype distributions with Hardy–Weinberg expectations allows the estimation of the degree of nonrandom mating within a population. A population's history of inbreeding will affect its response to additional inbreeding (Templeton and Read, 1984). Although long a focus of evolutionary biology, estimation of breeding structure is not a part of most investigations by conservation geneticists examining animals; however, it is often invoked as one reason why some plant populations have low genetic diversity (Dole and Sun, 1992; Soltis et al., 1992; Kress et al., 1994).

Population Structure

Next to the estimation of genetic diversity within populations, allozyme electrophoresis is most often used to characterize the genetic relationships among populations (Figure 6-2). Several authors have suggested that genetically similar populations be treated as the same population for management purposes (Smith et al., 1976; Vrijenhoek et al., 1985). Genetically differentiated populations are often suggested as candidates for special management consideration to prevent the loss of unique genetic variants (Dole and Sun, 1992; Leary et al., 1993). Estimates of relationships among populations based on allozymes often provide insight into the effects of past bottlenecks and range fragmentation on the genetic structure of a species (Leberg, 1991b, Stangel et al., 1992a; Ellsworth et al., 1994). Fishery biologists often use allozymes to evaluate stock structure and the effects of hatchery releases on population composition (Garcia de Leaniz et al., 1989; Hauser et al., 1991).

Hybridization

Because hybridization is a major threat to genetic stocks in some groups, such as freshwater fish (Deacon et al., 1979), it has received some attention from conservation geneticists (Figure 6-2). If there are a number of unique allozyme markers for each species associated with a hybrid complex, it is possible to assess the degree of hybridization. By using allozymes in conjunction with a maternally inherited marker such as mitochodrial DNA (mtDNA), it is often possible to determine the directionality of the hybrid matings (Leary et al., 1993). For example, some populations of native trout (*Salmo* spp.) exhibit species-specific mtDNA haplotypes while containing some allozyme alleles diagnostic of introduced species (Dowling and Childs, 1992). This pattern may indicate that introgression occurs through matings of males of introduced species with females of the native stock.

Paternity and Species Determination

Sometimes the ability to identify an individual or species with molecular techniques has applications in resource management, although this is not a major focus of conservation genetics research (Figure 6-2). If there are several polymorphic loci in a population, it is sometimes possible to determine the probability that a specific male is the father of an individual. Because allozymes are not as polymorphic as other molecules, their usefulness in paternity studies is often limited to the exclusion of individuals as the father, as well as to the identification of broods sired by multiple males.

Forensics can be of critical importance in stemming the illegal trade in wildlife. In investigations of illegal harvest and sale of wildlife species, allozyme electrophoresis can often be used to identify the species of an individual on the basis of a small sample of flesh (Dilworth and McKenzie, 1970). Allozymes have also been used to identify unique genetic forms such as unisexual vertebrates (Vrijenhoek et al., 1978). Because of the unique hemiclonal nature of these unisexual biotypes, they may be very susceptible to new environmental challenges such as exotic species (Leberg and Vrijenhoek, 1994).

Phylogenetic Systematics

The systematics of a group of species may also be of interest to conservation biologists. There have been several suggestions that cladistic analysis could be used to determine what taxa (Vane-Wright et al., 1991; Vogler and Desalle, 1994) or areas should be protected (Erwin 1991). Although allozymes have been used as characters in phylogenetic analyses (Murphy et al., 1990), there are few instances where systematics, above the level of intraspecific comparisons, has been applied to management issues (Figure 6-2).

CASE STUDIES

Endangered Topminnows

Allozyme electrophoresis has been used to assess genetic diversity and population structure in the endangered topminnow, *Poeciliopsis occidentalis* (Vrijenhoek et al., 1985). Once one of the most common fishes in southern Arizona, it has been reduced to about a dozen isolated populations because of habitat loss and introduction of exotic predators (Meffe et al., 1983). It is more widespread in Mexico, but the evolutionary relationships of the Mexican populations with those in the United States were not well understood. Characterizing the genetic relationships between populations and river drainages was critical to establishing a sound management program.

Assessing variation at 25 allozyme loci for 827 individuals for 21 populations, Vrijenhoek et al. (1985) found that the genetic diversity within the Arizona populations was very low compared to most topminnow populations in Mexico. Several of the populations in the Gila River of Arizona were fixed for alternate alleles that were polymorphic in populations from Mexico. The low genetic diversity and fixed allelic differences among populations were interpreted as evidence of bottlenecks and reduced gene flow in the Arizona populations (Vrijenhoek et al., 1985; Meffe and Vrijenhoek, 1988). Subsequent laboratory experiments found that only the Gila River population exhibiting allozyme polymorphism had higher brood sizes, survival, growth rates, and developmental stability than a population that was monomorphic for all loci (Quattro and Vrijenhoek et al., 1989). Following these studies, conservation biologists have begun to use the polymorphic population of topminnows in reintroduction programs; past reintroductions were made with fish from sources with no allozyme diversity (Simons et al., 1989).

Allozyme variation also revealed the occurrence of three distinctive forms of topminnows in the United States and Mexico (Vrijenhoek et al., 1985). Distributions of these forms correspond roughly to river drainages in southern Arizona and Mexico, which led Meffe and Vrijenhoek (1988) to argue for management of stream fishes as a hierarchical system of demes nested within connected drainages. A recent assessment of mtDNA variation in these same populations (Quattro et al., 1996) supported the existence of these genetically distinct units. The assessment of mtDNA restriction site variation, however, did not detect any genetic differences among the Gila River populations with fixed allozyme differences. The mtDNA survey also identified large genetic differences between other populations with similar allozyme frequencies. Disconcordance between allozyme and mtDNA data emphasizes the importance of using combinations of different molecular tools to characterize genetic variation in a species. Conservation geneticists should use caution when making management recommendations based on techniques that can only examine limited portions of the genome.

Wildlife Reintroductions

Although wildlife reintroduction programs are important conservation tools, little is known about their effects on genetic diversity (Leberg, 1990). Therefore, with a number of collaborators, I have examined allozyme variation in wild turkeys (*Meleagris gallopavo*) and white-tailed deer (*Odocoileus virginianus*), two species that were reintroduced in many parts of their original range. Our objectives were to determine whether reintroduced populations had lost genetic diversity during their establishment and whether the gene flow caused by wildlife translocations affected the genetic structure of the species (Leberg, 1991b; Stangel et al., 1992a; Leberg et al., 1994).

Over the short term, reductions in population size due to restoration programs and other management practices are more likely to result in changes in allele frequencies than in the fixation of many loci. Thus, evaluation of the effects of human actions on genetic diversity requires the examination of large numbers of individuals and loci. We initially screened turkey and deer populations for variation at 28 and 18 loci, respectively. We assessed allozyme variation for 713 turkeys from 22 populations and for 1476 deer from 21 populations. Without a technique as inexpensive and easy to apply as allozyme electrophoresis, it is unlikely that a genetic survey of this many individuals, loci, and populations would be possible.

As in many allozyme studies, most populations had only a few polymorphic loci (1–5 in turkeys, 4–9 in deer). Given the sampling error associated with these small numbers of loci, it would be hard to argue that a population had low allozyme diversity, relative to another population, because it had experienced a

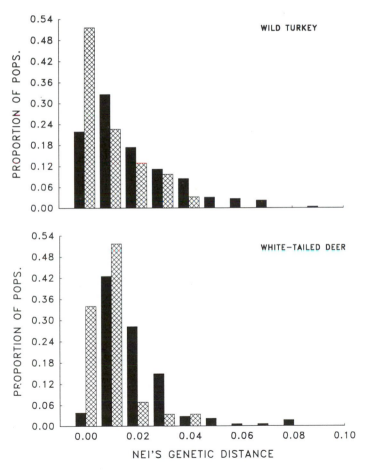

Figure 6-3 Distributions of genetic distances between pairs of 22 wild turkey and 21 white-tailed deer populations from the eastern United States. Populations were either associated (cross-hatched bars) or not associated by translocations (solid bars). A pair of populations was considered to be associated by a reintroduction if individuals were moved between them or if both populations were stocked with individuals from the same source population (redrawn from Leberg et al., 1994).

founder event (Leberg, 1992). On the other hand, by examining large numbers of populations and individuals it was possible to ask whether average heterozygosity was lower in reintroduced populations than in native populations. For both species, heterozygosity was significantly lower in the reintroduced populations (Leberg, et al., in preperation). The effect of this loss of genetic variation on population viability is unknown; however, associations between heterozygosity and traits related to fitness have been documented in both species (Chesser et al., 1982; Leberg 1994).

There is significant spatial subdivision in both the turkey and deer populations (Leberg et al., 1994); however, the relative influence of range fragmentation, founder events, population size reduction, and biogeography on the genetics of these populations is unclear (Leberg, 1991b; Ellsworth et al., 1994; Leberg et al., 1994). Reintroductions appear to have altered the genetic relationships among populations (Leberg et al., 1994). Populations that have had individuals translocated between them are more genetically similar than expected by chance alone (Figure 6-3). Genetic effects of translocations persisted many generations after populations were reestablished, suggesting that gene flow due to natural dispersal may be low in these species.

LIMITATIONS AND ADVANTAGES

Limitations

Many of the problems associated with the use of allozyme electrophoresis have been reviewed elsewhere (Harris and Hopkinson, 1976; Richardson et al., 1986; Murphy et al., 1990). These include the presence of null alleles, the inability to detect much of the variation occurring in the DNA sequence, and posttranslational modification of isozymes. Although these concerns are important, I will focus on problems associated with the application of allozymes to such conservation issues as estimation of genetic diversity, effective population size, gene flow, and population structure.

Selection. Many population genetic models assume that allele or genotype frequencies are not being strongly influenced by natural selection. For example, selection can bias estimates of effective population size based on heterozygosity (Mueller et al., 1985). Although some allozymes appear to be under selection (Watt 1977; Koehn et al., 1988), the degree to which the assumption of neutrality holds for allozyme (or other molecular) markers is generally unknown.

Sampling Error. Some of the principal problems related to estimates of genetic diversity based on allozyme electrophoresis involve error resulting from sampling insufficient numbers of loci and individuals. At least 40 loci should be sampled for one to have much confidence in heterozygosity estimates (Archie, 1985). Although this number is within the reach of most laboratories, many allozyme surveys do not meet this criterion (Table 6-1). Even if >40 loci are assayed, individual loci should not be considered monomorphic until a large sample of

Table 6-1   Number of Populations (N) and Mean Number of Individuals (n), Loci (L), and Polymorphic Loci (P) examined in 16 studies that used allozyme electrophoresis to estimate genetic diversity. Range of number of polymorphic loci detected in the populations surveyed are reported in parenthesis. All studies were published in *Conservation Biology* during 1991–1994.

Species	N	n	L	P	Study
Bensoniella oregona	6	23	22.5	0	Soltis et al. (1992)
Bull trout, *Salvelinus confluetus*	21	17	45	1.3(0–4)	Leary et al. (1993)
Butte County meadowforam, *Limanthes floccosa*	9	68	28	0.3(0–1)	Dole and Sun (1992)
Cotton-top tamarin, *Saquinus oedipus*	1	>75	41	2	Cheverud et al. (1994)
European bison, *Bison bonasus*	1	35	69	4	Hartl and Pucek (1994)
Harperella, *Ptilimnium nodosum*	14	13	13	1.7(0–4)	Kress et al. (1994)
Lactoris fernandeziana	12	7	22	0	Crawford et al. (1994)
Red-cockaded woodpecker, *Picoides borealis*	26	17	16	3(1–5)	Stangel et al. (1992b)
Gray wolf, *Canis lupus*	14	10	25	4.3(2–5)	Wayne et al. (1991)
Eastern barred bandicoot, *Perameles qunnii*	2	21	27	0	Sherwin et al. (1991)
Halocarpus bidwillii	16	35	20	3(0–5)	Billington (1991)
Plains bison, *Bison bison*	2	50	24	1(1)	McClenaghan et al. (1990)
Guam rail, *Rallus owstoni*	1	112	23	4	Haig et al. (1990)
Trifloium spp.	10	21	16	3.6(0–7)	Hickey et al. (1991)
Wiregrass, *Aristida striata*	15	11	25	17.5(15–20)	Walters et al. (1994)
Pentaclethra macroloba	12	78	14	3.8(2–5)	Hall et al. (1994)

individuals is examined (Sjorgren and Wyoni, 1994). Appropriate sample sizes for such analyses are likely to involve hundreds of individuals from a large population (Sjorgren and Wyoni, 1994) rather than the 10–40 examined in most studies (Table 6-1).

Attempts to estimate the effects of bottlenecks on genetic diversity also require a large number of polymorphic loci. Unfortunately, even when a large number of individuals and allozyme loci can be assayed, polymorphisms are often uncommon (Table 6-1). The majority of the loci sampled within most populations are monomorphic and provide no information on genetic relationships of populations or on temporal changes in genetic diversity within a population.

When comparing large numbers of populations or species, it is not unusual to observe decreases in multiple locus heterozygosity in populations experiencing bottlenecks (Leberg, 1992; Stangel et al., 1992a; Hartl and Pucek, 1994). The problem is that most managers want to know whether a specific population or species has lost genetic variability. In experiments with mosquitofish, it was impossible to use multiple-locus heterozygosity to accurately assess changes in genetic diversity due to severe bottlenecks. Because only 6–7 of the loci sampled were polymorphic, severe bottlenecks actually increased average heterozygosity in some cases (Leberg, 1992).

On the basis of interlocus variation in response to bottlenecks, I have estimated that 16–25 polymorphic loci would be needed to have an 80% chance of detecting the loss of heterozygosity from a bottleneck of six individuals, one generation in duration (Leberg, 1992). Many investigations are interested in detecting less severe bottlenecks, which would require examining even more polymorphic loci. Few studies in conservation genetics detect sufficient numbers of polymorphic loci to evaluate the effects of population size on heterozygosity (Table 6-1). This set of studies may be somewhat biased, as conservation biologists tend to study organisms with small population sizes. However, examination of a more comprehensive review of the literature (Nevo et al., 1984) indicates that less than 10% of all allozyme studies examine more than 10 polymorphic loci per population. Other measures of diversity, such as numbers of alleles per polymorphic locus and heterozygosity per polymorphic locus (Hartl and Pucek, 1994), are more sensitive to population reductions. Their use in the detection of past bottlenecks, however, is still limited by small numbers of polymorphic loci (Leberg, 1992).

Approaches for estimating population structure, effective population size, and gene flow are also affected by the sampling error due to small numbers of polymorphic loci. Although Slatkin (1985a) recommended that > 20 private alleles be used in the estimation of gene flow, many studies applying his method do not approach this number. Likewise, estimates of effective population size obtained with the temporal method (Waples, 1989) were very imprecise when based on seven polymorphic loci (Richards and Leberg, in press). Because most estimates of gene flow, effective population size, and genetic diversity are subject to potentially high and often unknown sampling error, conservation geneticists should be careful in not overstating their value as management tools.

Problems with sampling error are unlikely to be resolved by application of DNA markers to estimates of gene flow, effective population size, and genetic diversity. Although some of these molecules are more polymorphic than allozymes (Hughes and Queller, 1993), expense and labor requirements will often prohibit the examination of variation of large numbers of both individuals and loci. Until the appropriate controlled experiments and simulations have been conducted using both allozyme and DNA markers to determine their sensitivity to bottlenecks and other population processes, we can have only limited confidence in the conclusions of many studies. Geneticists should carefully point out potential limitations of molecular techniques to conservation biologists who will use their results to formulate management plans for endangered taxa.

Advantages

The existence of allozyme data for many taxa provides an opportunity for comparative studies that does not exist for other molecular markers. For example, Hartl and Pucek (1994) used the comparative approach to demonstrate that species that had experienced past reductions in population size have reduced levels of allozyme diversity. Surveying the existing literature also makes it possible to reduce costs of a new study by focusing on electrophoretic conditions used in other investigations of closely related species. Usually little modification of electrophoretic conditions is necessary for study of allozyme variation in taxonomically disparate organisms. Compared to other molecular techniques, costs and labor requirements associated with allozyme electrophoresis are low.

Many questions in conservation genetics are best addressed by the survey of genotypes at large numbers of nuclear loci that segregate independently. No molecular approach is as well suited for this task as allozyme electrophoresis. A typical survey of a natural population may characterize genetic variation at 15–50 allozyme loci (Table 6-1); a few studies have examined as many as 100 loci. Unlike some molecular markers, it is possible to identify genotypes at specific loci, simplifying the application of many existing models from population genetics to the analysis of allozyme data.

The low cost and ease of application of allozyme electrophoresis allow rapid examination of genotypes for hundreds or thousands of individuals. For example, I determined genotypes at seven polymorphic loci for >3000 mosquitofish (*Gambusia holbrooki*) in a 3-month period. Gauldie and Smith (1978) screened 2700 fish for one enzyme, and five species for 22 different enzymes in 7 and 10 days, respectively.

In conclusion, I expect that other techniques will soon replace allozyme electrophoresis as the approach of choice for some applications in conservation genetics. For example, DNA sequence analysis is rapidly becoming the most common method for estimating systematic relationships. The use of allozymes to assess genetic differences among individuals within populations will also decrease as molecular biologists continue to develop efficient methods of examining variation at hypervariable loci. The widespread use of allozyme electrophoresis in conservation biology will continue, however, because it provides the most inexpensive way to rapidly assess genetic diversity within and between populations using markers with straightforward inheritance patterns. In future studies, allozymes will be most effectively used as one of several molecular markers, each assessing variation in different portions of the genome. Such an approach provides our best hope of understanding the genetic structure of endangered taxa.

ACKNOWLEDGMENTS

I thank M.C. Hager, W. Jordan, R. Lance, and M. Mulvey for reading early drafts of this manuscript. Preparation of this manuscript was supported by National Science Foundation Grant DEB-AC09-76SR00-819 and the Louisiana Education Quality Support Fund.

REFERENCES

Allendorf, F.W. and R.F. Leary. 1986. Heterozygosity and fitness in natural populations of animals. In: M.E. Soulé, ed. Conservation biology: The science of scarcity and diversity, pp. 57–76. Sunderland, Mass.: Sinauer Associates.

Archie, J.W. 1985. Statistical analysis of heterozygosity data: Independent sample comparisons. Evolution 39: 623–627.

Ayala, F.J. 1968. Genotype, environment, and population numbers. Science 162: 1453–1459.

Bartley, D., M. Bagley, G. Gall, and B. Bentley. 1992. Use of linkage disequilibrium data to estimate effective size of hatchery and natural fish populations. Conserv. Biol. 6: 365–375.

Billington, H.L. 1991. Effect of population size on genetic variation in a dioecious conifer. Conserv. Biol. 5: 115–119.

Bonnell, M.L. and R.K. Selander. 1974. Elephant seals: Genetic variation and near extinction. Science 184: 908–909.

Borlase, S.C., D.A. Loebel, R. Frankham, R.K. Nurthen, D.A. Briscoe, G.E. Daggard. 1993. Modeling problems in conservation genetics using captive *Drosophila* populations: Consequences of equalization of family sizes. Conserv. Biol. 7: 122–131.

Briscoe, D.A., J.M. Malpica, A. Robertson, G. J. Smith, R. Frankham, R.G. Banks, and J.S.F. Barker. 1992. Rapid Loss of genetic variation in large captive populations of *Drosophila* flies: Implications for the genetic management of captive populations. Conserv. Biol. 6: 416–425.

Chesser, R.K., Smith, M.H., Johns, P.E., Manlove, M.N., Straney, D.O., and Baccus, R. 1982. Spatial, temporal, and age-dependent heterozygosity of beta-hemoglobin in white-tailed deer. J. Wildlife Management 46: 983–990.

Cheverud, J., E. Routman, C. Jaquish, S. Tardif, G. Peterson, N. Belfiore, and L. Forman. 1994. Quantitative and molecular genetic variation in captive Cotton-top Tamarins (*Saguinus oedipus*). Conserv. Biol. 8: 95–105.

Crawford, D.J., T.F. Stuessy, M.B. Cosner, D.W. Haines, D. Wiens, and P. Penaillo. 1994. *Lactoris fernandeziana* (Lactoridaceae) on the Juan Fernandez Islands: Allozyme uniformity and field observations. Conserv. Biol. 8: 277–280.

Deacon, J.E., G. Kobetich, J. Williams, and S. Conteras. 1979. Fishes of North America endangered, threatened, or of special concern: 1979. Fisheries 4: 29–44.

Dilworth, T.G. and J.A. McKenzie. 1970. Attempts to identify meat of game animals by starch gel electrophoresis. J. Wildlife Management 34: 917–921.

Dole, J.A. and M. Sun. 1992. Field and genetic survey of the endangered Butte County Meadowfoam—*Limnanthes floccosa* subsp. *californica* (Limnanthaceae). Conserv. Biol. 6: 549–558.

Dowling, T.E. and M.R. Childs. 1992. Impact of hybridization on a threatened trout of the southwestern United States. Conserv. Biol. 6: 355–364.

Ellsworth, D.L., R.L. Honeycutt, N.J. Silvy, J.W. Bickham, and W.D. Klimstra. 1994. White-tailed deer restoration to the southeastern United States: evaluation of mitochondrial DNA and allozyme variation. J. Wildlife Management 58: 686–697.

Erwin, T.L. 1991. An evolutionary basis for conservation strategies. Science 253: 750–752.

Garcia de Leaniz, E. Versforo, and A.D. Hawkins. 1989. Genetic determination of the contribution of stocked and wild Altantic salmon, *Salmo salar* L., to the angling fisheries in two Spanish rivers. J. Fish Biol. 35(A): 261–270.

Gauldie, R.W. and P.J. Smith. 1978. The adaptation of cellulose acetate electrophoresis to fish enzymes. Comp. Biochem. Physiol. 61B: 421–425.

Gilpin, M.E. and M.E. Soulé. 1986. Minimum viable populations: Processes of species extinctions. In: M.E. Soulé, ed. Conservation biology: The science of scarcity and diversity, pp. 19–34. Sunderland, Mass.: Sinauer Associates.

Haig, S.M., J.D. Ballou, and S.R. Derrickson. 1990. Management options for preserving genetic diversity: Reintroduction of Guam Rails to the wild. Conserv. Biol. 4: 290–300.

Hall, P., M.R. Chase, and K.S. Bawa. 1994. Low genetic variation but high population differentiation in a common tropical forest tree species. Conserv. Biol. 8: 471–482.

Harris, H. and D.A. Hopkinson. 1976. Handbook of enzyme electrophoresis in human genetics. Oxford, U.K.: North Holland.

Hartl, G.B. and Z. Pucek. 1994. Genetic depletion in the European Bison (*Bison bonasus*) and the significance of electrophoretic heterozygosity for conservation. Conserv. Biol. 8: 167–174.

Hauser, L., A.R. Beaumont, G.T.H. Marshall, and R.J. Wyatt. 1991. Effects of sea trout stocking on the population genetics of landlocked brown trout, *Salmo trutta* L., in the Conwy River system, North Wales, U.K. J. Fish Biol. 39(A): 109–116.

Hedrick, P.W., P.F. Brussard, F.W. Allendorf., J.A. Beardmore, and S. Orzack. 1986. Protein variation, fitness, and captive propagation. Zoo Biol. 5: 91–99.

Hickey, R.J., M.A. Vincent, and S.I. Guttman. 1991. Genetic variation in Running Buffalo Clover (*Trifolium stoloniferum* Fabaceae). Conserv. Biol. 5: 309–316.

Hill, W.G. 1981. Estimation of effective population size from data on linkage disequilibrium. Genet. Res. Cambridge 38: 209–216.

Hughes, C.R. and D.C. Queller. 1993. Detection of highly polymorphic microsatellite loci in a species with little allozyme polymorphism. Mol. Ecol. 2: 131–137.

Jordan, W.C., A.F. Youngson, and J.H. Webb. 1990. Genetic variation at the Malic Enzyme-2 locus and age at maturity in sea-run Altantic salmon (*Salmo salar*). Can. J. Aquat. Sci. 47: 1672–1677.

Koehn, W., J. Diehl, and M. Scott. 1988. The differential contribution of individual enzymes of glycolysis and protein catabolism to the relationship between heterozygosity and the growth rate of the coot clam, *Mulinia lateralis*. Genetics 118: 121–130.

Kress, W.J., G.D. Maddox, and C.S. Roesel. 1994. Genetic variation and protection priorities in *Ptilimnium nodosum* (Apiaceae), an endangered plant in the eastern United States. Conserv. Biol. 8: 271–276.

Leary, R.F., F.W. Allendorf, and K.L. Knudsen. 1987. Differences in inbreeding coefficients do not explain the association between heterozygosity at allozyme loci and developmental stability in rainbow trout. Evolution 41: 1413–1415.

Leary, R.F., F.W. Allendorf, S.H. Forbes. 1993. Conservation genetics of bull trout in the Columbia and Klamath River drainages. Conserv. Biol. 7: 856–865.

Leberg, P.L. 1990. Genetic considerations in the design of introduction programs. Trans. N. Am. Wildlife and Natural Resources Conf. 55: 609–619.

Leberg, P.L. 1991a. Effects of genetic variation on the growth of fish populations: Conservation implications. J. Fish. Biol. 37: (A) 193–195.

Leberg, P.L. 1991b. Effects of bottlenecks on the genetic divergence in populations of the wild turkey. Conserv. Biol. 5: 522–530.

Leberg, P.L. 1992. Effects of population bottlenecks on genetic diversity as measured by allozyme electrophoresis. Evolution 46: 477–494.

Leberg, P.L. 1993. Strategies for population reintroduction: Effects of genetic variability on population growth and size. Conserv. Biol. 7: 194–199.

Leberg, P.L. 1994. Genetic diversity, morphology, and demography of the wild turkey. Proc. Int. Union of Game Biology 21: 126–131.

Leberg, P.L. and R.C. Vrijenhoek. 1994. Variation among desert topminnows and their susceptibility to attack by exotic parasites. Conserv. Biol. 8: 419–424.

Leberg, P.L., M.H. Smith, and O.E. Rhodes. 1990. The association between heterozygosity and growth of deer fetuses is not explained by effects of the loci examined. Evolution 44: 254–259.

Leberg, P.L., P.E. Stangel, H.O. Hillestad, L.R. Marchinton, and M.H. Smith. 1994. Genetic structure of reintroduced wild turkey and white-tailed deer populations. J. Wildlife Management 58: 698–711.

Lesica, P. and F.W. Allendorf. 1992. Are small populations of plants worth preserving? Conserv. Biol. 6: 135–139.

Marsden, J.E., and B. May. 1984. Feather pulp: a nondestructive sampling technique for electrophoretic studies of birds. Auk 101: 173–175.

McClenaghan, L.R., J. Berger, and H.D. Truesdale. 1990. Founding lineages and genic variability in Plains Bison (*Bison bison*) from Badlands National Park, South Dakota. Conserv. Biol. 4: 285–289.

Meffe, G.K. and R.C. Vrijenhoek. 1988. Conservation genetics in the managment of desert fishes. Conserv. Biol. 2: 157–169.

Meffe, G.K., D.A. Hendrickson, W.L. Minckley, and J.N. Rinne. 1983. Factors resulting in decline of the endangered Sonoran topminnow *Poeciliopsis occidentalis* (Atheriniformes: Poeciliidae) in the United States. Biol. Conserv. 25: 135–159.

Mueller, L.D., B.A. Wilcox, P.E. Ehrlich, D.G. Heckel, and D.D. Murphy. 1985. A direct assessment of the role of genetic drift in determining allele frequency variation in populations of *Euphydryas editha*. Genetics 110: 495–511.

Murphy, R.W., J.W. Sites, Jr. D.G. Buth, and C.H. Haufler. 1990. Proteins I: Isozyme electrophoresis. In: D.M. Hillis and C. Moritz, eds. Molecular systematics, pp. 45–126. Sunderland, Mass.: Sinauer Associates.

Nevo, E., A. Beiles, and R. Ben-Shlomo. 1984. The evolutionary significance of genetic diversity: ecological, demographic and life history correlates. Lecture Notes in Biomathematics 53: 12–213.

O'Brien, S.J., D.E. Wildt, D. Goldman, C.R. Merril, and M. Bush. 1983. The cheetah is depauperate in genetic variation. Science 221: 459–461.

Packer, C., A.E. Pusey, H. Rowley, D.A. Gilbert, J. Martenson, and S.J. O'Brien. 1991. Case study of a population bottleneck: Lions of the Ngorongoro Crater. Conserv. Biol. 5: 219–230.

Quattro, J.M. and R.C. Vrijenhoek. 1989. Fitness differences in remnant populations of the endangered Sonoran topminnow. Science. 245: 976–978.

Quattro, J.M., P.L. Leberg, M.E. Douglas, and R.C. Vrijenhoek. 1996. Molecular evidence for a unique evolutionary lineage of endangered sonoran desert fish (genus *Poeciliopsis*). Conserv. Biol. 10: 128–135.

Richards, C. and P.L. Leberg. Evaluation of the temporal method for estimating population size from changes in allele frequencies. Conserv. Biol. [In press].

Richardson, B.J., P.R. Baverstock, and M. Adams. 1986. Allozyme electrophoresis: A handbook for animal systematics and population genetics studies. Sydney: Academic Press.

Robbins, L.W., D.K. Toliver, and M.H. Smith. 1987. Nondestructive methods for obtaining genotypic data from fish. Conserv. Biol. 3: 88–91.

Sherwin, W.B., N.D. Murray, J.A.M. Graves, and P.R. Brown. 1991. Measurement of genetic variation in endangered populations: Bandicoots (Marsupialia: Peramelidae) as an example. Conserv. Biol. 5: 103–108.

Simons, L.H., D.A. Hendrickson, and D. Papoulias. 1989. Recovery of the Gila topminnow: A success story? Conserv. Biol. 3: 11–15.

Sjogren, P. and P. Wyoni. 1994. Conservation genetics and detection of rare alleles in finite populations. Conserv. Biol. 8: 267–270.

Slatkin, M. 1985a. Rare alleles as indicators of gene flow. Evolution 39: 53–65.

Slatkin, M. 1985b. Gene flow in natural populations. Annu. Rev. Ecol. Syst. 16: 53–65.

Smith, M.H., H.O. Hillestad, M.N. Manlove, and R. Larry Marchinton. 1976. Use of population genetics data for the management of fish and wildlife populations. Trans. N. Am. Wildlife and Natural Resources Conf. 41: 119–132.

Soltis, P.S., D.E. Soltis, T.L. Tucker, and F.A. Lang. 1992. Allozyme variability is absent in the narrow endemic *Bensoniella oregona* (Saxifragaceae). Conserv. Biol. 6: 131–134.

Stangel, P.W., P.L. Leberg, and J.I. Smith. 1992a. Systematics and population genetics. In: J.G. Dickson, ed. The wild turkey: ecology and management. Harrisburg, Pa.: Stackpole Books.

Stangel, P.W., M.R. Lennartz, and M.H. Smith. 1992b. Genetic variation and population structure of red-cockaded woodpeckers. Conserv. Biol. 6: 283–292.

Stuwe, M. and K.T. Scribner. 1989. Low genetic variability in reintroduced Alpine ibex populations. J. Mammal. 70: 370–373.

Templeton, A.R. and B. Read. 1984. Factors eliminating inbreeding depression in a captive herd of Speke's gazelle. Zoo Biol. 3: 177–200.

Templeton, A.R., S.M. Davis, and B. Read. 1987. Genetic variability in a captive herd of Speke's Gazelle (*Gazella speki*). Zoo Biol. 6: 305–313.

Vane-Wright, R.I., C.J. Humphries, and P.H. Williams. 1991. What to protect?—Systematics and the agony of choice. Biol. Conserv. 55: 235–254.

Vogler, A.R. and R. Desalle. 1994. Diagnosing units of conservation management. Consev. Biol. 8: 354–363.

Vrijenhoek, R.C., R.A. Angus, and R.J. Schultz. 1978. Variation and anal structure in a unisexual fish. Am. Nat. 112: 41–55.

Vrijenhoek, R.C., M.E. Douglas, and G.K. Meffe. 1985. Conservation genetics of endangered fish populations in Arizona. Science 229: 400–402.

Walters, T.W., D.S. Decker-Walters, and D.R. Gordon. 1994. Restoration considerations for wiregrass (*Aristida stricta*): Allozyme diversity of populations. Conserv. Biol. 8: 581–585.

Waples, R.S. 1989. A generalized approach for estimating effective population size from temporal changes in allele frequency. Genetics 121: 379–391.

Waples, R.S. and D.J. Teal. 1990. Conservation genetics of Pacific Salmon I. Temporal changes in allele frequency. Conserv. Biol. 4: 144–156.

Watt, W.B. 1977. Adaptations at specific loci. I. Selection on phosphoglucose isomerase of *Colias* butterflies: Biochemical and population aspects. Genetics 87: 177–194.

Wayne, R.K., L. Forman, A.K. Newman, J.M. Simonson, and S.J. O'Brien. 1986. Genetic markers of zoo populations: morphology and electrophoretic assays. Zoo Biol. 5: 215–232.

Wayne, R.K., N. Lehman, D. Girman, P.J.P. Gogan, D.A. Gilbert, K, Hansen, R.O. Peterson, U.S. Seal, A. Eisenhawer, L.D. Mech, and R.J. Krumenaker. 1991. Conservation genetics of the endangered Isle Royal gray wolf. Conserv. Biol. 5: 41–51.

7

Regional Approaches to Conservation Biology: RFLPs, DNA Sequence, and Caribbean Birds

ELDREDGE BERMINGHAM, GILLES SEUTIN,
AND ROBERT E. RICKLEFS

The assessment of conservation priorities on a regional basis depends on an understanding of the genetic uniqueness and genetic diversity of local populations. Genetic surveys of regional biotas have become a practical goal only recently with the development of rapid, molecular methods of genetic evaluation. Because mitochondrial DNA (mtDNA) is homologous across animals, analysis of mtDNA variation permits objective, quantitative comparisons across independently evolving taxa and provides a particularly useful tool for regional biotic surveys. Although mtDNA has some unique attributes—maternal inheritance and rapid nucleotide substitution rates, to name two—insights obtained from mtDNA into phylogenetic relationship and genetic diversity are representative of a taxon's history and population biology, if not of the evolution of the nuclear genome. Thus, mtDNA-based assessments of regional faunas can provide information on genetic diversity that can inform conservation programs.

Archipelagos provide excellent opportunities for analyses of regional biotas, and island systems have long served as laboratories for the study of evolution and speciation (Lack, 1947; 1976; Mayr, 1963; MacArthur, 1972; MacArthur and Wilson, 1967; Carlquist, 1974; Abbott, 1980; Grant, 1986). Their suitability for this purpose arises from the fact that island populations are generally discrete and, under many circumstances, evolutionarily independent. That is, gene flow between islands is small relative to diversifying selection and genetic drift. Thus, once islands are colonized, each island population tends to follow its own evolutionary path, which may result in divergence, accumulation of behavioral or genetic isolating mechanisms, or even extinction. The analysis of genetic variation within and among island populations provides a powerful tool for assessing the historical development of biogeographic patterns and the processes responsible for these patterns (Rosen, 1976, 1978; Kluge, 1988).

One island chain that has been the focus of considerable attention from biogeographers is the Lesser Antilles (Ricklefs and Cox, 1972, 1978; Bond, 1979; Pregill, 1981; Rosen, 1985; Roughgarden and Pacala, 1989; Woods, 1989). We are currently measuring genetic diversity in populations of Lesser Antillean songbirds to determine whether colonization (founder) events and small population size (bottlenecks) erode genetic variability within populations, and whether genetic divergence between populations is more pronounced among taxonomically differentiated species or species groups, particularly those exhibiting distribution gaps or historical extinctions, than among widely dispersed taxa. Ecological studies have demonstrated that geographically restricted species tend to have narrow habitat distributions and low population densities where they do occur (Wilson, 1961; Greenslade, 1968, 1969; Ricklefs and Cox, 1972; Spitzer et al., 1993). These species, some island populations of which have already suffered extinction, appear vulnerable at present. We are assessing whether they also exhibit distinctive attributes of genetic variation within and between extant populations. Our genetic surveys of island bird species should offer practical guidelines for the development of regional conservation programs within the Caribbean, and our approach also may be generalizable to other groups of organisms in other localities.

With the potential now afforded by certain molecular techniques, particularly the analysis of DNA sequence divergence, it has become possible to measure genetic divergence between populations or genotypes directly and to infer the history of genetic changes with increased certainty (Bermingham and Avise, 1986; Avise et al., 1987; Bermingham et al., 1992; Avise, 1992; Seutin et al., 1993, 1994). While precise dating is not possible, molecular techniques do allow the establishment of an approximate chronology for the branch points in a phylogenetic tree of the genetic lineages of a population (Bermingham and Avise, 1986; Bermingham et al., 1992; Page, 1990, 1993; Bermingham and Lessios, 1993; Knowlton et al., 1993). One can use this chronology to test hypotheses concerning the pattern of island colonization, to ascertain the general course of evolutionary and ecological changes since colonization, and, by comparing the chronology to long-term variation in climate, to assess the role of physical factors in promoting these changes. Such historical reconstructions help us to understand the dynamics of evolutionary changes in species and of ecological changes in communities.

CONSERVATION IMPLICATIONS OF MOLECULAR SURVEYS OF REGIONAL BIOTAS

Molecular appraisals of genetic diversity within populations and genetic divergence between populations provide the following important information for the planning of conservation programs.

1. Molecular data allow one to identify genetically unique populations. Where a species is represented by many populations, as is often the case within groups of islands or island habitats, one may wish to direct conservation efforts toward the most distinctive of these—the population most differentiated genetically from the others (Vane-Wright et al., 1991; Krajewski, 1994). The rarest population may not be the most unique, and therefore not the most valuable in the currency of the total genetic diversity of the species. It will often be the case that, owing to the history and geography of an island group, genetically unique populations of several species will occur on the same island, allowing one to generalize the allocation of conservation efforts. Finally, genetic analyses can identify the most suitable source populations for augmenting faltering populations or for reintroductions.

2. Molecular techniques allow one to identify populations with reduced genetic diversity and, therefore, impaired capacity to respond to environmental change. Fewer genotypes mean fewer physiological varieties able to cope with climate or habitat change, and reduced diversity of genetically based disease-resistance mechanisms. All other things being equal, the genetically depauperate population will be the most vulnerable (least adaptable) to environmental change.

3. Molecular approaches allow one to determine genetic characteristics of populations that are vulnerable to extinction. Vulnerability often is revealed by historical records of extinction and gaps in the geographical distributions of species through island chains, such as the West Indies and Hawaiian Islands (Ricklefs and Cox, 1972). Ecological studies on West Indian birds have shown that where a population has gone extinct on one island, populations of the same species on other islands tend to be rare and locally distributed, hence vulnerable (Ricklefs and Cox, 1978). Genetic analyses of the remaining populations may indicate general characteristics of susceptible taxa, particularly with respect to genetic diversity within populations and degree of genetic differentiation between populations. Because isolated populations may diverge at a relatively constant molecular evolutionary rate, genetic differences between populations can indicate the relative age of the species within an island setting. As in the death of individual organisms, vulnerability to extinction may increase with the age of the taxon: historically recorded extinctions on islands have occurred with the highest frequency among the most taxonomically distinct (evolutionarily oldest?) species.

Although differentiation is most readily studied in island groups because of the discrete distribution of island populations, the genetic principle of population divergence could be extended easily to continental species as well. Thus, molecular techniques for assessing genetic variation hold the promise of producing generalizable characterizations of vulnerable populations as well as identifying particular islands, populations, and habitats requiring the greatest protection.

THE DATA, RFLPs OR DNA SEQUENCE?

The development of rapid and automated DNA sequencing might easily lead one to question the continued utility of restriction fragment length polymorphisms (RFLPs) for molecular systematic and conservation genetic analyses. After all, an improved understanding of DNA sequence evolution results from knowing the actual divergence between two homologous sequences as opposed to a restriction fragment or site-based estimate of DNA sequence differentiation. Furthermore, in comparisons across species, it is generally more informative to compare homologous DNA sequences than mapped restriction endonuclease sites of less certain homology. Thus, why not simply discard RFLP-based approaches to the study of evolutionary genetics?

It is useful to approach this question from a mtDNA perspective because mtDNA is the only DNA marker that has found wide application in the study of conspecific populations and their close relatives. The advantages that accrue to direct determination of mitochondrial nucleotide sequence are of sufficient significance that it would be disingenuous to compare RFLP-based analyses to sequence-based results without specifying the particular pragmatic needs of conservation planning. Rather, we hope to provide insights into complementary uses of RFLP and sequence analyses in evolutionary genetic studies. Among the few studies that directly compare the phylogenetic information revealed by restriction endonuclease studies of the entire mitochondrial genome, it is sometimes the case that RFLP data provide greater phylogenetic resolution than do nucleotide data (e.g., Edwards and Wilson, 1990; Geffen et al., 1992; Quinn et al., 1991). Nevertheless, the practical limitations of cross-taxa comparisons using RFLP data, coupled with the technological ease of nucleotide data collection, would soon diminish arguments based on apparent differences in the phylogenetic information content of RFLP versus sequence data. Even in the event that restriction site data were demonstrated to be less prone to homoplasy than nucleotide data, this finding would only suggest a different method for analyzing nucleotide data. After all, restriction-site sequences can be directly recovered and analyzed from nucleotide sequence data.

Another argument that might be advanced to support RFLP-based studies of mtDNA results from failure to distinguish nuclear pseudogenes from functional mtDNA genes. With the widespread use of "universal" mtDNA primers, it is conceivable that one could amplify and sequence a nuclear gene thinking all the while that the sequence represented a functional mitochondrial DNA target. Lopez and coworkers (1994) have documented the transposition of a 7.9 kilobase (kb) mtDNA fragment (named *Numt*) to a specific nuclear chromosome position in the domestic cat. In this case, two mitochondrial ribosomal genes, three complete protein coding genes (plus most of a fourth), and probably 13 transfer RNAs were carried into the nucleus. Excepting two insertions in the nuclear pseudogene NADH dehydrogenase 2 (ND2) relative to the mitochondrial ND2, there were no significant nucleotide sequence signals that obviously distinguish the nonfunctional nuclear *Numt* genes from their functional cytoplasmic mtDNA homologues. The unwitting inclusion of a nuclear pseudogene in an mtDNA-based analysis of phylogenetic relationship presents obvious problems, a point elaborated by Smith et al. (1992) in their characterization of a putative mitochondrial cytochrome *b* gene translocated to the nuclear genome of the South American rodent *Chroemys jelskii*. In contrast, it is unlikely that RFLP assays of the entire mitochondrial genome could suffer from analytical problems caused by molecule misidentification because restriction fragments should sum. However, even the recognition that masquerading genes are a serious problem in molecular systematics will not provide a sufficient rationale for preferring RFLP analyses under most circumstances.

Thus, have RFLP analyses outlived their usefulness to evolutionary and conservation genetics? The answer, we think, is no. Speed and cost become particularly important considerations when the power of statistical analysis depends on moderate to large sample size, and RFLP analyses can be considerably faster and less expensive than DNA sequencing. Better understanding of the genetic architectures of populations and closely related species will often require large-scale geographical and numerical sampling. Moderate to large sample sizes may also allow one to resolve the demographic history of a species—particularly whether population crashes are a prominent feature—in addition to its phylogenetic history. Thus, RFLP analyses and DNA sequencing, as well as other approaches (e.g., microsatellites), should be used to complement one another.

TECHNICAL DESCRIPTION, RFLPs ANALYSES

Although we are now following a mixed sequencing and RFLP strategy in the laboratory, our previous research has primarily used mtDNA RFLP analyses to investigate the evolutionary relationships of geographic populations and closely related species. Thus, in this chapter, we have been asked to center our technical focus on a brief description of RFLP methodologies. An excellent and more complete discussion of the laboratory techniques utilized in RFLP analyses can be found in Dowling et al. (1996) and numerical analyses are further covered by Swofford et al. (1996).

RFLP analyses utilize bacterial enzymes termed restriction endonucleases to identify nucleotide differences between DNA samples that have been extracted from the organisms to be studied. Restriction endonucleases recognize a specific sequence of nucleotides, typically four to six bases in length, and snip the DNA at a specific nucleotide (restriction site) within or near the recognition sequence. In bacteria these enzymes function as a kind of bacterial immune system and disable foreign, potentially deleterious nucleic acids. Molecular biologists siezed on these enzymes as a handy way to cut up DNA in a highly targeted manner, and population biologists (first and foremost, John Avise, Wes Brown, and Alan Wilson) determined that restriction endonucleases could reveal genetic polymorphisms (called restriction fragment length polymorphisms, hence RFLPs) that were extremely useful for evolutionary and population genetic studies.

Population biologists typically isolate a small sample of each study individual's entire genome (genomic DNA), although sometimes a portion of the genome is enriched for and extracted (e.g., mitochondrial DNA). In either case, aliquots of the resulting DNA are incubated and digested with restriction endonucleases which cut the DNA each time the enzyme encounters its particular recognition sequence. Investigators can choose among a large number of commercially available restriction enzymes, and price, availability, and tradition usually dictate which suite of restriction enzymes are used in a study. The more enzymes one uses and the shorter their recognition sequence (e.g., four- versus six-base cutters), the greater the proportion sampled of the study individual's genome. A typical vertebrate mtDNA study might use 10–20 five- and six-base cutters and sample 2–4% of the 16,000 nucleotide pair mitochondrial genome. Although recognizing many more sites on average, four-base restriction endonucleases have been less utilized by population biologists (but see later) owing to technical considerations (e.g., recovery of short restriction fragments when using filter hybridization techniques; difficulties associated with assigning fragment homology).

Once the DNA has been cut by restriction enzymes, it needs to be size-

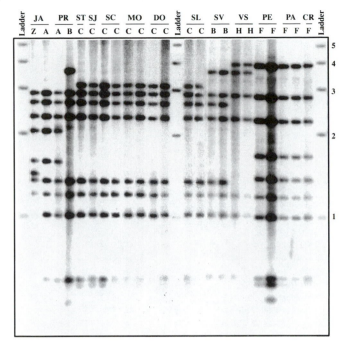

Figure 7-1 Restriction-fragment-length polymorphism analysis of the bananaquit, *Coereba flaveola*, from various locations around the Caribbean Basin, using the endonuclease *Stu*I. Two-letter codes at top refer to sampling localities presented in Table 7-1; single letters below refer to restriction site patterns. A 1-kilobase ladder was used as a molecular size marker; units of calibration are kilobases (kb).

fractionated and visualized in order to distinguish restriction fragment similarity and/or polymorphism among study organisms. Restriction fragments are separated electrophoretically and then visualized using specific stains (e.g., ethidium bromide) or radioactive tags exposed to film. In the latter case, restriction fragments can be directly labeled with radioactive nucleotides prior to electrophoresis (end-labeling) or the fragments can be hybridized with radioactively labeled probe DNA following electrophoresis, membrane transfer (Southern blotting) and denaturation (e.g., Figure 7-1). Filter hybridization and DNA staining share the disadvantage that small fragments (typically shorter than 500 nucleotides in length) are often not seen, because there is less DNA to hybridize or stain and/or because transfer efficiency declines with small fragments. End-labeling and DNA staining share the disadvantage that genomic DNA cannot be used because the dye or radioactive label is indiscriminate in its action; instead one must first isolate and purify the specific DNA region to be studied (e.g., ultrapure mtDNA or purified recombinant DNA).

Once restriction fragments are visualized, the data are ready to be scored. Data are generally scored in one of three ways: (1) Fragments are measured against a known size standard and the fragment's presence or absence is recorded for each individual surveyed (e.g., Figure 7-1); (2) using a double-digestion strategy (see Dowling et al., 1996), a genome (e.g., mtDNA, choloroplast), gene, or gene region

Table 7-1   Subspecies and Samples of the Bananaquit from the West Indies and Continental Locations around the Caribbean Basin

Subspecies	Location	Code	Sample Size	Haplotype Diversity	Nucleotide Diversity
flaveola	Jamaica	JA	10	0.889	0.003 74
bananivora	Hispaniola		2		
nectarea	Tortue		not sampled		
portoricensis	Puerto Rico	PR	25	0.853	0.002 13
sanctithomae	St. Thomas	ST	1		
	St. John	SJ	1		
newtoni	St. Croix	SC	10	0.378	0.000 54
bartholemica	Montserrat	MO	5		
	Guadeloupe	GU	6		
	Dominica	DO	19	0.790	0.002 74
martinicana	Martinique	MA	8		
	St. Lucia	SL	18	0.758	0.001 88
barbadensis	Barbados		8		
atrata	St. Vincent	SV	15	0.848	0.002 02
aterrima	Grenada	GR	17	0.544	0.000 80
luteola	Venezuela (Sucre)	VS	12	0.758	0.002 58
	Venezuela (Falcon)	VF	2		
	Venezuela (Federal)	VD	4		
mexicana	Panama (Bocas del Toro)	PA	8		
	Costa Rica	CR	1		
cerinoclunis	Panama (Pearl Islands)	PE	8		
Total			180		

is mapped for each restriction site and each individual is scored for the presence or absence of the physically mapped restriction site; (3) restriction sites are inferred using knowledge of gene/genome size and the transformation series of restriction fragment genotypes across individuals (see Bermingham, 1990). The last approach is only useful in studies of closely related populations and/or species where fragment homology is high and the genotype transformation series interconnect through single restriction site gains or losses. Restriction site analysis is preferred to restriction fragment analysis owing to the fact that restriction fragment data are not analytically independent. When analyzing restriction sites, each restriction site is recorded as a single character with two character states (present or absent). When analyzing restriction fragments, there are three fragment characters (each with a present/absent character state) associated with each restriction site. If the restriction site is present it divides one larger fragment (scored as absent) into two smaller fragments (scored as present).

The physical mapping of restriction sites is time-consuming and moderately expensive. In those studies where mapping would be necessary to correctly interpret restriction site homology, many investigators are now choosing direct sequence analysis in place of mapped restriction site analysis. Thus future uses of RFLP analysis in conservation studies will probably be restricted to situations where restriction sites can easily be inferred, have already been mapped, or can be directly determined from available nucleotide sequence. In our own research we are increasingly turning to RFLP analyses of PCR-amplified genes whose full nucleotide sequence we have previously determined for representative taxa (including geographic populations). This is a simple approach permitting rapid analysis of large numbers of individuals. In practice, we sequence a gene or gene region (e.g., mtDNA ATP synthase 6 and 8) and then use available software (e.g., MacVector) to search for all restriction sites across the taxa we have sequenced. After summarizing the extent of restriction site polymorphism across the gene region as well as the precise location of each restriction site, we can choose restriction enzymes (usually four-base cutters) which reveal particular polymorphisms of interest. In turn, we PCR amplify the gene region for large numbers of individuals, cut the DNAs with restriction enzymes, and visualize the resulting fragments on agarose or acrylamide gels after staining the gel with ethidium bromide. From our sequence analyses we can anticipate restriction fragment size and optimize our restriction enzyme choices such that the fragments are large enough to be visualized using DNA stains. Because the PCR process yields high concentrations of DNA, we can easily visualize fragments as small as 25 base pairs.

Now we turn to the numerical analyses of the data and, in particular, the analyses of nonrecombining genomes such as mtDNA. Molecular data analysis permits two classes of historical information to be recovered. First, these data permit a hypothesis of branching order to be deduced and, second, molecular data permit analyses of genetic distance that, if calibrated, can be used to estimate branching time in addition to branching order. Although controversial, it is the estimation of branching time that probably accounts for the seductive appeal of molecular approaches to phylogeny reconstruction. There are two general approaches to historical reconstruction. In the first approach, the phylogenetic history of contemporaneous DNA genotypes can be traced back through time by examining the derived characters (e.g., restriction sites, nucleotide states) that they share. This approach is generally referred to as cladistic analysis, and shared, derived characters are referred to as synapomorphies. The second approach uses genetic similarity or, alternatively, distance to infer the relationship among lineages. The rationale of the phenetic approach is that differences between lineages accumulate over time and that difference thereby indicates relative time since separation. The difference between phenetic and cladistic approaches is that phenetics ignores considerable information in individual characters, such as polarity (designation of character states, typically through comparison to an outgroup, as primitive or derived), homoplasy (parallel changes and convergences), and reversal (reversion to the ancestral state), in reconstructing phylogenetic hypotheses. Notwithstanding the important differences in the two approaches, we usually apply both methods to our data. When analyzing conspecific populations or closely related species, we have found that the trees resulting from phenetic and cladistic analyses are often congruent over much of the resulting topologies. For our cladistic analyses we rely on PAUP (Swofford, 1993); and MacClade (Maddison and Maddison, 1993); for our phenetic analyses we rely on NTSYS (Rohlf, 1993), and MEGA (Kumar et al., 1993).

Molecular data can also be used to provide insights into the demographic behavior and history of a species (Takahata and Palumbi, 1985; Slatkin and Maddison, 1989; Slatkin and Hudson, 1991; Excoffier et al., 1992; Hudson et al., 1992; Rogers and Harpending, 1992; Moritz, 1994; Shulman and Bermingham, 1995; McMillan and Bermingham, 1996). Thus, using either cladistic approaches (Slatkin and Maddison, 1989, 1990) or population genetic models (Takahata and Palumbi, 1985) one can estimate gene flow (Nm) and the genetic connectedness between contemporaneous populations. Sequence-based statistics such as Gst (Takahata and Palumbi, 1985) and AMOVA (utilizing the molecular information option; Excoffier et al., 1992) in conjuction with permutation tests permit one to test for population subdivision. Particularly useful is AMOVA which partitions genetic variance (1) within populations; (2) between populations within a region; and (3) between regions. Finally, the analysis of the distribution of pairwise differences between alleles (Slatkin and Hudson, 1991; Rogers and Harpending, 1992; Marjoram and Donnelly, 1994) can provide useful information regarding past population declines and expansions. Although we anticipate that analyses of demographic history will be particularly useful to conservation biologists, this is a nascent field and underlying models and their interpretation are controversial and still under development.

CASE STUDIES: EXAMPLES FROM CARIBBEAN BIRDS

Molecular data, specifically mtDNA RFLP analyses, have now been used to interpret the history of the streaked saltator (*Saltator albicollis*; Seutin et al., 1993), yellow warbler (*Dendroica petechia*; Klein, 1992), and bananaquit (*Coereba flaveola*; Seutin et al., 1994) in the West Indies. Using West Indian birds as examples, we suggest that RFLP analyses remain a cost-effective alternative to direct sequence analyses for analyzing large sample sizes, which are critical to a genetic description of a taxon's distribution, level of geographic subdivision, and diversity among geographic populations. In the future, cost reductions resulting from further automation of direct sequence analysis, and advances in the interpretation of sequence data, will erode the cost-related advantages of RFLPs.

Genetic Divergence, Phylogeographic Structure, and Taxonomic Distinctiveness

Our studies of Caribbean birds have revealed a degree of phylogeographic structuring and divergence among populations that is high by avian standards. Levels of mtDNA differentiation between populations of some species are typical of interspecific or even intergeneric differences among temperate bird species. Such marked genetic divergence among populations of neotropical birds has now been observed in species from a wide range of passerine families (Hackett and Rosenberg, 1990; Capparella, 1988; Escalante-Pliego, 1991; Peterson et al., 1992; Seutin et al., 1993; Seutin et al., 1994), suggesting that high levels of genetic divergence between populations may be a general feature of tropical species.

The most thoroughly documented example of phylogeographic structure in a neotropical bird is provided by the bananaquit (distribution shown in Figure 7-2),

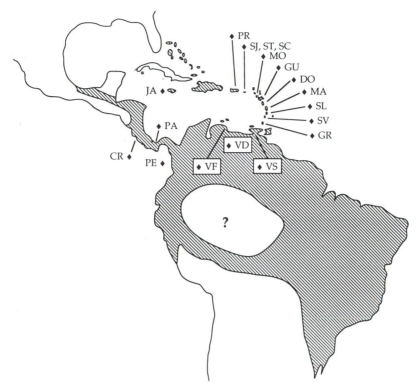

Figure 7-2 Geographic distribution of the bananaquit with collection localities indicated. (Locality abbreviations are presented in Table 7-1.)

for which mtDNA restriction-site differences defined six regional populations: Jamaica (JA), Central America (Costa Rica and Panama; hereafter CA), Venezuela (VE), southern Lesser Antilles (Grenada and St. Vincent; GSV), north-central Lesser Antilles and the U.S. Virgin Islands (St. Lucia to St. Thomas; LA), and Puerto Rico (PR). In most cases, the geographic structuring of mtDNA variation was well documented in both genetic distance and cladistic analyses (Figure 7-3). The mtDNA clade representing the 10 Jamaican individuals was the most divergent; six fixed restriction-site differences distinguished the Jamaican banana-quits from all others. In pairwise comparisons between Jamaican birds and all other samples, we observed a minimum of 15 restriction site changes, an estimated mean mtDNA sequence divergence (d_{xy}) of 0.029 (range 0.027–0.035). Three groups of populations representing Central America, northern South America, and the eastern Antilles (Puerto Rico to Grenada) were approximately equally differentiated among themselves (average $d_{xy} = 0.014$), suggesting that these independent lineages were established at about the same time. Within the eastern Antilles, the three geographic populations (PR, LA, and GSV) shared no mtDNA haplotypes among themselves or with other population groups, but the phylo-genetic relationships of the three Antillean groups could not be resolved un-ambiguously.

The geographical distribution of mtDNA variation often shows a poor

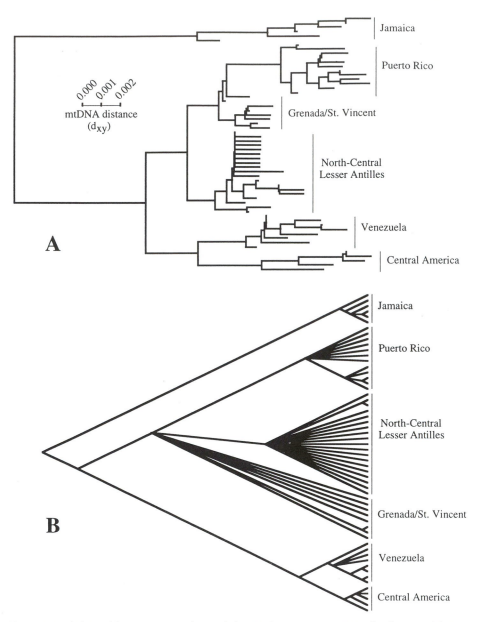

Figure 7-3 (A) Neighbor-joining analysis of the 58 bananaquit mtDNA haplotypes. (B) Strict consensus of unrooted Wagner parsimony trees (length 98; CI 0.62) showing the relationship among the 58 bananaquit mtDNA haplotypes.

congruence with the ranges of named subspecies, and we have concluded that taxonomic distinction provides only marginally useful estimates of genetic divergence in Antillean birds. Although fixed restriction-site differences separated the bananaquit subspecies *flaveola* (Jamaica) and *portoricensis* (Puerto Rico) from one another and from all other subspecies studied, other West Indian mtDNA groupings (GSV and LA) encompassed two and four subspecies, respectively. Whereas at least three restriction-site differences distinguished the single mtDNA haplotype carried by the eight Pearl Islands bananaquits (*C. f. cerinoclunis*) from the three haplotypes assayed in the eight birds from Bocas del Toro (*C. f. mexicana*), a minimum of five site differences was inferred between bananaquits from Bocas del Toro and our unique Costa Rica specimen, which all belong to the *mexicana* subspecies. Thus, mtDNA differentiation was higher among individuals within the *mexicana* subspecies than it was between this and the *cerinoclunis* subspecies. In the streaked saltator (*Saltator albicollis*) no significant mtDNA differentiation was observed between populations of named subspecies on mainland Panama (*isthmicus*) or the Pearl Islands (*speratus*), or between the two Lesser Antillean subspecies (*albicollis* and *guadelupensis*). Thus from a mtDNA perspective, subspecies designations do not provide a reliable guide to either the pattern or the depth of genetic distinctiveness between geographic populations of neotropical birds.

Levels of mtDNA sequence divergence between Antillean bird populations within species or species groups, as estimated by restriction-site analyses, ranged from 4% to 8% in *Myadestes genibarbis* and *Icterus* species-groups to zero in the monotypic *Tiaris bicolor* (unpublished results). Furthermore, to the degree that mtDNA in birds evolves in an approximately clocklike manner, it appeared that not all named subspecies within a group are historically and evolutionarily equivalent. This finding has important consequences for the study of biogeographic patterns, and we have pointed out elsewhere that cross-taxa biogeographic analyses would benefit from having phylogenetic branch points sorted by age (Bermingham et al., 1992). In developing the concept of the taxon cycle (Wilson 1961) for West Indian birds, Ricklefs and Cox (1972) characterized the relative ages of taxa in part by taxonomic distinction among island populations. Their estimate of age predicted ecological breadth and population density of island populations. With respect to the implications of the taxon cycle concept for conservation policies, it will be important to determine whether genetically designated ages sharpen or obscure these relationships.

mtDNA Diversity within Island and Mainland Populations

RFLP-based surveys of regional biotas are sufficiently rapid and inexpensive to provide a basis for studies of genetic diversity within populations. Genetic diversity of many West Indian island populations is similar to that observed among continental populations of passerines (0.0008–0.0060; Seutin et al., 1993), but some populations show evidence of founder effects. For example, *Tiaris bicolor* is virtually monomorphic throughout the West Indies, as are bananaquits on islands from Guadeloupe to the Virgin Islands (see below). In both the bananaquit and the yellow warbler, genetic diversity is correlated with island area and, presumably, population size (Klein, 1992; Seutin et al., 1994). In the bananaquit, the Jamaican population had both the highest mtDNA haplotype and the highest nucleotide diversity (0.89 and 0.0037, respectively), and St. Croix had the lowest (0.38 and 0.0005, respectively).

We observed reduced mtDNA variability in two groups of bananaquits. Eight individuals from Isla Chapera, in the Pearl Islands, Gulf of Panama, carried a single mtDNA haplotype. Isla Chapera has been isolated from the mainland of Panama for less than 10,000 years (Bartlett and Barghoom, 1973; Fairbanks, 1989) and supports a large population of bananaquits assigned to a subspecies endemic to the Pearl Archipelago (*Coereba flaveola cerinoclunis*; Wetmore et al., 1984). Reduced genetic variability in the Chapera population could be due either to recent colonization of the island by a small founder population or, more probably, to the loss of haplotype diversity through genetic drift.

Reduced mtDNA variability was also noted in the northern LA islands (Guadeloupe north to St. Croix). From the islands of Guadeloupe, Montserrat, and St. Croix, 90% of 21 birds carried the dominant LA mtDNA haplotype. In the southern LA islands (Dominica, Martinique, and St. Lucia), only 44% of 45 bananaquits carried the dominant LA haplotype. From a phylogenetic perspective, the bananaquit populations in the north-central Lesser Antilles represent a single mtDNA clade, yet from a demographic perspective it appeared that southern island populations have had a history quite distinct from that of the northern island populations (see below).

Patterns of Expansion and Dispersal Ability

Our saltator and bananaquit results provide examples of recent expansions of specific lineages within the ranges of old, established or extinct populations and suggest a pattern of alternation between geographical expansion and quiescence within archipelagoes. Although genetic analyses of additional insular bird species are required to assess the generality of this pattern, our data so far suggest that different populations of the same species may be in different phases of colonizing activity at a specific time.

For example, the limited geographic distribution of bananaquit mtDNA haplotypes found in Puerto Rico, Grenada and St. Vincent, Jamaica, and most continental locations contrasted sharply with the widespread distribution of north-central Antillean (LA) mtDNA haplotypes (Figure 7-4). The numerically predominant LA haplotype (LA1), observed in 39 of the 68 samples assayed from the region, was in high frequency or fixed on all islands from St. Lucia north to St. Croix (subspecies *martinicana*, *bartholemica*, and *newtoni*). If founder effects resulting in reduced within-population genotypic variability accompany colonization, the pattern of haplotypic diversity in the north-central Lesser Antillean islands suggests that the spread of *C. flaveola* within the Lesser Antilles occurred in two waves: an older one in the southern islands (St. Lucia, Martinique, Dominica), which nowadays show moderate levels of mtDNA variability, and a more recent progression, probably from Dominica, through the northern islands (Guadeloupe through St. Croix and the Virgin Islands), within which little mtDNA variation has yet appeared. Because Guadeloupe is the largest of the Lesser Antillean islands, its apparent reduced mtDNA variability can be explained most readily by a founder effect resulting from recent colonization or from a post-founding bottleneck.

A second example regards populations of the streaked saltator on the adjacent

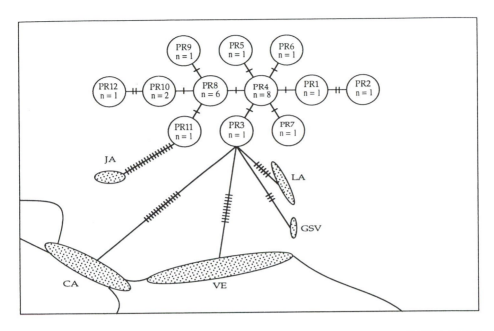

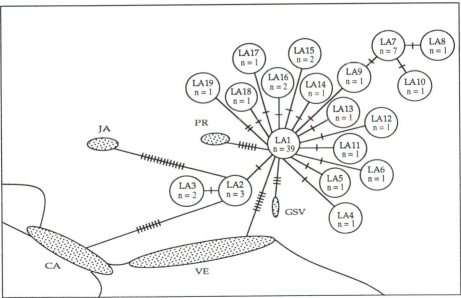

Figure 7-4 Relationships between mtDNA haplotypes within Puerto Rico (PR) and within the north-central Lesser Antilles (LA) identified on the basis of multiple restriction-site differences. The minimal distance between populations are also shown. Each mark represents the gain or loss of a restriction site.

117

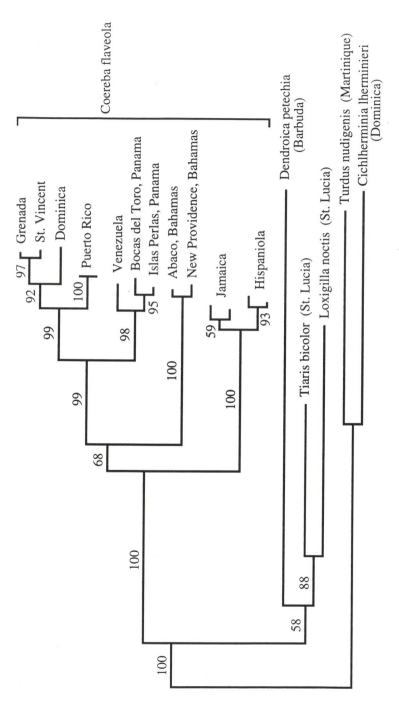

Figure 7-5 Neighbor-joining analysis of the complete mtDNA ATPase 6 gene for each of the major bananaquit clades identified by mtDNA RFLP analysis and outgroups.

islands of Dominica, Martinique, and St. Lucia which share most mtDNA restriction-site haplotypes and exhibit a maximum genetic divergence among haplotypes (d_{xy}) of 0.006. Recent derivation of the populations from a common ancestor, rather than a high level of migration between islands, is indicated by the limited Caribbean distribution of the species to the central part of the Lesser Antilles; with high levels of migration, one would expect further colonization of the island chain. The fact that the Antillean haplotypes differ from continental haplotypes of the same species by d_{xy} values exceeding 0.06 further implies low vagility. On some of the islands occupied by saltators, Klein (1992) observed two coexisting mtDNA haplotypes in the yellow warbler that differed by $d_{xy} = 0.014$, one of West Indian origin and the other of Venezuelan derivation. This observation suggests a more complex origin for Lesser Antillean populations of the yellow warbler, compared to those of the streaked salt-ator, featuring overlapping invasions from two genetically differentiated source areas.

The patterns described above document the idiosyncratic demographic his-tories of bananaquits, saltators, and yellow warblers in the Lesser Antilles. It is also the case, however, that prominent barriers to dispersal are evi-denced by congruence between phylogeographic patterns across species. For example, the water gap between the island of St. Vincent and St. Lucia marks either the end of a distribution (saltator, yellow warbler, *Icterus dominicensis* group) or a major genetic disjunction (bananaquit, *Vireo altiloguus*, *Loxigilla noctis*) in the species we have examined to date. The reason for this barrier is not apparent, and species that are assumed to have recently colonized the Lesser Antilles (e.g., *Tiaris bicolor*) do not show a phylogeographic break across this water gap.

The Origin of Bananaquits—Beyond the Reach of RFLP-based Analyses

Of the bananaquit populations that we sampled, that of Jamaica is the most distinctive, and it is almost equally differentiated from all other populations. However, without an outgroup we cannot determine whether the Jamaican population is part of an older West Indian taxon that is ancestral to all other bananaquit clades, or whether it represents an older invasion of the West Indies from established continental populations. We could not perform an outgroup analysis in our RFLP-based study of the bananaquit because we could not recover sufficient restriction endonuclease site homology between bananaquits and the many potential sister taxa. Figure 7-5 pictures the bananaquit phylogeny rooted through direct sequence analysis of representative bananaquit mtDNA clades identified by our RFLP study as well as a few of the possible sister taxa of bananaquits. Although the details of the analysis will be presented elsewhere, it is interesting to note that the basal position of the Jamaica, Hispaniola and Bahamas clades is well supported. Thus, at this level of analysis, it would appear that Central and South American continental populations (as well as Puerto Rico and the Lesser Antilles) were derived from older island populations.

CONCLUSIONS AND FUTURE DIRECTIONS

RFLP-based approaches to conservation genetics should remain attractive so long as costs militate against the direct sequence analysis of large sample sizes. RFLP analyses are a fast and convenient method of describing general patterns of population subdivision across the breadth of a species' distribution and can help focus more detailed genetic studies which might take advantage of other molecular approaches (particularly direct nucleotide sequencing and microsatellite analysis). Furthermore, reconstructing the demographic histories of species requires large sample sizes and suggests a long-term role for RFLP analyses in conservation genetics. Once a representative portion of a species' variation has been surveyed, one can identify a small subset of enzymes for extensive population genetic studies marked by better statistical power.

Surveying large numbers of individuals across moderate numbers of species with overlapping distributions should be a goal in conservation biology for both theoretical and applied reasons. On the theoretical side, species richness may be influenced more strongly by extrinsic biogeographical relationships and historical circumstances than by such intrinsic, local processes as competiton and predation (Ricklefs, 1987; Cornell and Lawton, 1992). In this context, Caribbean birds have been used to test the idea that local ecology imposes a ceiling on species richness: Terborgh and Faaborg (1980) argued for such a ceiling, while analyses by Ricklefs (1987) and Wiens (1989) suggested that proportional sampling of the regional West Indian avifauna explained local species richness patterns as well as or better than a local saturation model. Further tests of the dependence of local richness on regional richness will benefit from a taxonomy placed on a phylogenetic footing provided by molecular analyses.

On the practical side, molecular genetic analyses can provide a resonably rapid means for surveying regional biotic diversity. Indices of species richness, sometimes taking into account abundance, have been traditional measures of diversity. When used to make decisions regarding the preservation of biodiversity, however, it has been argued that these indices fail because they consider all species to be qualitatively equal or nearly equal. Erwin (1991), Vane-Wright et al. (1991) and others (Crozier, 1992; Faith, 1992; Weitzman, 1992; Solow et al., 1993; reviewed by Krajewski, 1994) have suggested that phylogenetic history and/or genetic divergence should be used in biodiversity indices to emphasize the phylogenetic and genetic distinctiveness of some populations compared to others. To the degree that this view is adopted by conservation biologists, molecular systematics will undoubtedly be called upon to provide objective measures of taxonomic distinctiveness. The resulting taxic diversity measures, when coupled to detailed knowledge of organismal distribution patterns, can be used to identify priority areas for conservation.

Although our studies of West Indian birds are far from complete, our results provide a useful introduction to the different types of conservation information yielded by RFLP and sequencing analyses. It might not be worthwhile, for example, to expend significant resources salvaging small bananaquit populations from the northern Lesser Antilles because they are genetically identical to those in

Guadeloupe. One would not want to repopulate the Virgin Islands with bananaquits from nearby Puerto Rico because they carry the wrong genotypes. Dominican yellow warblers are likely to have a unique mix of Antillean and South American genes. St. Vincent and Grenada are likely to be distinctive for many taxa and worthy of special attention.

ACKNOWLEDGMENTS

G. Seutin was supported by a Smithsonian Scholarly Studies postdoctoral fellowship and R. E. Ricklefs by a Regents' Fellowship from the Smithsonian Tropical Research Institute. Grants from the National Geographic Society and the Smithsonian Institution (Scholarly Studies Program, Abbott Fund, and the Tropical Research Institute's Molecular Systematics Program) made the field and laboratory work possible. We particularly acknowledge the support, assistance, and collecting permits provided by the island nations, departments, and territories of the Caribbean and thank the many people on those islands who have encouraged our work on West Indian birds.

REFERENCES

Abbott, I. 1980. Theories dealing with the ecology of landbirds on islands. Adv. Ecol. Res. 11: 329–371.

Avise, J.C. 1992. Molecular population structure and the biogeographic history of a regional fauna—A case history with lessions for conservation biology. Oikos 63: 62–76.

Avise, J.C., J. Arnold, R.M. Ball, E. Bermingham, T. Lamb, J.E. Neigel, C.A. Reeb, and N.C. Saunders. 1987. Intraspecific phylogeny: The mitochondrial DNA bridge between population genetics and systematics. Ann. Rev. Ecol. Syst. 18: 489–522.

Bartlett, A.S. and E.S. Barghoorn. 1973. Phytogeographic history of the Isthmus of Panama during the past 12,000 years: a history of vegetation, climate and sea-level change. In: A. Graham, ed. Vegetation and vegetational history of Northern Latin America, pp. 203–299. New York: Elsevier.

Bermingham, E. 1990. Mitochondrial DNA and the analysis of fish population structure. In: D. Whitmore, ed. Electrophoretic and isoelectric focusing techniques in fisheries management, pp. 197–221.. Boca Raton, Fla.: CRC Press.

Bermingham, E. and J.C. Avise. 1986. Molecular zoogeography of freshwater fishes in the southeastern United States. Genetics 113: 939–965.

Bermingham, E. and H.A. Lessios. 1993. Rate variation and mitochondrial DNA evolution as revealed by sea urchins separated by the Isthmus of Panama. Proc. Natl. Acad. Sci. U.S.A. 90: 2734–2738.

Bermingham, E., S. Rohwer, S. Freeman, and C. Wood. 1992. Vicariance biogeography in the Pleistocene and speciation in North American wood warblers: a test of Mengel's model. Proc. Natl. Acad. Sci. U.S.A. 89: 6624–6628.

Bond, J. 1979. Derivations of Lesser Antillean birds. Proc. Acad. Nat. Sci. Phila. 131: 89–103.

Capparella, A.P. 1988. Genetic variation in Neotropical birds: Implications for the speciation process. In: H. Ouellet, ed. Acta XIX Congressus Internationalis Ornithologici, pp. 1658–1664. Ottawa: University of Ottawa Press.

Carlquist, S.J. 1974. Island biology. New York: Columbia University Press.

Cornell, H.V. and J.H. Lawton. 1992. Species interactions, local and regional processes, and limits to the richness of ecological communities: a theoretical perspective. J. Animal Ecol. 61: 1–12.

Crozier, R.H. 1992. Genetic diversity and the agony of choice. Biol. Conserv. 61: 11–15.

Dowling, T.E., C. Moritz, J.D. Palmer, and L.H. Rieseberg. 1996. Nucleic acids III: Analysis of fragments and sites. In: D.M. Hillis, C. Moritz, and B.K. Mable, eds. Molecular systematics pp. 249–320. Sunderland, Mass.: Sinauer Associates.

Edwards, S.V. and A.C. Wilson. 1990. Phylogenetically informative length polymorphism and sequence variability in mitrochondrial DNA of Australian songbirds (*Pomatostomus*). Genetics 126: 695–711.

Erwin, T.L. 1991. An evolutionary basis for conservation strategies. Science 253: 758–761.

Escalante-Pliego, B.P. 1991. Genetic differentiation in yellowthroats (Parulinae: *Geothlypis*). In: Acta XX Congressus Internationalis Ornithologici pp. 333–343. Wellington: New Zealand Ornithological Congress Trust Board.

Excoffier, L., P.E. Smouse, and J.M. Quattro. 1992. Analysis of molecular variance inferred from metric distances among DNA haplotypes: Application to human mitochondrial DNA restriction data. Genetics 131: 479–491.

Fairbanks, R.G. 1989. A 17,000-year glacio-eustatic sea level record: influences of glacial melting rates on the Younger Dryas event and deep-ocean circulation. Nature 342: 637–642.

Faith, D.P. 1992. Conservation evaluation and phylogenetic diversity. Biol. Conserv. 61: 1–10.

Geffen, E., A. Mercure, D.J. Girman, D.W. Macdonald, and R.K. Wayne. 1992. Phylogeny of the fox-like canids: analysis of mtDNA restriction fragment, site, and cytochrome b sequence data. J. Zool. 228: 27–39.

Grant, P.R. 1986. Ecology and evolution of Darwin's finches. Princeton, N.J.: Princeton University Press.

Greenslade, P.J.M. 1968. Island patterns in the Solomon Islands bird fauna. Evolution 22: 751–761.

Greenslade, P.J.M. 1969. Land fauna: insect distribution patterns in the Solomon Islands. Phil. Trans. R. Soc., Ser. B 255: 271–284.

Hackett, S.J. and K.V. Rosenberg. 1990. Comparison of phenotypic and genetic differentiation in South American antwrens (Formicariidae). Auk 107: 473–489.

Hudson, R.R., D.D. Boos, and N.L. Kaplan. 1992. A statistical test for detecting geographic subdivision. Mol. Biol. Evol. 9(1): 138–151.

Klein, N.K. 1992. Historical processes and the evolution of geographic variation patterns in the yellow warbler (*Dendroica petechia*). Ph.D. dissertation, University of Michigan, Ann Arbor, Michigan.

Kluge, A.G. 1988. Parsimony in vicariance biogeography: a quantitative method and a Greater Antillean example. System. Zool. 37: 315–328.

Knowlton, N., L.A. Weigt, L.A. Solorzano, D.K. Mills, and E. Bermingham. 1993. Divergence in proteins, mitochondrial DNA, and reproductive compatability across the Isthmus of Panama. Science 260: 1629–1632.

Krajewski, C. 1994. Phylogenetic measures of biodiversity: A comparison and critique. Biol. Conserv. 69: 33–39.

Kumar, S., K. Tamura, and M. Nei. 1993. MEGA: Molecular evolutionary genetics analysis, version 1.0. University Park, Pa.: Pennsylvania State University.

Lack, D. 1947. Darwin's finches. Cambridge: Cambridge University Press.

Lack, D. 1976. Island biogeography illustrated by the land birds of Jamaica. Berkeley: University of California Press.

Lopez, J.V., N. Yuhki, R. Masuda, W. Modi, and S.J. O'Brien. 1994. *Numt*, a recent transfer and tandem amplification of mitochondrial DNA to the nuclear genome of the domestic cat. J. Mol. Evol. 39: 174–190.

MacArthur, R.H. 1972. Geographical ecology: Patterns in the distribution of species. New York: Harper and Row.

MacArthur, R.H. and E.O. Wilson. 1967. The theory of island biogeography. Princeton, N.J.: Princeton University Press.

McMillan, O. and E. Bermingham. 1996. The patterning of mtDNA variation in Dall's porpoise. Mol. Ecol. 5, 47–61.

Maddison, W. and D. Maddison. 1992. MacClade, v. 3.0. Sunderland, Mass.: Sinauer Associates.

Marjoram, P. and P. Donnelly. 1994. Pairwise comparisons of mitochondrial DNA sequences in subdivided populations and implications for early human evolution. Genetics 135: 673–683.

Mayr, E. 1963. Animal species and evolution. Cambridge, Mass.: Belknap Press.

Moritz, C. 1994. Defining evolutionary significant units for conservation. Trends Ecol. Evol. 9: 373–375.

Page, R.D.M. 1990. Temporal congruence and cladistic analysis of biogeography and cospeciation. System. Zool. 39: 205–226.

Page, R.D.M. 1993. Genes, organisms, and areas: The problem of multiple lineages. System. Biol. 42: 77–84.

Peterson, A.T., P. Escalante, P., and A. Navarro, S. 1992. Genetic variation and differentiation in Mexican populations of common bush-tanagers and chestnut-capped brush-finch. Condor 94: 244–253.

Pregill, G.K. 1981. An appraisal of the vicariance hypothesis of Caribbean biogeography and its application to West Indian terrestrial vertebrates. Syst. Zool. 30: 147–155.

Quinn, T.W., Shields, G.F., and A.C. Wilson. 1991. Affinities of the Hawaiian goose based on two types of mitochondrial DNA data. Auk 108: 585–593.

Ricklefs, R.E. 1987. Community diversity: relative roles of local and regional processes. Science 235: 167–171.

Ricklefs, R.E. and G.W. Cox. 1972. Taxon cycles in the West Indian avifauna. Am. Nat. 106: 195–219.

Ricklefs, R.E. and G.W. Cox. 1978. Stage of taxon cycle, habitat distribution, and population density in the avifauna of the West Indies. Am. Nat. 112: 875–895.

Rogers, A.R. and H. Harpending. 1992. Population growth makes waves in the distribution of pairwise genetic differences. Mol. Biol. Evol. 9: 552–569.

Rohlf, F. James. 1993. NTSYS-pc: Numerical taxonomy and multivariate analysis system, version 1.80. Setauket, New York: Exeter Software.

Rosen, D.E. 1976. A vicariance model of Caribbean biogeography. System. Zool. 24: 431–464.

Rosen, D.E. 1978. Vicariant patterns and historical explanation in biogeography. System. Zool. 27: 159–188.

Rosen, D.E. 1985. Geological hierarchies and biogeographic congruence in the Caribbean. Ann. Missouri Bot. Gard. 72: 636–659.

Roughgarden, J. and S. Pacala. 1989. Taxon cycles among *Anolis* lizard populations: review of the evidence. In: D. Otte and J.A. Endler, eds. Speciation and its consequences, pp. 403–432. Sunderland, Mass.: Sinauer Associates.

Seutin, G., J. Brawn, R.E. Ricklefs, and E. Bermingham. 1993. Genetic divergence among populations of a tropical passerine, the streaked saltator (*Saltator albicollis*). Auk 110: 117–126.

Seutin, G., N.K. Klein, R.E. Ricklefs, and E. Bermingham. 1994. Historical biogeography of the bananaquit (*Coereba flaveola*) in the Carribean region: A mitochondrial DNA assessment Evolution 48: 1041–1061.

Shulman, M.J. and E. Bermingham. 1995. Early life histories, ocean currents, and the population genetics of Caribbean reef fishes. Evolution 49: 897–910.

Slatkin, M. and R.R. Hudson. 1991. Pairwise comparison of mitochondrial DNA sequences in stable and exponentially growing populations. Genetics 129: 555–562.

Slatkin, M. and W.P. Maddison. 1989. A cladistic measure of gene flow inferred from the phylogenies of alleles. Genetics 123: 603–613.

Slatkin, M. and W.P. Maddison. 1990. Detecting isolation by distance using phylogenies of genes. Genetics 126: 249–260.

Smith, M.F., W.K. Thomas, and J.L. Patton. 1992. Mitochondrial DNA-like sequence in the nuclear genome of an akodontine rodent. Mol. Biol. Evol. 9: 204–215.

Solow, A.R., J.M. Broadus, and N. Tonring. 1993. On the measurement of biological diversity. J. Environ. Econ. Management 24: 60–68.

Spitzer, K., V. Novotny, M. Tonner, and J. Leps. 1993. Habitat preferences, distribution and seasonality of the butterflies (Lepidoptera, Papilionoidea) in a montane tropical rain forest, Vietnam. J. Biogeogr. 20: 109–121.

Swofford, D.L. 1993. PAUP: Phylogenetic analysis using parsimony. Version 3.1. Champaign, Ill.: Illinois Natural History Survey.

Swofford, D.L., G.J. Olsen, P.J. Waddell, and D.M. Hillis. 1996. Phylogenetic inference. In: D.M. Hillis, C. Moritz, and B.K. Mable, eds. Molecular systematics pp. 407–414. Sunderland, Mass.: Sinauer Associates.

Takahata, N. and S.R. Palumbi. 1985. Extranuclear differentiation and gene flow in the finite island model. Genetics 109: 441–457.

Terborgh, J.W. and J. Faaborg. 1980. Saturation of bird communities in the West Indies. Am. Nat. 116: 178–195.

Vane-Wright, R.I., C.J. Humphries, and P.H. Williams. 1991. What to protect?—Systematics and the agony of choice. Biol. Conserv. 55: 235–254.

Weitzman, M.L. 1992. On diversity. Q. J. Econ., 107: 363–405.

Wetmore, A., R.F. Pasquier, and S.L. Olson. 1984. The birds of the Republic of Panamá. Part 4. Passeriformes: Hirundinidae (swallows) to Fringillidae (finches). Smithsonian Misc. Coll., Vol. 150, Part 4. Washington, D.C.: Smithsonian Institution Press.

Wiens, J.A. 1989. The ecology of bird communities. Vol. 1. Foundations and patterns. Cambridge, Mass.: Cambridge University Press.

Wilson, E.O. 1961. The nature of the taxon cycle in the Melanesian ant fauna. Am. Nat. 95: 169–193.

Woods, C.A. (ed.). 1989. Biogeography of the West Indies. Past, present, and future. Gainesville, Fla.: Sandhill Crane Press.

8

The Use of Mitochondrial DNA Control Region Sequencing in Conservation Genetics

PIERRE TABERLET

One of the main challenges in conservation biology is to identify potential conservation units (or evolutionarily significant units) (Ryder, 1986; Woodruff, 1989) using a molecular approach. Such units correspond to sets of populations that have a recent common history, and should be managed separately in order to preserve their biological diversity. Usually, management programs are founded on current taxonomic assignments. At the subspecific or specific levels many of these assignments were proposed a long time ago, and were based on only a small number of specimens. Such a taxonomy, based solely on morphological traits, can provide two kinds of errors in the recognition of potential conservation units (Avise, 1989): recognition of groups showing little evolutionary differentiation (Laerm et al., 1982), or lack of recognition of distant forms (Avise and Nelson, 1989). Therefore, where current morphologically based taxonomy remains uncertain, molecular tools should be advocated. At the intraspecific level, as different populations may be geographically structured, potential conservation units can be identified using a phylogeographic approach that combines biogeographic and phylogenetic data (Avise et al., 1987; Avise, 1992; Dizon et al., 1992).

The analysis of animal mitochondrial DNA (mtDNA) polymorphisms represents the most commonly used means for revealing phylogenetic relationships among closely related species, and among populations of the same species (Avise and Lansman, 1983; Wilson et al., 1985; Avise, 1986; Moritz et al., 1987; Harrison, 1989). With few exceptions, the mitochondrial genome consists of a circular, double-stranded DNA molecule about 15 000 to 20 000 base pairs (bp) long, containing 13 protein-coding genes, 22 tRNA genes, 2 rRNA genes, and a noncoding segment about 1000 bp long called the control region that initiates replication and transcription. The control region is also called the "A-T rich region" for invertebrates, or the "D-loop region" for vertebrates. Mitochondrial DNA is transmitted without recombination predominantly through maternal lineages (e.g., Dawid and Blackler, 1972; Hutchison et al., 1974; Giles et al., 1980). Thus the entire mtDNA molecule can be considered a single genetic unit with multiple alleles, which can be used to trace maternal phylogenies. Two facts explain why mtDNA polymorphism is now so widely used in molecular phylogenetics: (1) mtDNA represents the best-known part of the animal genome,

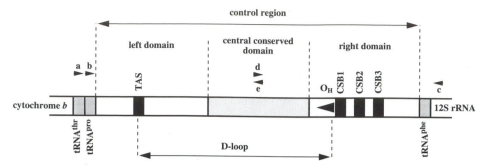

Figure 8-1 Schematic diagram of the mammalian mtDNA control region, showing the conserved central region and the more variable left and right domains. Orientation is 5′ and 3′ relative to the light strand. Locations of the primers which amplify the mtDNA control region in most vertebrate species are also indicated. The primers L15926, L16007, H00651 were designed by Kochler et al. (1989). L15926 and H00651 can be shortened by removing the restriction site on the 5′ part. The primers L16517 and H16498 were based on a sequence alignment proposed by Southern et al. (1988). (a) L15925: 5′-TCAAAGCTTACACCAGTCTTGTAAACC-3′. (b) L16007: 5′-CCCAAAGCTAAAATTCTAA-3′. (c) H00651: 5′-TAACTGCAGAAGGCTAGGACCAAACCT-3′. (d) L16517: 5′-CATCTGGTTCTTACTTCAGG-3′. (e) H16498: 5′-CCTGAAGTAAGAACCAGATG-3′.

and (2) it exhibits an evolutionary rate 5–10 times higher than single-copy nuclear genes (W.M. Brown et al., 1979, 1982); furthermore, the control region contains variable blocks that evolve about 4–5 times faster than the entire mtDNA molecule (Greenberg et al., 1983; Horai and Hayasaka, 1990; J.R. Brown et al., 1993). Thus, when studying closely related species or conspecific populations, the sequencing of this region can provide better resolution using less experimental effort.

In vertebrates, the control region can be divided into three domains (Figure 8-1): (1) the left domain (or 5′ end) containing one or more termination-associated sequences (TAS) where the synthesis of the nascent heavy strand pauses; (2) the central conserved region, C-G rich, which has been implicated in the regulation of the heavy strand replication; (3) the right domain (or 3′ end) that contains the site of initiation for heavy-strand replication (O_H), and two or three short conserved sequence blocks (CSBs) (Walberg and Clayton, 1981; G.G. Brown et al., 1986; Southern et al., 1988; Saccone et al., 1991). The displacement-loop (D-loop) represents the part of the control region spanning the site of initiation of heavy strand replication (O_H) to the termination-associated sequence (TAS), and corresponds to a three-stranded structure involving the two pre-existing strands plus the nascent heavy strand (Clayton, 1982, 1991). Despite their functional importance, the left and right domains evolve rapidly, both by length mutations and by base substitutions (Saccone et al., 1991), but due to long insertions/deletions the distribution of mutations in the control region can change among species. For example, when compared to the left domain the region surrounding CSB 1 is much more variable in shrews (Fumagalli et al., 1996) than in humans (Vigilant et al., 1989). A number of species exhibit tandem repeats, mainly in the TAS and CSB regions (see review for vertebrates in Hoelzel,

1993). Such repeats were found in all shrews species (Insectivora) examined so far (Fumagalli et al., 1996), and are quite common in Carnivora (Hoelzel, 1993).

Up to now, few studies involving a large number of control region sequences have been published. Obviously, the largest set of data concerns humans (e.g., W.M. Brown, 1983; Vigilant et al., 1989, 1991; Horai et al., 1993). In mammals, brown bears (Taberlet and Bouvet, 1994) and whales (C.S. Baker et al., 1993) have been studied extensively, and a few sequences are available for gorillas (Garner and Ryder, 1992). Some other papers deal with control region sequence variation in birds (Quinn, 1992; Edwards, 1993a,b; Wenink et al., 1993; A.J. Baker et al., 1994), in reptiles (Norman et al., 1994), and especially in fish (Meyer et al., 1990; Bernatchez et al., 1992; Fajen and Breden, 1992; Sturmbauer and Meyer, 1992, 1993; Bernatchez and Danzmann, 1993; J.R. Brown et al., 1993; Sang et al., 1994). There is no doubt that studies based on control region sequences will be carried out increasingly often in the short term in order to reveal either intraspecific phylogeographic structures or phylogenies between closely related species.

TECHNICAL DESCRIPTION

Sampling and DNA Extraction

The possibility of amplifying DNA via the polymerase chain reaction (PCR) (review in White et al., 1989; Erlich et al., 1991; Erlich and Arnheim, 1992) has revolutionized conservation genetics by allowing noninvasive sampling. Indeed, sampling represents one of the main difficulties when studying an endangered species throughout its entire geographic distribution. DNA extraction can be performed from fresh or alcohol preserved tissues using the classical proteinase K–phenol/chloroform–alcohol precipitation protocol (see, e.g., Kocher et al., 1989; Hillis et al., 1990), but such tissues are not always available. Recent advances in DNA extraction protocols allow the use of material that contains nanograms or even picograms of DNA. Hairs have been used as a source of DNA in humans (Higuchi et al., 1988), chimpanzees (Morin and Woodruff, 1992), gorillas (Garner and Ryder, 1992), and bears (Taberlet and Bouvet, 1992a,b; Taberlet et al., 1993). Plucked or shed feathers can be used for genetic analysis of birds (Taberlet and Bouvet, 1991), while bones and teeth represent another valuable source of DNA (Hagelberg and Sykes, 1989; Horai et al., 1989; Hänni et al., 1990). More recently, DNA has been amplified from bear droppings (Höss et al., 1992) and from fingernails (Kaneshige et al., 1992). Museum specimens can also provide DNA: mammal skins have been used (e.g., R.H. Thomas et al., 1989; W.K. Thomas et al., 1990; Wayne and Jenks, 1991), as well as bird feathers (Ellegren, 1991, 1994).

The possibility of using such a wide variety of tissues is connected with two DNA extraction procedures: (1) the "Chelex" protocol (Singer-Sam et al., 1989; Walsh et al., 1991), which is particularly suitable for hair samples, and (2) the "silica" protocol (Boom et al., 1990; Höss and Pääbo, 1993), which allows the efficient removal of PCR inhibitors from ancient tissues.

Primer Selection

When studying a species for which no sequence data for the areas flanking the control region are available, the first step is to try universal primers (Kocher et al., 1989). These primers, located on flanking genes (Figure 8-1), work for most vertebrate species and amplify the entire control region. However, if the DNA used as template is partially degraded, it may be necessary to amplify shorter fragments. A control region sequence alignment including distantly related mammalian species (Southern et al., 1988) allowed the design of two internal primers that work in almost all mammals (Figure 8-1) and in some fish. However, these primers do not work in birds, because the gene order around the control region is different (Desjardins and Morais, 1990). Figure 8-2 presents some known variations in genes flanking the control region. Table 8-1 gives references that

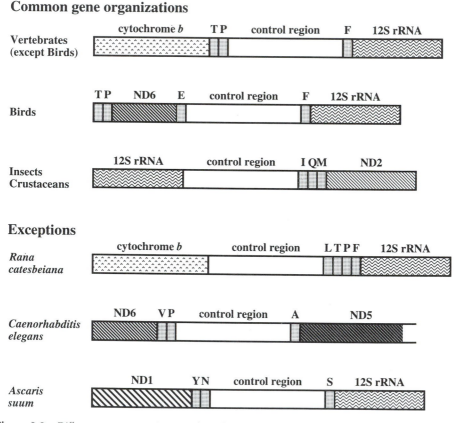

Figure 8-2 Different gene organizations found around the control region in animal mtDNA. The three commonly observed gene organizations are given, as well as three exceptions (Anderson et al., 1981; Clary and Wolstenholme, 1985; Roe et al., 1985; Desjardins and Morais, 1990; Okimoto et al., 1992, Yoneyama, 1987; Van Raay and Crease, 1994). Knowledge of the gene order around the control region is necessary to design new primers when no sequence data are available for the considered species.

Table 8-1 Selected papers where primer sequences directed to the mtDNA control region are given

Organism	References
Mammals	
Primate, *Macca fuscata*	Hayasaka et al. (1991)
Cetacea	Hoelzel et al. (1991)
Insectivora, *Sorex* spp.	Stewart and Baker (1994)
Chiroptera, *Nycticeius humeralis*	Wilkinson and Chapman (1991)
Rodentia, *Dipodomys panamintinus*	Thomas et al. (1990)
Birds	
Anseriformes, *Chen caerulescens*	Quinn (1992)
Charidriiformes, *Uria* spp.	Moum and Johansen (1992)
Charidriiformes	
Arenaria interpres, Calidris alpina	Wenink et al. (1994)
Calidris alpina	Wenink et al. (1993)
Passeriformes	
Pomatostomus temporalis	Edwards (1993b)
Sericornis spp.	Joseph and Moritz (1993)
Oscine passerines	Tarr (1995)
Reptiles	
Chelonids, *Chelonia mydas*	Normal et al. (1994)
Fish	
Acipenser transmontanus	Brown J.B. et al. (1993)
Cichlids	Meyer et al. (1990)
Poecilia reticulata	Fajen and Breden (1992)
Salvelinus fortinalis	Bernatchez and Danzmann (1993)
Anguilla japonica	Sang et al. (1994)
Insects	
Lepidoptera, *Jalemnus* spp.	Taylor et al. 1(993)

contain primers directed toward the control region for various organisms. These primers should work for related species.

If no primers have previously been used in related species, the alternative approach is to design new ones from the available sequences deposited in the GenBank or EMBL databases. In this case the process is (1) to find the gene order around the control region, and sequences for the species under analysis or another closely related species; (2) to align these sequences with the homologous genes of several more distantly related species in order to find the more conserved part of the two flanking regions; and (3) to design a new set of primers located on these conserved sequences according to the general rules of primer design (Erlich, 1989). The more highly conserved the sequence of the primers, the higher the probability they will work in related species.

DNA Amplification and Direct Sequencing of PCR Products

DNA amplification is carried out according to classical protocols (e.g., Erlich, 1989; Innis et al., 1990; Newton and Graham, 1994). To detect possible contamination during the extraction processes, or during the preparation of reagents for the polymerase chain reaction, each round of amplification must involve an extraction control (the template consists of a blank extract subjected to the same protocol as DNA extraction, but without tissue), and a PCR control (the template is replaced by water only). The PCR products can usually be directly sequenced following one of the alternative protocols already published (see review in Bevan et al., 1992).

Sequence Analysis

The sequences obtained are aligned, and then used either to estimate genetic distances or to infer phylogenetic trees. The description of the alternative methods of analysing the sequence data are far beyond the scope of this chapter (see review in Swofford and Olsen, 1990; Miyamoto and Cracraft, 1991; Penny et al., 1992).

However, as the pattern of nucleotide substitution in the control region is quite complicated, and the rate of substitution varies extensively among different sites, particular mathematical methods have recently been developed to estimate the number of nucleotide substitutions between such sequences (Tamura and Nei, 1993; Wakeley, 1993; Tamura, 1994). The method proposed by Tamura and Nei (1993) takes excess transitions, unequal nucleotide frequencies, and variation of substitution rates among different sites into account, and has been applied to human and chimpanzee data to find the age of the common ancestral mtDNA within each species.

CASE STUDIES

Conservation Units in the European Brown Bear

The brown bear (*Ursus arctos*) formerly occupied most of the European continent. Its present range has been dramatically reduced since the mid-1800s by habitat loss and excessive hunting in the past (Servheen, 1990). In western Europe, this species now exhibits a patchy geographic distribution (see Figure 8-3) with no possibility of reestablishment of a continuous habitat (Sørensen, 1990). The Pyrenean and the Trentino populations (localities A and E in Figure 8-3) may each number less than 10 individuals at present, and face the threat of extinction in the near future.

A research program was initiated in 1991 by the French Ministry of the Environment to manage the endangered population of brown bears in the Pyrenees. One of the goals of this program was to identify the potential conservation units at the European level, and then to determine which of the other nonendangered populations could provide a ready source of bears for reinforcement of the Pyrenean population.

The phylogeographic approach has been used to try to understand the recent history of the European brown bear, and to identify the potential conservation units (Taberlet and Bouvet, 1994). The phylogeography of this species was revealed by comparing the geographic distribution with the intraspecific phylogeny based on variation in mtDNA control region sequences. Sixty brown bears representative of the different European populations (Figure 8-3) were assayed for mtDNA polymorphisms. The DNA was extracted from hair samples preserved either dry or in 70% alcohol using the "Chelex" protocol (Walsh et al., 1991). The primers used for the PCR amplification were 5'-CTCCACTATCAGCACCCAAAG-3' (forward) and 5'-GGAGCGAGAAGAGGTACACGT-3' (reverse). Usually, a double-stranded DNA amplification (45 cycles), followed by single-stranded amplifications (30–35 cycles) and direct sequencing (using either the primers used for the amplification or other internal primers) were performed according to classical protocols (Kocher et al., 1989; Palumbi, 1991).

Sequencing of 272 base pairs of the control region enabled 16 different haplotypes to be distinguished. The most striking result is that these haplotypes exhibit a precise geographic pattern, with almost no local variation. The four Pyrenean bears have exactly the same sequence, as do the seven individuals from Dalarna (Sweden), the seven from Lapland (Sweden), and the 13 from Slovenia. This could be explained by very low dispersal habits of females (mtDNA is maternally inherited), and/or by the stochastic extinction of the primitive haplotypes in small populations.

Both the genetic distances and the phylogenetic tree (Figure 8-3) deduced from these sequences clearly indicate the presence of two main lineages in Europe that differ from each other by a mean pairwise genetic distance of 7.13% using the two-parameter model (Kimura, 1980). The eastern lineage is mainly represented by the large populations of Russia and Romania, while the western lineage includes several isolates that contain fewer than 100 individuals (Pyrenees, Cantabrian mountains, Trentino, Abruzzo; localities A, B, E, F respectively, in Figure 8-3). Based on the evolutionary rate of the homologous human sequence (Vigilant et al., 1989), these two main lineages have been estimated to have diverged about 0.85 million years ago. This crude estimate of an absolute date of divergence corresponds to the time since they had a common maternal ancestor, which must precede the time of the population divergence. It has to remain qualified due to uncertainties both in the genetic distance estimates and in the molecular clock calibration, which is based on humans. Independent of the phylogeographic study, a complete revision of the fossils of Eurasian bears also suggested (1) a split of *Ursus arctos* into two distinct lineages, one in Europe and one in Asia, and (2) a first occurrence of the brown bear in Europe about 1 million years ago. Therefore, the mtDNA divergence observed in the brown bear agrees with paleontological data, and the split between the eastern and western lineages, which occurred at an early stage in the history of the brown bear species, was probably due to geographic separation during the earlier Quaternary cold periods.

Furthermore, the western lineage itself appears to be organized into two clades (Figure 8-3). The first includes bears from Trentino, Abruzzo, Slovenia, Bosnia, Croatia, Greece, and Bulgaria (localities E to K in Figure 8-3), and may correspond to populations that have been isolated in a Balkan refugium during recent cold periods. The second clade is composed of bears from the Cantabrian and Pyrenean mountains, and from the south of Sweden and Norway (localities A to D in Figure 8-3). These last populations may have originated from an Iberian refugium, and would have been able to reach the south of Scandinavia via low-altitude land during the last postglacial warming. During the same period,

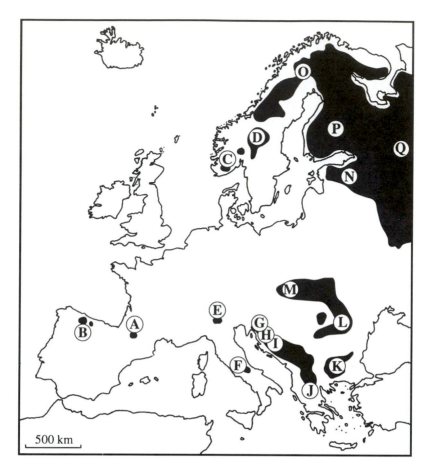

Figure 8-3 Phylogeography of the European brown bear. Geographic range and sample localities (top). For each locality, the number of individuals analysed and the collector name(s) (in parentheses) are: A, Pyranees, France (*n* = 4, J.-J. Camarra, C. Plisson, G. Caussimont, H. Laborde, M. Clemente, J. Herrero); B, Cantabrian mountains, Spain (*n* = 2, G. Palomero, A. Fernandez); C, Norway (*n* = 1; K. Elgmork); D, Dalarna, Sweden (*n* = 7, R. Franzén); E, Trentino, Italy (*n* = 1, F. Osti); F, Abruzzo, Italy (*n* = 4, G. Boscagli, H. Roth, L. Gentile); G, Slovenia (*n* = 13, M. Academic, D. Huber); H, Croatia (*n* = 2, D. Huber, A. Frkovic); I, Bosnia (*n* = 2, D. Huber); J, Greece (*n* = 1, D. Bousbouras); K, Bulgaria (*n* = 1, N. Spassov); L, Roumania (*n* = 4, L. Kalabér, N. Lestienne); M. Slovakia (*n* = 2, P. Hell); N, Estonia (*n* = 4, A. Allos, M. Koal); O, Lapland, Sweden (*n* = 7, R. Franzén); P, Finland (*n* = 2, E.S. Nyholm); Q, Karelia, Russia (*n* = 3, P. Danilov). All the samples were hairs, except for localities M and P (skin samples). Three individuals in locality A, and one in each of the localities C, E and J were assayed using hairs found in the field.

the ice sheets of the Alps would have constituted a barrier toward the north and the west for the bears of the Balkan refugium, as would the presence of bears of the eastern lineage toward the east. These two barriers could explain the fact that

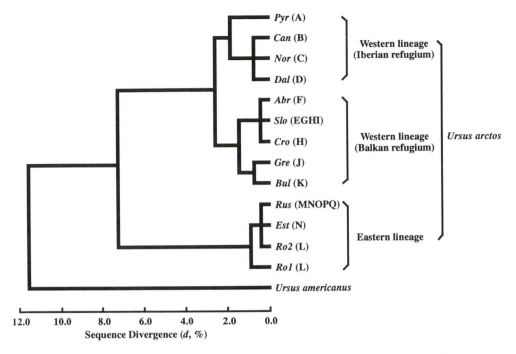

Figure 8-3 (*cont.*) UPGMA (unweighted pair group method with arithmetic mean) phenogram summarizing mtDNA relationships among the 13 haplotypes of European brown bear and 1 haplotype of American black bear (outgroup). The phenogram was deduced from a matrix of sequence divergences estimated using the two-parameter model (Kimura, 1980). The cophenetic correlation is 0.985.

bears of the Balkan refugium were unable to colonize the north of Europe: when !the ice sheets of the Alps melted, the north of Europe was already occupied by bears of another lineage. This scenario could be tested by studying historical and fossil samples of brown bears throughout the historical range.

The identification of the potential conservation units can be deduced from the phylogeographic data. The first possibility is to include all European populations of brown bears in the same conservation unit, but this choice must be rejected due to the clear genetic break that separates the eastern and the western lineages. The second possibility is to consider each of these two main lineages as two separate conservation units. However, in this case, the geographic structuring corresponding to the Iberian and the Balkan refugia is not taken into account. Since some transplantations can lead to the irretrievable loss of the historical genetic record (Avise, 1992), a third possibility is preferred: populations originating from the Balkan and the Iberian refugia (Figure 8-3) should be considered as two distinct potential conservation units.

This suggestion has clear consequences for management policy: (1) bears from Romania, Russia, and Lapland are not suitable to reinforce the threatened isolates of the western lineage, and further studies will be necessary in order to delimit the exact border between the eastern and the western lineages, both in central Europe and in Scandinavia; (2) in the short term, the management efforts must focus on the most endangered populations, i.e., those originating from the Iberian refugium.

In practice, the reinforcement of the Pyrenean population is difficult to achieve. The other populations originating from the Iberian refugium may not be able to act as a ready source of bears. The population from the Cantabrian mountains is endangered, and the population from the southern part of Scandinavia lives in an environment quite different from that found in the Pyrenees. As it becomes obvious that the Pyrenean population has no chance of independent survival, the two potential conservation units of the western lineage could pragmatically be pooled into a single operational conservation unit, and animals from the Balkan refugium (Slovenia–Croatia–Bosnia and Bulgaria) might ultimately be used for the reinforcement.

Worldwide Population Structure of Humpback Whales

To evaluate the influence of commercial hunting on genetic variation of humpback whales (*Megaptera novaeangliae*), a worldwide survey of mtDNA control region sequence variation was carried out by C.S. Baker et al. (1993). Indeed, before international protection was afforded in 1966, hunting exploitation lowered the number of living humpback whales from a maximum of 125 000 to less than 5000. This species is distributed in three oceanic populations: the North Atlantic, the North Pacific, and the southern Oceans. Within each population, in spite of significant migratory movements, humpback whales form relatively discrete subpopulations that are not separated by any obvious geographic barrier.

Tissue samples from 3 oceanic populations, 6 stocks, and 13 regional habitats were analysed. A 283 bp region located in the left domain of the mtDNA control region revealed 37 haplotypes and 33 variable sites among the 90 humpback whale individuals studied. The two most divergent haplotypes differ by 15 transitions and one transversion (i.e, 5.65%). The average divergence among the 37 haplotypes is 3.00%, about 10 fold-higher than the average divergence found by RFLP for the entire mtDNA.

A phylogenetic analysis of these 37 haplotypes revealed three major clades, each predominant in a different ocean. Within each major clade monophyletic groups were found that originated from an ocean that was different from the origin of the other haplotypes in the major clade. The simplest interpretation of this phylogeographic structure would be an ancient divergence in the three oceans, and then six migration events between oceans in order to explain the presence of these monophyletic groups within major clades.

To estimate the age of the three major humpback whale clades, C.S. Baker et al. (1993) tried to calibrate the divergence rate of the control region in whales. The divergence dates determined from the fossil record were compared to the pairwise sequence divergences between the three families of baleen whales. A sequence divergence of 0.7–1.0% per million years was found. Based on this estimate, the three major clades of humpback whales would have diverged about 5 million years ago. The current mtDNA polymorphism seems to indicate that humpback whales did not undergo a major loss of genetic diversity during the recent bottleneck due to the commercial whaling industry.

Population Structure of Green Turtles

Marine turtles occupy a vast geographic range. Their life histories involve first a pelagic oceanic phase, and then seasonal movements between feeding habitats and distant breeding sites on sandy beaches. Information on the spatial and temporal distribution of marine turtles can be deduced (1) from long-term mark–recapture experiments, and (2) from the use of appropriate genetic markers. Previous mtDNA restriction fragment length polymorphism (RFLP) analyses suggest that mtDNA represents a good genetic marker for inferring the genetic structure of marine turtles. However, the slower rate of mtDNA evolution in Testudines (Avise et al., 1992) limits the resolution of RFLP studies, and so Norman et al. (1994) proposed, using mtDNA control region sequencing, to recognize genetically and demographically discrete stocks of green turtles (*Chelonia mydas*) in the Indo-Pacific region.

As the universal primer for the control region (Kocher et al., 1989) resulted in poor amplification, a new set of primers located on the flanking cytochrome *b* and 12S rRNA gene sequences was designed to amplify the entire control region. A sequencing primer located within the "F" conserved sequence block (in the central conserved region) was used to obtain a control region sequence of about 200 bp in two turtle species. Two new primers were designed from these sequences, and allowed the characterization of a variable 400 bp segment in the green turtle and the subsequent design of two additional primers flanking this variable region. Alignment of sequences from 15 individuals representing 12 rookeries revealed 41 variable sites that define 14 distinct haplotypes (mean sequence divergence: 3.67%). This variable segment of the control region is evolving about eight times faster than the average rate deduced from RFLP studies on entire mtDNA. The distribution of seven variable restriction sites along the sequence alignment allows the identification of eight distinct haplotypes (as opposed to the 14 haplotypes identified by sequencing). The study was continued using RFLP analyses on PCR products (256 individuals assayed), and using denaturing gradient gel electrophoresis (DGGE) in order to reveal polymorphism in two restriction types (106 individuals assayed).

The results obtained clearly confirm the existence of a number of discrete breeding populations of green turtles within the Indo-Pacific region. This strong geographic structuring suggests that most major rookeries are genetically and demographically discrete, and that they should be managed separately.

LIMITATIONS AND MOST APPROPRIATE APPLICATIONS

The main advantage of mtDNA control region sequencing is that it extends the resolution of this molecule between 5 and 10 times when addressing intraspecific questions. The three case studies presented above clearly demonstrate the relevance of geographic variation in mtDNA control region sequences for revealing cryptic conservation units. Furthermore, the local polymorphism can also reveal past demographic events: for example, in a large population, a lack of variation in control region sequences indicates a recent bottleneck. Concerning endangered species, the possibility of use of a noninvasive sampling technique represents another major advantage of this PCR-based method.

However, mtDNA control region sequencing presents some drawbacks. First, DNA sequencing is an expensive and time-consuming technique, but the total cost of a study can be significantly reduced by combining DNA sequencing with other less expensive methods which can reveal nucleotide polymorphisms. The case study concerning the green turtle (Norman et al., 1994) represents a good example of such a combination. These other techniques can be PCR-RFLPs if polymorphic restriction sites occur; single-strand conformation polymorphism (SSCP) (e.g., Hayashi, 1991); or denaturing gradient get electrophoresis (DGGE) (e.g., Sheffield et al., 1990; Lessa, 1992).

Second, the results obtained by mtDNA control region sequencing stem from only one maternally inherited locus. If males disperse more than females, as it is often the case in mammals, the genetic structure observed may be less strongly correlated with the geographic distribution observed by screening nuclear genes which are transmitted by both sexes. Furthermore, mtDNA introgression can occur between conspecific populations or between closely related species (Ferris et al., 1983; Carr et al., 1986; Tegelström, 1987; Lehman et al., 1991). Such an introgression can hide the real phylogenetic relationships. For these two reasons, it would be desirable to confirm the results obtained by mtDNA control region sequencing by also studying nuclear loci.

Third, the molecular evolution of the mitochondrial control region is not well understood, and estimates of evolutionary rate are questionable. For example, the two main lineages of brown bear (see above) are estimated to have diverged about 0.85 million years ago. This estimate is very rough and should be qualified because (1) there are uncertainties in the genetic distance estimates, and (2) the molecular clock calibration used is based on humans. Actually, the problems of molecular clock calibration arise from large differences in mutation rates among nucleotide sites (Tamura and Nei, 1993). As a consequence, saturation is quickly reached, even at the intraspecific level (L. Fumagalli and P. Taberlet, unpublished data), and the estimation of a reliable genetic distance becomes difficult, compromising any temporal scale inference based on the molecular clock concept. Thus, attempts to estimate evolutionary rates based on sequences that are too distantly related will give a significant underestimate owing to saturation at nucleotide sites exhibiting high mutation rates. This bias could be an explanation of the very low evolutionary rate calculated by C.S. Baker et al. (1993) for humpback whales, which was based on the comparison of species which diverged more than 5 million years ago. Another way of calibrating the mtDNA control region molecular clock could consist of a comparison, among closely related individuals, of sequence variations between the control region and another region such the cytochrome *b* gene for which the molecular evolution is better understood (Irwin et al., 1991).

In conclusion, mtDNA control region sequencing can be a powerful method for investigating population structure at the intraspecific level when the resolution of other methods is not accurate enough to reveal a sufficient level of poly-morphism. The study of nuclear loci can be used as a complement to avoid the bias of using only one maternally inherited genetic marker.

REFERENCES

Anderson, S., A.T. Bankier, B.G. Barrell, M.H.L. DeBruijn, A.R. Coulson, J. Drouin, I.C. Eperon, D.P. Nierlich, B.A. Roe, F. Sanger, P.H. Schreier, A.J.H. Smith, R. Staden, and I.G. Young. 1981. Sequence and organization of the human mitochondrial genome. Nature 290: 457–465.

Avise, J.C. 1986. Mitochondrial DNA and the evolutionary genetics of higher animals. Phil. Trans. R. Soc. Lond. B 312: 325–342.

Avise, J.C. 1989. A role for molecular genetics in the recognition and conservation of endangered species. Trends Ecol. Evol. 4: 279–281.

Avise, J.C. 1992. Molecular population structure and the biogeographic history of a regional fauna: a case history with lessons for conservation biology. Oïkos 63: 62–76.

Avise, J.C. and R.A. Lansman. 1983. Polymorphism of mitochondrial DNA in populations of higher animals. In: M. Nei and R.K. Koehn, eds. Evolution of genes and proteins, pp. 147–164. Sunderland, Mass.: Sinauer.

Avise, J.C. and W.S. Nelson. 1989. Molecular genetic relationship of the extinct dusky seaside sparrow. Science 243: 646–648.

Avise, J.C., J. Arnold, R.M. Ball, E. Bermingham, T. Lamb, J.E. Neigel, C.A. Reeb, and N.C. Saunders, 1987. Intraspecific phylogeography: the mitochondrial DNA bridge between population genetics and systematics. Annu. Rev. Ecol. Syst. 18: 489–522.

Avise J.C., B.W. Bowen, T. Lamb, A.B. Meylan, and E. Bermingham. 1992. Mitochondrial DNA evolution at a turtle's pace: evidence for low genetic variability and reduced microevolutionary rate in the Testudines. Mol. Biol. Evol. 9: 457–473.

Baker C.S., A. Perry, J.L. Bannister, M.T. Weinrich, R.B. Anernethy, J. Calambokidis, J. Lien, R.H. Lambersten, J. Urba Ramirez, O. Vasquez, P.J. Clapham, A. Alling, S.J. O'Brien, and S.R. Palumbi. 1993. Abundant mitochondrial DNA variation and world-wide population structure in humpback whales. Proc. Natl. Acad. Sci. U.S.A. 90: 8239–8243.

Baker, A.J., T. Piersma, and L. Rosenmeier. 1994. Unraveling in the intraspecific phylo-geography of knots Calidris canutus: a progress report on the search of genetic markers. J. Ornithol. 135: 599–608.

Bernatchez, L. and R.G. Danzmann. 1993. Congruence in control-region sequence and restriction-site variation in mitochondrial DNA of brook charr (Salvelinus fontinalis Mitchill). Mol. Biol. Evol. 10: 1002–1014.

Bernatchez, L., R. Guyomard, and F. Bonhomme. 1992. DNA sequence variation of the mitochondrial control region among geographically and morphologically remote European brown trout Salmo trutta populations. Mol. Ecol. 1: 161–173.

Bevan, I.S., R. Rapley, and M.R. Walker. 1992. Sequencing of PCR-amplified DNA. PCR Methods Applic. 1: 222–228.

Boom, R., C.J.A. Sol, M.M.M. Salimans, C.L. Jansen, P.M.E. Wertheim-van Dillen, and J. Van Der Noordaa. 1990. Rapid and simple method for purification of nucleic acids. J. Clin. Microbiol. 28: 495–503.

Brown, W.M. 1983. Evolution of animal mitochondrial DNA. In: M. Nei and R.K. Koehn, eds. Evolution of genes and proteins, pp. 62–88. Sunderland, Mass.: Sinauer.

Brown, W.M., M.J. George, and A.C. Wilson. 1979. Rapid evolution of animal mito-chondrial DNA. Proc. Natl. Acad. Sci. U.S.A. 76: 1967–1971.

Brown, W.M., E.M. Prager, A. Wang, and A.C. Wilson. 1982. Mitochondrial DNA sequences of primates: tempo and mode of evolution. J. Mol. Evol. 18: 225–239.

Brown, G.G., G. Gadaleta, G. Pepe, C. Saccone, and E. Sbisa. 1986. Structural conservation and variation in the D loop-containing region of vertebrate mitochondrial DNA. J. Mol. Biol. 192: 503–511.

Brown, J.R., A.T. Beckenbach, and M.J. Smith. 1993. Intraspecific DNA sequence variation of the mitochondrial control region of white sturgeon (*Acipenser transmontanus*). Mol. Biol. Evol. 10: 326–341.

Carr, S.M., S.W. Ballinger, J.N. Derr, L.H. Blankenship, and J.W. Bickman. 1986. Mitochondrial DNA analysis of hybridization between sympatric white-tailed deer and mule deer in west Texas. Proc. Natl. Acad. Sci. U.S.A. 83: 9576–9580.

Clary, D.O. and D.R. Wolstenholme. 1985. The mitochondrial DNA molecule of *Drosphila yakuda*: nucleotide sequence, gene organization, and genetic code. J. Mol. Evol. 22: 252–271.

Clayton, D.A. 1982. Replication of animal mitochondrial DNA. Cell 28: 693–705.

Clayton, D.A. 1991. Replication and transcription of vertebrate mitochondrial DNA. Annu. Rev. Cell Biol. 7: 453–478.

Dawid, I.B. and A.W. Blacker. 1972. Maternal and cytoplasmic inheritance of mtDNA in *Xenopus*. Dev. Biol. 29: 152–161.

Desjardins, P. and R. Morais. 1990. Sequence and gene organization of the chicken mitochondrial genome. A novel gene order in higher vertebrates. J. Mol. Biol. 212: 599–634.

Dizon, A.E., C. Lockyer, W.F. Perrin, D.P. Demaster, and J. Sisson. 1992. Rethinking the stock concept: a phylogeographic approach. Conserv. Biol. 6: 24–36.

Edwards, S.V. 1993a. Long-distance gene flow in cooperative breeder detected in genealogies of mitochondrial DNA sequences. Proc. R. Soc. Lond. B 252: 177–185.

Edwards, S.V. 1993b. Mitochondrial gene genealogy and gene flow among island and mainland populations of a sedentary songbird, the grey-crowned babbler (*Pomatstomus temporalis*). Evolution 47: 1118–1137.

Ellegren, H. 1991. DNA typing of museum birds. Nature 354: 113.

Ellegren, H. 1994. Genomic DNA from museum bird feathers. In: B. Herrmann and S. Hummel, eds. Ancient DNA, pp. 211–217. New York: Springer-Verlag.

Erlich, H.A. 1989. PCR technology. New York: Stockton Press.

Erlich, H.A. and N. Arnheim. 1992. Genetic analysis using the polymerase chain reaction. Annu. Rev. Genet. 26: 479–506.

Erlich, H.A., D. Gelfand, and J.J. Sninsky. 1991. Recent advances in the polymerase chain reaction. Science 252: 1643–1651.

Fajen, A. and F. Breden. 1992. Mitochondrial DNA sequence variation among natural populations of the trinidad guppy, *Poecilia reticulata*. Evolution 46: 1457–1465.

Ferris, S.D., R.D. Sage, C.-M. Huang, J.T. Nielsen, U.W. Rittel, and A.C. Wilson. 1983. Flow of mitochondrial DNA across a species boundary. Proc. Natl. Acad. Sci. U.S.A. 80: 2290–2294.

Fumagalli, L., P. Taberlet, L. Favre, and J. Hausser. 1996. Origin and evolution of homologous repeated sequences in the mitochondrial DNA control region of shrews. Mol. Biol. Evol. 13: 31–46.

Garner, K.J. and O.A. Ryder. 1992. Some applications of PCR to studies in wildlife genetics. In: H.D.M. Moore, W.V. Holt and G.M. Mace, eds. Biotechnology and the conservation of genetic diversity, pp. 167–181. New York: Oxford University Press.

Giles, R.E., H. Blanc, H.M. Cann, and D.C. Wallace. 1980. Maternal inheritance of human mitochondrial DNA. Proc. Natl. Acad. Sci. U.S.A. 77: 6715–6719.

Greenberg, B.D., J.E. Newbold, and A. Sugino. 1983. Intraspecific nucleotide sequence variability surrounding the origin of replication in human mitochondrial DNA. Gene 21: 33–49.

Hagelberg, E. and B. Sykes. 1989. Ancient bone DNA amplified. Nature 342–485.

Hänni, C., V. Laudet, M. Sakka, A. Bègue, and D. Stéhlein. 1990. Amplification de fragments d'ADN mitochondrial à partir de dents et d'os humains anciens. C.R. Acad. Sci. (III) 310: 365–370.

Harrison, R.G. 1989. Animal mitochondrial DNA as a genetic marker in population and evolutionary biology. Trends Ecol. Evol. 4: 6–11.

Hayasaka, K., T. Ishida, and S. Horai. 1991. Heteroplasmy and polymorphism in the major noncoding region of mitochondrial DNA in Japanese monkeys: association with tandemly repeated sequences. Mol. Biol. Evol. 8: 399–415.

Hayashi, K. 1991. PCR-SSCP: a simple and sensitive method for detection of mutations in the genomic DNA. PCR Methods Applic. 1: 34–38.

Higuchi, R., C.H. VonBeroldingen, G.F. Snesbaugh, and H.A. Erlich. 1988. DNA typing from single hairs. Nature 332: 543–546.

Hillis, D.M., A. Larson, S.K. Davis, and E.A. Zimmer. 1990. Nucleic acids III: sequencing. In: D.M. Hillis and C. Moritz, eds. Molecular systematics, pp. 318–370. Sunderland, Mass.: Sinauer.

Hoelzel, A.R. 1993. Evolution by DNA turnover in the control region of vertebrate mitochondrial DNA. Curr. Biol. 3: 891–895.

Hoelzel, Z.R., J.M. Hancock, and G.A. Dover. 1991. Evolution of the cetacean mitochondrial D-loop region. Mol. Biol. Evol. 8: 475–493.

Horai, S. and K. Hayasaka. 1990. Intraspecific nucleotide sequence differences in the major noncoding region of human mitochondrial DNA. Am. J. Hum. Genet. 46: 828–842.

Horai, S., K. Hayasaka, K. Murayama, N. Wate, H. Koite, and N. Nakai. 1989. DNA amplification from ancient human skeletal remains and their sequence analysis. Proc. Jpn. Acad. B 65: 229–233.

Horai, S., R. Kondo, Y. Nakagawa-Hattori, S. Hayashi, S. Sonoda, and K. Tajima. 1993. Peopling of the Americas, founded by four major lineages of mitochondrial DNA. Mol. Biol. Evol. 10: 23–47.

Höss, M. and S. Pääbo. 1993. DNA extraction from Pleistocene bones by a silica-based purification method. Nucleic Acids Res. 21: 3913–3914.

Höss, M., M. Kohn, S. Pääbo, F. Knauer, and W. Schröder. 1992. Excrement analysis by PCR. Nature 359: 199.

Hutchinson, C.A.I., J.E. Newbold, and S.S.E. Potter M.H. 1974. Maternal inheritance of mammalian mitochondrial DNA. Nature 251: 536–538.

Innis, M.A., D.H. Gelfand, J.J. Sninsky, and T.J. White. 1990. PCR protocols. San Diego: Academic Press.

Irwin, D.M., T.D. Kocher, and A.C. Wilson. 1991. Evolution of the cytochrome *b* gene of mammals. J. Mol. Evol. 32: 128–144.

Joseph L. and C. Moritz. 1993. Phylogeny and historical aspects of the ecology of eastern Australian scrubwrens *Sericornis* spp.—evidence from mitochondrial DNA. Mol. Ecol. 2: 161–170.

Kaneshige, T., K. Takagi, S. Nakamura, T. Hirasawa, M. Sada, and K. Uchida. 1992. Genetic analysis using fingernail DNA. Nucleic Acids Res. 20: 5489–5490.

Kimura, M. 1980. A simple method for estimating evolutionary rate of base substitution through comparative studies of nucleotide sequences. J. Mol. Evol. 16: 111–120.

Kocher, T.D., W.K. Thomas, A. Meyer, S.V. Edwards, S. Pääbo, F.X. Villablanca, and A.C. Wilson. 1989. Dynamics of mitochondrial DNA evolution in animals: amplification and sequencing with conserved primers. Proc. Natl. Acad. Sci. U.S.A. 86: 6196–6200.

Laerm, J., J.S. Patton, J.C. Avise, and R.A. Lansman. 1982. Genetic determination of the status of an endangered species of pocket gopher in Georgia. J. Wildlife Management 46: 513–518.

Lehman, N., A. Eisenhawer, K. Hansen, L.D. Mech, R.O. Peterson, P.J.P. Gogan, and R.K. Wayne. 1991. Introgression of coyote mitochondrial DNA into sympatric North American gray wolf populations. Evolution 45: 104–119.

Lessa, E.P. 1992. Rapid surveying of DNA sequence variation in natural populations. Mol. Biol. Evol. 9: 323–330.

Meyer, A., T.D. Kocher, P. Basasibwaki, and A. C. Wilson. 1990. Monophylectic origin of Lake Victoria cichlid fishes suggested by mitochondrial DNA sequences. Nature 347: 550–553.

Miyamoto, M.M. and J. Cracraft, eds. 1991. Phylogenetic analysis of DNA sequences. New York: Oxford University Press.

Morin, P.A. and D.S. Woodruff. 1992. Paternity exclusion using multiple hypervariable microsatellite loci amplified from nuclear DNA of gair cells, In: R.D. Martin, A.F. Dixson and E.J. Wickings, eds. Paternity in primates: Genetic tests and theories, pp. 63–81. Basel: Karger.

Moritz, C., T.E. Dowling, and W.M. Brown. 1987. Evolution of animal mitochondrial DNA: relevance for population biology and systematics. Annu Rev. Ecol. Syst. 18: 269–292.

Moum, T. and S. Johansen. 1992. The mitochondrial NADH dehydrogenase subunit 6 (ND6) gene in murres: relevance to phylogeneric and population studies among birds. Genome 35: 903–906.

Newton, C.R. and A. Graham. 1994. PCR. Oxford, U.K.: BIOS Scientic Publishers.

Norman, J.A., C. Moritz, and C.J. Limpus. 1994. Mitochondrial DNA control region polymorphisms: genetic markers for ecological studies of marine turtles. Mol. Ecol. 3: 362–373.

Okimoto, R., J.L. Macfarlane, D.O. Clary, and D.R. Wolstenholme. 1992. The mito-chondrial genomes of two nematodes, *Caenorhabditis elegans* and *Ascaris suum*. Genetics 130: 471–498.

Palumbi, S.R. 1991. The simple fool's guide to PCR. Honolulu. University of Hawaï, Department of Zoology.

Penny, D., M.D. Hendy, and M.A. Steel. 1992. Progress with methods for constructing evolutionary trees. Trends Ecol. Evol. 7: 73–79.

Quinn, T.W. 1992. The genetic legacy of mother goose (*Chen caerulescens caerulescens*) maternal lineages. Mol. Ecol. 1: 105–117.

Roe, B.A., D.-P. Ma, R.K. Wilson, and J.F.-H. Wong. 1985. The complete nucleo-tide sequence of the *Xenopus laevis* mitochondrial genome. J. Biol. Chem. 260: 9759–9774.

Ryder, O.A. 1986. Species conservation and systematics: the dilemma of subspecies. Trends Ecol. Evol. 1: 9–10.

Saccone, C., G. Pesole, and E. Sbisa. 1991. The main regulatory region of mammalian mitochondrial DNA: structure–function model and evolutionary pattern J. Mol. Evol. 33: 83–91.

Sang, T.-K., H.-Y. Chang, C.-T. Chen, and C.-F. Hui. 1994. Population structure of the Japanese eel, *Anguilla japonica*. Mol. Biol. Evol. 11: 250–260.

Servheen, C. 1990. The satus and conservation of the bears of the world. Int. Conf. Bear Res. Management Monogr. Series 2: 1–32.

Sheffield, V.C., D.R. Cox, and R.M. Myers. 1990. Identifying DNA polymorphisms by denaturing gradient gel electrophoresis. In: M.A. Innis, D.H. Gelfand, J.J. Sninski and T.J. White, eds. PCR protocols, a guide to methods and applications, pp. 206–218. San Diego: Academic Press.

Singer-Sam, J., R.L. Tanguay, and A.D. Riggs. 1989. Use of Chelex to improve the PCR signal from a small number of cells. Amplifications 3: 11.

Sørensen, O.J. 1990. The brown bear in Europe in the mid 1980's. Aquilo Ser. Zool. 27: 3–16.

Southern, S.O., P.J. Southern, and A.E. Dizon. 1988. Molecular characterization of a cloned dolphin mitochondrial genome. J. Mol. Evol. 28: 32–42.

Stewart, D.T. and A.J. Baker. 1994. Pattern of sequence variation in the mitochondrial D-loop region of shrews. Mol. Biol. Evol. 11: 9–21.

Sturmbauer, C. and A. Meyer. 1992. Genetic divergence, speciation and morphological stasis in a lineage of African cichlid fishes. Nature 358: 578–581.

Sturmbauer, C. and A. Meyer. 1993. Mitochondrial phylogeny of the endemic mouth-brooding lineages of cichlid fishes from Lake Tanganyika in Eastern Africa. Mol. Biol. Evol. 10: 751–768.

Swofford, D.L. and G.J. Olsen. 1990. Phylogeny reconstruction. In: D.M. Hillis and C. Moritz, eds. Molecular systematics, pp. 411–501. Sunderland, Mass.: Sinauer.

Taberlet, P. and J. Bouvet. 1991. A single plucked feather as a source of DNA for bird genetic studies. Auk 108: 959–960.

Taberlet, P. and J. Bouvet. 1992a. Bear conservation genetics. Nature 358: 197.

Taberlet, P. and J. Bouvet. 1992b. Génétique de l'Ours brun des Pyrénées (Ursus arctos): premiers résultats. C.R. Acad. Sci. (III) 314: 15–21.

Taberlet, P. and J. Bouvet. 1994. Mitochondrial DNA polymorphism, phylogeography, and conservation genetics of the brown bear (Ursus arctos) in Europe. Proc. R. Soc. Lond. B 255: 195–200.

Taberlet, P., H. Mattock, C. Dubois-Paganon, and J. Bouvet. 1993. Sexing free-ranging brown bears Ursus arctos using hairs found in the field. Mol. Ecol. 2: 399–403.

Tamura, K. 1994. Model selection in the estimation of the number of nucleotide substitutions. Mol. Biol. Evol. 11: 154–157.

Tamura, K. and M. Nei. 1993. Estimation of the number of nucleotide substitutions in the control region of mitochondrial DNA in humans and chimpanzees. Mol. Biol. Evol. 10: 512–526.

Tarr, C.L. 1995. Primers for amplification and determination of mitochondrial control-region sequences in oscine passerines. Mol. Ecol. 4: 527–529.

Taylor, M.F.J., S.W. McKechnie, N. Pierce, and M. Kreitman. 1993. The lepidoteran mitochondrial control region: structure and evolution. Mol. Biol. Evol. 10: 1259–1272.

Tegelström, H. 1987. Transfer of mitochondrial DNA from the northern red-backed vole (Clethrionomys rutilus) to the bank vole (C. glareolus). J. Mol. Evol. 24: 218–227.

Thomas, R.H., W. Schaffner, A.C. Wilson, and S. Pääbo. 1989. DNA phylogeny of the extinct marsupial wolf. Nature 340: 465–467.

Thomas, W.K., S. Pääbo, F.X. Villablanca, and A.C. Wilson. 1990. Spatial and temporal continuity of kangaroo rat populations shown by sequencing mitochondrial DNA from museum specimens. J. Mol. Evol. 31: 101–112.

Van Raay, T.J. and T.J. Crease. 1994. Partial mitochondrial DNA sequence of the crustacean Daphnia pulex. Curr. Genet. 25: 66–72.

Vigilant, L., R. Pennington, H. Harpending, T.D. Kocher, and A.C. Wilson. 1989. Mitochondrial DNA sequences in single hairs from a sourthern African population. Proc. Natl. Acad. Sci. U.S.A. 86: 9350–9354.

Vigilant, L., M. Stoneking, H. Harpending, K. Hawkes, and A.C. Wilson. 1991. African populations and the evolution of human mitochondrial DNA. Science 253: 1503–1507.

Wakeley. 1993. Substitution rate variation among sites in hypervariable region 1 of human mitochondrial DNA. J. Mol. Evol. 37: 613–623.

Walberg, M.W. and D.A. Clayton. 1981. Sequence and properties of the human KB cell and mouse L cell D-loop regions of mitochondrial DNA. Nucleic Acids Res. 9: 5411–5421.

Walsh, P.S., D.A. Metzger, and R. Higuchi. 1991. Chelex 100 as a medium for simple extraction of DNA for PCR-based typing forensic material. BioTechniques 10: 506–513.

Wayne, R.K. and S.M. Jenks. 1991. Mitochondrial DNA analysis implying extensive hybridization of the endangered red wolf *Canis rufus*. Nature 351: 565–568.

Wenink, P.W., A.J. Baker, and M.G.J. Tilanus. 1993. Hypervariable-control-region sequences reveal global population structuring in a long-distance migrant shorebird, the dunlin (*Calidris alpina*). Proc. Natl. Acad. Sci. U.S.A. 90: 94–98.

Wenink, P.W., A.J. Baker, and M.G.J. Tilanus. 1994. Mitochondrial control-region sequences in two shorebird species, the turnstone and the dunlin, and their utility in population genetic studies. Mol. Biol. Evol. 11: 22–31.

White, T.J., N. Arnheim, and H.A. Erlich. 1989. The polymerase chain reaction. Trends Genet. 5: 185–189.

Wilkinson, G.S. and A.M. Chapman. 1991. Length and sequence variation in evening bat D-loop mtDNA. Genetics 128: 607–617.

Wilson, A.C., R.L. Cann, S.M. Carr, M. George, U.B. Gyllensten, K.M. Helmbychowski, R.G. Higuchi, S.R. Palumbi, E.M. Prager, R.D. Sage, and M. Stoneking. 1985. Mitochondrial DNA and two perspectives on evolutionary genetics. Biol. J. Linn. Soc. 26: 375–400.

Woodruff, D.S. 1989. The problems of conserving genes and species. In: D. Western and M.C. Pearl, eds. Conservation for the twenty-first century, pp. 76–88. New York: Oxford University Press.

Yoneyama, Y. 1987. The nucleotide sequence of the heavy and light strand replication origins of the *Rana catesbeiana* mitochondrial genome. Nippon Ika Daigaku Zasshi 54: 429–440.

9

Chloroplast DNA Sequencing to Resolve Plant Phylogenies Between Closely Related Taxa

LUDOVIC GIELLY AND PIERRE TABERLET

When morphological data do not allow clear identification of endangered plant taxa, the first step consists of defining the potential conservation units by establishing a molecular phylogeny among the relevant species, subspecies, or populations. Analysis of chloroplast DNA (cpDNA) sequence variations represents one of the most powerful tools used by scientists to infer phylogenetic relationships of plants (Palmer, 1987; Palmer et al., 1988; Clegg and Zurawski, 1992; Clegg et al., 1991). Within seed plants, the *rbc*L gene, which encodes the large subunit of ribulose-1,5-bisphosphate carboxylase/oxygenase (RUBISCO), has been sequenced in all major taxonomic groups. The resulting database greatly aided studies of plant phylogenies (Palmer et al., 1988; Clegg et al., 1991; Chase et al., 1993).

Based on *rbc*L sequences, phylogenies were successfully obtained at the family level (e.g., Zurawski et al., 1984; Soltis et al., 1990; Wilson et al., 1990; Jansen et al., 1992; Bousquet et al., 1992b; Michaels et al., 1993; Morgan and Soltis, 1993), and also at higher taxonomic levels (e.g., Bousquet et al., 1992a; Gaut et al., 1992; Chase et al., 1993 and references therein). However, in the Zingiberales, the utility of *rbc*L is limited to the interordinal or intrafamilial level (Smith et al., 1993). Although some phylogenies were obtained at lower taxonomic (inter- and intrageneric) levels (Conti et al., 1993; Gadek and Quinn, 1993; Kron and Chase, 1993; Price and Palmer, 1993; Soltis et al., 1993; Xiang et al., 1993), the *rbc*L gene reaches, in some instances, its lower limits of resolution at these taxonomic levels (Doebley et al., 1990; Gaut et al., 1992; Kim et al., 1992; Soltis et al., 1993; Xiang et al., 1993). However, numerous cases of intraspecific variation, revealed mainly by restriction fragment length polymorphism (RFLP), have already been reported (reviewed by Soltis et al., 1992).

Noncoding regions of cpDNA tend to evolve more rapidly than coding regions owing to the accumulation of insertions/deletions at a rate at least equal to nucleotide substitutions (Curtis and Clegg, 1984; Wolfe et al., 1987; Zurawski and Clegg, 1987; Clegg et al., 1991). Analysis of such regions can extend the resolution

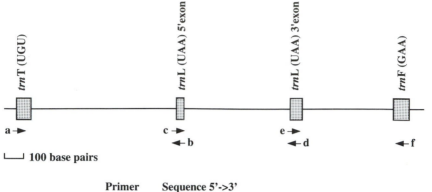

Primer	Sequence 5'->3'
a	CATTACAAATGCGATGCTCT
b	TCTACCGATTTCGCCATATC
c	CGAAATCGGTAGACGCTACG
d	GGGGATAGAGGGACTTGAAC
e	GGTTCAAGTCCCTCTATCCC
f	ATTTGAACTGGTGACACGAG

Figure 9-1 Positions, directions, and sequences of universal primers used to amplify three noncoding regions of cpDNA. Arrows point toward the 3' ends of the primers. The length of the noncoding regions corresponds to the tobacco sequence (from Taberlet et al., 1991).

Table 9-1 Species on Which Universal Primers for Three Noncoding Regions of cpDNA Were Tested

Branch	Family	Species	Primers[a]			
Algae	Laminariaceae	*Laminaria digitata*		c & d		
	Codiaceae	*Codium tomentosum*		c & d		
Bryophytae	Amblystegiaceae	*Acrocladium cuspidatum*	a & b	c & d	e & f	a & f
	Lunulariaceae	*Lunularia cruciata*	a & b	c & d	e & f	
Pteridophytae	Thelypteridaceae	*Thelypteris palustris*	a & b	c & d		
	Equisetaceae	*Equisetum arvense*	a & b	c & d	e & f	
Gymnosperms	Gingkoaceae	*Gingko biloba*	a & b	c & d	e & f	
	Abietaceae	*Pinus nigra*	a & b	c & d	e & f	a & f
Angiosperms	Magnoliaceae	*Magnolia* sp.	a & b	c & d		
	Ranunculaceae	*Aconitum* sp.		c & d	e & f	
	Salicaceae	*Salix babylonica*	a & b	c & d	e & f	
	Caryophyllaceae	*Saponaria officinalis*	a & b	c & d		
	Rosaceae	*Rosa canina*		c & d		
	Fabaceae	*Robinia pseudacacia*	a & b	c & d	e & f	a & f
	Compositae	*Bellis perennis*	a & b	c & d		
	Cyperaceae	*Carex elata*	a & b	c & d	e & f	
	Poaceae	*Phalaris arundinacea*	a & b	c & d		

[a] The column "primers" indicates the primer sets for which amplifications were attempted and successfully obtained. The abbreviations of the different primers are as in Figure 9-1.

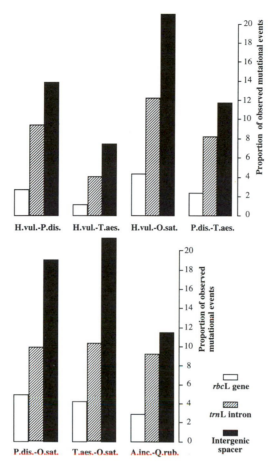

Figure 9-2 Comparisons for seven pairs of species of the proportion of observed mutational events in *rbc*L, the *trn*L intron and the intergenic spacer between the *trn*L (UAA) 3′ exon and the *trn*F (GAA), showing higher sequence divergences of noncoding regions of cpDNA than those of *rbc*L. The proportion of observed mutational events was estimated following O'Donnell (1992): proportion of mutational events = $[(TS + TV + ID)/L] \times 100$, where TS = number of observed transitions, TV = number of observed transversions, ID = number of observed insertions/deletions, L = TS + TV + ID + number of sites showing the same nucleotide). Species abbreviations: A. inc., *Alnus incana*; O. sat., *Oryza sativa*; H. vul., *Hordeum vulgare*; P. dis., *Puccinellia distans*; Q. rub., *Quercus rubra*; T. aes., *Triticum aestivum* (from Gielly and Taberlet, 1994b).

of the molecule, and permit one to address plant conservation genetics by assessing phylogenetic relationships at lower taxonomic levels.

Accordingly, we designed six universal primers for the amplification of three noncoding regions of cpDNA via the polymerase chain reaction (Taberlet et al., 1991). These zones, located in the large single-copy region, are: (1) an intergenic spacer between *trn*T (UGU) and the *trn*L (UAA) 5′ exon; (2) the *trn*L (UAA) intron; and (3) another intergenic spacer between the *trn*L (UAA) 3′ exon and *trn*F (GAA) (Figure 9-1).

To find out whether these primers were universal, we attempted to amplify DNA from various plant species representing all the major plant lineages. The results are reported in Table 9-1. The primers for the *trn*L (UAA) intron work for all the species tested, and PCR products of the two spacers were successfully obtained for each attempt. It was also possible to amplify the whole region (*trn*T–*trn*F) for three species, suggesting that the 1.5 kb long fragment obtained could be used for RFLP analysis using 4 bp restriction endonucleases (Taberlet et al., 1991). It appears that the *trn*L (UAA) intron and the intergenic spacer between *trn*L (UAA) and *trn*F (GAA) evolve more than three time faster than the *rbc*L gene (Figure 9-2).

TECHNICAL DESCRIPTION

Extraction. Total DNA extractions can be carried out using a modified version of the cationic hexadecyl trimethyl ammonium bromide (CTAB) protocol as described by Doyle and Doyle (1990), modified for the polymerase chain reaction. The isopropanol precipitation is replaced by a centrifugal dialysis (e.g., Microcon 100 000 MWCO low-binding filter units, Amicon, Beverly, Mass.) (Taberlet et al., 1991). To prevent possible contamination from foreign DNA, it is advisable to perform the whole procedure in Eppendorf tubes. A procedure using CTAB also allows extraction of DNA from herbarium specimens (Rogers and Bendich, 1985).

Amplification and Sequencing. Double-stranded DNA amplifications are performed via PCR according to classical protocols (e.g., Erlich, 1989; Innis et al., 1990; Newton and Graham, 1994). Extreme care must also be taken concerning the amplification procedure in order to avoid contamination of the PCR mixture (Kwok, 1990). The PCR products can usually be directly sequenced following one of the alternative protocols that have already been published (see review in Bevan et al., 1992).

Data analysis. The high frequency of length mutations in cpDNA noncoding regions permits only a parsimony approach for phylogeny assessments, since no reliable genetic distance can be estimated from both nucleotide substitutions and insertions/deletions. The different mutational events (substitutions as well as insertions/deletions) are coded in a matrix of unordered multistate characters; the nucleotide stretch corresponding to one (or two overlapping) insertion/deletion(s) being considered as a single site (Gielly and Taberlet, 1994b).

CASE STUDIES

Phylogeny of Section *Megalanthe* Gaudin (Genus *Gentiana* L.; Gentianaceae)

The genus *Gentiana* L., with 361 species distributed into 15 sections, is the largest of the family Gentianaceae (Ho and Liu, 1990). It exhibits a subcosmopolitan distribution, ranging from Europe to temperate zones of Asia and North America for most species (some species have tropical distributions). The major center of distribution is located in Asia, with 14 sections and 312 species (299 endemics) represented, corresponding to 93.3% and 86.4% of the world total, respectively. The main area of origin for the genus is thought to be represented by the mountains of southwestern China, in which the highest species concentration is found (190 species, 98 endemics, 11 sections). A second important center of distribution for the genus occurs in Europe, with 27 species (17 endemics) belonging to 8 sections. One is strictly endemic (section *Megalanthe* Gaudin) and another is almost endemic (section *Gentiana*) (Ho and Liu, 1990). Section *Megalanthe* was first considered at the morphological level as a very homogeneous group. This morphological similarity within the section led Bonnier (1935) to consider only a single collective species, *G. acaulis* L., encompassing three subspecies. More recently, the section was divided into seven species with similar morphologies, but it was suggested that the rank of subspecies might be more appropriate (Tutin et al., 1972). Table 9-2 summarizes the taxa included in this study, as well as their origin. *Gentiana ligustica* and *G. dinarica* (Italy) were sampled in the herbarium of Neuchâtel University (Switzerland). *Gentianella ciliata* was used as an outgroup.

In the multiple alignment of the **trmL** (UAA) intron for the eight taxa of Gentianaceae (all the species of section *Megalanthe* and the outgroup, data not shown), two deletions of several bases occur at the same position in *G. acaulis*, *G. angustifolia*, *G. ligustica* and *G. occidentalis* sequences, suggesting that such events can be considered as phylogenetically informative (Böhle et al., 1994; Gielly and

Table 9-2 Taxa included in the Phylogeny of Section *Megalanthe* (Following Flora Europaea; Tutin et al., 1972). Several species were sampled in botanical gardens. Where possible, the origin of the cultivated plant is indicated (in parentheses)

Genus	Section	Species (Abbreviation)	Origin (Locality of Sampling)	Country
Gentiana L.	*Megalanthe*	*G. clusii* Perr. & Song.	? (Botanical Garden Geneva)	Switzerland
	Gaudin	*G. occidentalis* Jakowatz	? (Alpine Garden Lautaret)	France
		G. ligustica R. de Vilmorin & Chopinet	Monjoie (Ligurian Alps)	Italy
		G. acaulis L.	Törbel, Valais canton (Botanical Garden Geneva)	Switzerland
		G. alpina Vill.	? (Alpine Garden Lautaret)	France
		G. dinarica G. Beck	Vhru Bora, Sarajevo	Bosnia
		G. dinarica G. Beck	Monti Sibillini	Italy
		G. angustifolia Vill.	? (Alpine Garden Lautaret)	France
Gentianella Moench	*Crossopetalae* (Froelich) Pritchard.	*G. ciliata* (L.) Borkh. subsp. *ciliata*	Lautaret pass	France

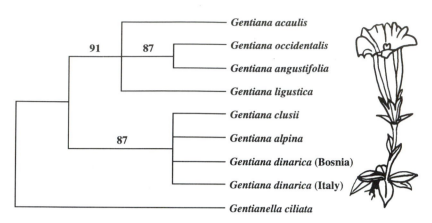

Figure 9-3 Phylogenetic relationships within the section *Megalanthe* of the genus *Gentiana*. Bootstrap 50% majority rule consensus tree of the eight trees retained by the branch-and-bound search of PAUP (tree length, 44; consistency index, 0.932). Numbers indicate bootstrap values (2000 replications).

Taberlet, 1994b; van Ham et al., 1994; Gielly and Taberlet, 1996). Intraspecific variation was found between origins of *G. dinarica* (Bosnia and Italy). We assessed phylogenetic relationships between taxa via a parsimony approach using a PAUP 3.1.1 (Swofford, 1993) branch-and-bound search. This procedure resulted in eight most-parsimonious trees. The data set was then submitted to a bootstrap (2000 replications) analysis (Felsenstein, 1985). The bootstrap 50% majority rule consensus tree has a consistency index of 0.932 (Figure 9-3). The eight trees produced by PAUP always showed two well-separated groups of species (*G. acaulis–G. angustifolia–G. ligustica–G. occidentalis*, and *G. alpina–G. clusii–G. dinarica*). Note that each group contains calcifuge (*G. acaulis* and *G. alpina*) and calcicole species (*G. angustifolia*, *G. clusii*, *G. dinarica*, *G. ligustica*, and *G. occidentalis*). From this we can deduce that the soil preference has evolved independently in the two clades.

Results presented here demonstrate the ability to use noncoding sequences of cpDNA to resolve phylogenies between closely related taxa. Even in the section *Megalanthe* of the genus *Gentiana*, where morphological characters are tenuous, the different taxa can be unambiguously distinguished by sequencing the *trnL* (UAA) intron.

Evolution of Insular Woodiness in the Genus *Echium* L. (Boraginaceae)

Böhle et al. (1994) recently investigated the history of the genus *Echium* in order to obtain insights into the evolution for woody and herbaceous habits. This circummediterranean–west Asian plant is subdivided into two distincts groups: (1) roughly 30 herbaceous continental species, and (2) 30 Macaronesian species (28 woody, 2 herbaceous) endemic to the Canary Islands, the Madeira archipelago, the Cape Verde archipelago, the Azores, and portions of the northwest African coast. Do the Macaronesian species represent a monophyletic group encompassing both woody and herbaceous taxa?

The first step consisted in the sequencing of three noncoding regions (see

Figure 9-1) for three herbaceous continental (*Echium asperrimum* Lam., *E. russicum* Gmel., *E. vulgare* L.), and one woody insular species (*E. simplex* DC.). Nucleotide divergence among the above species is about 1%. Two insertions/deletions of 5 bp and 9 bp distinguish the three continental and the insular species. The second step was to check the occurence of these two insertions/deletions for 24 *Echium* species. The insular species, whether woody or herbaceaous, appear to be monophyletic for these two characters, and therefore could be the result of a single invasion.

LIMITATIONS AND MOST APPROPRIATE APPLICATIONS

The sequencing of noncoding regions of cpDNA represents one of the most powerful tools for assessing phylogenetic relationships among closely related taxa, and consequently this approach can be used successfully in conservation genetics. The efficiency of DNA amplification and direct sequencing of PCR products permits the use of very small amounts of tissue (less than 1 mg), even preserved in herbaria. Therefore, this noninvasive method is particularly suitable for studying endangered plants, for which it is impossible to obtain enough tissues for standard RFLP analysis.

The sequencing represents the most expensive and time-consuming approach, but no alternative methods are available when the level of sequence variation is too low to be detected by RFLP. A solution to reduce the cost of the analysis could be to proceed first to a screening by sequencing several taxa to discover phylogenetically informative markers, and then to carry out a PCR-RFLP analysis for all taxa.

The use of noncoding sequences of cpDNA for the establishment of plant phylogenies can be criticized because of the presence of length mutational events, which can cause some problems with sequence alignments. The more distant the taxa, the more doubtful the alignment. For example, it is impossible to achieve an alignment between monocots and dicots (Gielly and Taberlet, 1994b). Even within a single family (Euphorbiaceae), it was difficult to align *trn*L (UAA) intron sequences of *Euphorbia* sp. and *Mercurialis* sp. (Gielly and Taberlet, 1994a). Consequently, the phylogenetic utility of these non-coding regions is limited to intrafamilial or intrageneric level. Furthermore, the occurrence of insertions/deletions does not allow reliable estimates of genetic distances, and thus only parsimony approaches for inferring phylogenetic relationships are applicable.

Several cases have been reported where cpDNA evolutionary rates differ among lineages. Within seed plants, extensive variation has been reported (Bousquet et al., 1992a; Frascaria et al., 1993; Gielly and Taberlet, 1994a). In monocots, the overall substitution rate for the *rbc*L locus is about 5 times higher in grasses than in palms (Gaut et al., 1992). Chloroplast DNA evolutionary rates also vary among more related taxa, at the family level (Doebley et al., 1990; Bousquet et al., 1992b); and at the genus level (Gielly and Taberlet, 1996). Consequently, because of the conservative mode of evolution of cpDNA and its variability of evolutionary rates in seed plants, it is difficult to predict whether the sequencing of noncoding regions will reveal enough informative sites for the establishment of a reliable phylogeny among closely related species of a given genus. For this reason, it is highly recommended that preliminary studies be conducted to assess the level of variability among the considered taxa. The *trn*L (UAA) intron seems to be well suited for such a preliminary study. The primers are universal enough to amplify double-stranded DNA from a wide taxonomic range of plant species without designing specific primers, and the relatively small size of this region does not require an internal primer for sequencing. This procedure also enables detection of long series of T (or A) bases, which sometimes prevent the achievement of the whole sequence.

If the analysis of the noncoding regions presented here does not reveal enough information, phylogenetic relationships could be resolved by one of the following. (1) The sequencing of other regions of cpDNA that may evolve at a rate equal to that of the *trn*L intron, for example the coding region *mat*K (Johnson and Soltis, 1994; Steele and Vilgalys, 1994), or non-coding regions such as the intergenic spacer between *rbc*L and *atp*B (Golenberg et al., 1993; Spichiger et al., 1993; Manen et al., 1994; Savolainen et al., 1994). (2) The sequencing of the internal transcribed spacers (ITS1 and ITS2) of nuclear ribosomal DNA (nrDNA), which can be a valuable tool for assessing plant phylogenetic relationships (Hamby and Zimmer, 1992, and references therein).

REFERENCES

Bevan, I.S., R. Rapley, and M.R. Walker. 1992. Sequencing of PCR-amplified DNA. PCR Methods Applica. 1: 222–228.

Böhle, U.R., H.H. Hilger, R. Cerff, and W.F. Martin. 1994. Non-coding chloroplast DNA for plant molecular systematics at the infrageneric level. In: R. DeSalle, G.P. Wagner, B. Schierwater and B. Streit, eds. Molecular approaches to ecology and evolution, pp. 341–403. Basel: Birkhauser Verlag.

Bonnier, G. 1935. Flore complète de France, Suisse et Belgique. Paris: Librairie générale de l'enseignement.

Bousquet, J., S.H. Strauss, A.H. Doerksen, and R.A. Price. 1992a. Extensive variation in evolutionary rate of *rbc*L gene sequences among seed plants. Proc. Natl. Acad. Sci. U.S.A. 89: 7844–7848.

Bousquet, J., S.H. Strauss, and P. Li. 1992b. Complete congruence between morphological and *rbc*L-based molecular phylogenies in birches and related species (Betulaceae). Mol. Biol. Evol. 9: 1076–1088.

Chase, M.W., D.E. Soltis, R.G. Olmstead, D. Morgan, D.H. Les, B.D. Mishler, M.R. Duvall, R.A. Price, H.G. Hills, Y.-L. Qiu, K.A. Kron, J.H. Rettig, E. Conti, J.D. Palmer, J.R. Manhart, K.J. Sytsma, H.J. Michaels, W.J. Kress, K.G. Karol, W.D. Clark, M. Hedren, B.S. Gaut, R.K. Jansen, K.-J. Kim, C.E Wimpee, J.F. Smith, G.R. Furnier, S.H. Strauss, Q.-Y. Xiang, G.M. Plunkett, P.S. Soltis, S.M. Swensen, S.E. Williams, P.A. Gadek, C.J. Quinn, L.E. Eguiarte, E. Golenberg, G.H.J. Learn, S.W. Graham, S.C.H. Barrett, S. Dayanandan, and V.A. Albert. 1993. Phylogenetics of seed plants: an analysis of nucleotide sequences from the plastid gene *rbc*L. Ann. Missouri Bot. Gard. 80: 528–580.

Clegg, M.T. and G. Zurawski. 1992. Chloroplast DNA and the study of plant phylogeny. In: P.S. Soltis, D.E. Soltis and J.J. Doyle, eds. Molecular systematics of plants, pp. 1–13. New York: Chapman and Hall.

Clegg, M.T., G.H. Learn, and E.M. Golenberg. 1991. Molecular evolution of chloroplast DNA. In: R.K. Selander, A.G. Clark and T.S. Whittam, eds. Evolution at the molecular level, pp. 135–149. Sunderland, Mass.: Sinauer.

Conti, E., A. Fischbach, and K.J. Sytsma. 1993. Tribal relationships in Onagraceae: implications from *rbc*L sequence data. Ann. Missouri Bot. Gard. 80: 672–685.

Curtis, S.E. and M.T. Clegg. 1984. Molecular evolution of chloroplast DNA sequences. Mol. Biol. Evol. 1: 291–301.

Doebley, J., M.L. Durbin, E.M. Golenberg, M.T. Clegg, and D.P. Ma. 1990. Evolutionary analysis of the large subunit of carboxylase (*rbc*L) nucleotide sequence among the grasses (Graminae). Evolution 44: 1097–1108.

Doyle, J.J. and J.L. Doyle. 1990. Isolation of plant DNA from fresh tissue. Focus 12: 13–15.

Erlich, H.A. 1989. PCR technology. New York: Stockton Press.

Felsenstein, J. 1985. Phylogenies and the comparative method. Am. Nat. 125: 1–15.

Frascaria, N., L. Maggia, M. Michaud, and J. Bousquet. 1993. The *rbc*L gene sequence from chestnut indicates a slow rate of evolution in the Fagaceae. Genome 36: 668–671.

Gadek, P.A. and C.J. Quinn. 1993. An analysis of relationships within the Cupressaceae sensu stricto based on *rbc*L sequences. Ann. Missouri Bot. Gard. 80: 581–586.

Gaut, B.S., S.V Muse, W.D. Clark, and M.T. Clegg. 1992. Relative rates of nucleotide substitutions at the *rbc*L locus of monocotyledonous plants. J. Mol. Evol. 35: 292–303.

Gielly, L. and P. Taberlet. 1994a. Chloroplast DNA polymorphism at the intrageneric level: implications for the establishment of plant phylogenies. C.R. Acad. Sci. (Life Science) 317: 885–692.

Gielly, L. and P. Taberlet. 1994b. The use of chloroplast DNA to resolve plant phylogenies: non-coding versus *rbc*L sequences. Mol. Biol. Evol. 11: 769–777.

Gielly, L. and P. Taberlet. 1996. A phylogeny of the European gentians inferred from chloroplast *trn*L (UAA) intron sequences. Bot. J. Linn. Soc. 120: 57–75.

Golenberg, E.M., M.T. Clegg, M.L. Durbin, and J. Doebley. 1993. Evolution of a noncoding region of the chloroplast genome. Mol. Phyl. Evol. 2: 52–64.

Hamby, R. K. and E.A. Zimmer. 1992. Ribosomal RNA as a phylogenetic tool in plant systematics. In: P.S. Soltis, D.E. Soltis and J.J. Doyle, eds. Molecular systematics of plants, pp. 50–91. New York: Chapman and Hall.

Ho, T.-N. and S.-W. Liu. 1990. The infrageneric classification of *Gentiana* (Gentianaceae). Bull. Br. Mus. Nat. Hist. (Bot.) 20: 169–192.

Innis, M.A., D.H. Gelfand, J.J. Sninsky, and T.J. White. 1990. PCR protocols. San Diego: Academic Press.

Jansen, R.K., H.J. Michaels, R.S. Wallace, K.-J. Kim, S.C. Keeley, L.E. Watson, and J.D. Palmer. 1992. Chloroplast DNA variation in the Asteraceae: phylogenetic and evolutionary implications. In: P.S. Soltis, D.E. Soltis and J.J. Doyle, eds. Molecular systematics of plants, pp. 252–279. New York: Chapman and Hall.

Johnson, L. A. and D.E. Soltis. 1994. *mat*K DNA sequences and phylogenetic reconstruction in Saxifragaceae sensu stricto. Syst. Bot. 19: 143–156.

Kim, K.-J., R.K. Jansen, R.S. Wallace, H.J. Michaels, and J.D. Palmer. 1992. Phylogenetic implications of *rbc*L sequence variation in the Asteraceae. Ann. Missouri Bot. Gard. 79: 428–445.

Kron, K.A. and M.W. Chase. 1993. Systematics of the Ericaceae, Empetraceae, Epacridaceae and related taxa based upon *rbc*L sequence data. Ann. Missouri Bot. Gard. 80: 735–741.

Kwok, L. 1990. Procedures to minimize PCR-product carry-over. In: M.A. Innis, D.H. Gelfand, J.J. Sninski and T.J. White, eds. PCR protocols, a guide to methods and applications, pp. 142–145. San Diego: Academic Press.

Manen, J.-E., A. Natali, and F. Ehrendorfer. 1994. Phylogeny of Rubiaceae-Rubieae inferred from the sequences of a cpDNA intergene region. Plant Syst. Evol. 190: 195–211.

Michaels, H.J., K.M. Scott, R.G. Olmstead, T. Szaro, R.K. Jansen, and J.D. Palmer. 1993. Interfamilial relationships of the Asteraceae: insights from *rbc*L sequence variation. Ann. Missouri Bot. Gard. 80: 742–751.

Morgan, D.R., and D.E. Soltis. 1993. Phylogenetic relationships among members of Saxifragaceae sensu lato based on *rbc*L sequence data. Ann. Missouri Bot. Gard. 80: 631–660.

Newton, C.R. and A. Graham. 1994. PCR. Oxford, U.K.: BIOS Scientific Publishers.

O'Donnell, K. 1992. Ribosomal DNA internal transcribed spacers are highly divergent in the phytopathogenic ascomycete *Fusarium sambucinum* (*Gibberella pulicaris*). Curr. Genet. 22: 213–220.

Palmer, J.D. 1987. Chloroplast DNA evolution and biosystematic uses of chloroplast DNA variation. Am. Nat. 130: S6–S29.

Palmer, J.D., R.K. Jansen, H.J. Michaels, M.K. Chase, and J.R. Manhart. 1988. Chloroplast DNA variation and plant phylogeny. Ann. Missouri Bot. Gard. 75: 1180–1206.

Price, R.A. and J.D. Palmer. 1993. Phylogenetic relationships of the Geraniaceae and Geraniales from *rbc*L sequences comparisons. Ann. Missouri Bot. Gard. 80: 661–671.

Rogers, S.O. and A.J. Bendich. 1985. Extraction of DNA from milligram amounts of fresh, herbarium and mummified plant tissues. Plant Mol. Biol. 5: 69–76.

Savolainen, V., J.-E. Manen, E. Douzery, and R. Spichiger. 1994. Molecular phylogeny of families related to the Celastrales based on *rbc*L 5' flanking sequences. Mol. Phyl. Evol. 3: 27–37.

Smith, J.E., W.J. Kress, and E.A. Zimmer. 1993. Phylogenetic analysis of the Zingiberales based on *rbc*L sequences. Ann. Missouri Bot. Gard. 80: 620–630.

Soltis, D.E., P.S. Soltis, M.T. Clegg, and M. Durbin. 1990. *rbc*L sequence divergence and phylogenetic relationships in Saxifragaceae sensu lato. Proc. Natl. Acad. Sci. U.S.A. 87: 4640–4644.

Soltis, D.E., P.S. Soltis, and B.G. Milligan. 1992. Intraspecific chloroplast DNA variation: systematic and phylogenetic implication. In: P.S. Soltis, D.E. Soltis and J.J. Doyle, eds. Molecular systematics of plants, pp. 117–150. New York: Chapman and Hall.

Soltis, D.E., D.R. Morgan, A. Grable, P.S. Soltis, and R. Kuzoff. 1993. Molecular systematics of Saxifragaceae sensu stricto. Am. J. Bot. 80: 1056–1081.

Spichiger, R., V. Savolainen, and J.-F. Manen. 1993. Systematic affinities of Aquifoliaceae and Icacinaceae from molecular data analysis. Candollea 48: 459–464.

Steele, K. P. and R. Vilgalys. 1994. Phylogenetic analyses of Polemoniaceae using nucleotide sequences of the plastid gene *mat*K. Syst. Bot. 19: 126–142.

Swofford, D.L. 1993. Phylogenetic analysis using Parsimony Version 3.1.1. Champaign, Ill.: Illinois Natural History Survey.

Taberlet, P., L. Gielly, G. Pautou, and J. Bouvet. 1991. Universal primers for amplification of three non-coding regions of chloroplast DNA. Plant Mol. Biol. 17: 1105–1109.

Tutin, T.G., V.H. Heywood, N.A. Burges, D.M. Moore, D.H. Valentine, S.M. Walters, and D.A. Webb. 1972. Flora Europaea: Diapensiaceae to Myoporaceae. Cambridge, U.K.: Cambridge University Press.

Van Ham, R.C.H.J., H. 't Hart, T.H.M. Mes, and J.M. Sandbrink. 1994. Molecular evolution of noncoding regions of the chloroplast genome in the Crassulaceae and related species. Curr. Genet. 25: 558–566.

Wilson, M.A., B. Gaut, and M.T. Clegg. 1990. Chloroplast DNA evolves slowly in the palm family (Arecaceae). Mol. Biol. Evol. 7: 303–314.

Wolfe, K.H., W.-H. Li, and P.M. Sharp. 1987. Rates of nucleotide substitution vary greatly among plant mitochondrial, chloroplast and nuclear DNAS. Proc. Natl, Acad. Sci. U.S.A. 84: 9054–9058.

Xiang, Q.-Y., D.E. Soltis, D.R. Morgan, and P.S. Soltis. 1993. Phylogenetic relationships of *Cornus* L. sensu lato and putative relatives inferred from *rbc*L sequence data. Ann. Missouri Bot. Gard. 80: 723–734.

Zurawski, G. and M.T. Clegg. 1987. Evolution of higher-plant chloroplast DNA-encoded genes: implications for structure–function and phylogenetic studies. Ann. Rev. Plant Physiol. 38: 391–418.

Zurawski, G., M.T. Clegg, and A.H.D. Brown. 1984. The nature of nucleotide sequence divergence between barley and maize chloroplast DNA. Genetics 106: 735–749.

10

Reconstructing Population History Using PCR–Restriction Site Data

NICHOLAS GEORGIADIS

Many recent advances in genetic technology have had almost instantaneous applications to conservation, as this volume attests. Examples include a growing variety of genetic markers, accessible with the polymerase chain reaction (PCR), that are suitable for forensic detection, and the definition of systematically inclusive "units" for conservation at population and species levels (Morin et al., 1994, Moritz 1994, Volgler and de Salle, 1994). While many of these markers are informative in the sense that their presence or absence is diagnostic, or that degrees of phenetic similarity can be derived from the combined use of multiple markers, relatively few provide *genealogically* informative data in the sense that phylogenetic (as opposed to phenetic) relationships can be reconstructed for sets of alleles. Data that are suited to the derivation of genealogies include DNA sequence and restriction site data, but not isozyme, restriction fragment, minisatellite, microsatellite or random amplified polymorphic DNA (RAPD) data.

Genealogical approaches are of great potential value because they reveal more explicit details than do patterns of allele frequencies about the recent history of populations and species, and the evolutionary processes by which they were structured. This retrospective information becomes useful when populations must be managed or manipulated, for instance by translocating individuals between populations, restoring locally extinct populations, or, indeed, reconstructing communities. Managers might choose, for example, not to translocate individuals between populations that are known to be genealogically distinct (comprising reciprocally monophyletic lineages; Moritz, 1994), which would imply long-term isolation and possible local adaptation. Alternatively, managers might deliberately increase genetic variation among the founders of a locally restored species by selecting individuals from populations that are known to be genetically different (in terms of allele frequencies) but not genealogically distinct.

Although in their infancy, theoretical approaches to the analysis of genealogical information are advancing rapidly, particularly at the intraspecific level (reviewed by Hudson, 1990). Unfortunately, however, genealogically informative data are among the most expensive and time-consuming to produce and, at least for the foreseeable future, funding realities dictate that speed and cost-effectiveness will continue to rank among the leading factors that limit the role of genetics in conservation. The principal goal of this chapter is to describe a method for generating genealogically informative data that is fast, cost-effective, and does not require radioactive isotopes: the mapping of restriction sites within large DNA fragments that have been amplified by PCR, in this case within fragments of mitochondrial DNA (mtDNA). A secondary goal is to outline several approaches that have been used to make inferences about population history from genealogical data, with references to the relevant literature.

TECHNICAL DESCRIPTION

The method described here was originally developed to address an applied conservation problem: Can genetic markers be used to trace the tusks of African elephants (*Loxodonta africana*) to their points of geographic origin, and thereby form the basis of a forensic method to regulate an ivory trade? The method had to be simple, robust, and applicable to samples of degraded DNA. The genetic structure of elephant populations proved unsuitable for tracing every individual tusk (Georgiadis et al., 1994), but the method provided important insights into the maternal history of elephant populations, and has been similarly applied to other mammals (e.g., Templeton and Georgiadis, 1995) and fish (Cronin et al., 1993). Methods for screening restriction fragment variation in large-sized PCR products have been developed independently in many laboratories, but the method for mapping restriction sites that is described here was originally devised by John Patton and myself at Washington University in St. Louis, and refined by Juan Carlos Morales at Texas A&M University. I will confine the description of the method to a sample of up to 20 individuals, representing, say, a single population, to be screened for restriction site variation within a specific segment of mtDNA. I have found this to be a convenient number to process as a single batch, and subsequent batches, for example representing different populations, are treated identically.

The description of lab procedures is divided into three phases. First, a large (1700–2500 bp), homologous segment of mtDNA is amplified from each sample using PCR. Second, those segments are screened for restriction fragment variation using a battery of restriction enzymes (mostly four-base cutters), and the fragments are visualized directly after electrophoresis in agarose. Third, after all samples in the entire survey have been screened in this manner, the total number of distinct haplotypes is enumerated, and the presence/absence of restriction sites in each haplotype is determined.

Phase 1. DNA Extraction and Amplification

Although it is possible to amplify DNA directly from a raw sample (e.g., with "Genereleaser"), or using a quick DNA extraction method (e.g., Walsh et al.,

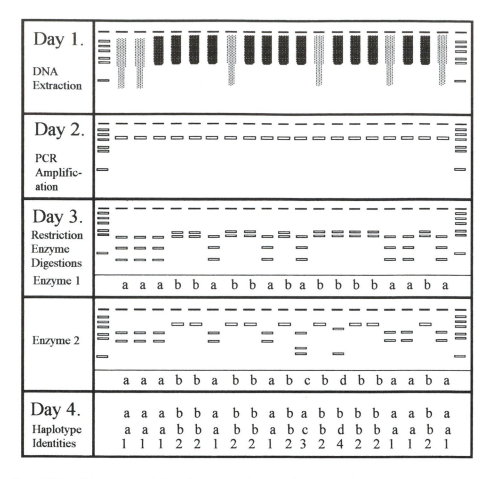

Figure 10-1 Schematic depiction of agarose gel electrophoresis products at progressive stages of restriction fragment screening. Day 1: Extracted genomic DNA of various degrees of degradation. Day 2: PCR amplification of a large segment of mitochondrial DNA from each of the samples of genomic DNA in the first panel. Day 3: Digestion of amplified segments by different restriction enzymes. Within each panel (gel), each unique fragment profile is scored with a unique letter. Day 4: Digestion profile scores for each restriction enzyme are aligned and each unique sequence of letters is assigned a number. These correspond to haplotypes with distinct DNA sequences. DNA size standards appear at the extreme left and right ends within each panel.

1991), I have achieved the most consistent results using the standard proteinase-K–phenol-chloroform extraction method (Manniatis et al., 1989). After total genomic extraction (Figure 10-1, day 1), a segment of about 2450 base pairs that spans the ND 5–6 region of mtDNA is PCR-amplified from each sample in a total volume 65 µl (day 2). Because this is a relatively large segment of DNA, amplification is more likely to succeed with well-preserved DNA. Even so, I have consistently amplified this region from DNA that was relatively degraded, such as from dried skins or scrapings from the base of elephant tusks. Primers were designed from mitochondrial sequences within the glutamine

$$(5'\text{-TTACAACGATGGGTTTTTCAT}^G/_A\text{TCA-3'})$$

and leucine

$$(5'\text{-AATAGTTTATCC}^A/_G\text{TTGGTCTTAGG-3'})$$

tRNA regions that are conserved among mammals and amphibians. Optimum PCR conditions for these primers in elephants were found to be as follows:

DNA	20–100 ng, e.g., in 2 µl H_2O
Salt	50 mmol/l KCl, 4.5 mmol/l $MgCl_2$, 10 mmol/l Tris-HCl, pH 9.0
Primers	9.4 µmol/l
dNTPs	30 µmol/l
Taq polymerase	0.35 U

Forty cycles were used at the following temperatures and times: melting at 94°C for 45 s, annealing at 55°C for 40 s, and extension at 72°C for 150 s, with 4 s added to the extension phase in each cycle. To check the quality of the amplification, 2 µl of each product was visualized in a 0.8% (w/v) agarose minigel, submersed in 1 × TEA buffer after straining with ethidium bromide.

Phase 2. Restriction Enzyme Digestion and Electrophoresis

Amplified segments can be screened for sequence variation using up to 18 restriction enzymes per batch (Figure 10-1, day 3). For example, the following 10 restriction enzymes were used on elephant samples: *BsaJI, DpnI, HaeIII, HhaI, HinfI, MseI, MspI, RsaI*, and *Sau 96I* (four-base cutters), and *SspI* (a six-base cutter). Additional possibilities are: *AvaII, BstNI, HincII, NciI, NlaIII*, and *TaqI*. For AT-rich regions of mtDNA (e.g., control region), *MseI* was found to cut too many times to enable discrimination in the gel, and may be substituted by the six-base cutters *AseI* and *DdeI*. *AluI* tended to yield partial digestions, even in high concentrations.

Digestions are performed in 96-well microtiter plates, using 2–5 µl of amplified DNA (depending on amplification quality) and 1.5 U of enzyme in a total volume of 30 µl. Fewer units may be used for some enzymes. I have found that less than 0.5 U per digestion is required for *Sau96I*, and more than 2 U of *SspI* is required to prevent partial digestion. All digestions are done simultaneously. First, prepare bulk cocktails for each restriction enzyme, containing all ingredients of the digestion mixture (except DNA). Then dispense the amplified DNA into each well directly from the PCR tube (there is no need to remove mineral oil if it was used in the PCR reaction). Aliquots of digestion mixture are then added to each well, the plates are sealed with "Saranwrap" and placed in a 37°C oven for at least 2 h. Enzymes having an optimal reaction temperature >37°C (e.g., *BsaJI, TaqI*) cut sufficiently well at 37°C that there is no need to prepare separate plates for these enzymes.

Once digestion is complete, plates may be frozen until needed, but it is convenient to set up the digestions in the morning and, while they are incubating, prepare the electrophoresis gels. After addition of 3 μl of 10 × loading dye (1 mg/ml Bromophenol Blue, 0.01 mol/l EDTA, and 40% (w/v) sucrose) to each well, the contents are loaded into 2% agarose electrophoresis gels that are submersed in 1 × TEA buffer. Gel molds are 20 × 13.5 × 0.5 cm (140 ml of agarose solution per gel), each fitted with up to three combs. The combs should ideally have one more "tooth" than there are samples in a given batch, with one lane on the gel occupied by the DNA size standard. I use 0.4 μg φX DNA, digested with HaeIII as a DNA fragment size standard in one well per comb. Run the enzymes in alphabetical order, with all the digestions from a given enzyme occupying one row (comb) in the gel, and each enzyme labeled by the number of the well in which the DNA standard is placed. Up to three gel molds can be stacked vertically in a given electrophoresis rig, so that digestions from up to nine restriction enzymes can be run per rig. The amperage is set at 100–150 mA, and gels take 2–4 h to run.

Digestion fragments are visualized and photographed directly under UV after staining with ethidium bromide. Gels are scored by assigning a letter to each unique digestion pattern for a given enzyme (Figure 10-1, day 3). This procedure is repeated for all samples in the survey. *But beware*: It is easy to unwittingly assign the same letter to digestion profiles on separate gels that appear to be very similar, but do in fact differ by at least one restriction site. Therefore, it is essential to run "lineup" gels to verify the identities of digestion profiles, with representatives from each batch run side-by-side on the same gel.

Once all samples have been screened with all enzymes, the letters that correspond to each enzyme digestion profile are aligned for each sample in order to assign haplotype identities (Figure 10-1, day 4). Unique haplotypes correspond to unique sequences of letters; each sample is assigned its appropriate haplotype identity; and the frequencies of each haplotype in each population are tallied. With regard to logistics, by far the most critical step is the PCR amplification. If a good product can be expected in 95% of attempted amplifications, it should be possible for one person to screen a batch of up to 20 samples, from DNA extraction, through screening with up to 18 restriction enzymes, to haplotype assignment, in 3 days. This rate of screening can be increased if batches are staggered to start on succeeding days. Costs of phase II may be cut by reusing the microtiter plates after thoroughly rinsing and drying, by reusing the TEA buffer at least once, and by reusing the agarose once (but beware of boiling agarose that has been infused with ethidium bromide).

Phase 3. Restriction Site Mapping

To avoid repetition, it is important to complete the restriction fragment screening process (phase II) before proceeding with site mapping (phase III). The goal of this phase is to divine the presence and absence of restriction sites within each of the haplotypes identified in phase II. To achieve this, it is not necessary to know the relative positions of all restriction sites along the amplified segment (that is, the physical map). Rather, it is sufficient to know only whether a given restriction site is present or absent in a given haplotype. This distinction saves much work because the presence of most restriction sites can be deduced simply by comparing fragment sizes among digestion profiles. Choose a representative sample for each unique digestion profile of each restriction enzyme that was used in phase II. Digest–10 µl of the original amplification product with the appropriate enzyme, and run them side-by-side in submersed agarose gels, similar to those used in phase II but of higher concentration (2.5% (w/v agarose), and with only one comb per gel. Use ϕX DNA digested with *Hae*III as a DNA fragment size standard in at least two lanes per gel, and run out the DNA across the entire gel. Photograph the gel with transparent plastic rulers (marked with a millimeter or smaller scale) superimposed on the lanes with the DNA size standards. Using those known fragment sizes as standards for calibration, the unknown sizes of the restriction fragments in each digestion profile can be estimated to within 5–10 bp. The sum of the fragment sizes within each profile should equal the known size of the undigested PCR product. For each profile there should be one fewer restriction site than there are fragments. With practice, the presence or absence of sites can be deduced from this fragment sizing process for over 90% of all digestion profiles.

PARTIAL DIGESTION

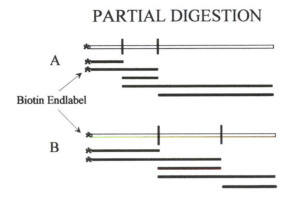

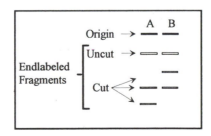

Figure 10-2 Schematic depiction of restriction site mapping by partial digestion of PCR-amplified mtDNA segments that have been end-labeled with biotin (asterisk). Upper panel: Undigested segments (open rectangles) representing haplotypes (A and B) that share a restriction site for a given enzyme but each of which characterized by a unique site (vertical bars). Fragments resulting from partial digestion of each haplotype are represented by filled rectangles. Lower panel: Schematic representation of restriction sites that appear on x-ray film due to photo-luminescence (open rectangles) for each of the two halotypes.

Two additional steps may be necessary to identify the presence/absence of sites that cannot be deduced from fragment size comparisons alone. First, information from partial digestions of fragments, run beside complete digestions using the same enzyme on fragment sizing gels, can often resolve ambiguous sites. Second, those sites that remain unresolved must be physically mapped, using a final procedure that is elegant but can be time-consuming (up to one week). The principle is illustrated in Figure 10-2, and described more fully by Morales et al. (1993). Briefly, samples representing the appropriate haplotypes are PCR-amplified as above, but with one of the primers end-labeled with biotin. *Partial* digestions using the appropriate restriction enzyme are performed on about 10 µl of the PCR product, and the fragments are electrophoresed in fragment sizing gels as described above. The fragments are then Southern blotted onto a nylon membrane, and exposed to autoradiograph film using a bioluminescent method (e.g. "Photogene" from BRL Life Technologies, Inc.), according to the manufacturer's specifications. Only those fragments that are end-labeled with biotin will be illuminated, and the order of the restriction sites can be read, much like the bases on a sequencing gel, from one end of the DNA region to the other (Figure 10-2). Application of this method to a 2450 bp segment in 270 African elephant mtDNA samples, and using 10 restriction enzymes, resulted in screening the equivalent of 194 base pairs for genetic variation in each sample, and the identification of 10 mitochondrial haplotypes (Georgiadis et al., 1994).

HISTORICAL RECONSTRUCTIONS: METHODS AND CASE STUDIES

Several approaches have recently been proposed for inferring population history by overlaying information about the geographical dispersion of alleles upon an allelic genealogy (see also Kreitman and Hudson 1991; Excoffier and Smouse, 1994). Two will be briefly described here, and both begin by deriving the most likely genealogical relationships among alleles or haplotypes. In the method described by Slatkin and Maddison (1989) (see also Hudson et al., 1992) a "maximum parsimony" genealogy (requiring the minimum number of evolutionary steps) is inferred from characters that are judged to be both shared and derived (e.g., Swofford 1993). The minimum number of migration events (s) that must have occurred over evolutionary time to yield the observed geographical dispersion of haplotypes, *given their genealogical relationships*, is then counted. The rate of gene flow between populations (Nm, where N is the effective population size and m the proportion of migrants per generation, assuming an island model), and its confidence intervals, are estimated from s using software available from M. Slatkin. Using Monte Carlo simulations, Slatkin and Maddison (1990) showed that the same approach can also be used to test for isolation by distance (Wright, 1946): Nm is estimated between all *pairs* of sampling locations, and these estimates are plotted against the geographical distances (d) between pairs of locations on log–log scales. The data are consistent with a model of isolation by distance if $\ln(Nm)$ declines with $\ln(d)$ with a slope of -0.5 (a two-dimensional model) to -1.0 (a one-dimensional model).

Applied to African savanna elephants, restriction fragment analysis of mtDNA showed that populations are markedly subdivided in terms of haplotype frequencies (Georgiadis et al., 1994). On their own, the haplotype frequency data implied that additional genetic screening would yield diagnostic mitochondrial restriction sites for elephants in specific locations, and that these could be applied to the forensic regulation of an ivory trade. Restriction site analysis, however, found no diagnostic (fixed) mitochondrial markers between even distant populations (> 2000 km). Moreover, a genealogical analysis using the method of Slatkin and Maddison (1989) showed that diagnostic restriction sites were unlikely to be found by further screening. While populations are isolated by distance at the regional level (within eastern Africa and within southern Africa), the analysis suggested that a relatively recent range expansion event has occurred across the entire sampled region. Maternal gene flow between elephant populations appears to have been too protracted to permit tracing of all ivory pieces to geographical regions that are sufficiently small for ivory trade regulation. Because females are more philopatric than males, we would not expect nuclear genetic markers to be more geographically localized than mitochondrial markers.

By providing insights that were not revealed by haplotype frequency data alone (see Edwards, 1993), the Slatkin and Maddison method constitutes an important advance in the reconstruction of intraspecific history from patterns of genetic variation, but it has several limitations. First, it is based on theory that was derived for *inter*specific analyses, in which taxa are assumed to be related by a strictly bifurcating genealogy, and ancestral taxa (those that would occupy internal nodes of the genealogy) are assumed to be extinct. Neither of these assumptions necessarily applies to *intra*specific genealogies (Crandall et al., 1994). Second, intraspecific alleles are commonly distinguished by few sites and, consequently, genealogies that are based on the identification of shared, derived characters are not supported by high bootstrap or decay indices. Third, the Slatkin and Maddison (1989) analysis requires only that the presence or absence of an allele is known in a population—information implicit in allele frequencies is not utilized. Last, statistical inferences are hard to draw in the test for isolation by distance because data in the pairwise analysis of populations are not independent.

A complementary approach to the analysis of population history has been explicitly designed to resolve these problems by Templeton et al. (1995, and references therein). This method is computationally intensive, and a detailed description is not appropriate here, but it differs from the Slatkin and Maddison (1989) approach in important respects. For example, genealogies are not inferred from shared, derived characters, but based on the probability that haplotypes are parsimoniously linked within a haplotype tree, depending on the number of sites that they share, relative to the number of sites by which they differ (Templeton et al., 1987, 1992). Thus the entire data set, not just polymorphic sites, is used to derive the tree. All trees that are judged on this basis to have a greater than 95% probability of being true are included in the set of plausible, alternative trees. Crandall (1994) has shown that this algorithm performs as well as or better than more traditional maximum parsimony methods with data from closely related (intraspecific) taxa. An additional distinction of the Templeton (1995) method is that gene flow is not assumed to be the *only* historical process that can have molded current population structure (accordingly, rates of gene flow are not explicitly quantified). Rather, recurrent processes such as gene flow can be distinguished from such historical events as population range expansion, long-distance colonization, or ancestral range fragmentation in a spatially explicit fashion. It is possible, indeed likely, for several or all of these to have occurred at different times and places in the history of a given species.

The Templeton et al. (1995) method was applied to PCR–restriction site data from a comparative study of three wild bovid species that occupy African savannas (buffalo *Syncerus caffer*, impala *Aepyceros melampus*, and blue wildebeest *Connochaetes taurinus*; Templeton and Georgiadis 1995). The species have strongly contrasting habitat and dietary preferences and, although they once occupied large regions in eastern and southern Africa, they exist today in fragmented geographical ranges. The question arises: To what extent, and by what factors, were populations of these species *naturally* subdivided before their ranges were fragmented by humanity? Within a relatively small region in northern Tanzania, sharp local disjunctions in habitat type and relief separate savanna ecosystems in ways that provide a natural experiment in which the effects of these factors on population genetic structure can be largely uncoupled by strategic sampling (details are described in Templeton and Georgiadis, 1995).

Seventeen individuals of each species were sampled at four or more locations, and samples were analyzed as described above, but in this study *two* mtDNA segments were amplified and screened for restriction site variation each individual: the 2450 bp segment spanning the ND 5–6 region that was used in the African elephant study, and a 1730 bp segment spanning the control region of mtDNA (primer sequences 5'-TACACTGGTCTTGTAAACC-3' and 5'-ATCGATTACAGAACAGGCTCCTC-3'). After combining results from both segments, the equivalent of approximately 400 base pairs were screened for genetic variation within each individual.

Similar numbers of mitochondrial haplotypes were found in all species, and all were significantly subdivided, but more so in wildebeest (24 haplotypes, $H_{st} = 0.16$), than in impala (25 haplotypes, $H_{st} = 0.1$) and buffalo (28 haplotypes, $H_{st} = 0.08$). Here, H_{st} is analogous to F_{st} and was calculated by the method described by Hudson et al. (1992). This analysis of haplotype frequencies detected quantitative differences in population structure between these bovid species, but the genealogical analysis revealed marked qualitative differences as well. For example, no fixed restriction site differences were detected between buffalo populations, even those separated by more than 2000 km. The genetic structure of buffalo populations was apparently not affected by geographic relief or by habitat disjunctions, and was entirely consistent with the interpretation of a history of isolation by distance. By contrast, at least five fixed restriction sites differentiate wildebeest populations in eastern and southern Africa, with corresponding genetic distances between those populations that fall within the range that is characteristic of different *species* (1.4%; Hey, 1991). A significant fragmentation event was detected between populations that are separated by just 50 km of closed canopy forest associated with the Ngorongoro Highlands and Rift Valley in northern Tanzania. Despite their mobility, wildebeest populations are markedly subdivided by extrinsic factors such as habitat disjunctions. Because the wildebeest is the most mobile, and the buffalo the most philopatric of these species, the results show that neither of these intrinsic factors strongly affects population structure. Rather, the structures and histories of these bovid species appear to be influenced primarily by interactions between intrinsic and extrinsic factors that are fundamentally ecological (i.e. habitat preference/avoidance and habitat juxtaposition). Other behavioral factors (e.g., philopatry and mobility) and abiotic factors (e.g., geographical relief and distance) seem to play a secondary role.

This brief description is intended to underscore the utility of data that afford genealogical interpretations, compared to analyses that are based solely on allele frequencies. At least in this example, it might be argued that allele frequencies alone would provide an inadequate basis for management or manipulation policies. The genealogical approach emphasizes evolutionary processes and ecological continuity. By understanding these processes, we can more effectively conserve, even reconstruct, the integrity of populations and communities.

LIMITATIONS

Foremost among the properties of mtDNA that make it so useful for reconstructing history are that it does not recombine and that it evolves rapidly (Avise et al., 1987; Avise, 1994). Because it is maternally inherited, however, analyses of mitochondrial DNA are informative only about maternal history—they are uninformative about the influence of male dispersal on population genetic structure. For species in which males have different dispersal tendencies from females (they are rarely identical; e.g., Cronin 1991), patterns of nuclear or paternal gene flow are likely to differ at least quantitatively from maternal patterns (Birky et al., 1989). While it is unlikely that a geographical or habitat boundary would inhibit gene flow in one sex and not the other, more complete descriptions of species or population histories require information from nuclear or Y chromosome-linked regions of DNA. No such regions are yet known that are as genealogically informative or as convenient to assay as is mtDNA, but a rapidly growing array of genomic markers are available that complement mitochondrial data in this respect (many are described in this volume). For example, microsatellite (Bishop et al., 1994) and major histocompatibility markers (van Eijk et al., 1992; Schwaiger et al., 1994) that have been developed for domestic bovids can be applied to their wild relatives. Analytical methods for historical reconstruction that accommodate recombination (Templeton et al., 1992), or that combine mitochondrial and nuclear data in historically informative ways (Arnold, 1993), are under development.

An assumption that is implicit in most current methods that attempt to reconstruct history from genetic data is that the focal loci are not only selectively neutral but also unlinked to selected loci. Several tests for neutrality have been proposed (e.g., Tajima, 1989; see Rand et al., 1994), and these often fail to confirm the assumption of neutrality, even for mtDNA (Rand et al., 1994; Ballard and Kreitman, 1994).

The extent to which selection affects historical reconstructions is currently under active investigation. One confounding consequence of selection, particularly of balancing or frequency-dependent selection, is that alleles may be shared between populations not as a consequence of gene flow (or of homoplasy) but owing to the persistence of ancestral polymorphisms (e.g., Larson et al., 1984; this can also occur by chance). Gene flow between populations might be mistakenly inferred where none in fact exists. In such cases, it should be remembered that the Statkin and Maddison (1989) method estimates gene flow over the *entire* history defined by the genealogy, and those estimates do not necessarily reflect current conditions. It may be expedient not to estimate *Nm* between those populations for which the Templeton et al. (1995) method detected a significant fragmentation event. Finally, populations might have become isolated from each other, and adapted to local conditions, too recently to have been marked by mutations within the gene region under investigation. It is important, therefore, to focus on loci that are evolving at a rate that is appropriate to the time-scale of the process in question.

CONCLUSION

Maintaining the functional and historical integrity of populations and ecosystems over the long term will increasingly demand informed intervention (Grumbine, 1994) based on our cumulative biological wisdom. The means to at least partially reconstruct population history from genealogically informative data are now sufficiently fast and inexpensive to be applied to conservation in a timely fashion. Because these methods can be informative about natural evolutionary and ecological processes long *after* populations have become unnaturally fragmented, we expect their utility to increase. We also expect these methods to improve rapidly, to the extent that the technical and theoretical limitations to reconstructing history from the genetic record will be subordinated by a much more serious limitation: the loss of the record itself. An obvious cause of this loss is the extinction of whole populations. Loss of genetic variation due to human-induced population "bottlenecks" and genetic drift is less serious in this context, since partial reconstruction of the past is possible as long as some genetic variation persists among populations, even those that have been unnaturally fragmented. A much more insidious threat to the genetic record is the undocumented translocation of individuals by humans from one population to another. If those individuals, or their descendants, are sampled for historical analysis, the natural record is confounded. Because large mammals are important economically, their populations have been and will increasingly be the target of such manipulations, particularly in Europe, the United States and South Africa (Greig, 1979). As a consequence, historical reconstructions are now impossible in many species for which such insights are in greatest demand. It is not too soon to apply these retrospective methods as broadly as possible.

REFERENCES

Arnold, J. 1993. Cytonuclear disequililbria in hybrid zones. Ann. Rev. Ecol. Syst. 24: 521–554
Avise, J.C. 1994. Molecular markers, natural history and evolution. New York: Chapman and Hall.

Avise, J.C., J. Arnold, R.M. Ball et al. 1987. Intraspecific phylogeography: the mitochondrial bridge between population genetics and systematics. Ann. Rev. Ecol. Syst. 18: 489–522.

Ballard, J.W.O. and M. Kreitman. 1994. Unraveling selection in the mitochondrial genome of *Drosophila*. Genetics 138: 757–772.

Birky, C.W. Jr., P. Fuerst, and T. Muruyama 1989. Organelle gene diversity under migration, mutation, and drift: Equilibrium expectations, approach to equilibrium, effects of heteroplasmic cells, and comparison to nuclear genes. Genetics 121: 613–627.

Bishop, M.D., S. M. Kappes, J. W. Keele, R. T. Stone et al. 1994. A genetic linkage map for cattle. Genetics 136: 619–639.

Crandall, K.A., A.R. Templeton, and C.F. Sing. 1994. Intraspecific cladogram estimation: accuracy at higher levels of divergence. Syst. Biol. 43: 222–235.

Crandall, K.A., 1994. Intraspecific phylogenetics: problems and solutions. In: R.W. Scotland, D. J. Siebert, and D.M. Williams, eds. Systematics Association Special Volume No. 52, pp. 273–297. Oxford, U.K.: Clarendon Press.

Cronin, M.A. 1991. Mitochondrial and nuclear genetic relationships of deer *Odocoileus* spp. in western North America. Can. J. Zool. 69: 1270–1279.

Cronin, M.A., W.J. Spearman, R.L. Wilmot, J.C. Patton, and J.W. Bickman, 1993. Mitochondrial DNA variation in Chinook salmon *Oncorhynchus tshawytscha*, and Chum salmon *O. keta* detected by restriction enzyme analysis of polymerase chain reaction (PCR) products. Can. J. Fish. Aquat. Sci. 50: 708–715.

Edwards, S.V. 1993. Mitochondrial gene genealogy and gene flow among island and mainland populations of a sedentary songbird, the grey-crowned babbler (*Pomatostomus temporalis*). Evolution 47: 1118–1137.

Excoffier, L. and P.E. Smouse. 1994. Using allele frequencies and geographic subdivision to reconstruct gene trees within a species: molecular variance parsimony. Genetics 136: 343–359.

Georgiadis, N.J., L.L. Bischof, A.R. Templeton, J.C. Patton, W. Karesh, and D. Western. 1994. Structure and history of African elephant populations. I: Eastern and Southern Africa. J. Hered. 85: 100–104.

Greig, J.C. 1979. Principles of genetic conservation in relation to wildlife management in southern Africa. S. Afr. J. Wildlife Management 9: 57–78.

Grumbine, R.E. 1994. What is ecosystem management? Conserv. Biol. 8: 27–38.

Hey, J. 1991. The structure of genealogies and the distribution of fixed differences between DNA sequence samples from natural populations. Genetics 128: 831–840.

Hudson, R.R. 1990. Gene genealogies and the coalescent process. Oxford Surveys in Evolutionary Biology 7: 1–44.

Hudson, R.R., M. Slatkin, and W.P. Maddison 1992. Estimation of levels of gene flow from DNA sequence data. Genetics 132: 583–589.

Kreitman, M. and R.R. Hudson. 1991. Inferring the evolutionary histories of the Adh and Adh. dup loci in *Drosophila melanogaster* from patterns of polymorphism and divergence. Genetics 127: 565–582.

Larson, A., D.B. Wake, and K.P. Yanev. 1984. Measuring gene flow among populations having high levels of genetic fragmentation. Genetics 106: 293–308.

Manniatis, T., E.F. Fritsch, and J. Sambrook. 1989. Molecular cloning, a laboratory manual. Cold Spring Harbor, N.Y.: Cold Spring Harbor Laboratory.

Morales, J.C., J.C. Patton, and J.W. Bickham. 1993. Partial endonuclease digestion mapping of restriction sites using PCR amplified DNA. PCR Methods Appl. 2: 228–233.

Morin, P.A., J.J. Chakraborty, L. Lin, J. Goodall, and D.S. Woodruff. 1994. Kin selection, social structure, gene flow, and the evolution of chimpanzees. Science 265: 1193–1201.

Moritz, C. 1994. Defining "Evolutionary significant units" for conservation. Trends Ecol. Evol. 9: 373–374.

Rand, D.M., M. Dorfmann, and L.M. Kann. 1994. Neutral and non-neutral evolution of *Drosophila* mitochondrial DNA. Genetics 138: 741–756.

Schwaiger, F.W., E. Weyers, J. Bitkamp, A.J. Ede, A. Crawford, and J.T. Epplen. 1994. Interdependent MHC-DRB exon-plus-intron evolution in Artiodactyls. Mol. Biol. Evol. 11: 239–249.

Slatkin, M. and W.P. Maddison 1989. A cladistic measure of gene flow inferred from the phylogenies of alleles. Genetics 122: 957–966.

Solatin, M. and W.P. Maddison. 1990. Detecting isolation by distance using phylogenies of genes. Genetics 126: 249–260.

Swofford, D. 1993. PAUP, version 3.1.1. Phylogenetic analysis using parsimony. User's manual. Champaign, Ill.: Illinois Natural History Survey.

Tajima, F. 1989. Statistical method for testing the neutral mutation hypothesis by DNA polymorphism. Genetics 123: 585–595.

Templeton, A.R. and N.J. Georgiadis 1996. A landscape approach to conservation genetics: conserving evolutionary processes in African bovids. In: J. Avise and J. Hamrick, eds. Conservation genetics: Case studies from nature, pp. 398–430. New York: Chapman and Hall [In press].

Templeton, A.R., Boerwinkle, E., and Sing, C.F. 1987. A cladistic analysis of phenotypic associations with haplotypes inferred from restriction endonuclease mapping. I. Basic theory and an analysis of alcohol dehydrogenase activity in *Drosophila*. Genetics 117: 343–351.

Templeton, A.R., K.A. Crandall, and C.F. Sing. 1992. A cladistic analysis of phenotypic associations with haplotypes inferred from restriction endonuclease mapping and DNA sequence data. III. Cladogram estimation. Genetics 132: 619–633.

Templeton, A.R., E.J. Routman, and C.A. Phillips. 1995. Separating population structure from population history: a cladistic analysis of the geographical distribution of mitochondrial DNA haplotypes in the Tiger Salamander, *Ambystoma tigrinum*. Genetics [In press].

van Eijk, M.J.T., J.A. Stewart-Haynes, and H.A. Lewin. 1992. Extensive polymorphism of the Bo-LA–DRB3 gene distinguished by PCR-RFLP. Animal Genet. 23: 483–496.

Vogler, A.P. and R. de Salle. 1994. Diagnosing units of conservation management. Conserv. Biol. 8: 354–363.

Walsh, P.S., D.A. Metzger, and R. Higuchi. 1991. Chelex 100 as a medium for simple extraction of DNA for PCR-based typing from forensic material. Biotechniques 10: 506–513.

Wright, S. 1946. Isolation by distance under diverse systems of mating. Genetics 31: 39–59.

11

The Use of PCR-Based Single-Stranded Conformation Polymorphism Analysis (PCR-SSCP) in Conservation Genetics

DEREK J. GIRMAN

The use of molecular markers to investigate the population structure of endangered or threatened species has become a key component in the genetic management of endangered species (see chapter 1). The identification of variable genetic loci has allowed the analysis of DNA sequence data to become one of the most useful measures of genetic differentiation, genetic variability, and evolutionary relatedness for conservation-oriented analysis at or below the species level (Wayne and Jenks, 1991; Wenink et al. 1993; Gottelli et al., 1994; Rosel et al., 1994; Taberlet and Bouvet, 1994; Taylor et al., 1994). In addition, the development of PCR (polymerase chain reaction) has allowed specific regions of the genome to be quickly isolated and amplified, so that the products of this relatively simple procedure can be readily sequenced (Saiki et al., 1988). The use of PCR technology is also advantageous because it provides a way to access DNA from a wide variety of sample sources including skin, hair, and feces which allows researchers to use noninvasive sampling methods (Karesh et al., 1987; Hoss et al., 1992; see chapter 18). With increased use of remote sampling, researchers can greatly increase the sample size of their studies, thereby increasing the strength of their analyses. However, with increased sample sizes (often hundreds of individuals), the task of generating DNA sequences from every individual in the study becomes labor-intensive, time-consuming, and costly. To facilitate the identification of DNA sequence genotypes in a more efficient manner, a number of DNA screening techniques have been developed that allow one to distinguish sequence differences among individuals without having to sequence each individual directly. Among the range of techniques that are available, PCR-SSCP (polymerase chain reaction combined with the analysis of single-stranded conformation polymorphisms) has emerged as a favored technique for screening DNA sequences in a wide variety of applications in molecular biology (Desgeorges et al., 1993; Hunter et al., 1993; Pietravalle et al., 1994). For conservation geneticists this technique is particularly useful because it is simple and inexpensive, requires only standard

DNA sequencing equipment, and can be conducted without the use of radioactive isotopes (Orita et al., 1989a; Yap and McGee, 1992; Hongyo et al., 1993; Lessa and Applebaum, 1993). Furthermore, SSCP analysis is a very sensitive technique allowing one to detect even single base pair changes with consistent reliability (Hayashi 1992; Fan et al., 1993).

TECHNICAL DESCRIPTION

Alternative DNA Screening Methods

The DNA screening techniques discussed here all make use of PCR, as this is the most efficient method for isolating and amplifying the genetic regions of interest. The techniques consist of protocols that make use of radioisotopes as well as protocols that avoid the use of radioisotopes. In addition, the available DNA screening techniques come with a range of applicability and equipment/reagent costs.

Heteroduplex Analysis. The analysis of heteroduplex molecules is an inexpensive and simple technique for determining whether a sample contains one or two genotypes at a given locus (White et al., 1993; Shen et al., 1993; Wilken et al., 1993). This technique involves the denaturation of PCR products generated from the genetic region of interest. The single strands are allowed to reanneal and are then separated according to size on a denaturing acrylamide gel. If more than one allele is present, some heteroduplex molecules will be formed between the different alleles. Mispairing between different allelic strands will cause the heteroduplex molecule to migrate at a different rate through the gel matrix. Homozygous individuals will have a single band of DNA for a given locus while heterozygous individuals will have multiple bands (Figure 11-1). In addition to identifying samples with multiple genotypes, heteroduplex analysis can be used to distinguish the presence of mitochondrial DNA duplications in the nuclear genome (Lessa and Applebaum, 1993). Furthermore, one can identify genotype differences between unknown samples and a reference sample by simply denaturing and mixing both samples prior to running on the gel. Finally, with the use of chemical cleavage of mismatched heteroduplex molecules, this technique becomes extremely sensitive in detecting mutations. However, the addition of these biochemical steps greatly increases the skill and effort necessary to complete the analyses (Hayashi, 1992). In its basic form, heteroduplex analysis has several advantages: it is simple, it uses standard DNA sequencing equipment, and can be conducted without radioactivity. Although heteroduplex analysis is effective for identifying homozygotes and heterozygotes, it is not practical for screening large numbers of individuals for genotype identification.

Denaturing Gradient Gel Electrophoresis (DGGE). The use of denaturing gradient gels has proven to be a highly sensitive technique for identifying sequence mutations. Unlike heteroduplex analysis, DGGE allows one to distinguish among

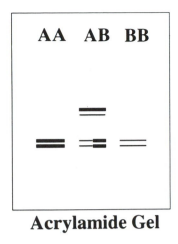

Acrylamide Gel

Figure 11-1 The separation of heteroduplex molecules from homoduplex molecules on an acrylamide gel in heteroduplex analysis.

alternative genotypes, including alternative homoduplex molecules. DGGE uses acrylamide gels with a narrow gradient of denaturing properties beginning at the top of the gel and increasing linearly toward the bottom of the gel (Figure 11-2). The denaturing properties are created chemically, generally using urea and formamide. As double-stranded PCR products run through the gel matrix, they will eventually reach a point in the gel where they will begin to denature; thereafter, their progression through the gel matrix will be greatly impeded. PCR products with different sequences will have slightly different associations owing to the base pairing of the two strands, and they will denature at different points in the denaturing gradient (Figure 11-2). Alternative genotypes are easily distinguished by their positions in the DGGE gel after 2–6 h of electrophoresis. The denaturation gradient range is initially determined by running an acrylamide gel with a wide gradient range extending across the gel (perpendicular to the path of the sample migration). PCR products from a single sample are loaded across the entire gel. One will see a sharp alteration in fragment location in the gradient where the progress of products begin to be noticeably inhibited. Once the concentration of denaturants at this inflection point is determined, standard DGGE gels with a more narrowly defined gradient range can be used to readily identify genotypes. Perpendicular calibration gels should be run for each new genetic region and for each new species being analyzed. Used in combination with heteroduplex analysis, this protocol has very strong resolving power. Also, with the addition of a 40 base pair (bp) GC clamp attached to one of the primers, PCR products up to to 500 base pairs in length can readily be analyzed (Sheffield et al., 1989). Finally, specific alleles can be cut out of a gel and directly sequenced, adding yet another advantage to DGGE that is lacking in heteroduplex analysis alone.

Although DGGE offers a rigorous and sensitive method of screening individuals for genotype differences, it is notably more complicated than heteroduplex analysis or SSCP analysis. In addition, it requires considerable initial investment

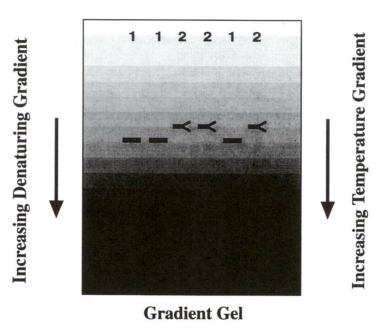

DGGE
Denaturing Gradient
Gel Electrophoresis

TGGE
Temperature Gradient
Gel Electrophoresis

1   GATTAGGATCCGAT**C**CGATCG**T**AGCTGAT
       CTAATCCTAGGCTA**G**GCTAGC**A**TCGACTA

2   GATTAGGATCCGAT**T**CGATCG**C**AGCTGAT
       CTAATCCTAGGCTA**A**GCTAGC**G**TCGACTA

Increasing Denaturing Gradient

Increasing Temperature Gradient

1   1   2   2   1   2

Gradient Gel

Figure 11-2   The separation of haploid alleles (e.g., mitochondrial DNA) with either DGGE or TGGE. Increasing shading indicates an increase in the denaturing properties of the gel. The sequences of the two alleles are given.

in specialized equipment such as gradient makers for the construction of the gradient gels and aquarium apparatus to maintain the necessary running conditions (Myers et al., 1986).

Temperature Gradient Gel Electrophoresis (TGGE).   TGGE is based on the same basic principle as DGGE with the exception that it uses a temperature gradient rather than a chemical gradient (Figure 11-2). Here, a temperature gradient is established that increases linearly from the top of the gel to the bottom. As each PCR product migrates through this gel it will reach a point at which the gel temperature will cause denaturation to begin and the migration of the product will be impeded (Figure 11-2). Like DGGE, TGGE requires the use of perpendicular calibration gels, is quite sensitive to mutations, can be used effectively in combination with heteroduplex analysis, and allows for the direct sequencing of alleles (Wartell et al., 1990; Hinney et al., 1994). However, TGGE requires the use of temperature-controlled water circulators to create the desired temperature gradient. This requires a large investment in specialized equipment that is estimated to be twice as expensive as the DGGE apparatus, and the technique does not make use of standard sequencing equipment as does heteroduplex analysis and SSCP analysis (Lessa and Applebaum, 1993).

Allele-Specific Oligonucleotide (ASO) Hybridization. This method allows one to identify specific sequence changes when the genotypes have been previously determined (Saiki et al., 1989). ASO hybridization is applicable only for short stretches of sequence and requires the construction of allele-specific probes for each allele (Hayashi et al., 1992). This makes the approach rather limited for large-scale population analyses where many alleles may be present and where one generally would like to quickly screen several hundred base pairs of sequence. However, certain steps in ASO can be automated such that many individuals can be screened at once (Saiki et al., 1989). In situations where individuals with specific alleles must be identified among many individuals (e.g., among fish stocks), this technique can be simple, fast, and relatively inexpensive to apply.

PCR-SSCP

Background

The analysis of SSCPs generated from PCR products has been in use since the late 1980s (Orita et al., 1989a,b). Since that time it has been used in a variety of molecular genetic applications and has undergone a number of protocol modifications. Originally, the technique involved amplifying the target sequence with radioactively labeled primers that were quickly labeled in a simple kinase reaction and then used in a PCR reaction (Orita et al., 1989a; Saiki et al., 1988). The PCR products were then denatured and the single-stranded products were subjected to electrophoresis in a non-denaturing polyacrylamide gel. Under these conditions the single-stranded DNA molecules would fold up upon themselves such that they would migrate through the gel according to both size and conformation (Figure 11-3). Molecules with different sequences are expected to have different secondary structures resulting in different mobilities through the gel.

The sensitivity of PCR-SSCP depends on how sequence mutations affect the conformations of single-stranded molecules. In general, the sensitivity of the method tends to decrease with increasing fragment length (Figure 11-4) (Hayashi, 1991). Estimates of the detection efficiency of SSCP analysis have suggested that single base pair changes will be detected 99% of the time for 100–300 bp fragments (Lessa and Appelbaum, 1993), $>90\%$ of the time for 350 bp fragments (Fan et al., 1993), and $>80\%$ of the time for 400 bp fragments (Hayashi, 1991). This would suggest that SSCP analysis is most efficient for DNA sequence surveys with PCR products <350 bp in length. There is currently no evidence for any detection preference based on the position of mutations in the fragment (Fan et al., 1993).

SSCP
SINGLE STRANDED
CONFORMATION POLYMORPHISMS

1 GATTAGGATCCGAT**C**CGATCG**T**AGCTGAT
CTAATCCTAGGCTA**G**GCTAGC**A**TCGACTA

2 GATTAGGATCCGAT**T**CGATCG**C**AGCTGAT
CTAATCCTAGGCTA**A**GCTAGC**G**TCGACTA

Non-denaturing
Acrylamide Gel

Figure 11-3   The separation of SSCPs from a haploid locus (e.g., mitochondrial DNA) on anondenaturing gel according to conformational differences. Examples of DNA sequence genotypes and their hypothetically folded molecules are depicted.

PCR-SSCP can be conducted with or without the use of radioisotopes. Nonradioactive techniques incorporate the use of fluorescent molecules, silver staining, or ethidium bromide to detect the single-stranded fragments (Hayashi, 1992; Yap and McGee, 1992; Hongyo et al., 1993). In addition, the PCR products, once separated on an SSCP gel, can be recovered for direct sequencing and genotype confirmation (Suzuki et al., 1991).

Methods

There are a variety of PCR-SSCP protocols currently available. They can generally be divided into two processes. The first set of methods involves labeling the PCR

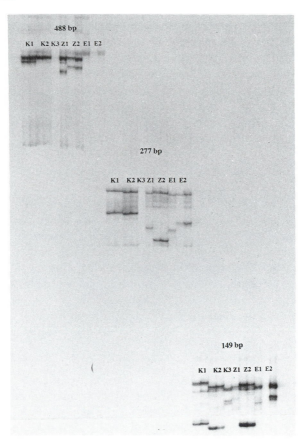

Figure 11-4 A comparison of nine African wild dogs examined for control region sequence differences using SSCP. Three different PCR product sizes were run simultaneously in a nondenaturing gel at 10°C (40 W) for 2.5 h. Genotypes and size of PCR products are depicted in the figure.

product, either before or during the PCR reaction, for later detection. Here, one can label the primers with either a radioactive isotope or flourescein molecule such that any PCR products made with these primers will be easily detected through autoradiography or with automated sequencing equipment (Orita et al., 1989a; Makino et al, 1992). Alternatively, a radioactive nucleotide could be incorporated into the PCR product during the PCR reaction and the strands detected with autoradiography (Dean et al., 1990; Dean and Gerrard, 1991). The second set of methods does not rely on any prelabeling of the PCR products; rather they involve staining the SSCPs directly with either a silver nitrate solution or with ethidium bromide (Dockhorn-Dworniczak et al., 1991; Mohabeer et al., 1991; Hongyo et al., 1993).

Whichever method is chosen for detection of the SSCPs, the PCR products require electrophoresis through a 5–10% nondenaturing polyacrylamide gel. These gels can be cast and run in standard DNA sequencing apparatus. However, there are a number of conditions that dramatically effect the clarity and detection sensitivity of SSCP genotypes. Clearly, the most important factor for successfully identifying SSCPs is the gel temperature during electrophoresis. SSCP gels are run at much cooler temperatures than standard sequencing gels in order to allow the single-stranded products to remain folded in their conformation. This is accomplished either with a commercially available apparatus specially designed for SSCP, with a fan unit to cool the gel, or by running the gel on standard equipment in a cold room or refrigeration unit. It is important to keep the gel temperature constant during electrophoresis. In addition, SSCP gels can be run at lower power to reduce the temperature of the gel (10–40 watts).

The running buffer concentration can also be altered to improve the sharpness of the bands in the SSCP gel. Generally, a $0.5 \times$ TBE buffer (45 mmol/l Tris, 45 mmol/l boric acid, 4 mmol/l EDTA) is recommended (Spinardi et al., 1991). In addition, 5–10% glycerol can be added to the gel when running at warmer temperatures (e.g., 25°C) to allow for the detection of certain mutation types (Spinardi et al., 1991). Finally, using polyacrylamide gels with small proportions of bisacrylamide (1.5–2%) can reduce cross-linking, thereby improving results (Hayashi, 1991). The combination of conditions that will distinguish all possible alleles depends on the method of identifying SSCPs in the gel. However, the conditions should be optimized for each genetic region and each species being studied.

Applications

PCR-SSCP has been used widely in medical molecular genetic research. The technique has been invaluable in the detection of mutations in studies of many genetic diseases. For example a number of cancer-related mutations have been identified with PCR-SSCP (e.g., Bafico et al., 1993; Mashiyama et al., 1991; Mazars et al., 1991). PCR-SSCP has also been used in the detection of genes for hereditary diseases such as cystic fibrosis (Dean et al., 1990; DesGeorges et al., 1993), phenylketanuria (Dockhorn-Dworniczak et al., 1991), and Tay–Sachs disease (Ainsworth et al., 1991). Genetic mapping has also been aided by the use of PCR-SSCP, which is proving to be easier and more cost-effective than Southern blotting methods (Beier, 1993; Hunter et al., 1993). Finally, considering the great number of polymorphisms that can be detected using PCR-SSCP, it becomes clear that this technique is useful in many areas of molecular genetic research (e.g., Lench et al., 1994; Pietravalle et al., 1994).

The uses in conservation genetics generally concern the rapid screening of large numbers of individuals for DNA sequence differences within and among populations. Ideally, the DNA sequences that are screened consist of highly variable coding or noncoding regions of the genome that can be isolated and amplified through a PCR reaction. Regions that are of interest include intron regions between transcribed exons (see chapter 2), portions of the highly variable major histocompatibility complex (MHC) (Bannai et al., 1993; Pietravalle et al., 1994; see chapter 14), or variable coding regions (e.g., Lessa, 1992; see chapter 13).

One commonly analyzed region is the mitochondrial DNA control region (e.g., Vigilant et al., 1991; Wenink et al., 1993; Edwards 1993; Rosel et al., 1994;

Taberlet and Bouvet, 1994). The high rate of mutation and lack of recombination makes this region extremely useful for population level analyses (see chapter 8). The use of PCR-SSCP greatly facilitates the screening of control region sequence genotypes, thereby reducing the overall effort and cost of analysis of this region (Yap and McGee, 1992). Although the use of PCR-SSCP in conservation genetics is relatively new, there are a few studies that provide examples of how the use of a DNA screening technique like PCR-SSCP can provide an efficient means of analyzing the structure of populations.

CONSERVATION GENETICS OF THE AFRICAN WILD DOG

The African wild dog (*Lycaon pictus*) is the largest canid in Africa and the only extant member of the genus *Lycaon*. Historically it existed in nearly every habitat in sub-Saharan Africa, being excluded only from the rainforest and extreme desert. African wild dogs are obligate carnivores with a unique hunting strategy and a highly developed social system (Estes and Goddard, 1967; Fuller and Kat, 1990). Unlike some other large carnivores whose numbers have increased substantially in Africa in recent years, African wild dog numbers have declined dramatically, particularly in western and eastern Africa (Fanshawe et al., 1991; Kingdon, 1977) and are now recognized as endangered by the World Conservation Union (Ginsberg and Macdonald, 1990). Currently, African wild dog numbers seem to have stabilized in some parts of southern Africa, where they largely occur in protected areas (Fanshawe et al., 1991). However, the persistence of wild dog packs requires a sufficient prey base distributed over large tracts of undisturbed land. For example, home range sizes of African wild dog packs vary from 500 km^2 to 2000 km^2 (Frame et al. 1979; Fuller and Kat, 1990; Reich 1978). With such high mobility we would expect wild dog populations to show little substructure throughout their range (Lehman and Wayne, 1991). However, the recent dramatic declines seen in eastern Africa may have caused a reduction in the relative genetic variability in that region compared to more stable populations in southern Africa.

In an initial study of African wild dogs from both eastern and southern Africa, we discovered several important features concerning the genetic structure of wild dogs (Girman et al., 1993). We examined approximately 80 samples from populations in Masai Mara, Kenya; Serengeti National Park, Tanzania; Hwange National Park, Zimbabwe; and Kruger National Park, Republic of South Africa (Figure 11-5). Through the analysis of restriction fragment length polymorphisms (RFLPS) from the mitochondrial genome we determined that African wild dog populations in eastern Africa are distinct from those in southern Africa (Girman et al., 1993). In addition, we found that the levels of mitochondrial DNA variability in eastern African populations are comparable to those found in other more stable populations. This is important because genetic variability is thought to be crucial for the long term persistence of any population or species (Gilpin and Soule, 1986).

Considering the potential dispersal abilities of African wild dogs, these results were somewhat surprising and had important conservation implications. Because

Figure 11-5 A map showing the localities of the populations sampled for the analysis of the genetic structure of African wild dogs.

captive stocks of African wild dogs appear to be solely from southern African stock, conservation strategies in eastern Africa have evolved from the notion of relocating and transplanting wild dogs from southern Africa into eastern Africa, to an emphasis on captive breeding wild dogs from eastern Africa in order to preserve the naturally occurring distinct populations. Since eastern and southern populations had what appeared to be distinct and long-term evolutionary histories, they could be considered separate evolutionary units (Ryder, 1986). These populations could now be treated as separate management units for the purposes of conservation (Moritz, 1994).

After the completion of this initial study we continued to collect genetic samples from these and other populations. Our sample size increased to approximately 200 genetic samples from wild dog populations in the Masai Mara, Kenya; Serengeti National Park, Tanzania; Selous, Tanzania; Okavango Delta, Botswana; Hwange National Park, Zimbabwe; Etosha National Park, Namibia; and Kruger National Park, Republic of South Africa (Figure 11-5). We examined these samples through the analysis of 402 bp of DNA sequence from region I of the mitochondrial DNA control region. However, the prospect of directly sequencing 200 samples led us to seek a more efficient method of screening these samples for sequence genotypes. Among the possible DNA screening techniques we chose to use PCR-SSCP for several reasons. First, SSCP had the sensitivity necessary to identify the genotypes that had been identified through direct sequencing. Second, we were able to easily determine the usefulness of the technique using existing DNA sequencing equipment in the laboratory. Using DGGE or TGGE would have required an additional investment of thousands of dollars. Finally, PCR-SSCP is simple and fast and could thus be easily and quickly evaluated as to its usefulness in this study.

For our analysis we chose to use the radioactively labeled primer method of Orita et al. (1989a). This method allowed the use of small reaction volumes in our PCR reactions and provided the detection efficiency necessary to carry out our survey of control region sequences. In addition, this method allowed access to DNA from poorly preserved skin and fecal samples that often consisted of very small amounts of partially degraded DNA. Standard PCR reactions were carried out in 10 μl or 25 μl volumes using PCR primers end-labeled with $[\gamma\text{-}^{32}P]ATP$ in a standard polynucleotide kinase reaction (Orita et al., 1989a). Two sets of PCR primers were included in each PCR reaction to produce one 296 bp fragment and one 256 bp fragment. The PCR products were added to a loading buffer (98% formamide, 20 mmol/l EDTA, 0.05% xylene cyanol, and 0.05% bromophenol blue) in 1:10 ratio of product to loading buffer. This mixture (2 μl) was loaded on to a 6% nondenaturing polyacrylamide (19:1 acrylamide–bisacrylamide) gel and run at 10°C (approximately 40 W in a 4°C cold room) for approximately 2 h. The gels were then exposed to x-ray film and the resultant bands from undiagnosed samples were compared to known control samples loaded on the same gel (Figure 11-6). Two individuals per genotype from each gel were then directly sequenced from the original PCR product for confirmation of the genotype.

The results of this study provided a contrast to our initial RFLP results. There was a distinction in the presence of genotypes between populations in

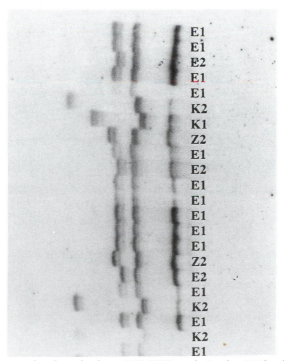

Figure 11-6   An example of results from PCR-SSCP analysis of a 256 bp fragment of mrDNA control region sequence. The mitochondrial DNA genotypes are given in the figure.

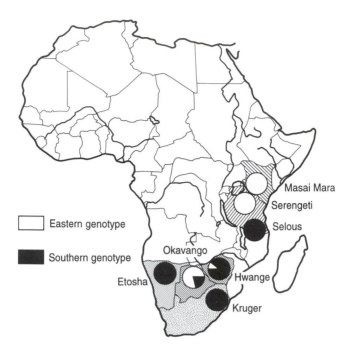

Figure 11-7 A map depicting the relative amount of eastern African or southern African mitochondrial DNA control region DNA sequence genotypes in each population. Populations in Hwange National Park and the Okavango Delta region show a mixture of eastern and southern genotypes.

the Masai Mara/Serengeti region and populations in Etosha and South Africa that supports the initial notion of population subdivision between eastern and southern African populations (Figure 11-7). In addition, phylogenetic analysis of the genotypes still results in two monophyletic clades. The sequence divergence between any two individuals from each clade is approximately 3–6% compared to 0.3–1.5% for any two individuals within either clade. However, analysis of populations in the Okavango Delta, Hwange National Park, and the Selous Game Preserve show that eastern and southern African populations are not completely subdivided. Eastern African genotypes were found mixed with southern African genotypes in Okavango and in Hwange. Thus, a mixing zone appears to exist in this south/central African region (Figure 11-7). In addition, genotypes found in the Selous were most closely related to a southern African genotype found only in Kruger Park (Girman, unpublished data). These results suggest that although there may have been a separation of eastern and southern populations at some point in the history of African wild dogs, there has recently been some gene flow between these regions. Current data suggest that wild dogs from eastern Africa have migrated down into Botswana and Zimbabwe while wild dogs from South Africa have made their way up the coast into southern Tanzania.

Of the 200 wild dogs analyzed for 402 bp of mitochondrial DNA control region sequence, 75 were directly sequenced. All 75 individuals that were directly sequenced were correctly identified for their genotype by the PCR-SSCP technique. Furthermore, by using PCR-SSCP to analyze the 125 individuals that were not directly sequenced, we were able to use five kinase reactions and three nondenaturing gels in place of 250 sequencing reactions and 21 sequencing gels. The use of PCR-SSCP allowed us to greatly reduce the amount of time, effort, and money expended on this project while obtaining an accurate account of the mitochondrial genotypes present among African wild dog genetic populations. The results of this analysis permitted us to map more precisely the genetic subdivision in African wild dogs, leading to the development of better management strategies.

The use of PCR-SSCP has been demonstrated to be a powerful technique for the detection of mutations in many fields of molecular genetic research. However, its potential use in conservation genetics has only just begun to be exploited. In addition to the many examples in humans and the study of African wild dogs described here, PCR-SSCP should be easily adaptable to the study of any species and any genetic region where sequence differences can be identified among DNA fragments 100–300 bp in length. In our laboratory alone, we have successfully used PCR-SSCP to identify mitochondrial DNA control region sequence differences in populations of California sea lions (*Zalophus californicus*) (Maldonado et al., 1995), raccoons (*Procyon lotor*), and coatis (*Nasua nasua*) (Girman, unpublished data) with no modifications in protocol. With its simplicity and cost-effectiveness, PCR-SSCP currently offers the most efficient way to quickly and accurately identify DNA sequence genotypes and offers the researcher a range of both radioactive and nonradioactive protocols to best suit their existing laboratory conditions and equipment.

ACKNOWLEDGMENTS

I thank R.K. Wayne and F. Hertel for their assistance on this project. P.W. Kat, M.G.L. Mills, J.R. Ginsberg, J.W. McNutt, M. Borner, V. Wilson, J.H. Fanshawe, L. Scheepers, and S. Creel contributed valuable genetic samples. M. Matocq and K. Koepfli provided valuable comments on the manuscript. L.M. Lau and K.H. Mamiya provided technical assistance. Support for this work was provided by E. and L. Cekanski and J.R. and P.L. Girman. Additional support was provided by a USPHS National Research Award (GM-07104) to D.J.G.

REFERENCES

Ainsworth, P.J., L.C. Surh, and M.B. Coulter-Mackie. 1991. Diagnostic single strand conformational polymorphism, (SSCP): a simplified non-radioisotopic method as applied to a Tay-Sachs B1 variant. Nucleic Acids Res. 19: 405–406.

Bafico, A., L. Varesco, L. De Benedetti, M.A. Caligo, V. Gismondi, S. Sciallero, H. Aste, G.B. Ferrara, and G. Bevilacqua. 1993. Genomic PCR-SSCP analysis of the metastasis associated NM23-HI (NME1) gene: a study on colorectal cancer. Anticancer Res. 13: 2149–2154.

Bannai, M., T. Mazda, K. Tokunaga, and T. Juji. 1993. DNA single-strand conformation polymorphism method to distinguish DR4 alleles. Lancet 341: 769.

Beier, D.R. 1993. Single-strand conformation polymorphism (SSCP) analysis as a tool for genetic mapping. Mammal. Genome 11: 627–631.

Dean, M. and B. Gerrard. 1991. Helpful hints for the detection of single-stranded conformation polymorphisms. Biotechniques 10: 8.

Dean, M., M. White, J. Amos, B. Gerrard, C. Stewart, K. Khaw, and M. Leppert. 1990. Multiple mutations in highly conserved residues are found in mildly affected cystic fibrosis patients. Cell 61: 863–870.

Desgeorges, M., P. Boulot, P. Kjellberg, G. Lefort, M. Rolland, J. Demaille, and M. Claustres. 1993. Prenatal diagnosis for cystic fibrosis using SSCP analysis. Prenat. Diagn. 13: 147–148.

Dockhorn-Dworniczak, B., B. Dworniczak, L. Brommelkamp, J. Bulles, J. Horst, and W.W. Bocker. 1991. Non-isotopic detection of single strand conformation polymorphism (PCR-SSCP): a rapid and sensitive technique for the diagnosis of phenylketonuria. Nucleic Acids Res. 19: 2500.

Edwards, S.V. 1993. Mitochondrial gene genealogy and gene flow among island and mainland populations of a sedentary songbird, the grey-crowned babbler (*Pomatostomus temporalis*). Evolution 47: 1118–1137.

Estes, R.D. and J. Goddard. 1967. Prey selection and hunting behavior of the African wild dog. J. Wildlife Management 31: 52–70.

Fan, E., D.B. Levin, B.W. Glickman, and D.M. Logan. 1993. Limitations in the use of SSCP analysis. Mutat. Res. 288: 85–92.

Fanshawe, J.H., L.H. Frame, and J.R. Ginsberg. 1991. The wild dog—Africa's vanishing carnivore. Oryx 25: 137–144.

Frame, L.H., J.R. Malcolm, G.W. Frame, and H. Van Lawick. 1979. Social organisation of African wild dogs (*Lycaon pictus*) on the Serengeti Plains, Tanzania. Z. Tierpsychol. 50: 225–249.

Fuller, T.K. and P.W. Kat. 1990. Movements, activity, and prey relationships of African wild dogs (*Lycaon pictus*) near Aitong, southwestern Kenya. Afr. J. Ecol. 28: 330–350.

Gilpin, M.E. and M.E. Soule. 1986. Minimum viable populations: process of species extinction. In: M.E. Soule, ed. Conservation biology: the science of scarity and diversity, pp. 19–34. Sunderland, Sinauer.

Ginsberg, J.R. and D.M. Macdonald. 1990. Foxes, wolves, and jackals: an action plan for the conservation of canids. Gland, World Conservation Union.

Girman, D.J., P.W. Kat, M.G.L. Mills, J.R. Ginsberg, M. Bomer, V. Wilson, J.H. Fanshawe, C. Fitzgibbon, L.M. Lau, and R.K. Wayne. 1993. Molecular genetic and morphologic analyses of the African wild dog (*Lycaon pictus*). J. Hered. 84: 450–459.

Gottelli, D., C. Sillero-Zubiri, G.D. Applebaum, M.S. Roy, D.J. Girman, J. Garcia-Moreno, E.A. Ostrander, and R.K. Wayne. 1994. Molecular genetics of the most endangered canid: the Ethiopian wolf *Canis simensis*. Mol. Ecol. 3: 301–312.

Hayashi, K., 1991. PCR-SSCP: a simple and sensitive method for detection of mutations in the genomic DNA. PCR Methods Applic. 1: 34–38.

Hayashi, K. 1992. PCR-SSCP: A method for detection of mutations. GATA 9: 73–79.

Hinney, A., C. Durr, C. Luckenbach, and H. Ritter. 1994. TGGE and HIEF—a comparison of two methods in the detection of carriers of the Z mutation of the alpha-1-antitrypsin gene. Hum. Genet. 93: 571–574.

Hongyo, T., G.S. Buzard, R.J. Calvert, and C.M. Weghorst. 1993. Cold SSCP—a simple rapid and non-radioactive method for optimized single-strand conformation polymorphism analyses. Nucleic Acids Res. 21: 3637–3642.

Hoss, M., M. Kohn, S. Paabo, F. Knauer, and W. Schroder. 1992. Excremental analysis of PCR. Nature 359: 199–201.

Hunter, K.W., M.L. Watson, J. Rochelle, S. Ontiveros, D. Munroe, M.F. Seldin, and D.E. Housman. 1993. Single-strand conformational polymorphism (SSCP) mapping of the mouse genome: integration of the SSCP, microsatellite, and gene maps of mouse chromosome 1. Genomics 18: 510–519.

Karesh, W.B., F. Smith, and H. Frazier-Taylor. 1987. A remote method for obtaining skin biopsy samples. Conserv. Biol. 3: 261–262.

Kingdon, J. 1977. Eastern African mammals, vol. IIIA. Chicago, Ill.: University of Chicago Press.

Lehman, N. and R.K. Wayne. 1991. Analysis of coyote mitochondrial genotype frequencies: estimation of the effective number of alleles. Genetics 129: 4405–4416.

Lench, N.J., A.H. Brook, and G.B. Winter. 1994. SSCP detection of a nonsense mutation in exon 5 of the amelogenin gene (AMGX) causing X-linked amelogenesis imperfects (AIH1). Hum. Mol. Genet. 3: 827–828.

Lessa, E.P. 1992. Rapid survey of DNA sequence variation in natural populations. Mol. Biol. Evol. 9: 323–330.

Lessa, E.P. and G. Applebaum. 1993. Screening techniques for detecting allelic variation in DNA sequences. Mol. Ecol. 2: 119–129.

Makino, R., H. Yazyu, Y. Kishimoto, T. Sekiya, and K. Hayashi. 1992. F-SSCP. fluorescence-based polymerase chain reaction-single-strand conformation polymorphism (PCR-SSCP) analysis. PCR Methods Applic. 2: 10–13.

Maldonado, J.E., F. Orta Davila, B.S. Stewart, E. Geffen, and R.K. Wayne. 1995. Intraspecific differentiation in California sea lions (*Zalphus californicus*) from Southern California and the Gulf of California. Marine Mammal Sci. 11: 46–58.

Mashiyama, S., Y. Murakami, T. Yoshimoto, T. Sekiya, and K. Hayashi. 1991. Detection of p53 gene mutations in human brain tumors by single-strand conformation polymorphism analysis of polymerase chain reaction products. Oncogene 6: 1313–1318.

Mazars, R., P. Pujol, T. Maudelonde, P. Jeanteur, and C. Theillet. 1991. p53 mutations in ovarian cancer: a late event? Oncogene 6: 1685–1690.

Mohabeer, A.J., A.L. Hiti, and W.J. Martin. 1991. Non-radioactive single strand conformation polymorphism (SSCP) using the Pharmacia Phastsystem. Nucleic Acids Res. 19: 3154.

Moritz, C. 1994. Applications of mitochondrial DNA analysis in conservation: a critical review. Mol. Ecol. 4: 401–411.

Myers, R.M., T. Maniatis, and L.S. Lerman. 1986. Detection and localization of single base changes by denaturing gradient gel electrophoresis. Methods Enzymol. 155: 501–527.

Orita, M., Y. Suzki, T. Sekiya, and K. Hayashi. 1989a. Rapid and sensitive detection of point mutations and DNA polymorphisms using the polymerase chain reaction. Genomics 5: 874–879.

Orita, M., H. Iwahana, H. Kanazawa, K. Hayashi, and T. Sekiya. 1989b. Detection of polymorphisms of human DNA by gel electrophoresis and single-strand conformation polymorphisms. Proc. Natl. Acad. Sci. U.S.A. 86: 2766–2770.

Pietravalle, F., J. Tkaczuk, M. Thomsen, A. Cambon-Thomsen, E. Ohayon, and M. Abbal. 1994. Evaluation of HLA-DPB incompatibility by PCR-SSCP in the choice of a bone marrow donor. Transplant. Proc. 26: 238.

Reich, A. 1978. The behavior and ecology of the African wild dog (*Lycaon pictus*) in the Kruger National Park (Ph.D. thesis). New Haven, Conn.: Yale University.

Rosel, P.E., A.E. Dizon, and J.E. Heyning. 1994. Genetic analysis of sympatric morphotypes of common dolphins (genus *Delphinus*). Marine Biol. 119: 159–167.

Ryder, O.A. 1986. Species conservation and systematics: the dilemma of subspecies. Trends Ecol. Evol. 1: 9–10.

Saiki, R.K., D.H. Gelfand, S. Stoffel, S.J. Scharf, R.H. Higuchi, G.T. Hom, K.B. Mullis, and H.A. Erlich. 1988. Primer-directed amplification of DNA with a thermostable DNA polymerase. Science 239: 487-491.

Saiki, R.K., P.S. Walsh, C.H. Levenson, and H.A. Erlich. 1989. Genetic analysis of amplified DNA with immobilized sequence-specific oligonucleotide probes. Proc. Natl. Acad. Sci. U.S.A. 86: 6230–6234.

Sheffield, V.C., D.R. Cox, and R.M. Myers. 1989. Attachment of a 40-base-pair GC rich sequence (GC clamp) to genomic DNA fragments by polymerase chain reaction results in improved detection of single base changes. Proc. Natl. Acad. Sci. U.S.A. 86: 232–236.

Shen, M.H., P.S. Harper, and M. Upadhyaya. 1993. Neurofibromatosis type-1 (NFI)—the search for mutations by PCR-heteroduplex analysis on hydrolink gels. Hum. Mol. Genet. 2: 1861–1864.

Spinardi, L., R. Mazars, and C. Theillet. 1991. Protocols for an improved detection of point mutations by SSCP. Nuceic Acids Res. 19: 4009.

Suzuki, Y., T. Sekiya, and K. Hayashi. 1991. Allele-specific polymerase chain reaction: a method for amplification and sequence determination of a single component among a mixture of sequence variants. Anal. Biochem. 192: 82–84.

Taberlet, P. and J. Bouvet. 1994. Mitochondrial DNA polymorphism, phylogeography, and conservation genetics of the brown bear *Ursus arctos* in Europe. Proc. R. Soc. Lond. B. 255: 195–200.

Taylor, A.C. , W.B. Sherwin, and R.K. Wayne. 1994. The use of simple sequence loci to measure genetic variation in bottlenecked species: the decline of the northern hairy-nosed wombat (*Lasiorhinus krefftii*). Mol. Ecol. 4: 277–290.

Vigilant, L., M. Stoneking, H. Harpending, K. Hawkes, and A.C. Wilson. 1991. African populations and the evolution of human mitochondrial DNA. Science 253: 1503–1507.

Wartell, R.M., S.H. Hosseini, and C.P. Moran. 1990. Detecting base substitutions in DNA fragments by temperature gradient gel electrophoresis. Nucleic Acids Res. 18: 2699–2705.

Wayne, R.K. and S.M. Jenks. 1991. Mitochondrial DNA analysis implying extensive hybridization of the endangered red wolf *Canis rufus*. Nature 351: 565–568.

Wenink, P.W., A.J. Baker, and M.G.J. Tilanus. 1993. Hypervariable-control-region sequences reveal global population structuring in a long-distance migrant shorebird, the dunlin (*Calidris alpina*). Proc. Natl. Acad. Sci. U.S.A. 90: 94–98.

White, M.B., M. Carvalho, D. Derse, S.J. O'Brien, and M. Dean. 1992. Detecting single base substitutions as heteroduplex polymorphisms. Genomics 12: 301–306.

Wilken, D.J., K.E. Kpeivnikar, and D.H. Cohn. 1993. Heteroduplex analysis can increase the informativeness of PCR-amphfied VNTR markers—application using a marker tightly linked to the COL2A1 gene. Genomics 15: 372–375.

Yap, E.P.H. and J.O.D. McGee. 1992. Non-isotopic SSCP detection in PCR products by ethidium bromide staining. Trends Gen. 8: 49–54.

12

Application of Chloroplast DNA Restriction Site Studies for Conservation Genetics

JAVIER FRANCISCO-ORTEGA, ROBERT K. JANSEN,
ROBERTA J. MASON-GAMER, AND ROBERT S. WALLACE

Mitochondrial DNA (mtDNA) has been used widely in systematic and conservation genetic studies of animals (e.g., Wayne et al., 1992; Moritz, 1994; O'Brien, 1994; O'Ryan et al., 1994). The mitochondrial genome of plants has not been used for these kind of studies, however, because of the high incidence of structural rearrangements and low levels of sequence variation (Palmer, 1992). Although chloroplast DNA (cpDNA) variation is used widely in systematics (Palmer et al., 1988; Olmstead and Palmer, 1994), few if any restriction site studies have focused specifically on the establishment of strategies for biodiversity conservation or the understanding of the biology of threatened species. There has been recent interest in exploring evolutionary and populational questions using the chloroplast genome (reviewed in Soltis et al., 1992a; Rieseberg and Brunsfeld, 1992) which have direct application in conservation genetic studies, and these fall into three categories: (1) origin of hybrids and polyploids; (2) patterns of genetic diversity within species; and (3) phylogenetic relationships within and among closely related species.

The chloroplast genome is the most widely used macromolecule for evolutionary and phylogenetic studies in plants (Palmer et al., 1988; Clegg and Zurawski, 1992; Olmstead and Palmer, 1994). It is well suited for these types of investigations, in part because of the ease of isolating large quantities of cpDNA from small amounts of leaf material, as well as its highly conserved structure and slow rate of evolution among land plants (Palmer, 1991; Clegg et al., 1994). Complete sequences of six chloroplast genomes are now available from a wide diversity of plants, including an alga (*Euglena*, Hallick et al., 1993), a bryophyte (*Marchantia*, Ohyama et al., 1986), a conifer (*Pinus*, Wakasugi et al., 1994), and three flowering plants (rice, Hiratsuka et al., 1989; tobacco, Shinozaki et al., 1986; beech drops, Wolfe et al., 1992), and these have provided a wealth of information on both structural and sequence evolution. This information has facilitated the use of this molecule for evolutionary studies at a variety of taxonomic levels ranging from intraspecific to among phyla.

Chloroplast DNA is a circular molecule that varies in size between 120 and

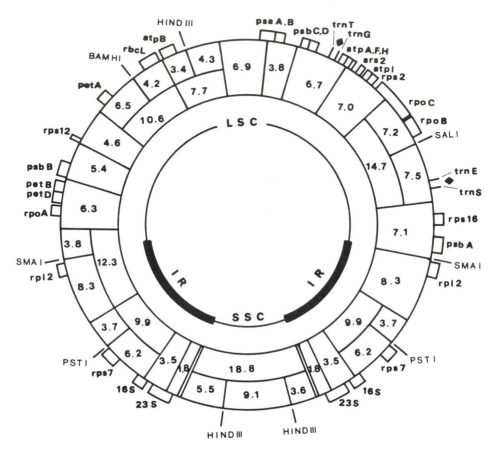

Figure 12-1 Restriction map of lettuce chloroplast DNA showing the location of 22 cloned restriction fragments and selected genes. All restriction sites are for *Sac*I, except where indicated. The two heavy lines indicate the extent of the inverted repeat (IR), and LSC and SSC indicate the large and small single-copy regions, respectively. Solid diamonds within the 6.7 and 7.5 kb fragments indicate the approximate location of the endpoints of the 22 kb inversion. Abbreviations for gene names follow Shiniozaki et al. (1986). All fragment sizes are given in kilobases. (Redrawn from Jansen and Palmer, 1987, 1988.)

217 kilobase pairs (kb), although most plants have genomes of approximately 160 kb (Palmer, 1991, Sugiura, 1992). Its most prominent feature is a large repeated sequence with two copies in opposite orientations, separating the remaining portion of the molecule into a large single-copy region (LSC) and a small single-copy (SSC) region (Figure 12-1). In general the chloroplast protein-coding regions evolve at a similar rate to plant mitochondrial genes and approximately five times slower than plant nuclear genes (Clegg and Zurawski, 1992; Palmer, 1991). The frequency of restriction site changes in different regions of the chloroplast genome has been examined in several plant groups. Jansen and Palmer (1987) demonstrated that the inverted repeat (IR) is the most highly conserved region, and this has been confirmed by sequence comparisons (Wolfe et al., 1987). Restriction site comparisons in the genus *Krigia* (Kim et al., 1992a) identified four

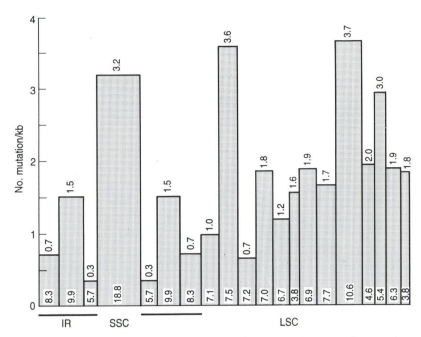

Figure 12-2   Histogram summarizing the location of 252 restriction site changes detected in different regions of the *Krigia* chloroplast genome from Kim et al. (1992a). The regions are indicated by the cloned *SacI* restriction fragments (Jansen and Palmer, 1987, 1988; Figure 12-1). The solid bar indicates the position and extent of the inverted repeat in lettuce. The numbers at the top of each bar indicate the average number of restriction site changes per kilobase of the genome.

rapidly changing regions (Figure 12-2). All of these variable regions include long spacer sequences or open reading frames (Figure 12-1). A similar pattern of restriction site changes was observed in two other studies (Johansson and Jansen, 1993; Whitton et al., 1995). Information on the frequency of restriction site changes in different portions of the chloroplast genome enables the selection of conserved regions of cpDNA for studies at higher taxonomic levels (Downie and Palmer, 1992, 1994) and rapidly evolving regions for investigations at lower levels (Mason-Gamer et al., 1995).

Another important feature of the chloroplast genome is its mode of inheritance. In most plants ($\sim$75%) the genome exhibits uniparental inheritance, paternal in conifers and maternal in angiosperms (reviewed in Sears. 1980; Corriveau and Coleman, 1988; Harris and Ingram, 1991). Early studies of plastid inheritance suggested that the genome was predominantly maternally inherited; however, recent studies have detected paternal and biparental inheritance and variation in mode of plastid inheritance within species (Cruzan et al., 1993; Sewell et al., 1993; Mason et al., 1994).

In this chapter we briefly describe the methodology for cpDNA restriction site studies and present three examples from our current research on the use of this approach at lower taxonomic levels, especially in conservation genetics. We will show that restriction site comparisons of the chloroplast genome, mainly by utilizing variable regions of the genome and frequent-cutting restriction enzymes, hold great potential for understanding patterns of genetic variation in rare plants. We will also discuss the merits of restriction site analyses versus DNA sequencing and their future utility in conservation genetics.

GENERAL METHODOLOGY

Restriction site studies of cpDNA can be performed primarily by two different approaches. The most common is based on Southern hybridization methods, which involve digestion of DNA with of a battery of 4–8 bp restriction enzymes, separation of DNA fragments via agarose electrophoresis, transfer of DNA to a filter membrane, and hybridization of labeled cpDNA probes to the membranes. Chloroplast genomes from a wide diversity of plants have been cloned and are available for use (Palmer, 1986; Jansen and Palmer, 1987; Chase and Palmer, 1989). The second method takes advantage of polymerase chain reaction (PCR) technology. This involves amplification of specific regions of the chloroplast genome, digestion of PCR fragments with restriction enzymes, and comparison of fragment patterns by agarose gel electrophoresis (Liston, 1992; Arnold, 1993; Cruzan and Arnold, 1993; Cruzan et al., 1993; McCauley 1994). This approach has the advantage of requiring much less plant material and in many instances even museum collections can be utilized (Jansen et al., 1996; Loockerman and Jansen, 1996). A number of specific primers have been developed and are useful for studying chloroplast genomes from a wide diversity of plant taxa (Arnold et al., 1991; Taberlet et al., 1991; Liston, 1992; Demeure et al., 1995). In spite of this advantage, the second approach is of limited utility for intraspecific studies because it examines such a small proportion of the chloroplast genome.

EXAMPLES OF APPLICATIONS OF RESTRICTION SITE APPROACH AT INTER- AND INTRASPECIFIC LEVELS

Origin of Polyploids in *Microseris*

Chloroplast DNA restriction site studies, usually in combination with similar studies of the nuclear ribosomal DNA (rDNA), have been used widely to study the origin and evolution of polyploids. This is a very common phenomenon in plants; it has been estimated that 95% of ferns and 47% of angiosperms are of polyploid origin (Grant, 1981). Chloroplast DNA studies have been used (1) to determine the parentage of polyploids; (2) to demonstrate multiple origins of both auto- and allopolyploids; and (3) to provide new insights into the evolution of polyploids (reviewed in Soltis and Soltis, 1992). Below we describe one example of the utility of cpDNA restriction site data for examining polyploid evolution in the flowering plant genus *Microseris* (Asteraceae).

Microseris comprises 16 species of annual and perennial herbs, native primarily

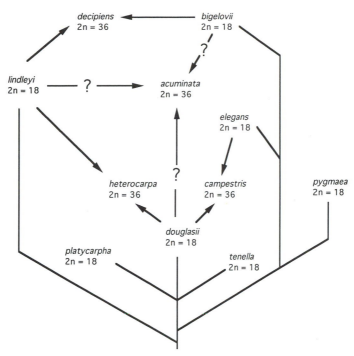

Figure 12-3 Proposed parentage of allotetraploids in *Microseris* based on morphological characters, chromosome numbers, and experimental crosses. (Redrawn from Chambers, 1955.)

to the western United States, with two disjunct species in Australia/New Zealand and Chile (Chambers, 1955). The genus contains five tetraploid species, four of which are placed in the annual subgenus *Microseris*. Chambers (1955) used morphological characters, geographic distribution, chromosome numbers, and experimental hybridizations to propose the parentage of the four annual tetraploids (Figure 12-3). Recent cpDNA and morphological studies (Wallace and Jansen, 1990; Jansen et al. 1991) indicated that one of the diploid species, *M. lindleyi*, should be recognized as the distinct monotypic genus *Uropappus*.

Wallace and Jansen (1995) examined chloroplast and nuclear rDNA variation in 70 populations of the annual species to determine the origin of the allotetraploid species *M. heterocarpa* and *M. decipiens*. Four cpDNA restriction site changes were identified to distinguish the diploid species (Table 12-1). Furthermore, additivity of nuclear rDNA restriction fragment patterns of the putative parents in the alloploids (Table 12-1) confirmed their parentage as suggested by Chambers (1955). The 70 populations were scored for their plastid genotypes and this enabled the identification of the maternal parent of each allotetraploid. The paternal parent of both allotetraploids is *M. lindleyi*, a species that has recently been segregated into the monotypic genus *Uropappus* because it is more closely related to the genera *Agoseris* and *Nothocalais* (Jansen et al., 1991). The involvement of the diploid *M. lindleyi* in the origin of these allotetraploids is enigmatic because of its distant relationship to the other diploid parents. Another exciting result of this study is the multiple origin of the allotetraploid *M. heterocarpa*. The plastid

Table 12-1   Chloroplast DNA Genotypes of Annual *Microseris*. Presence of a restriction site
= +, absence = −

Taxon	Observed Plastid Genotypes				Nuclear rDNA Fragments	
	*Hae*II 6.9	*Eco*RI 7.0	*Dra*I 12.3	*Eco*RV 18.8	9.6	7.1 + 1.5 + 1.0
Microseris douglasii (DC.)	+	−	+	+	+	
Sch.-Bip. (2N)	+	+	−	−	+	
	+	−	−	−	+	
	+	−	+	−	+	
Microseris heterocarpa	+	+	−	−	+	+
(Nutt.) K. Chambers (4N)[a]	+	−	−	−	+	+
	+	−	+	−	+	+
Microseris lindleyi (DC.) Gray (2N)[b]	−	−	−	−		+
Microseris decipiens K. Chambers (4N)[c]	+	−	+	+	+	+
Microseris bigelovii (Gray) Sch.-Bip. (2N)	+	−	+	+	+	

Source: Wallace and Jansen (1995).

[a] *Stebbinsoseris heterocarpa* (Nutt.) K. Chambers.

[b] *Uropappus lindleyi* (DC.) Nutt.

[c] *Stebbinsoseris decipiens* (K. Chambers) K. Chambers.

genotypes observed within the maternal parent *M. douglasii* showed four distinct combinations of restriction site patterns, three of which were also found in the examined populations of the tetraploid (Table 12-1). Thus, there is evidence for at least three independent origins of allopolyploidy in *M. heterocarpa*. Chloroplast DNA restriction site studies in several other plant groups have revealed additional examples of the multiple origin of polyploids (reviewed in Soltis and Soltis, 1992).

Patterns of Chloroplast DNA Variation in a Progenitor–Derivative Species Pair in *Coreopsis*

Studies of population differentiation in animals have employed restriction site variation extensively in the mitochondrial genome (reviewed in Wilson et al., 1985: Avise, 1986, Moritz et al., 1987). In contrast, similar studies in plants have relied primarily on allozyme data (Hamrick, 1989) because the organelle genomes were considered to be too conserved for population studies. Several recent studies, however, have suggested that intraspecific cpDNA diversity is high enough for examinination of population-level issues, including population differentiation (Neale et al., 1989; Soltis et al., 1989, 1991; Fenster and Ritland, 1992; Kim et al., 1992b; Hong et al., 1993; Petit et al., 1993; Byrne and Moran, 1994; Dong and Wagner, 1994; Mason-Gamer et al., 1995), gene flow (McCauley, 1994), hybridization (Whittemore and Schaal, 1991), introgression (reviewed in Rieseberg and Brunsfeld, 1992), and polyploidy (reviewed in Soltis et al., 1992b). Below we summarize the results of a populational study in the flowering plant genus

Coreopsis, which is especially relevant to the application of cpDNA markers for conservation genetics because one of the species examined is a recent derivative, and both are narrowly restricted geographically.

Coreopsis nuecensoides and *C. nuecensis* (Asteraceae) are narrowly distributed species endemic to Texas. Although their geographic ranges overlap, mixed populations are unknown (Smith, 1974). The two species are very similar morphologically and are primarily distinguished from each other by the presence of hairs on the phyllaries of *C. nuecensis*. The species are intersterile and also differ with regard to chromosome number, with *C. nuecensis* having $n = 9$ or 10 and *C. nuecensis* with $n = 6$ or 7. Morphological, chromosomal (Smith, 1974), and allozyme (Crawford and Smith, 1982) data suggested that *C. nuecensis* is a recent derivative of *C. nuecensoides*.

A cpDNA restriction site analysis of *Coreopsis* was performed to test this putative progenitor–derivative species pair hypothesis (Mason-Gamer et al., unpublished results). DNA variation was examined by the standard Southern hybridization approach for 113 individuals from eight populations using six 4 bp restriction enzymes. The cpDNA comparisons identified 16 variable restriction sites and 13 distinct haplotypes in the two species (Table 12-2). Haplotypes differed by between one and six restriction site changes, or 0.037–0.220% sequence divergence. A phylogenetic analysis (Figure 12-4) of the haplotypes supports the hypothesis that *C. nuecensis* is a derivative of *C. nuecensoides*. Furthermore, the

Table 12-2   Chloroplast DNA Restriction Site Haplotypes[a] Detected in *Coreopsis nuecensoides* F.B. Smith and *C. nuecensis* Heller

	Restriction Enzymes				
	*Alu*I 2 4 7	*Hae*III 6 6	*Hinf*I 3 6	*Mbo*I 2 4 4 4 7	*Msp*I 4 5 5 5
Haplotypes					
1	1 1 0	0 1	1 1	1 0 1 1 1	0 1 0 1
2	1 0 1	0 1	1 1	1 0 0 1 1	0 1 0 1
3	1 0 1	0 1	1 1	1 0 0 1 1	0 0 0 1
4	1 1 1	0 1	1 1	1 0 1 0 1	0 1 0 1
5	0 1 1	0 1	1 1	1 0 1 1 1	0 1 0 1
6	1 1 0	0 1	1 1	1 0 1 1 1	1 1 1 1
7	1 1 1	0 0	1 1	1 0 1 1 1	0 1 0 1
8	1 1 1	0 1	0 1	1 0 1 1 1	0 1 0 1
9	1 1 1	0 1	1 1	1 0 1 1 0	0 1 0 1
10	1 1 1	1 1	1 1	1 0 1 1 0	0 1 0 1
11	1 1 1	0 1	1 0	1 0 1 1 1	0 1 0 1
12	1 1 1	0 1	1 1	0 1 1 1 1	0 1 0 0
13	1 1 1	0 1	1 1	0 1 1 1 1	0 1 0 1
C. grandiflora[b]	1 1 1	0 1	1 1	0 1 1 1 1	0 1 0 1

Source: Mason-Gamer et al., unpublished.

[a] 0 = absence of site; 1 = presence of site. Numbers below each enzyme refer to regions in which restriction site changes were found in terms of the lettuce cpDNA fragments used a probes (Figure 12-1; Jansen and Palmer 1987, 1988): 2 = 4.6 kb; 3 = 5.4 kb; 4 = 6.4 kb; 5 = 6.9 kb; 6 = 7.5 kb; 7 = 18.8 kb.

[b] *Coreopsis grandiflora* Sweet was used as the outgroup in cladistic analyses (Figure 12-4).

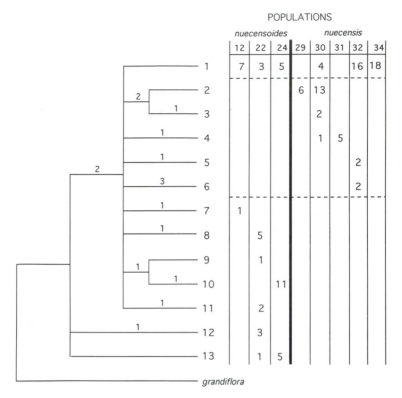

Figure 12-4 Single most parsimonious tree of the 13 cpDNA haplotypes of *Coreopsis necen-soides* and *C. nuecensis* based on 16 restriction site changes. The tree has a length of 16 and a consistency index of 1.0. Numbers at the tips of the branches indicate specific haplotypes (Table 12-2). Numbers along branches indicate the number of changes in each lineage. Numbers at the top of each column on the right indicate distinct populations and the numbers within each column show how many of each of the haplotypes appear in each population.

reduced haplotype and nucleotide diversity of *C. nuecensis* is consistent with this hypothesis.

This study demonstrates the utility of cpDNA restriction site comparisons for population studies. This is in spite of previous comparisons of cpDNA variation in *Coreopsis* (Crawford et al., 1990) which found very little differentiation among species in section *Coreopsis*. The use of more frequent-cutting restriction enzymes, the examination of only the most variable regions of chloroplast genome, and modifications of agarose gel electrophoresis conditions (higher-percentage gels and longer electrophoresis times) allowed the detection of additional cpDNA variation.

Phylogenetic Relationships in *Argyranthemum*

Argyranthemum (Asteraceae: Anthemideae) is the largest endemic genus of the Macaronesia. This biogeographical region is situated in the Atlantic Ocean between latitudes 15° and 40° N and comprises the archipelagos of Azores, Madeira, Desertas, Selvagens, Canaries and Cape Verde (Figure 12-5). At least

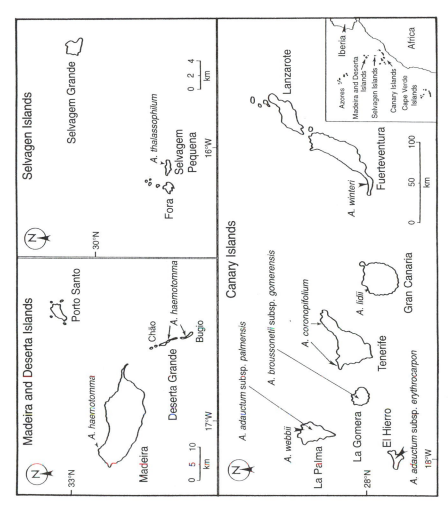

Figure 12-5 The Macaronesian islands and the distribution of the 10 rare taxa of *Argyranthemum*.

Table 12-3   Summary of cpDNA Restriction Site (Francisco-Ortega et al., 1996b) and Isozyme (Francisco-Ortega et al., 1996c) Data for the *Argyranthemum* Taxa with the Highest Priority for Genetic Conservation. U_{RS} = number of unique mutations detected from cpDNA restriction site, U_I = number of unique allozymes; H_O = gene diversity from isozyme data

Taxon	Island	cpDNA Lineage	U_{RS}	U_I	H_O
A. adauctum (Link) Humphries subsp. *erythrocarpon* (Svent.) Humphries	El Hierro	Lineage 2	0	0	0.1108
A. adauctum subsp. *jacobifolium* (Sch. Bip.) Humphries	Gran Canaria	Lineage 1	0	0	0.0411
A. adauctum subsp. *palmensis* A. Santos	La Palma	Lineage 2	1	0	0.1431
A. broussonetii (Pers.) Humphries subsp. *gomerensis* Humphries	La Gomera	Lineage 1	1	0	0.0842
A. coronopifolium (Willd.) Humphries	Tenerife	Lineage 2	1	0	0.0613
A. haemotomma (Lowe) Lowe	Deserta Grande and Madeira	Lineage 3	0	1	0.0819
A. lidii Humphries	Gran Canaria	Lineage 1	0	0	0.0536
A. thalassophilum (Svent.) Humphries	Selvagem Pequena	Lineage 3	2	1	0.0091
A. webbi Sch. Bip.	La Palma	Lineage 2	1	0	0.1177
A. winteri (Svent.) Humphries	Fuerteventura	Lineage 2	0	0	0.0554

one different species of *Argyranthemum* is endemic to Madeira, Selvagem Pequena and each of the Canaries. In addition *A. haemotomma* is found both in the islands of Deserta Grande and of Madeira. The genus comprises 23 species and 15 subspecies (Humphries, 1976a; Borgen, 1980; Rustan, 1981; Santos-Guerra, 1983) and each of these taxa is restricted to a particular ecological zone of the islands. Ten taxa are considered endangered, with few populations and individuals known in the wild (Table 12-3, Figure 12-5; Francisco-Ortega et al., 1996a).

A cladistic study of cpDNA restriction site data revealed that the genus comprises three major lineages (Francisco-Ortega et al. 1996b; Figure 12-6). One of these lineages is restricted to the Portuguese islands of Madeira, Desertas, and Selvagens, whereas the other two are exclusive to the Canaries. Relationships were poorly resolved among these three groups and the current sectional classification of *Argyranthemum* (Humphries, 1976a) was not supported. The first lineage of Canarian species was formed by taxa of the islands of Tenerife, La Palma, La Gomera, and Gran Canaria, whereas the second had endemics from all the islands except Gran Canaria.

Five of the endangered taxa had unique restriction site changes (Table 12-3). These data were compared with results obtained from isozyme studies (Francisco-Ortega et al., 1996c) that were valuable in terms of giving an overall estimation of the genetic variation, which ranged between 0.091 for the extremely rare *A. thalassophilum* to 0.1431 for *A. adauctum* subsp. *adauctum*. However, unique alleles were identified for only two of the endangered taxa (i.e., *A. thalassophilum* and *A. haemotomma*). In this instance, cpDNA restriction site studies detected more

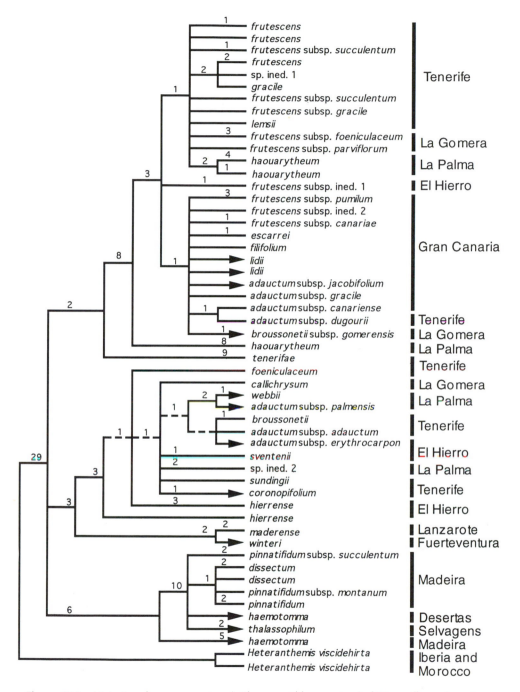

Figure 12-6 Majority rule consensus tree (with compatible groupings) of 54 equally parsimonious trees. The tree topology is identical to one of these 54 most parsimonious tress (tree length = 137 steps; consistency index excluding autopomorphies = 0.823; retention index = 0.971). Dashed lines indicate branches which collapse in the strict consensus tree. Number of changes along each branch is also given. Rare taxa are indicated with arrows.

unique variants in the endangered taxa than allozyme studies, underscoring the potential of these types of studies in conservation genetics.

Chloroplast DNA markers of the rare species can be extremely useful for tracing the key biological features of these taxa, mainly related to their reproductive biology, and for evaluating to what extent hybridization with other species could be a threat for their genetic conservation. Hybridization may play a very important role in species extinction, especially in oceanic islands where species are known to have weak crossing barriers (Levin et al., 1996). The fact that we detected some unique restriction site changes in the widespread taxa also indicates the potential use of cpDNA in conservation studies of this genus in relation to introgression and hybridization.

CONCLUSIONS AND PROSPECTS

Although the chloroplast genome has been used primarily for phylogenetic comparisons at or above the species level, several recent studies have demonstrated its utility for population biology (Soltis et al., 1989, 1991; Whittemore and Schaal, 1991; Fenster and Ritland, 1992; Kim et al., 1992b; Hong et al., 1993; Petit et al., 1993: Byrne and Moran, 1994; Dong and Wagner, 1994; Mason-Gamer et al., 1995; Neale, 1986). It is not possible to predict the levels of genetic variation that may be found by restriction site comparisons within species. However, the utilization of frequent-cutting enzymes in conjunction with the examination of variable regions of the chloroplast genome (Figure 12-2) will often yield variation.

The Southern hybridization approach detects more variation in restriction site comparisons than the PCR approach because it is possible to screen more restriction sites by the former approach. A PCR-based strategy is generally limited to sampling several hundred sites in one or a few fragments up to 4 kb in size. In contrast, the hybridization approach can sample several thousand sites from the entire genome. The development of more rapid DNA sequencing methods, especially when combined with the use of denaturing gradient gel electrophoresis (Abrams and Stanton, 1992), may make this approach more attractive for population studies of cpDNA variation in the near future. However, a restriction site strategy is currently cheaper and easier and provides sufficient variation in many plant groups.

Establishment of conservation genetic strategies must be based on an understanding of the biological features of the rare species. In this review we have discussed several examples of the utility of cpDNA restriction site variation for determining the origin of polyploids and for examining patterns of genetic diversity both within and between populations and species. Although we have not found any examples of the application of this genome for conservation biology studies, the intraspecific variation detected in many groups clearly indicates that this genome has much potential to provide useful markers for examining genetic variation of rare and endangered plants.

Hybridization and backcrossing between rare and widespread taxa is one important threat to some rare species. This is particularly true in oceanic islands where, in most cases, congeneric taxa have evolved rapidly and have not developed reproductive isolating mechanisms (Crawford et al., 1987). In these situations, offspring of interspecific crosses are often fertile and hybrid swarms are relatively common in those areas where the distribution range of the insular endemics are sympatric (Rieseberg et al., 1989). Habitat destruction following human intervention has been reported extensively in many volcanic islands (Heywood, 1979; Parsons, 1981; Schofield, 1989; Mueller-Dombois and Loope, 1990). Such destruction has meant that, at present, closely related taxa which formerly were isolated can hybridize in such a way that the rare species can be genetically absorbed by the common ones and their genetic integrity can thus be lost (Levin et al., 1996). In the Canary Islands *Argyranthemum coronopifolium* has almost vanished from most of the northwest part of Tenerife as a result of extensive hybridization with the weedy and extremely common *A. frutescens* (Humphries, 1976b; Brochmann, 1984). The identification of unique cpDNA markers (Table 12-3) in combination with ecological and morphological studies can be extremely useful in tracing the loss of some genetic combinations through introgression and hybridization, and in following gene flow from widespread into rare species.

Events such as genetic drift and close inbreeding could lead to a reduction of genetic variation in those cases where population size and population numbers decrease dramatically. In the present review we have shown that there are instances in which the existence of cpDNA polymorphisms within species could lead to partitioning of the genetic variation both among and within populations. This would facilitate (1) the estimation of how the genetic structure of the populations of a particular species is affected by a decline in the number of individuals, and (2) what modifications occur for the overall levels of genomic variation of a taxon once habitat fragmentation and destruction lead to a discontinuous distribution. This approach has been utilized with large animals, where restriction site polymorphism in mtDNA have been useful in identifying cases of inbreeding depression after habitat partition and population decline in gray wolves (Wayne et al., 1992), black rhinoceros (Ashley et al., 1990; O'Ryan et al., 1994), and cheetahs (O'Brien, 1994).

The application of cpDNA restriction site data for phylogenetic reconstructions has provided many new insights in plants. There are numerous examples in which the phylogenetic analysis of restriction site changes has led to an evolutionary reinterpretation of classifications (e.g., Jansen et al., 1991; Spooner et al., 1993). A phylogenetic approach is valuable for the establishment of conservation priorities because it identifies species that are closely related. This approach will lead to the recognition of "evolutionary significant units" (Ryder, 1986), which will avoid the problems of assigning priorities for conservation solely on the basis of taxonomic classifications. Phylogenetic partitioning of biodiversity will provide information concerning evolutionary uniqueness of each species, and it will indicate the hierarchical structure of rarity at both the infra- and supraspecific levels (Avise, 1989; Vane-Wright et al., 1991). Restriction site analysis of the chloroplast genome has been used in this context on numerous occasions. Notable examples include the silversword alliance (Asteraceae) in the Hawaiian Islands (Baldwin et al., 1990), *Dendroseris* (Asteraceae) in the Juan Fernandez Islands (Crawford et al., 1992), *Argyranthemum* in the Macaronesian Islands (Franciso-Ortega et al., 1996b), the North American representatives of *Streptanthus*

(Brassicaceae) (Mayer and Soltis, 1994), and *Helianthus* (Asteraceae) (Rieseberg et al., 1991). All of these groups have at least one rare taxon, and the interspecific phylogenies generated from cpDNA restriction site data will be of great value in the implementation of management strategies.

Trees generated from cladistic analyses of restriction site changes could be combined with biogeographical data to establish conservation priorities (Vane-Wright et al., 1991). This utilization has recently been discussed by Moritz (1994), who proposed that it is in the field of molecular ecology and conservation that the utilization of the "phylogeographical" data generated by molecular methods could be of great value. This could be applied in plants to define management units in a geographic scale (Moritz, 1994).

In conclusion, restriction site studies of cpDNA have provided a wealth of new data that are relevant to conservation genetics. Phylogenetic analyses at the interspecific level have assisted in the identification of significant evolutionary units which could be the focus of conservation strategies. Population studies of cpDNA variation have also revealed much useful information concerning the pattern and amount of genetic diversity within rare species. In spite of a recent shift to DNA sequencing methods, we believe that the restriction site approach to studying cpDNA variation will continue to have a significant impact on plant conservation genetics.

ACKNOWLEDGMENTS

This research was supported by grants from the National Science Foundation to R.K.J. (DEB-9318279) and to R.J.M. (BSR9020171), and by the Ministerio de Educación y Ciencia, Spain, to J.F.O. (PG92 42044506). Our gratitude to A. Prather (University of Texas) for his valuable comments of an early version of the manuscript.

REFERENCES

Abrams, E.S. and Stanton, V.P. 1992. Use of denaturing gel electrophoresis to study conformational transitions in nucleic acids. Methods Enzymol. 212: 71–104.

Arnold, M.L. 1993. *Iris nelsonii* (Iridaceae): origin and genetic composition of a homoploid hybrid species. Am. J. Bot. 80: 577–583.

Arnold, M.L., C.M. Buckner, and J.J. Robinson. 1991. Pollen-mediated introgression and hybrid speciation in Louisiana irises. Proc. Natl. Acad. Sci. U.S.A. 88: 1398–1402.

Ashley, M.V., D.J. Melnick, and D. Western. 1990. Conservation genetics of the black rhinoceros (*Diceros bicornis*), I: evidence from the mitochondrial DNA of three populations. Conserv. Biol. 4: 71–77.

Avise, J.C. 1986. Mitochondrial DNA and the evolutionary genetics of higher animals. Phil. Trans. R. Soc. London B 312: 325–342.

Avise, J.C. 1989. A role for molecular genetics in the recognition and conservation of endangered species. Trends Ecol. Evol. 4: 279–281.

Baldwin, B.G., D.W. Kyhos, and J. Dvorak. 1990. Chloroplast DNA evolution and adaptive radiation in the Hawaiian silversword alliance (Asteraceae-Madiinae). Ann. Missouri Bot. Gard. 77: 96–109.

Borgen, L. 1980. A new species of *Argyranthemum* (Compositae) from the Canary Islands. Norwegian J. Bot. 19: 149–170.

Brochmann, C. 1984. Hybridization and distribution of *Argyranthemum coronopifolium* (Asteraceae-Anthemideae) in the Canary Islands. Nordic J. Bot. 4: 729–736.

Byrne, M. and G.F. Moran. 1994. Population divergence in the chloroplast genome of *Eucalyptus nitens*. Heredity 73: 18–28.

Chambers, K.L. 1955. A biosystematic study of the annual species of *Microseris*. Contrib. Dudley Herb. 4: 207–312.

Chase, M.W. and J.D. Palmer. 1989. Chloroplast DNA systematics of lilioid monocts: resources, feasibility, and an example from the Orchidaceae. Am. J. Bot. 76: 1720–1730.

Clegg, M.T. and G. Zurawski. 1992. Chloroplast DNA and the study of plant phylogeny: present status and future prospects. In: P.S. Soltis, D.E. Soltis and J.J. Doyle, eds. Molecular systematics of plants, pp. 1–13. New York: Chapman and Hall.

Clegg, M.T., B.S. Gaut, G.H. Learn, and B.R. Morton. 1994. Rates and patterns of chloroplast DNA evolution. Proc. Natl. Acad. Sci. U.S.A. 91: 6795–6801.

Corriveau, J.L. and Coleman A.W. 1988. Rapid screening method to detect potential biparental inheritance of plastid DNA and results for over 200 angiosperm species. Am. J. Bot, 75: 1443–1458.

Crawford, D.J. and E.B. Smith. 1982. Allozyme variation in *Coreopsis nuecensoides* and *C. nuecensis* (Compositae), a progenitor-derivative species pair. Evolution 36: 379–386.

Crawford, D.J., R. Whitkus, and T.F. Stuessy.1987. Plant evolution and speciation on oceanic islands. In: K. Urbanska, ed. Patterns of differentiation in higher plants, pp. 183–199. London: Academic Press.

Crawford, D.J., J.D. Palmer, and M. Kobayashi. 1990. Chloroplast DNA restriction site variation and the phylogeny of *Coreopsis* section *Coreopsis* (Asteraceae). Am. J. Bot. 77: 552–558.

Crawford, D.J., T.F. Stuessy, M.E. Cosner, D.W. Haines, M. Silva, and M. Baeza. 1992. Evolution of the genus *Dendroseris* (Asteraceae: Lactuceae) on the Juan Fernandez Islands: evidence from chloroplast and ribosomal DNA. Syst. Bot. 17: 676–682.

Cruzan, M.B. and M.L. Arnold. 1993. Ecological and genetic associations in an *Iris* hybrid zone. Evolution 47: 1432–1445.

Cruzan, M.B., M.L. Arnold, S.E. Carney, and K.R. Wollenberg. 1993. cpDNA inheritance in interspecific crosses and evolutionary inference in Louisiana irises. Am. J. Bot. 80: 344–350.

Demesure, B., N. Sodzi, and J. Petit. 1995. A set of universal primers for amplification of polymorphic non-coding regions of mitochondrial and chloroplast DNA in plants. Mol. Ecol. 4: 129–131.

Dong, J. and D.B. Wagner. 1994. Paternally inherited chloroplast polymorphism in *Pinus*: estimation of diversity and population subdivision, and tests of disequilibrium with a maternally inherited mitochondrial polymorphism. Genetics 136: 1187–1194.

Downie, S. R. and J.D. Palmer. 1992. Restriction site mapping of the chloroplast DNA inverted repeat: a molecular phylogeny of the Asteridae. Ann. Missouri Bot. Gard. 79: 266–284.

Downie, S.R. and J.D. Palmer. 1994. A chloroplast DNA phylogeny of the Caryophyllales based on structural and inverted repeat restriction site variation. Syst. Bot. 19: 236–252.

Fenster, C.B. and K. Ritland. 1992. Chloroplast DNA and isozyme diversity in two *Mimulus* species (Scrophulariaceae) with contrasting mating systems. Am. J. Bot. 79: 1440–1447.

Francisco-Ortega, J., A. Santos-Guerra, R. Mesa-Coello, E. González-Feria, and D.J. Crawford. 1996a. Genetic resource conservation of the endemic genus *Argyranthemum* Sch. Bip. (Asteraceae: Anthemideae) in the Macaronesian islands. Genet. Resour. Crop. Evol. 43: 33–39.

Francisco-Ortega, J., R.K. Jansen, and A. Santos-Guerra. 1996b. Chloroplast DNA evidence of colonization, adaptive radiation and hybridization in the evolution of the Macaronesian flora. Proc. Natl. Acad. Sci. U.S.A. [In press].

Francisco-Ortega, J., D.J. Crawford, A. Santos-Guerra, and J.A. Carvalho. 1996c. Isozyme differentiation in the endemic genus *Argyranthemum* Sch. Bip. (Asteraceae: Anthemideae) in the Macaronesian islands. Plant Syst. Evol. 93: 4085–4090.

Grant, V. 1981. Plant speciation, 2d ed. New York: Columbia University Press.

Hallick, R.B., L. Hong, R.G. Drager, M.R. Favreau, A. Monfort, B. Orsat, A. Spielmann, and E. Stutz. 1993. Complete sequence of *Euglena gracilis* chloroplast DNA. Nucleic. Acids Res. 21: 3537–3544.

Hamrick, J.L. 1989. Isozymes and the analysis of genetic structure in plant populations. In: D.E. Soltis and P.S. Soltis, eds. Isozymes in plant biology, pp. 87–105. Portland, Oreg.: Dioscorides.

Harris, S.A. and R. Ingram. 1991. Chloroplast DNA and biosystematics: the effects of intraspecific diversity and plastid transmission. Taxon 40: 393–412.

Heywood, V.H. 1979. The future of islands floras. In: D. Bramwell, ed. Plants and islands, pp. 431–441. London: Academic Press.

Hiratsuka J., H. Shimada, R. Whittier, T. Ishibashi, M. Sakamoto, M. Mori, C. Kondo, Y. Honji, C.-R. Sun, B.-Y. Meng, Y.-Q. Li, A. Kanno, Y. Nishizawa, A. Hirai, K. Shinozaki, and M. Sugiura. 1989. The complete sequence of the rice (*Oryza sativa*) chloroplast genome: intermolecular recombination between distinct tRNA genes accounts for a major plastid DNA inversion during the evolution of the cereals. Mol. Gen. Genet. 217: 185–194.

Hong, Y., V.V. Hipkins, and S.H. Strauss. 1993. Chloroplast DNA diversity among trees, populations, and species in the California closed-cone pines (*Pinus radiata*, *Pinus muricata* and *Pinus attenuata*). Genetics 135: 1187–1196.

Humphries, C.J. 1976a. A revision of the Macaronesian genus *Argyranthemum* Webb ex Schultz Bip. (Compositae-Anthemideae). Bull. Br. Mus. (Nat. Hist.) Bot. 5: 1–240.

Humphries, C.J. 1976b. Evolution and endemism in *Argyranthemum* Webb ex Schultz Bip. (Compositae: Anthemideae). Bot. Macar. 1: 25–50.

Jansen, R.K. and J.D. Palmer. 1987. Chloroplast DNA from lettuce and *Barnadesia* (Asteraceae): structure, gene localization, and characterization of a large inversion. Current Genet. 11: 553–564.

Jansen, R.K. and J.D. Palmer. 1988. Phylogenetic implications of chloroplast DNA restriction site variation in the Mutisieae (Asteraceae). Am. J. Bot. 75: 751–764.

Jansen, R.K., R.S. Wallace, K.-J. Kim, and L.L. Chambers. 1991. Systematic implications of chloroplast DNA variation in the subtribe Microseridinae (Asteraceae: Lactuceae). Am. J. Bot. 78: 1015–1027.

Jansen, R.K., D.J. Lookerman, and H.-G. Kim. 1996. DNA sampling from herbarium material: a current perspective. In: D.A. Metsger, ed. Managing the modern herbarium [In press].

Johansson, J.T. and R.K. Jansen. 1993. Chloroplast DNA variation and phylogeny of the Ranunculaceae. Plant Syst. Evol. 187: 29–49.

Kim, K.-J., B.L. Turner, and R.K. Jansen. 1992a. Phylogenetic and evolutionary implications of interspecific chloroplast DNA variation in *Krigia* (Asteraceae-Lactuceae). Syst. Bot. 17: 449–469.

Kim, K.-J., R.K. Jansen and B.L. Turner. 1992b. Evolutionary implications of intraspecific chloroplast DNA variation in dwarf dandelions (*Krigia*; Asteraceae). Am. J. Bot. 79: 708–715.

Levin, D.A., J. Francisco-Ortega, and R.K. Jansen. 1996. Hybridization and the extinction of rare plant species. Conserv. Biol. 10: 1–7.

Liston, A. 1992. Variation in the chloroplast genes *rpo*C1 and *rpo*C2 of the genus *Astragalus* (Fabaceae): evidence from restriction site mapping of a PCR-amplified fragment. Am. J. Bot. 79: 953–961.

Loockerman, D.J. and R.K. Jansen. 1996. The use of herbarium material for DNA studies. In: T.F. Stuessy, ed. Sampling the green world. New York: Columbia University Press [In press].

Mason, R.J., K.E. Holsinger, and R.K. Jansen. 1994. Biparental inheritance of the chloroplast genome in *Coreopsis* (Asteraceae). J. Heredity 85: 171–173.

Mason-Gamer, R.J., K.E. Holsinger, and R.K. Jansen. 1995. Chloroplast DNA haplotype variation within and among populations of *Coreopsis grandiflora* (Asteraceae). Mol. Biol. Evol. 12: 371–381.

Mayer, M.S. and P.S. Soltis 1994. The evolution of serpentine endemics: a chloroplast DNA phylogeny of the *Streptanthus glandulosus* complex (Cruciferae). Syst. Bot. 19: 557–574.

McCauley, D.E. 1994. Contrasting the distribution of chloroplast DNA and allozyme polymorphism among local populations of *Silene alba*: implications for studies of gene flow in plants. Proc. Natl. Acad. Sci. U.S.A. 91: 8127–8131.

Moritz, C. 1994. Applications of mitochondrial DNA analysis in conservation: a critical review. Mol. Ecol. 3: 401–411.

Moritz, C., T.E. Dowling, and W.M. Brown. 1987. Evolution of animal mitochondrial DNA: relevance for population biology and systematics. Annu. Rev. Ecol. Syst. 18: 269–282.

Mueller-Dombois, D. and L.L. Loope. 1990. Some unique ecological aspects of oceanic islands ecosystems. In: J.E. Lawesson, O. Hamann, G. Rogers, G. Reck and H. Ochoa, eds. Botanical research and management in Galapagos, pp. 21–27. Monographs in Systematic Botany from the Missouri Botanical Garden, vol. 32. St Louis, Mo.: Missouri Botanical Garden.

Neale, D.B., M.A. Saghai-Maroof, R.W. Allard, Q. Zhang, and R.A. Jorgensen. 1986. Chloroplast DNA diversity in populations of wild and cultivated barley. Genetics 120: 1105–1110.

O'Brien, S.J. 1994. A role for molecular genetics in biological conservation. Proc. Natl. Acad. Sci. U.S.A. 84: 9054–9058.

Ohyama, K., H. Fukuzawa, T. Kohchi, H. Shirai, T. Sano, S. Sano, K. Umesono, Y. Shiki, M. Takeuchi, Z. Chang, S. Aota, H. Inokuchi, and H. Ozeki. 1986. Chloroplast gene organization deduced from complete sequence of liverwort *Marchantia polymorpha* chloroplast DNA. Nature 322: 572–574.

Olmstead, R.G. and J.D. Palmer. 1994. Chloroplast DNA systematics: a review of methods and data analysis. Am. J. Bot. 81: 1205–1224.

O'Ryan, C., J.R.B. Flamand, and E.H. Harley. 1994. Mitochondrial DNA variation in black rhinoceros (*Diceros bicornis*): conservation management implications. Conserv. Biol. 8: 495–500.

Palmer, J.D. 1986. Isolation and structural analysis of chloroplast DNA. Methods Enzymol. 118: 167–186.

Palmer, J.D. 1991. Plastid chromosomes: structure and evolution. In: I.K. Vasil, ed. The molecular biology of plastids, Vol. 7A, Cell culture and somatic cell genetics of plants, pp. 5–53. New York: Academic Press.

Palmer, J.D. 1992. Mitochondrial DNA in plant systematics: applications and limitations.

In: P.S. Soltis, D.E. Soltis and J.J. Doyle, eds. Molecular systematic of plants, pp. 36–49. New York: Chapman and Hall.

Palmer, J.D., R.K. Jansen, H.J. Michaels, M.W. Chase, and J. W. Manhart. 1988. Chloroplast DNA variation and plant phylogeny. Ann. Missouri Bot. Gard. 75: 1180–1206.

Parsons, J.J. 1981. Human influence in the pine and laurel forests of the Canary Islands. Geograph. Rev. 71: 253–271.

Petit, R.J., A. Kremer, and D.B. Wagner. 1993. Geographic structure of chloroplast DNA polymorphisms in European oaks. Theor. Appl. Genet. 87: 122–128.

Rieseberg, L.H. and S. J. Brunsfeld. 1992. Molecular evidence and plant introgression. In: P.S. Soltis, D.E. Soltis and J. Doyle, eds. Molecular systematics of plants, pp. 151–176. New York: Chapman and Hall.

Rieseberg, L.H., S. Zona, L. Aberbom, and T.D. Martin. 1989. Hybridization in the island endemic, *Catalina mahogany*. Conserv. Biol. 3: 52–58.

Rieseberg, L.H., S.M. Beckstrom-Stemberg, A. Liston, and D.M. Arias. 1991. Phylogenetic and systematic inferences from chloroplast DNA and isozyme variation in *Helianthus* sect. *Helianthus* (Asteraceae). Syst. Bot. 16: 50–76.

Rustan, Ø.H. 1981. Infraspecific variation in *Argyranthemum pinnatifidum* (Lowe) Lowe. Bocagiana 55: 2–18.

Ryder, O.A. 1986. The species problem and conservation: the dilemma of subspecies. Trends Ecol. Evol. 1: 9–10.

Santos-Guerra, A. 1983. Vegetación y flora de La Palma. Santa Cruz de Tenerife, Canary Islands, Spain: Interinsular Canaria.

Schofield, E.K. 1989. Effects of introduced plants and animals on island vegetation: examples from the Galapagos archipelago. Conserv. Biol. 3: 227–238.

Sears, B. 1980. Elimination of plastids during spermatogenesis and fertilization in the plant kingdom. Plasmid 4: 233–255.

Sewell, M.M., Y.-L. Qiu, C.R. Parks, and M.W. Chase. 1993. Genetic evidence for trace paternal transmission of plastids in *Liriodendron* and *Magnolia* (Magnoliaceae). Am. J. Bot. 80: 854–858.

Shinozaki, K., M. Ohme, M. Tanaka, T. Wakasugi, N. Hayashida, T. Matsubayashi, N. Zaita, J. Chunwongse, J. Obokata, K. Yamaguchi-Shinozaki, C. Ohto, K. Torazawa, B.-Y. Meng, M. Sugita, H. Deno, T. Kamogashira, K. Yamada, J. Kusuda, F. Takaiwa, A. Kato, N. Tohdoh, H. Shimada, and M. Sugiura. 1986. The complete nucleotide sequence of the tobacco chloroplast genome: its gene organization and expression. EMBO J 5: 2043–2049.

Smith, E.B. 1974. *Coreopsis nuecensis* and a related new species from southern Texas. Brittonia 26: 161–171.

Soltis, D.E. and P.S. Soltis. 1992. Chloroplast DNA and nuclear rDNA variation: insights into autopolyploid and allopolyploid evolution. In: P.S. Soltis, D.E. Soltis, and J.J. Doyle, eds. Molecular systematics of plants, pp. 97–117. New York: Chapman and Hall.

Soltis, D.E., P.S. Soltis, T.A. Ranker, and B.D. Ness. 1989. Chloroplast DNA variation in a wild plant, *Tolmiea menziesii*. Genetics 121: 819–826.

Soltis, D.E., M.S. Mayer, P.S. Soltis, and M. Edgerton. 1991. Chloroplast-DNA variation in *Tellima grandiflora*. Am. J. Bot. 78: 1379–1390.

Soltis, D.E., P.S. Soltis, and B.G. Milligan. 1992a. Intraspecific chloroplast DNA variation: systematic and phylogenetic implications. In: P.S. Soltis, D.E. Soltis, and J.J. Doyle, eds. Molecular systematics of plants, pp. 117–150. New York: Chapman and Hall.

Soltis, P.S., J.J. Doyle, and D.E. Soltis. 1992b. Molecular data and polyploid evolution in plants. In: P.S. Soltis, D.E. Soltis, and J.J. Doyle, eds. Molecular systematics of plants, pp. 177–201. New York: Chapman and Hall.

Spooner, D.M., G.J. Anderson, and R.K. Jansen. 1993. Chloroplast DNA evidence for the interrelationships of tomatoes, potatoes, and pepinos (Solanaceae). Am. J .Bot. 80: 676–688.

Sugiura, M. 1992. The chloroplast genome. Plant Mol. Biol. 19: 149–168.

Taberlet, P., L. Gielly, P. Guy, and J. Bouvet. 1991. Universal primers for amplification of three non-coding regions of chloroplast DNA. Plant. Mol. Biol. 17: 1105–1109.

Vane-Wright, R.I., C.J. Humphries, and P.H. Williams. 1991. What to protect?—Systematics and the agony of choice. Biol. Conserv. 55: 235–254.

Wakasugi, T., J. Tsudzuki, S. Ito, T. Nakashima, T. Tsudzuki, and M. Sugiura. 1994. Loss of all *ndh* genes as determined by sequencing the entire chloroplast genome of the black pine *Pinus thunbergii*. Proc. Natl. Acad. Sci. U.S.A. 91: 9794–9798.

Wallace, R.S. and R.K. Jansen. 1990. Systematic implications of chloroplast DNA variation in the genus *Microseris* (Asteraceae: Lactuceae). Syst. Bot. 15: 606–616.

Wallace, R.S. and R.K. Jansen. 1995. DNA evidence for multiple origins of intergeneric allopolyploids in annual *Microseris* (Asteraceae). Plant Syst. Evol. 198: 253–265.

Wayne, R.K., N. Lehman, M.W. Allard, and R.L. Honeycutt. 1992. Mitochondrial DNA variability of the gray wolf: genetic consequences of population decline and habitat fragmentation. Conserv. Biol. 6: 559–569.

Whittemore, A.T. and D.A. Schaal. 1991. Interspecific gene flow in sympatric oaks. Proc. Natl. Acad. Sci. U.S.A. 76: 2540–2544.

Whitton, J., R.S. Wallace, and R.K. Jansen. 1995. Phylogenetic relationships and patterns of character change in the tribe Lactuceae (Asteraceae) based on chloroplast DNA restriction site variation. Can. J. Bot. 73: 1058–1073.

Wilson, A.C., R.L. Cann, S.M. Carr, M.L. George, U.B. Gyllenstein, K.M. Helm-Bychowski, R.G. Higuchi, S.R. Palumbi, E.M. Prager, R.D. Sage, and M. Stoneking. 1985. Mitochondrial DNA and two perspectives on evolutionary genetics, Biol. J. Linn. Soc. 26: 375–400.

Wolfe, K.H., W.- H. Li, and P.M. Sharp. 1987. Rates of nucleotide substitution vary greatly among plant mitochondrial, chloroplast, and nuclear DNAs. Proc. Natl. Acad. Sci. U.S.A. 84: 9054–9058.

Wolfe, K.H., C.W. Morden, and J.D. Palmer. 1992. Function and evolution of a minimal plastid genome from a non photosynthetic parasitic plant. Proc. Natl. Acad. Sci. U.S.A. 89: 10648–10652.

13

A PCR Approach to Detection of Malaria
in Hawaiian Birds

REBECCA L. CANN, ROBERT A. FELDMAN, LEILA AGULLANA,
AND LEONARD A. FREED

Organisms that cause disease are recognized as having an influential role in the population dynamics of host species and in the communities to which host species belong (Dobson and Hudson, 1986). Within host species, individuals may vary in tolerance or resistance to disease, with consequences for survival and reproductive competition (van Riper et al., 1986). Within communities, species may vary in the extent to which individuals tolerate or resist disease, with consequences for interspecific competition and community structure (Scott, 1988). Disease is of special concern for endangered species, and the journal *Conservation Biology* highlighted this concern (May, 1988). Unfortunately, our recognition of the potential importance of disease often exceeds our ability to accurately document its role in natural populations and communities and assess the significance it has for the maintenance of intraspecific genetic variation. While prevalence is a notoriously oversimplified index (Ewald, 1994), part of the larger problem is the methodological difficulty in accurately estimating prevalence of disease as a first step.

Molecular genetics may provide an approach to a more comprehensive study of disease by increasing the sensitivity of diagnosis and defining the pool of susceptibles. Here we illustrate this approach using a polymerase chain reaction (PCR)-based test that can detect the DNA of an avian malarial blood parasite (*Plasmodium*) in part of the total DNA extracted from a blood sample from the host bird. Traditional approaches for studying malaria are reviewed to point out the need for more sensitive diagnoses. Then the basis of the PCR test for *Plasmodium*, which can be generalized to other blood pathogens, is presented. Results are presented of the operational use of the test with Hawaiian birds, many of which are highly susceptible to *Plasmodium* (van Riper et al., 1986). Diverse issues that can be addressed with the PCR test, especially those germane to conservation biology, are discussed.

REVIEW OF DIAGNOSES FOR PLASMODIUM

Traditional investigation of *Plasmodium* in birds involves microscopy (Atkinson and van Riper, 1991). A small sample of blood is smeared across a slide to form a monolayer. The smear is then fixed in alcohol to keep the red blood cells intact, and a stain such as Giemsa is applied to the smear, causing the nucleic acids on the slide to become darkly stained. For birds with nucleated erythrocytes, cells with more than one dark stain are scored as infected. Microscopy also enables the species of blood parasite and stage of infection to be identified, and the level of parasitemia (number of infected cells per 10 000) to be calculated.

The smear/microscopy approach, however, has several limitations. One set includes the quality of the smear and the expertise and motivation of the microscopist. Variation in these factors decreases the reliability of estimates of prevalence and the identity of pathogenic strains or species among studies. There is another set of limitations that are associated with sampling error, and that increase in importance with lower levels of parasitemia. Typical coverages of slides by microscopy entail 10 000 to 25 000 cells (Garnham, 1966), or less than 0.06% of the estimated 41 million cells available (Feldman et al., 1995). At low levels of parasitemia, corresponding to early, later, or the latent stages of infection, the uneven dispersion of rare infected cells on the slide may result in the absence of such cells in a random sample of fields (Garnham, 1966). A study of human malaria found that significant differences in detection resulted from microscopic reexamination of the same slides (Barker et al., 1989).

A further complication is the problem of using blood smears to diagnose individuals as disease-free. The blood of smear-negative birds has produced malarial infections in other individuals (Herman, 1968). Smear-negative birds subjected to an immunosuppressant have also developed malarial infections (Nakamura et al., 1984). Such findings indicate that birds previously infected may serve as reservoir hosts, even though they appear disease-free.

Molecular approaches have been employed to look at proteins and nucleic acids associated with human malaria. In principle, these could also be used with birds. The popular ELISA technique detects antibodies that are produced in response to surface antigens and other molecular markers of *Plasmodium* and has potential to reveal previous exposure or historical infections within an individual who has cleared the pathogen from its blood stream. In combination with a technique that can detect latent infections, the method may distinguish historical from current infections. However, following infection, there is a lag period before the development of circulating immune response, when serological assays for infection would be expected to be negative by the ELISA technique.

Autoradiographic blot tests for the small nuclear subunit ribosomal genes of *Plasmodium* have proven superior to microscopy and ELISA tests (Barker et al., 1989; Waters and McCutchan, 1989) because they combine molecular sensitivity and do not require a long waiting period before confirming diagnosis. However, blot tests also require a threshold level of product to be present, even with autoradiography. The first PCR-based diagnostic for birds using ribosomal gene targets revealed a sensitivity that exceeded that of the DNA blot tests for human malaria by an order of magnitude (Feldman et al., 1995), in accord with the comparisons of smear versus PCR and blot sensitivity by vaccine researchers studying *falciparum* malaria in Papua New Guinea (Felger et al., 1994).

DESIGN OF THE PCR TEST

The PCR test for malaria in birds requires a design that permits detection of *Plasmodium* DNA sequences not found in the avian genome. One possibility, following the biomedical approach to developing a vaccine against malaria in humans, would be to emphasize the circumsporozoite and merozoite proteins. DNA sequences, of 22 known alleles for the extensively studied *P. falciparum* merozoite surface antigen 2 were available as of 1993 (Felger et al., 1993). Their extraordinary mutation rate, when combined with the phenomenon of sexual recombination, may have important consequences for the evolution of virulence (Hughes, 1993; Ewald 1994). However, this variability suggests that the sequences are likely to be unreliable in a general test for all species and strains of *Plasmodium*. The problem is compounded by the fact that over 54 species of *Plasmodium* are thought to occur in birds alone (Herman, 1968).

The limited sequence information available that may be used to design general *Plasmodium* PCR primers comes from studies of their ribosomal genes (Unnasch and Wirth, 1983; Gunderson et al., 1987; Waters and McCutchan 1989; Waters et al., 1991). In particular, the small nuclear subunit (18S) ribosomal gene has known structure and function (Neefs et al., 1990). Some progress appears to have been made in designing different sets of primers to detect differences in sequence between humans, mosquitoes, *Plasmodium*, and even strains of *Plasmodium* (Mathiopoulos et al., 1993). However, no data were presented in that paper to indicate the generality of the test to birds or other vertebrates. The creation of a general diagnostic primer, rather than a series of specialized primers, takes advantage of a size difference between homologous regions of the gene in *Plasmodium* and vertebrates. Such a region exists in domain 3 between nucleotides 1367 and 1945 of the 18S rRNA gene sequence reported for *Plasmodium berghei*, known to infect rodents (Gunderson et al., 1987).We have concentrated on these sequences in *Plasmodium* to design a relatively inexpensive diagnostic test for malaria in birds (Feldman et al., 1995).

Development of a PCR-based diagnostic that is accurate and quick depends on sequence information which can identify unique regions of a genome, present and/or expanded in the *Plasmodium* but absent in the avian host. In order for it to be inexpensive, one should not need to sequence the region in each bird tested but simply see a size difference in migration of amplified DNA fragments on an agarose minigel. We compared sequences in domain 3 from a variety of *Plasmodium* and vertebrate species. We then designed primers which anneal to highly conserved regions of secondary structure but amplify a region which is about 200 bases longer in the parasite than in all vertebrates analyzed (Figure 13-1).

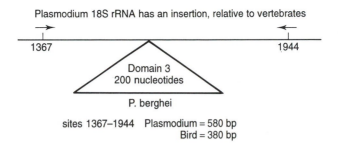

Figure 13-1 Insertion in *Plasmodium* 18S rRNA relative to vertebrates. Numbers refer to nucleotide numbers for *Plasmodium berghei* (Gunderson et al., 1987). The locations of the PCR primers are indicated by the arrows.

METHODS

Birds used in our studies are caught in mist nets, color-banded, measured, inspected for visible signs of health, photographed, bled with a sterile 26.5-gauge needle, and then released. Total genomic DNA is prepared from about 50 μl of whole blood, using a simple method that works well with the nucleated red blood cells of birds (Quinn and White, 1987). If the bird is infected with *Plasmodium*, some fraction of the blood cells, depending on the stage of infection, will yield DNA from the parasite as well as that of the host and any other infectious agents present.

Our primers (5'-GCATGGCCGTTTTTAGTTCGTGAAT-3; 5'-TATCTTT-CAATCGGTAGGAGCGACG-3') are normally used in a 25 μl reaction containing about 40 ng of total genomic DNA, with standard concentrations of deoxynucleotides and *Taq* DNA polymerase (50 mmol/l KCl, 10 mmol/l Tris-HCl, pH 8.3, 30 mmol/l $MgCl_2$, 0.01% sterile gelatin, 1.25 mmol/l of each dNTP, 1 mmol/l of each primer, and 0.625 units *Taq*). Amplification conditions we have found to be ideal in this system are 30 cycles at 94°C for 40 s, 48°C for 2 min, and 72°C for 45 s, followed by a final extension at 72°C for 10 min. Half of this reaction is electrophoresed on a 2% agarose–Tris–borate EDTA (TBE) gel, stained with ethidium bromide, and photographed.

Negative controls are samples containing all reagents, where distilled water replaces the volume normally contributed by bird DNA. Good laboratory practices standard with PCR include the use of aerosol-resistant pipette tips, positive displacement pipettes, physical separation of pre- and post-PCR samples, and UV irradiation of reagents, in order to avoid contamination.

RESULTS AND DISCUSSION

A typical run is shown in Figure 13-2. Some of the bands show reduced mobility in the agarose matrix, indicative of larger DNA fragments. These have been identified by extensive sequence analysis as DNA fragments from *Plasmodium* ribosomal genes. On this simple gel, the fragments match the relative mobility of a

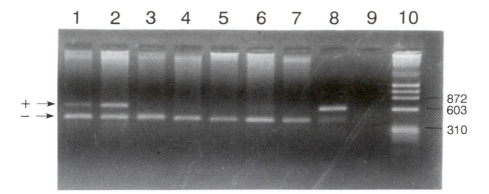

Figure 13-2 Diagnostic PCR gel. Lanes 1–7 are tested samples; lane 8 is a positive control (*P. gallinaceum*) DNA; and lane 9 is a negative control (double-distilled water). The arrow indicates the presence of *Plasmodium* rDNA bands at 580 bp and bird rDNA bands at 380 bp. Lanes 1 and 2 contain positive test samples. Lane 10 is a molecular weight marker containing standard phage lambda DNA restriction fragments of known size.

positive control made from purified *Plasmodium* DNA that was passaged through duck cells (some trace amounts of duck ribosomal gene amplification are also visible). The smaller molecular weight bands visible in each sample lane reflect the amplified fragment of the bird's own ribosomal gene, which we confirmed by DNA sequencing.

Application of the test to new species of hosts and parasites may require verification by sequencing of amplified DNA from the putative host and parasite bands.

Sensitivity of PCR Test

We have conducted extensive dilution experiments and quantitative competitive PCR (Piatek et al., 1993) to establish limits of sensitivity. Feldman et al. (1995) contains details of these experiments and adjustments for the number of rRNA genes in *Plasmodium* and birds. Under noncompetitive conditions, the test is sensitive to 2.12 parasites per reaction. Under competitive conditions, the test is capable of detecting as few as 8 parasites per 10 000 avian red blood cells. Initial conditions are very important in competitive PCR, and the smaller bird fragment has an early advantage, so further work may reveal even greater sensitivity when hot start conditions are routinely employed (Piatek et al., 1993). In addition, the sensitivity is dependent on the number of rRNA genes in birds as well as *Plasmodium*. Refinements in estimating genome equivalents of ribosomal genes in our PCR reactions will also alter the estimates of sensitivity (Feldman et al., 1995).

We have direct and indirect comparisons with blood smears, using standards of detection employed by expert microscopists. Dr. Carter Atkinson, head of the National Wildlife Health Center, Hawaii Field Station, provided us with two sets of samples from birds from around 4000 feet elevation at Hawaii Volcanoes National Park, Island of Hawaii. In each of these sets the PCR test correctly identified as positive each of the samples that had been so identified by smear.

The PCR test identified additional birds as being positive, with a 67% increase in prevalence over that estimated by smear in one sample (Feldman et al., 1995). Further, as documented below, we detected high prevalence of malaria in introduced bird species at Lyon Arboretum (500 feet elevation) in Manoa Valley, Oahu. Previous work within a year at that site, in which smears were analysed, did not detect malaria in the same species, although the vector responsible (*Culex quinquefasciatus*) is present year round (R. Nakamura, personal communication).

Application of PCR Test to Hawaiian Birds

The PCR test was considered for its potential contribution to the conservation of Hawaiian forest birds. The unique avifauna of Hawaii has been undergoing accelerated extinction since historical times (Scott et al., 1986; Freed et al. 1987), with almost half of the remaining bird species currently listed as endangered. An initial extinction peak coincided with the arrival in the islands of the first indigeneous human populations about 1500 years ago, but exploitation of bird populations was at subsistence levels for food and decoration (Holt, 1985). Disease is now thought to be the principal cause of modern extinction events (Warner, 1968; van Riper et al., 1986), and malaria is one of the two diseases (the other is caused by a pox virus transmitted by the same mosquito vector) thought to be important in restricting the range of many species to habitats above 5000 feet. The lack of knowledge about basic prevalence levels of *Plasmodium*, insect vectors, and strain genetic diversity present in the multitude of bird species, especially those in the low-elevation forests, limits more successful conservation efforts in Hawaii (van Riper et al., 1986). Pressure to step up captive rearing efforts has also increased the urgency of having appropriate diagnostic tools in place, as animals are moved between islands and habitats.

Using the PCR diagnostic, we have been able to identify the presence of malaria in high-elevation habitats on Maui and the Island of Hawaii that have been protected for endangered forest birds and have begun to identify significant reservoirs of the disease in lower-elevation habitats on other Hawaiian islands. We have also used the PCR test to explore the limits of parasitemia detectable in small blood samples taken nondestructively from endangered species. Finally, we are studying the temporal fluctuation of infected individuals in native and introduced bird populations.

We now have data from four main islands which give us information about the prevalence of malaria at different elevations and in different bird species. More than 150 species of birds have been introduced to Hawaii since 1850 (Stone, 1989). Some species have been introduced multiple times, and in their study of the properties of successful invaders Moulton and Pimm (1986) record that 50 species were introduced in various combinations 84 times on seven of the main Hawaiian islands. About 70% of all bird introductions to Hawaii have been successful. Many conservation biologists suspect that these alien species are important disease reservoirs that threaten the existence of endemic Hawaiian birds. We find some evidence to support their assertions, especially in low-elevation habitats where introduced species outnumber native ones.

Table 13-1 shows results of our spring 1994 sampling of birds from Oahu's

Table 13-1 Malaria in Oahu's Birds. Site: Lyon Arboretum, Hawaii Section, Manoa (18 April–17 May 1994)

	N	Number positive	Percentage positive
Common amakihi	4	0	0
White-rumped shama	8	6	75
Northern cardinal	3	0	0
Common waxbill	5	0	0
Red-whiskered bulbul	5	4	80
Red-vented bulbul	1	1	100
Common myna	2	0	0
Nutmeg mannikin	11	4	36
Japanese white-eye	10	0	0
Red-billed leiothrix	1	0	0
Total	50	15	30

Manoa valley, a low-elevation site located in an area of high rainfall with plentiful mosquitoes. Only two native forest birds, the common amakihi (*Hemignathus virens*; Oahu subspecies) and the apapane (*Himatione sanguinea*), occur here. Introduced species, such as the white-rumped shama (*Copsychus malabaricus*), red-whiskered bulbul (*Pycnonotus jocosus*), red-vented bulbul (*Pycnonotus cafer*), and nutmeg mannikin (*Lonchura punctulata*), that make up a significant fraction of the passerine birds normally seen on Oahu, are commonly infected with *Plasmodium*. Four specimens of common amakihi, a Hawaiian honeycreeper, tested negative. We assumed that these birds might be positive, because of the high number of mosquitoes present and the increased susceptibility that the native species supposedly have to malaria.

A working hypothesis to investigate is that only *Plasmodium*-tolerant or even resistant common amakihi can survive in such an environment. Van Riper et al. (1986) discovered a population of common amakihi (Island of Hawaii subspecies) that resisted challenge experiments. Both behavioral and genetic adaptations have probably taken place, in the face of over 150 years of natural selection for improved ability to cope with this disease. Common amakihis may provide the best model system for investigating the evolution of resistance and tolerance to malaria in Hawaiian honeycreepers. If knowledge of mouse genetics is any clue, natural (preimmune) resistance to *Plasmodium* infections can be controlled by a few loci, all so far located outside the major histocompatibility complex (Malo and Skamene, 1994).

Middle-elevation birds, between 3000 and 4400 feet elevation, were sampled on Kauai and Maui. The island of Kauai is the most isolated of the high islands in the archipelago. We have tested a small number of birds from Kauai's central plateau at the edge of the Alakai swamp. Our screen revealed that the Elepaio (*Chasiempsis sandwichensis*), a native flycatcher, appeared in the summer of 1992 to be the focus of an epizootic (Table 13-2). Kauai occupies a special position in the radiation of Hawaii's birds, with 14 endemic forest species known there historically. Unfortunately, there is no forest habitat above 5000 feet, so tolerance

Table 13-2 Malaria in Kauai's Birds. Site: Alakai Swamp, Pihea Trail (9 July–August 1992)

	N	Number positive	Percentage positive
Common amakihi	16	0	0
Apapane	4	0	0
Iiwi	5	0	0
Anianiau	8	0	0
Kauai creeper	1	0	0
Elepaio	11	4	37
Japanese white-eye	15	0	0
White-rumped shama	1	0	0
Northern cardinal	1	0	0
Melodious laughing-thrush	1	0	0
Total	63	4	6.3

or resistance to malaria, or eradication of the *Culex* mosquito vector, are the only hopes for management actions possible to deal with this and perhaps other diseases. West Maui is also limited in elevation, and here two native apapane (*Himatione sanguinea*), as well as one introduced Japanese white-eye (*Zosterops japonicus*) tested positive for malaria (Table 13-3). *Culex* mosquitos are known to exist at these elevations throughout the year (van Riper et al., 1986).

Sampling efforts at upper elevations on Maui and Hawaii are especially important because they are the centers of the remaining distribution of endangered forest birds on these islands (tables in Feldman et al., 1995). On the slopes of Maui's Haleakala volcano (Hanawi Natural Area Reserve, almost 7000 feet), individuals of five native and introduced species (including one endangered species) tested positive (7 of 159 birds), and some birds appeared to acquire their infections as residents on this reserve (Feldman et al., 1995). We inferred this because of capture and sightings of color-banded birds over a 6-month period in 1992 that spanned the prebreeding and postbreeding periods. Although almost twice as many birds were tested in our prebreeding sample of February 1992 (102 versus the postbreeding July 1992 sample of 57), we were able to document that the endangered crested honeycreeper (*Palmeria dolei*), the common amakihi (Maui

Table 13-3 Malaria in West Maui's Birds. Site: Kaulalewelewe Mauka and Pu'u Kukui Trail, Nakalaua (Spring 1994)

	4400 ft			2800–3000 ft		
	N	Number positive	Percentage positive	N	Number positive	Percentage positive
Common amakihi	5	0	0	0	0	0
Apapane	3	0	0	3	2	67
Iiwi	1	0	0	1	0	0
Japanese white-eye	8	1	12.5	3	0	0
Total	17	1	6	7	2	29

subspecies), and the red-billed leiothrix (*Leiothrix lutea*) were infected in July, while apapane (Maui subspecies) and Maui creeper (*Paroreomyza montana*) tested positive in February.

Birds sampled from Hakalau Forest National Wildlife Refuge on the island of Hawaii, at elevations between 5200 and 6200 feet, provide additional evidence that infections can be acquired at upper elevations (Feldman et al., 1995). Three species of native birds had some positive individuals (common amakihi, apapane, and omao (*Myadeste obscurus*), a native thrush). The apapane may have contracted the disease at mid or lower elevations while tracking floral blooms for nectar, but the other two species are year-round residents. Intriguing questions are raised by the 571 samples of eight native and six introduced species, in which only 12 individuals from three native (nonendangered) species tested positive. How long can infected mosquitos, which may be blown upslope by strong tradewinds, survive to feed on resident birds? Does high susceptibility to malaria result in rapid death or debilitation that would prevent such birds from being captured? Does seasonal flocking behavior that mixes resident and nonresident birds increase the likelihood of malarial transmission? These are some tough questions that must be answered in assessing the viability of upper elevation populations of endangered forest birds.

FURTHER CONSIDERATIONS AND CONCLUSIONS

The PCR test presented here and in Feldman et al. (1995) is the most sensitive and reliable technique for investigating avian malaria. The positive and negative controls that are part of each reaction guard against false conclusions generated by operator error or contamination. Ambiguous cases (barely discernible *Plasmodium* band, multiple bands) can be easily and cheaply re-run or evaluated directly within new PCR machines that assess the presence of the amplified bands by fluorimetry. The test is inexpensive in terms of supplies; a PCR machine costs less than a high-quality microscope for histological work, and numerous samples can be analyzed during a given run, saving time as well as money. It will be useful for massive screening of blood samples for studying epizootics, for research involving challenge experiments aimed at identifying tolerant and resistant individuals, and for refining disease categories to differentiate between those individuals that are developing malaria (prepatent), those with acute infections (patent), and those with low levels after an acute phase (subpatent or latent). Here we discuss some extensions of the approach that may increase its research value even more.

The quantitative competitive PCR admits a stochastic element to the sensitivity of the test, based on initial starting conditions (low target number, preferential extension of smaller fragments, etc.). The implication is that a PCR-negative test may itself be erroneous because of the initial condition realized during the test. This stochasticity limits the confidence a prudent investigator should place in a single negative PCR test, but it also gives the PCR test potential to estimate the level of parasitemia when multiple reactions from the same bird are run. We suggest that samples be replicated several times during the same PCR run. The proportion of replicates that are positive should ultimately reflect the level of parasitemia. High levels would generate a high proportion of positives. Very low levels would generate only a single or a few positives. Indeed, this application of the PCR test may be the surest way of documenting that an individual bird is free of malaria or that a previously infected bird has completely cleared malaria from its system. Such precision is also crucial to evaluating hypotheses about the correlation between the plumage of bright tropical birds and parasite levels, especially resident species compared to migrants (Zuk, 1991).

The current version of the PCR test also has potential to identify malarial strains. The *Plasmodium* rRNA gene fragment can be sequenced and the sequences can be compared from different locations in Hawaii that vary in the types of introduced species present. Humans with *falciparum* malaria often show mixed-strain infections with PCR, with consequences for the development of acquired immunity. Strain identification may be the first step toward recognizing that the problem with malaria in Hawaii is a global one, given the diversity of geographical regions from which birds have been introduced. A new set of primers, ones targeting more variable regions of the rRNA gene than those amplified by this set of primers, or even different genes, would be advantageous for later fine-scale identification of strains. Some progress in this area is reported by Mathiopoulos et al. (1993).

We also anticipate a second generation of primers for the basic PCR test. These would target a region that was of the same size in both *Plasmodium* and birds but which differed in the sequence, so that a unique restriction site is present in one group but not the other. Amplified products could simply be digested with a restriction endonuclease and run on an agarose gel. This would obviate the competitive disadvantage the current test has owing to the bias against amplification of the parasite's larger DNA fragment in the same reaction. In so doing, it would provide the most accurate estimates of prevalence (but not parasitemia) at the lowest cost. In order to be truly accurate, however, this type of test requires much more knowledge about the DNA sequence variability than currently exists for any species of bird.

Conservation genetics can contribute new technologies, novel insights, and proactive attitudes to the community of biologists concerned with the progressive loss of biodiversity. Forest birds in Hawaii can never truly be protected from disease vectors, even though they represent some of the most geographically isolated populations on this planet. Accurate diagnostic tools are the first step to identifying host-resistant genotypes, discovering their frequency in natural populations, and constructing management strategies that boost their contributions to the gene pool of threatened species.

ACKNOWLEDGMENTS

We thank J. Bennett, M. Burt, S. Fretz, P. Hart, J. Lepson, G. Massey, R. Peck, J. Rohrer, S. Santos, B. Thorsby, and E. VanderWerf for blood samples; E. Laxson, D. Horie, and L. Hanakahi for DNA extractions; and C. Atkinson for smear results, blood samples, and discussion. Plasmodium DNA samples used for positive controls were prepared by Yvonne

Ching and provided by M. Skoldager of the Baltimore Zoological Society and D. Kaslow at the National Institutes of Health. This work was made possible by a J.D. and C.T. MacArthur Foundation grant to L. Freed, R. Cann, and S. Conant. Support for L. Agullana was provided by a grant from the Howard Hughes Medical Institute through the Undergraduate Biological Sciences Education Program. Y. Ching also contributed to the early development of the PCR test, supported in part by grants from the Hawaiian Audubon Society and the Pew Charitable Trust. We also thank H. Carson, J. Hunt, B. Nakamura, and S. Palumbi for additional advice and discussion.

REFERENCES

Atkinson, C.T. and C. van Riper III. 1991. Pathogenicity and epizootiology of avain haematozoa: *Plasmodium, Leucocytozoon*, and *Haemoproteus*. In: J.E. Loye and M. Zuk, eds. Bird–parasite interactions, pp. 19–48. New York: Oxford University Press.

Barker, R.H., L. Suebsaeng, W. Rooney, and D.F. Wirth. 1989. Detection of *Plasmodium falciparum* infection in human patients: a comparison of the DNA probe method to microscopic diagnosis. Am. J. Trop. Med. Hy. 41: 266–272.

Dobson, A.P. and P.J. Hudson. 1986. Parasites, disease, and the structure of ecological communities. Trends Ecol. Evol. 1: 11–15.

Ewald, P.W. 1994. The evolution of infectious disease. New York: Oxford University Press.

Feldman, R.A., L.A. Freed, and R.L. Cann. 1995. A PCR test for avian malaria in Hawaiian birds. Mol. Ecol. 4: 663–673.

Felger, I., L. Tavul, and H.-P. Beck. 1993. *Plasmodium falciparum*: a rapid technique for genotyping the merozoite surface protein 2. Exp. Parasitol. 77: 372–375.

Felger, I., L. Tavul, S. Kabintik, V. Marshall, B. Genton, M. Alpers, and H.P. Beck. 1994. *Plasmodium falciparum*: extensive polymorphism in merozoite surface antigen 2 alleles in an area with endemic malaria in Papua New Guinea. Exp. Parasitol. 79: 106–116.

Freed, L.A., S. Conant, and R.C. Fleischer. 1987. Evolutionary ecology and radiation of Hawaiian passerine birds. Trends Ecol. Evol. 2: 196–203.

Garnham, P.C.C. 1966. Malaria parasites and other Haemosporidia. Oxford: Blackwell Scientific.

Gunderson, J.H., M.L. Sogin, G. Wollett, M. Hollingdale, V.F. De la Crus, A.P. Waters, and T.F. McCuchan. 1987. Structurally distinct, stage-specific ribosomes occur in *Plasmodium*. Science 238: 933–937.

Herman, C.M. 1968. Blood protozoa of free-living birds. Symp. Zool. Soc. London 24: 177–195.

Holt, J.D. 1985. The art and featherwork in Old Hawaii. Topgallant Publishing Co., Honolulu.

Hughes, A.L. 1993. Coevolution of immunogenic proteins of *Plasmodium falciparum* and the host's immune system. In: N. Takahata and A.G. Clark, eds. Mechanisms of molecular evolution, pp. 109–127. Sunderland, Mass.: Japan Scientific Societies Press/Sinauer Associates.

Malo, D. and E. Skamene. 1994. Genetic control of host resistance to infection. Trends Genet. 10: 365–371.

Mathiopoulos, K., M. Bouare, G. McConkey, and T. McCutchan. 1993. PCR detection of *Plasmodium* species in blood and mosquitoes. In: D. Persing, T. Smith, F. Tenover, and T. White, eds. Diagnostic molecular microbiology, pp. 462–467. Washington D.C.: American Society for Microbiology/Mayo Clinic.

May, R. 1988. Conservation and disease. Conserv. Biol. 2: 28–30.

Moulton, M.P. and S.L. Pimm. 1986. Species introductions to Hawaii. In: H.A. Mooney and J.A. Drake, eds. Ecology of biological invasions of North America and Hawaii, pp. 231–249. New York: Springer-Verlag.

Nakamura, R., D.M. Berger, E.M.L. Chang, W. Hansen, and A.Y. Miyahara. 1984. A survey of infections, diseases, and parasites of Hawaiian forest birds. Workshop on Avian Diseases in Wild Birds on Pacific Islands, Honolulu, Dec. 13–14, 1984.

Neefs, J.M., V. de Peer, L. Hendriks, and R. De Wachter. 1990. Compilation of small ribosomal subunit RNA sequences. Nucleic Acids Res. 18(s): 2237–2317.

Piatek Jr. M., K.C. Luk, B. Williams, J.D. Lifson. 1993. Quantitative competitive polymerase chain reaction for accurate quantification of HIV DNA and RNA species. Biotechniques 14(1): 70–80.

Quinn, T.W. and B.N. White. 1987. Identifications of restriction fragment-length polymorphisms in genomic DNA of the lesser snow goose (*Anser caerulescens caerulescens*). Mol. Biol. Evol. 4: 126–143.

Scott, M.E. 1988. The impact of infection and disease on animal populations: implications for conservation biology. Conserv. Biol. 2: 40–56.

Scott, J.M., S. Mountainspring, F.L. Ramsey, and C.B. Kepler. 1986. Forest bird communities of the Hawaiian Islands: their dynamics, ecology, and conservation. Studies in Avian Biology 9. Berkeley, Calif.: Cooper Ornithological Society.

Stone, C.P. 1989. Non-native land vertebrates. In: C.P. Stone and D.B. Stone, eds. Conservation biology in Hawaii, pp. 89–95. Honolulu.: University of Hawaii Cooperative National Park Resources Unit.

Unnasch, T.R. and D.F. Wirth. 1983. The avian malaria *Plasmodium lophorae* has a small number of heterogeneous ribosomal RNA genes. Nucleic Acids Res. 11(23): 8443–8459.

van Riper III, C., S.G. van Riper, M.L. Goff, and M. Laird. 1986. The epizootiology and ecological significance of malaria in Hawaiian land birds. Ecol. Monog. 56: 327–344.

Warner, R.E. 1968. The role of introduced diseases in the extinction of endemic Hawaiian avifauna. The Condor 70: 101–120.

Waters, A.P. and T.F. McCutchan. 1989. Rapid, sensitive diagnosis of malaria based on ribosomal RNA. The Lancet 8651: 1336–1343.

Waters, A.P., D.G. Higgins, and T.F. McCutchan. 1991. *Plasmodium falciparum* appears to have arisen as a result of lateral transfer between avian and human hosts. Proc. Natl. Acad. Sci. U.S.A. 88: 3140–3144.

Zuk, M. 1991. Parasites and bright birds: new data and a new prediction. In: J.E. Loye and M. Zuk, eds. Bird–parasite Interactions, pp. 317–327. New York: Oxford University Press.

14

Polymorphism of Genes in the Major Histocompatibility Complex (MHC): Implications for Conservation Genetics of Vertebrates

SCOTT V. EDWARDS AND WAYNE K. POTTS

Genes of the major histocompatibility complex (MHC) are emerging as important paradigms in the study of natural selection at the molecular level (Hedrick, 1994). Several decades of immunological and molecular genetic research have revealed that MHC genes play a critical role in the mounting of an immune response by vertebrate hosts to foreign pathogens. Additionally, more recent research in mice suggests that MHC genes influence individual odors in an allele-specific fashion and that these odors are used in mate choice and kin recognition (reviewed in Potts and Wakeland, 1993; Brown and Ecklund, 1994). The extraordinary polymorphism of MHC genes at the molecular level provides the basis of their ability to perform both of these diverse functions. This extreme polymorphism, and the nature of the forces generating and maintaining it, is the basis for recent claims that MHC genes should play a disproportionate role in the design of programs to conserve genetic diversity in captive populations and endangered species (Hughes, 1991a). Both the promise and the controversy surrounding the precise use of MHC genes in conservation genetics lie in our detailed, albeit incomplete, knowledge of proximate and ultimate mechanisms of selection on MHC genes in nature; we suspect that MHC genes have roles in disease resistance and reproductive success, yet there is considerable disagreement among biologists concerning basic mechanisms underlying these roles. In this regard, issues surrounding the use of MHC markers in conservation genetics are relevant to larger issues in conservation genetics concerning the importance of genetic diversity in general (Harcourt, 1992; Caro and Laurensen, 1994). In this chapter we review the data for MHC as a case example of these larger issues.

THE MHC FOR CONSERVATION BIOLOGISTS

Structure and Function

The MHC is one of three major multigene families contained within the immunoglobulin superfamily of metazoans (Klein, 1986; Hood and Hunkapiller,

Table 14-1 Characteristics of Mammalian MHC Class I and Class II genes

Character	Class I	Class II
Tissue distribtion	All nucleated cells	Immune system cells specialized for uptake and presentation of extracellular antigens (e.g., macrophages and B-cells)
Source site of antiegn	Intracellular	Extracellular
Number of loci (mouse/human)	3/3	2/5
Presents antigen to:	Cytotoxic T-lymphocyte (CTL or Tc)	Helper T-lymphocyte (T_h)
Function	Activates CTLs for killing of infected cells	Activates Th-cells which in turn activates appropriate B-cells for antibody production and appropriate CTLs for further proliferation beyond class I activation alone
Number of potentially polymorphic chains	1	2
Pattern of long-term evolution	Frequent gene deletion, duplication and death	More stable than class I

1991). In mammals, the MHC spans about 3500 kilobases of DNA and contains several hundred genes (Trowsdale et al., 1991; Trowsdale 1993). However, when one speaks of "MHC genes," one is usually referring to the few members of two classes in the MHC known as class I and II (Table 14-1). The class I and II molecules of mammals are the best-studied MHC genes and usually number about 2 to 3 in each class, although different species and haplotypes within species differ in these numbers. Class I and II MHC molecules are receptors that bind fragments of foreign proteins ("antigen" or "peptides") that have been truncated ("processed") into short amino acid chains (9–20 residues) by cellular machinery (Table 14-1; Figure 14-1). For any given MHC, a small subset of these peptides bind to a part of the MHC molecule called the antigen-binding site (ABS). These bound peptides are taken to the cell surface and "presented" to two important components of the vertebrate immune system, cytotoxic and helper T-cells; binding by T-cells to the peptide–MHC complex initiates the adaptive immune response (Figure 14-1). Recent studies of peptides purified from intact class I and II molecules from humans and mice reveal the importance of specific amino acid positions in the peptides which act as "anchor positions" for binding to particular MHC alleles, lending support to the idea that different MHC molecules bind different universes of peptides (reviewed in Germain and Margulies, 1993).

Although discovered prior to the human MHC (known as HLA), the MHC of chickens is much less thoroughly studied (Briles et al., 1948; Kaufman et al., 1990; Kaufman, 1995). As for other nonmammalian vertebrates, the overall function of the MHC in chickens (the "B" complex) is thought to be similar to that in mammals, but it is already clear that there are significant differences between avian and mammalian MHCs. For example, the B complex contains a whole family of receptor genes (B-G genes) with no known function or known homologues in

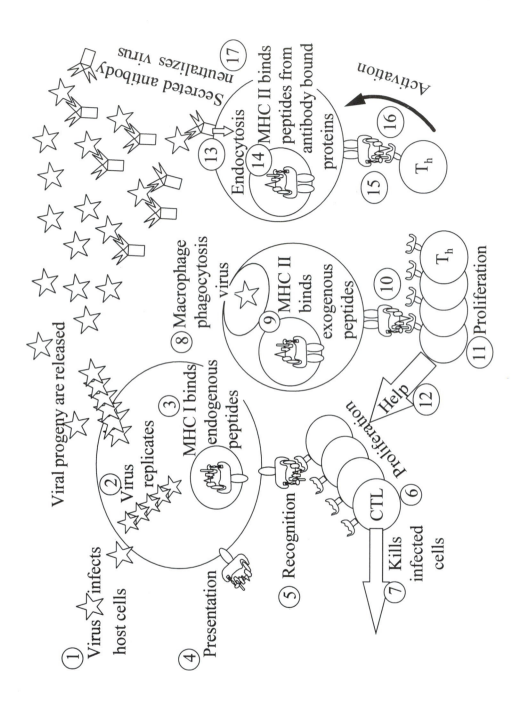

mammals (Kaufman and Salamonsen, 1992). Also, unlike any other vertebrate MHC, B-complex genes were recently shown to reside in at least two independent linkage groups (Briles, et al. 1993; Miller et al. 1994); thus it is almost certain that further comparative work within each vertebrate class will yield surprises, as has been the case with other compartments of the immune system (McCormack et al., 1991).

MHC molecules are currently known only from vertebrates, thus limiting use of MHC polymorphisms to the conservation genetics of this group. Nonetheless, a number of immunoglobulin superfamily molecules occur in invertebrates (Marchalonis and Schluter, 1990), and plants possess self–nonself recognition systems to which MHC molecules may in many respects be analogous (Dangle, 1992; Potts and Wakeland, 1993). Because of their importance in inbreeding avoidance and pathogen surveillance, such molecules may ultimately prove useful in the conservation genetics of these taxa as well.

POLYMORPHISM OF MHC GENES

MHC genes are the most polymorphic functional genes in vertebrates and possess a unique combination of features underlying this extraordinary diversity (reviewed in Hedrick, 1994). The most important conclusion that can be drawn from an analysis of each of these features is that some form of natural selection, specifically balancing selection, is required to explain each one. One of the first surprising features of MHC genes and molecules detected by immunologists was the extreme

---

Figure 14-1 (opposite)   Outline of critical events and features of MHC-dependent immune recognition in response to a viral infection. Development of self/non-self discrmination by the adaptive immune system: Self/non-self discrimination is primarily achieved when T-cells that recognize MHC + self peptides are eliminated during T-cell maturation in the thynus. Consequently, activated T-cells (normally) only recognize MHC + non/-self peptides. Infection: Virus infects host cells (1) and replicates (2); viral progeny are released and new host cells are infected. Within infected host cells foreign and self proteins are being degraded by antigen processing pathways and MHC class I molecules are binding a subset of the resulting peptides (3). Activation of cytotoxic T-lymphocytes (CTL): MHC class I molecules bind a small subset of these peptide fragments from endogenously synthesized proteins (both viral and self) and present them on the surface of infected cells (4). Cytotoxic T-lymphocytes (CTL) that recognize (bind) MHC + viral peptides (5) proliferate (6) and begin killing infected cells (7). Activation of T-helper cells that enhance CTL activity: Macrophages at the site of infection phagocytose virus (8) and a subset of the resulting peptide fragments (derived from exogenous viral proteins) bind to MHC class II molecules (9) and are presented. T-helper (T$_H$) cells that recognize class II MHC + viral peptide fragments presented by these macrophages (10) respond by proliferating themselves (11) and also by enhancing the proloferation of nearby activated CTL via cytokine release (12). Activation of antibody producing B-cells: B-cells whose cell surface antibody binds to free virus. Endocytoses this virus (13), and a subset of the resulting peptide fragments (derived from these exogenous proteins) are bound (14) and presented by MHC class II molecules. T-helper cells that recognize class II MHC + viral peptide fragments presented by these B-cells (15) respond by proliferating themselves and also stimulating the presenting B-cells to proliferate (16) and release soluble antibodies that neutralize virus (17), CTL killing of infected cells and antibody neutralization of free virus continues until the infection is cleared.

number of alleles, which, for some MHC genes is known to exceed 100—a number too high to be explained by most neutral scenarios (Potts and Wakeland, 1990). This feature combined with suspiciously uniform allele frequencies and the excess of MHC heterozygotes in human populations, was among the first indications that some form of balancing selection was occurring in the MHC region (Hedrick and Thomson, 1983; Klitz et al., 1986; Potts and Wakeland, 1990; Hedrick et al., 1991; Hedrick, 1994). However, recent discoveries of MHC variation in isolated human populations suggest that the effects of balancing selection are not so strong as to override significant effects of genetic drift (Watkins et al., 1992; Belich et al., 1992; Titus-Trachtenberg et al., 1994).

Selection Favors Changes in the Antigen Binding Site

The diversity of MHC genes is largely, although not completely, confined to codons forming the ABS, that part of mature MHC molecules that binds foreign peptides. Hughes and Nei (1988, 1989) found that rates of nonsynonymous (amino acid-changing) substitutions were higher than rates of synonymous (non-amino acid-changing) substitutions in codons encoding the ABS, indicating that mutations that change the binding properties of MHC molecules often have a selective advantage when they arise. This pattern has been observed at all classical MHC genes for which one can infer by homology or otherwise the codons encoding the ABS (e.g., Sato et al., 1993). The extreme sequence divergence between MHC alleles is another hallmark of MHC polymorphism. For example, the mean divergence between most pairs of MHC loci can approach 10%, and some allelic pairs have diverged by as much as 30% (Gyllensten and Erlich, 1989; She et al., 1991). It is now known that both the extreme age of MHC allelic lineages maintained by balancing selection (Takahata, 1990, 1991; Takahata and Nei, 1990; Takahata et al., 1992; Klein et al., 1993) and the rapidity with which interallelic and interlocus gene conversion generates new variability (Melvold and Kohn, 1990; Kuhner et al., 1991; Gyllensten et al., 1991; She et al., 1991) cause this high sequence divergence. The age of MHC lineages within some species approaches 30 million years—a point illustrated dramatically by the pattern of relatedness of alleles in long-diverged species. In rodents and primates, pairs of MHC alleles found in distantly related species are often more related than all pairs within species (McConnell et al., 1988; Figueroa et al., 1988; Lawlor et al., 1988; Gyllensten and Erlich 1989; Gyllensten et al., 1990; She et al., 1990); phylogenetic trees of MHC alleles are in general much deeper than trees of the species from which the alleles are derived, or the trees of other nuclear genes for that matter (Klein, 1987; Golding and Felsenstein, 1990; Golding, 1993; Takahata, 1993). In addition, recent gene conversion events in and around the ABS codons have elevated the standing level of MHC diversity in several species, including humans (Erlich and Gyllensten, 1991; Belich et al., 1992; Watkins et al., 1992; Zoorob et al., 1993). Thus, both recombination and the long persistence times of MHC alleles make them unreliable indicators of the true ancestry of organismal lineages, suggesting that conservation plans focused on the geography of evolutionarily significant units (ESUs; Moritz, 1994) should rely on other, neutral markers for such purposes.

WHAT SELECTIVE MECHANISMS MAINTAIN MHC GENETIC DIVERSITY?

Models Based on Infectious Disease

Most models in which infectious disease plays a role in maintaining MHC genetic diversity, and in which there are immunological risks imposed by low MHC variability, are based on the idea that pathogens can evade those components of the immune response that are dependent on MHC molecules ("MHC-dependent" immune recognition or immunity), leading to a molecular evolutionary arms race between MHC molecules and pathogens (Doherty and Zinkenagel, 1975; O'Brien et al., 1985; O'Brien and Evermnann, 1988; Potts and Slev, 1995). In principle, pathogen variants that arise with mutations in peptides that are presented by host MHC molecules may escape MHC-dependent immunity because either MHC molecules no longer bind the variant peptide or T-cells no longer bind the peptide–MHC complex owing to "holes in the T-cell repertoire" (Figure 14-1). Frequency-dependent selection arises when most pathogens evade the most common host MHC genotypes, a situation in which common MHC alleles become disfavored and rare alleles become favored (Clarke and Kirby, 1966; Bodmer, 1972; Howard, 1991; Slade and McCallum, 1992). That pathogens do evolve in response to MHC alleles of the host is evident in experiments in which the mutations characterizing recent pathogen variants have arisen precisely in those portions of peptides that are presented by MHC molecules (Phillips et al., 1991; De Campos et al., 1993). Other types of MHC-pathogen interactions leading to MHC diversity have also been examined recently (Bertoletti et al., 1994; Klenerman et al., 1994).

Pathogen evasion can also give rise to heterozygote advantage for MHC genes because each MHC allele will be resistant to one set of pathogens and susceptible to another. In this model, resistances and susceptibilities are largely determined by past and ongoing pathogen evasion events. For example, specific MHC haplotypes proved resistant to pathogen infection in the case of Marek's disease, malaria, and other pathogens (Briles et al., 1987; Bacon, 1987; Lamont et al., 1987; Wakelin and Blackwell, 1988; Hill et al., 1991). MHC-linked *resistance* (via successful peptide presentation and T-cell recognition) is known to be dominant to susceptibility, so in theory individuals heterozygous for MHC alleles enjoy the combined set of resistances for both alleles and will in theory be more disease-resistant than either homozygote (Wakeland et al., 1989). One caveat to this hypothesis is that normally each allele expressed in an individual is responsible for deletion of a subset of T-cells during ontogeny of the immune system (see Figure 14-1), so it is possible that heterozygotes could suffer a net loss of overall immune recognition! This type of argument is usually used to explain why the duplication of MHC genes has not proceeded to the extent of other components of the immune system, such as the genes for T-cell receptors and antibodies (Figure 14-1), a situation that would make allelic diversity at individual MHC loci unnecessary (Hughes and Nei, 1988; Takahata, 1994).

Maintenance of MHC Diversity without Pathogen Evolution

Although theoretically attractive, pathogen evasion of MHC-dependent immunity has only indirect empirical support and may play only a minor role in MHC evolution (Tiwari and Terasaki, 1985). However, MHC diversity can in principle be maintained without evolution of invading pathogens: MHC heterozygotes will enjoy increased disease resistance simply because they can present a wider variety of antigens, completely independent of the occurrence of pathogen evasion events (Takahata and Nei, 1990; Hughes and Nei, 1992). Again there is no strong empirical support for this hypothesis and, as described above, it is possible that heterozygotes actually present fewer antigens than homozygotes. However, other models leading to frequency-dependent selection (Andersson et al., 1987b), in which the parasite can incorporate host MHC molecules and thereby immunize other hosts with any of the same MHC alleles, do have indirect support from immunization trials for SIV in macaques (Stott, 1991) and corroborating in vitro studies (Arthur et al., 1992).

Models Based on MHC-Based Mating Preferences

If MHC homozygosity is deleterious owing to increased susceptibility to infectious disease, then one would predict the evolution of reproductive mechanisms that allowed parents to preferentially produce MHC-heterozygous, disease-resistant offspring. Such mating preferences have in fact been experimentally demonstrated in house mice (*Mus musculus*), in which MHC-dissimilar mates are preferred both under laboratory conditions (Yamazaki et al., 1976, 1978, 1988, Egid and Brown, 1989) and in seminatural populations (Potts et al., 1991). These MHC-based disassortative mating preferences are diversity-maintaining and the strength of selection imposed by them was sufficient to account for the majority of MHC diversity observed in natural populations (Potts et al., 1991; Hedrick, 1992). Recent reports of MHC-based mating patterns in humans (Ober et al., 1993) suggest that this trait may have some generality in mammals and possibly other vertebrates. Olfactory mechanisms whereby individuals could evaluate the MHC genotypes of prospective mates have been convincingly demonstrated in both house mice (Yamaguchi et al., 1981) and rats (*Rattus norvegicus*, Singh et al., 1987). Although the precise molecular mechanisms are still unknown, MHC genes influence individual odors in an allele-specific fashion. Mutations in a single MHC gene alter the odor of those individuals (Yamazaki et al., 1983). Thus, the extreme genetic diversity of MHC antigen binding sites results not only in extensive variation in patterns of antigen presentation, but also in an extensive array of MHC-specific odor types.

The avoidance of inbreeding is an alternative, but not mutually exclusive, function for the evolution of MHC-based disassortative mating preferences (Brown, 1983; Partridge, 1988; Uyenoyama, 1988; Potts and Wakeland, 1990, 1993; Alberts and Ober, 1993; Brown and Ecklund, 1994). The extreme genetic diversity of MHC genes coupled with the olfactory ability to discriminate MHC-mediated odor types by rodents (Yamaguchi et al., 1981; Singh et al., 1987) and humans (Gilbert et al., 1986) makes it a potentially powerful system for recognizing and avoiding mating with kin (Getz, 1981; Alberts and Ober, 1993; Brown, 1983; Potts and Wakeland, 1993; Partridge, 1988; Brown and Echlund 1994; Potts et al., 1994). For example, by avoiding mating with prospective mates who carry one or more alleles identical to those found in ones' own parents, all full- and half-sib matings and half of all cousin matings will be avoided (Potts and Wakeland, 1993). Furthermore, there is data indicating that MHC genes are being used as a kin recognition marker in contexts other than inbreeding avoidance (Manning et al., 1992). Finally, the only other genetic system exhibiting all the extreme genetic features of MHC genes are plant self-incompatibility genes (Potts and Wakeland, 1993), whose role in inbreeding avoidance is uncontested. This suggests that disassortative mating patterns could in principle contribute to, if not account for, the patterns of diversity observed at MHC genes.

MHC-Based Selective Fertilization or Abortion

One would expect selective fertilization or abortion mechanisms favoring zygotes that were heterozygous at MHC genes to evolve for the same reasons as discussed above for mating preferences. The data supporting the existence of such mechanisms in mammals is equivocal and has been recently reviewed (Alberts and Ober, 1993).

MHC VARIATION AND CONSERVATION GENETICS

Hypotheses explaining MHC diversity focus on either disease or reproduction (see above); both have fundamental implications for the welfare and longevity of small populations and hence are of fundamental importance to conservation biologists. The dramatic evidence for selection on MHC genes makes them of interest far beyond their use strictly as unusually polymorphic markers to keep track of lineages or measure overall heterozygosity. But does this constellation of unusual features of MHC genes warrant elevating their status in genetic studies of endangered species and captive breeding programs as suggested by Hughes (1991a) in a provocative article in *Conservation Biology*? In principle, the answer is irrefutable: *yes*, special attention should be given to genes affecting viability, fecundity, and mating. In practice, however, the situation is much less straight-forward. The problem is twofold: First, *there is no empirical data indicating an unambiguous link between either MHC heterozygosity or MHC genotype of mates and the probability of successful reproduction* (Hill et al., 1991; Potts and Wakeland, 1993). Second, *any captive breeding scheme designed around a single linkage group will inevitably lose variation at other potentially important loci* (Vrijenhoek and Leberg, 1991; Gilpin and Wills, 1991; Miller and Hedrick, 1991). Hughes' (1991a) plea for a "radical reorientation" of captive breeding programs was based on a decade of population genetic research indicating selection for diversity at MHC loci and the known ability of MHC molecules to bind immunogenic peptides. However, whereas some workers insist that the balancing selection driving MHC polymorphisms must be overdominant selection (heterozygote advantage), many others argue that other mechanisms are involved (e.g., Wills, 1991; Slade and McCallum, 1992). The intricacy and redundancy of the vertebrate immune system alone should discourage simplistic translations of MHC heterozygosity or monomorphism into a detailed immunological phenotype of an endangered species (e.g., Parham, 1991). Furthermore, the number of associations of HLA haplotypes and *autoimmune* diseases (at least in humans) far exceeds the number involving *infectious* disease; thus MHC variability is not by any means a panacea, and some haplotypes are overtly detrimental (Tiwari and Teraski, 1985; Potts and Wakeland, 1993). Understanding the precise mechanisms of molecular diversification is a prerequisite for constructing detailed conservation schemes around MHC loci. Such understanding is, unfortunately, very difficult to acquire and is likely to come about only after experimentation with natural and seminatural populations and rigorous controls.

Conserve Heterozygotes or Haplotypes?

In an ideal world in which our knowledge of MHC diversity was perfect, differences in hypothesized forces governing MHC diversity could translate into differences in practices for conservation biologists: the simplest models for the maintenance of MHC diversity suggest different schemes for short-term captive breeding or selection of individuals for reintroduced populations. If it were known that heterozygote advantage was the primary form of selection (a simple symmetrical overdominance model in which fitness is not conditioned on particular alleles; Hedrick, 1994), then a captive breeding scheme in which heterozygosity and allelic diversity is maximized would be favored (Hughes, 1991a). When choosing individuals to found new populations for reintroduction, heterozygous individuals could be chosen over homozygotes. By contrast, if it were known that particular rare alleles or haplotypes, and not heterozygosity per se, conferred greater fitness to individuals in a given place or time, as predicted by models of frequency-dependent selection, then preservation or reintroduction of these alleles might be a priority—at least for the short term. This latter scheme would result in substantially lowered allelic diversity at both MHC loci and at other loci in the genome in the short term in the same way that selection for particular haplotypes in nature would lower this observed level of polymorphism and heterozygosity (Hill et al., 1994). Indeed, Hedrick (1994) and Hedrick and Miller (1994) have shown that this type of management protocol would erode genome-wide diversity drastically compared to one based on preservation of overall levels of genetic diversity (see also Vrijenhock and Leberg, 1991; Gilpin and Wills, 1991; Miller, 1995). Another problem with the latter scheme is that under frequency-dependent selection, a "fit" allele today must have a lower fitness tomorrow; the same logic applies to fitness of alleles in particular localities. The unpredictability inherent in pathogen–MHC interactions necessitates a constantly evolving conservation scheme, one that would change on time-scales coincident with those driving the coevolution of fitness and particular alleles. In our ideal world both overdominant and frequency-dependent scenarios suggest conservation protocols minimizing loss of MHC diversity in the long term. Unfortunately, current knowledge of the forces maintaining MHC diversity is not half so detailed for any vertebrate as to allow prescription of equally detailed conservation plans based solely on MHC.

Role of Mating Preferences in Conservation Genetics

Most of the authors participating in the MHC debate in *Conservation Biology* sparked by Hughes' (1991a) article (Vrijenhoek and Leberg, 1991; Gilpin and Wills, 1991), or subsequent discussions of the role of MHC in conservation genetics (Caro and Laurensen, 1994), ignored recent findings indicating the existence of MHC-based mating preferences and the use of MHC alleles as possible kin recognition markers in semi-natural populations (Potts et al., 1991; Manning et al., 1992). In principle, however, these two findings have dramatic implications for captive breeding programs. Say it was known that an endangered mammal used MHC-mediated olfactory cues to discriminate between possible mates and avoid inbreeding. Then by allowing MHC genetic diversity to become low we would be robbing a threatened species of an evolved mechanism to avoid inbreeding. If a species evolved an MHC-based inbreeding avoidance system in its evolutionary history, then it would be unwise to allow essential components of such a system, in this case MHC diversity, to be lost. The functional diversity of MHC genes for mating preferences could be lost in a few generations of inattention to this problem by managers of threatened species. It would be illogical to ignore the self-incompatibility system in the management of a threatened plant species; in so far as MHC genes function in ways similar to plant histocompatibility genes, the same logic could apply. In this case, appropriate management or breeding techniques would ensure MHC-dissimilarity among captive individuals or prospective founders of new populations and that sufficient variability was available for effective operation of an MHC-based inbreeding avoidance system (Manning et al., 1992; Brown and Ecklund, 1994).

Extrapolating from seminatural populations of mice to a typical endangered vertebrate may well be risky, and the rules and mechanisms by which mice process the information from MHC genes are still in question and may be different or nonexistent in other vertebrates. Contrary to textbook wisdom, many birds possess and use well-developed olfactory senses (Waldvogel, 1989), but we know nothing about their potential interactions with the avian MHC. It is our obligation at least to ask what the consequences would be were such genetic discrimination occurring in endangered mammals, and whether such consequences have fitness costs large enough to warrant attention. Could the inappropriate "social conditions" underlying low reproductive rates in captive cheetah populations have a component in MHC variation (Caro, 1993)? Could the success of programs designed to maximize hybridity (Templeton and Read, 1983; Templeton, 1987) be due in part to MHC genes? As our knowledge of important genetic systems becomes more precise, managers will be better able to assess the impacts of genes like those in the MHC on conservation goals; "conserving genetic diversity" in species already endangered may come to possess a logic ultimately more gratifying than the logic of preserving diversity at neutral markers throughout the genome that are only surrogates of functional, fitness-related diversity.

CORRELATES AND TAXONOMIC DISTRIBUTION OF LOW MHC VARIABILITY

Case Studies of MHC Monomorphism

The most frequently mentioned cause for concern for species exhibiting low MHC variability is their possible increased susceptibility to epizootics (e.g., Hughes, 1991a). Some of the most thorough case studies in conservation genetics have voiced such a concern (O'Brien et al., 1985; O'Brien and Everman, 1988). Despite the usually abundant polymorphism at MHC genes, there are a number of species in which MHC genes exhibit low or no detectable polymorphism. Table 14-2 summarizes the known cases falling into this category, the causes of the low polymorphism hypothesized by the authors of those studies, and whether 14 significant health problems are known in each species. *None* of the studies referenced in Table 14-2 identified causes of low variability experimentally (see below), and none has established causal relationships between low MHC variability and disease susceptibility. Unfortunately, there is no unambiguous link between low intraspecific MHC variation and disease susceptibility (Caro and Laurenson, 1994), just as there is no link between typical levels of MHC polymorphism and "population health"; the recent viral-induced decline of a Serengeti population of lions (Morell, 1994)—population with average levels of MHC variability (Yuhki and O'Brien, 1990)—is a case in point. Perhaps the most celebrated case for a link between MHC variability and disease susceptibility is in cheetahs (O'Brien et al., 1985); nonetheless, cheetahs are low in variability throughout the genome (Yuhki and O'Brien, 1990), and other loci could be responsible for the numerous susceptibilities to disease recorded in cheetahs (Munson, 1993; Everman et al., 1993).

Two further points emerge from Table 14-2. First, even if the hypothesized cause of low MHC variation were correct in each case, no single cause could explain low MHC variation across taxa. Hypothesized causes range from the loss of polymorphism due to low pathogen loads in particular environments (e.g., southern elephant seals, naked mole-rats) or reproductive biologies (tamarins) to nonadaptive loss due to population bottlenecks (cheetahs, Asiatic lions). Second, low variability at MHC loci does not necessarily entail low variability at non-MHC loci (e.g., some marine mammals, beavers). Importantly, no study conducted at the level of nucleotide sequences has reported complete lack of variability, whereas several studies utilizing RFLP approaches have failed to detect any variability (see below). Most of the instances of low variability involve highly expressed class I loci, which are known sometimes to originate from ancestral loci exhibiting low polymorphism (Watkins et al., 1990b) or to undergo gene duplication much more frequently than class II loci (Rada et al., 1990; Hughes, 1991b; Nei and Hughes, 1991); both of these phenomena could influence the level of polymorphism detected in contemporary populations and could cloud understanding of the significance of MHC diversity.

Table 14-2 Low MHC Polymorphism in Mammalian Species and Its Hypothesized Correlates

Species (MHC Class I and/or II)	Method of Assay[a]	Low Variation at Other Loci?	Documented Health Problems?	Hypothesized Cause	Low Polymorphism Maladaptive?	Referen
Large-bodied terrestrial mammals						
Cheetah (I) (*Aconyx jubatus*)	RFLP	Yes	Yes	Bottleneck	Yes	1–3, 1
Asiatic lions (I) (*Panthera leo persica*)	RFLP	Yes		Bottleneck	Yes	4, 5
Small-bodied terrestrial mammals						
Naked mole-rat (I) (*Heterocephalus glaber*)	RFLP	Yes	No	Stable niche	No	6–8
Syrian hamster (I) (*Mesocricetus auratus*)[b]	Ser	?	No	Solitary, fewer parasites	Np	9
Helgoland Island mice (I) (*Mus musculus helgolandicus*)	Ser	?	No	Bottleneck	No	10
Balkan mole rats (I, II) (4 *Spalax* spp.)	RFLP	?	No	Bottleneck	No	11
Beaver (I, II) (*Castor fiber*)	RFLP	Yes/No	No	Reduced selection	No	12
Marine mammals						
Southern elephant seal (I, II) (*Mirounga leonina*)	RFLP	No	Yes	Fewer parasites	No	13
Sei whale (II) (*Balaenoptera borealis*)	RFLP	No	No	Fewer parasites	No	14
Fin whale (II) (*Balaenoptera physalus*)	RFLP	No	No	Fewer parasites	No	14
Beluga whale (II) (*Delphinapterus leucas*)	Seq	Yes	Yes	Reduced selection	No	19
Primates						
Cotton-top tamarin (I) (*Saguinus oedipus*)	Seq	?	Yes	Bone marrow chimaeras, recent origin of gene	Yes/No[c]	15–1

[a] Method of assay: RFLP, restriction fragment length polymorphisms; Ser, serology (antibodies); Seq, nucleotide sequencing.

[b] Class I alleles of the Syrian hamster have been shown to be polymorphic by sequence analysis (Watkins 1990c) and is included here to augment the variety of causes of low polymorphism hypothesized by researchers.

[c] Low polymorphism was hypothesized to facilitate rejection of fetuses sharing bone marrow, but is also thought to be associated with susceptibility to viruses.

References: 1, O'Brien et al. (1985); 2, Yuhki and O'Brien (1990); 3, Everman et al. (1993); 4, Wildt et al. (1987); 5, O'Brien et al. (1987); 6, Faulkes et al. (1990); Reeve et al. (1990); 8, Honeycutt al. (1991); 9, Streilein and Duncan (1983); 10, Figueroa et al. (1986); 11, Nizetic et al. (1988); 12, Ellegren et al. (1993); 13, Slade (1992); 14, Trowsdale et al. (1989); 15, Watkins et al. (1990a);. Watkins et al. (1990b); 17, Hughes (1991); 18, Munson (1993); 19, Murray et al. (1995).

MINING THE VARIABILITY OF MHC GENES

Immunological Techniques

The techniques for isolating and characterizing MHC genes have evolved over the past three decades just as those for studying other genetic markers (reviewed in Parham, 1992). MHC molecules and allotypes were originally isolated with antibodies directed toward various conserved and variable regions of molecules. It was immediately clear from such methods that MHC molecules were polymorphic, and immunobiological methods are still of great value in isolating MHC molecules from novel species (Kaufman et al., 1990).

Restriction Fragment Analyses

RFLP techniques have been used extensively to characterize MHC diversity. A number of studies have successfully used cDNA probes from humans (Andersson et al., 1987a; Slade, 1992) or mice (Faulkes et al., 1990; Yuki and O'Brien, 1990) to reveal genomic variation in natural populations of other species via Southern hybridizations. Since estimates of nucleotide diversity via DNA sequencing are usually higher than when assayed via RFLPs, it is almost certain that the Southern blot approach using heterologous probes will usually underestimate the actual nucleotide diversity and heterozygosity. (The extent of this underestimation is not known for MHC genes, and most workers utilizing RFLPs have argued that it is negligible.) In natural populations it is also rarely known whether the observed RFLPs occur in functionally important regions of the MHC genes (e.g., exons 2 and 3 for class I or exon 2 for class II). Detailed conclusions about pathogen resistance or susceptibility based on RFLP variation alone would be at best inaccurate and at worst misleading.

Occasionally the use of probes from distantly related organisms can give ambiguous results. To determine patterns of realized reproductive success in red-winged blackbirds (*Agelaius phoeniceus*) by means of highly polymorphic markers, Gibbs et al. (1990) compared banding patterns of putative parents and chicks generated by Southern analysis using a mouse MHC class II cDNA probe. However, the polymorphism exhibited by these blots was extremely high even for an MHC locus, and it was unclear whether the bands generated by this probe represented bona fide MHC variation rather than polymorphism at an anonymous locus (Gibbs et al., 1991). (Surprisingly, use of probes from the chicken B complex failed to give reliable hybridization signals.) Nonetheless, probes from the human MHC were used to detect polymorphisms in putative MHC genes in chickens (Anderson et al., 1987a). Thus, while the use of heterologous MHC probes in Southern hybridizations has been extremely useful, caution should be exercised.

Hybridization probes from the focal species or a congener have frequently been used to detect MHC variation (e.g., Hala et al., 1988; Miller et al., 1988). We have developed a simple method utilizing PCR for rapidly generating probes

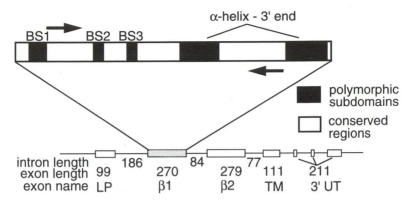

Figure 14-2 Position of degenerate PCR primers between polymorphic subdomains of the second exon of a chicken class II β-chain gene. Primers are indicated by bold arrows above the second exon (Edwards et al., 1995a). Approximate location of polymorphic subdomains in mammals appear in black in the enlarged second exon (above). The structure of a typical chicken class II β gene with lengths (in base pairs) of introns and exons are according to Xu et al. (1988). LP = leader peptide, TM = transmembrane, 3' UT = 3' untranslatedsequences.

specific for class II β-chain MHC genes for use in natural populations (Figure 14-2). By using degenerate primers we have amplified MHC class II β-chain genes from birds and crocodiles (Edwards et al., 1995a,b). These primers amplify approximately 210 bp of exon 2 and are targeted toward some of the same conserved regions of exon 2 that primer pairs in other studies have utilized (Gyllensten et al., 1990; Slade et al., 1994). Probes made from such PCR products cloned in TA-cloning vectors (Invitrogen) reveal abundant and specific MHC polymorphisms when used in Southern hybridizations (Edwards et al., 1995a), and are likely to prove useful in population surveys of natural populations.

PCR and Nucleotide Sequencing

Human HLA genes were one of the first to be amplified by the polymerase chain reaction (PCR; Horn et al., 1988), and all of the methods currently used to conduct population surveys of HLA genes are based on PCR. Such methods include sequencing of cloned PCR products amplified from cDNA (Ennis et al., 1990) and nonradioactive typing via hybridization with allele-specific oligonucleotides (ASOs) fixed to a nylon membrane (Saiki et al., 1989; Erlich and Bugawan, 1993). All of these methods rely on databases not only of primary structures of MHC molecules, but (for oligonucleotide typing) allelic variability in representative samples. Population surveys utilizing PCR sequencing or ASOs have shown that, while most of the major lineages of HLA molecules delineated by serology have survived scrutiny at the level of nucleotide sequences, some serological designations are inadequate reflections of differences at the molecular level (Parham, 1991). In a few cases, MHC molecules that were deemed monomorphic using serological or RFLP techniques have been found to be polymorphic, albeit at a low level, upon DNA sequencing (e.g., Syrian hamster, Watkins et al., 1990c; cheetah, Yuhki and O'Brien, 1994).

Determining variability at the nucleotide level for all relevant regions from specific MHC genes of most vertebrates can be difficult, and is usually attacked by constructing cDNA or genomic libraries (e.g., Ono et al., 1993). A drawback of sequencing studies using only primers directed toward conserved regions in exon 2 (Figure 14-2) is that at least one functionally important portion of exon 2 (β-strand 1 or BS1) will not be contained within the amplified product. Thus, what might appear to be similar alleles for the 210 bp amplified region might in fact be different alleles had exon 2 been sequenced in its entirety. To overcome this, we recommend utilizing the sequence data from the amplified portion of exon 2 to design further experiments to clone flanking exons. We have successfully used a recently described version of ligation-assisted primer amplification, termed ligation-anchored PCR (LA-PCR; Troutt et al., 1992), to obtain sequences, and hence primers, 5' to exon 2. Sequences downstream of exon 2 can be obtained via conventional reverse-transcription PCR (RT-PCR) using primers targeted to conserved blocks in exon 4 (Kawasaki, 1990; Edwards et al., 1995b). A disadvantage of using either of the above techniques on endangered species is the need to isolate mRNA from tissue, usually from spleen in the case of class II genes (Table 14-1). However, methods are available for isolating sufficient mRNA from leukocytes to permit amplification of class I and II and MHC genes from approximately 0.5 ml of blood (J. Kaufman, personal communication).

It is already possible to amplify the entire exon 2 of class II genes of certain taxonomic groups using primers placed in flanking introns (e.g., Zoorob et al., 1993), and we suspect that PCR primers of wider taxonomic utility will be designed within the next few years that will amplify relevant exons of MHC genes in their entirety. Although some class II genes are so closely related that primers often amplify one or a few loci, it is often easy to distinguish cloned PCR products from different loci based on sequence motifs. Nonetheless, for studies employing DNA sequencing, is particularly important to have some criterion for stating that the sequences analyzed come from orthologous (identical by descent) loci, rather than from multiple loci. Such criteria, although well established in mammals, will be improved for other vertebrates through comparative studies of MHC gene organization.

MHC POLYMORPHISMS IN ENDANGERED SPECIES: POPULATION BIOLOGY FOR IMMUNOLOGISTS

A consensus is emerging that many important aspects of the dynamics of MHC evolution and of MHC-linked disease susceptibilities will almost certainly depend on analysis of relatively undisturbed human populations or natural populations of nonhuman vertebrates (e.g., Hill et al., 1991; Klein et al., 1993). Focus on natural populations removes many recent influences perturbing the action of natural selection, such as recent worldwide admixture of human populations and access to medical facilities in developed countries. Although the experimental data are lacking, it is plausible that MHC variation could have been intimately involved with some of the most dramatic examples of disease- or viral-induced selection in natural populations of vertebrates (e.g., Hawaiian birds, *Myxomatosis* in Australian rabbits, etc.; O'Brien and Evermann, 1988). That the diseases mediating such selection in some cases are known to interact with MHC genes in humans makes such inferences even more compelling (Hill et al., 1991). We expect that many of the technical and analytical methods for identifying MHC-linked susceptibilities in model organisms will be useful in identifying such susceptibilities in natural and captive populations of vertebrates.

Our own studies are directed toward determining whether basic life history attributes are associated with MHC heterozygosity or particular alleles in both natural and seminatural populations. One such study is focused on the cooperatively breeding Florida scrub jay (*Aphelocoma caerulescens caerulescens*), a federally threatened subspecies of the scrub jay, which is widespread in western North America and Mexico. The total number of Florida scrub jays was recently estimated at 7000–10 000, and the ancient xeric scrub communities with which they are tightly associated are disappearing at an alarming rate owing to development and cessation of the frequent natural fires (Fitzpatrick et al., 1991). Several aspects of this subspecies are well suited to analysis of MHC correlates of life history variation and reproductive success: (1) The population at Archbold Biological Station has been the subject of one of the most detailed long-term studies of any songbird in North America; (2) this population is known to have undergone an episodic decline due to an epidemic in the late, 1970s; and (3) there are differences in reproductive success between groups that cannot be explained by differences in habitat quality or demographic variables (Woolfenden and Fitzpatrick, 1984). Unravelling the intricate relationship between MHC poly-morphisms and reproductive success—a relationship which all the models for MHC polymorphism assume—will require large sample sizes (Hill et al., 1991) and, in some cases, long-term genetic and demographic data. The generation of knowledge pertaining to the function of MHC molecules in natural populations—including humans—will undoubtedly require a two-way communication between evolutionary biologists and immunologists (Hedrick, 1994). Monitoring of varia-bility at MHC genes and other loci of immunological importance in captive and reintroduced populations could provide fertile ground for such communica-tion.

CONCLUSION

There are abundant hints that the variation harbored in the vertebrate MHC could be of exceptional importance to conservation geneticists. Given the role of MHC genes in immunity and reproduction, it is likely that conserving MHC diversity will contribute to the vigor and longevity of populations in the long term. However, for most species, the immediate consequences of loss of MHC diversity are uncertain, and we lack the detailed knowledge required to predict these consequences in any one species. Considerable basic research should be devoted to evaluating the importance of MHC diversity in enough experimental and non-endangered natural populations that the importance of preserving MHC variability in endangered species can be ranked alongside other conservation genetic goals. As long as the roles of the MHC in disease resistance, reproductive efficiency, and inbreeding avoidance remain viable possibilities, conserving MHC diversity should be an important consideration for conservation biologists.

ACKNOWLEDGMENTS

We thank E. Titus-Trachtenberg, P. Hedrick, M. Miller, and C. Moritz for helpful discussion and/or sharing of unpublished manuscripts. R. Ruff and P. Hedrick provided helpful comments on the manuscript. This work was supported by an Alfred P. Sloan Fellowship in Molecular Evolution to S.V.E. and support from the National Science Foundation to W.K.P. and from the National Institutes of Health to W.K.P. and E.K. Wakeland.

REFERENCES

Alberts, S.C. and C. Ober, 1993. Genetic variability in the major histocompatibility complex: a review of non-pathogen-mediated selective mechanisms. Yearbook Phys. Anthropol. 36: 71–89.

Andersson, L., C. Lundburg, L. Rask, B. Gissel-Nielsen, and M. Simonsen. 1987a. Analysis of class II genes of the chicken MHC (B) by use of human DNA probes. Immunogenetics 26: 79.

Andersson, L., S. Paabo, and L. Rask, 1987b. Is allograft rejection a clue to the mechanism promoting MHC polymorphism? Immunol. Today 8: 206–209.

Arthur, L.O., J.W. Bess, R. Sowder, et al. 1992. Cellular proteins bound to immuno-deficiency viruses: implications for pathogenesis and vacancies. Science 258: 1935–1938.

Bacon, L.D. 1987. Influence of the major histocompatibility complex on disease resistance and productivity. Poultry Sci. 66: 802–811.

Belich, P.M., J.A. Madrigal, W.H. Hildebrand, J. Zemmour, R.C. Williams, R. Luz, M.L. Petzl-Erler, and P. Parham. 1992. Unusual HLA-B alleles in two tribes of Brazilian Indians. Nature 357: 326–329.

Bertoletti, A., A. Sette, F.V. Chiasri, A. Penna, M. Levrero, M. De Carli, F. Flacciadori, and C. Ferrari, 1994. Natural variants of cytotoxic epitopes are T-cell receptor antagonists for antiviral cytotoxic T-cells. Nature 369: 407–410.

Bodmer, W.F. 1972. Evolutionary significance of the HLA system. Nature 237: 139–145.

Briles, W. E., W.H. McGibbon, and M.R. Irwin. 1948. Studies of the time of development of cellular antigens in the chicken. Genetics 33: 97.

Briles, W.E., R.W. Briles, and H.A. Stone, 1987. Resistance to a malignant lymphoma in chickens is mapped to a subregion of the major histocompatibility (B) complex. Science 219: 977.

Briles, W.E., R.M. Goto, C. Auffrey, and M.M. Miller, 1993. A polymorphic system related to but genetically independent of the chicken major histocompatibility complex. Immunogenetics 37: 408–414.

Brown, J.L. 1983. Some paradoxical goals of cells and organisms: the role of the MHC. In: D.W. Pfaff, ed. Ethical questions in brain and behavior, pp. 111–124. New York: Springer Verlag.

Brown, J.L. and A. Eklund, 1994. Kin recognition and the major histocompatibility complex: an integrative review. Am. Nat. 143: 435–461.

Caro, T.M. 1993. Behavioral solutions to breeding cheetahs in captivity: insights from the wild. Zoo Biol. 12: 19–30.

Caro, T.M. and M.K. Laurenson. 1994. Ecological and genetic factors in conservation: a cautionary tale. Science 263: 485–486.

Clarke, B. and D.R.S. Kirby. 1966. Maintenance of histocompatibility polymorphisms. Nature 211: 999–1000.

Dangle, J.L. 1992. The major histocompatibility complex a la carte: are there analogies to plant disease resistance genes on the menu? Plant J. 2: 3–11.

De Campos, P.O., Gavioli, R., Zhang, Q.J., et al. 1993. HLA-A11 epitope loss isolates of Epstein–Barr virus from a highly A11+ population. Science 260: 98–100.

Doherty, P.C. and R.M. Zinkernagel. 1975. Enhanced immunological surveillance in mice heterozygous at the H-2 gene complex. Nature 256: 50–52.

Edwards, S.V., M. Grahn, and W.K. Pohs. 1995a. Dynamics of *mhc* evolution in birds and crocodilians: amplification of class II gene with degenerate primers. Mol. Ecol. 4: 719–729.

Edwards, S.V., E.K. Wakeland, and W.K. Potts. 1995b. Contrasting histories of avian and mammalian *mhc* genes revealed by class II B sequences from songbirds. Proc. Natl. Acad. Sci. U.S.A. 92: 12200–12204.

Egid, K. and J.L. Brown. 1989. The major histocompatibility complex and female mating preferences in mice. Animal Behav. 38: 548–549.

Ellegren, H., G. Hartman, M. Johansson, and L. Anderson. 1993. Major histocompatibility complex monomorphism and low levels of DNA fingerprinting variability in a reintroduced and rapidly expanding population of beavers. Proc. Natl. Acad. Sci. U.S.A. 90: 8150–8153.

Ennis, P.D., J. Zemmour, R.D. Slater, and P. Parham. 1990. Rapid cloning of HLA-A, B cDNA by using the polymerase chain reaction: frequency and nature of errors produced in amplification. Proc. Natl. Acad. Sci. U.S.A. 87: 2833–2837.

Erlich, H.A. and T. Bugawan. 1993. Analysis of HLA class II polymorphism using polymerase chain reaction. Arch. Pathol. Lab. Med. 117: 482–485.

Erlich, H.A. and U.B. Gyllensten, 1991. Shared epitopes among HLA class II alleles: gene conversion, common ancestry and balancing selection. Immunol. Today 12: 411–414.

Everman, J.F., M.K. Laurenson, A.J. McKeirman, and T.M. Caro. 1993. Infectious disease surveillance in captive and free-living cheetahs: an integral part of the species survival plan. Zoo Biol. 12: 125–135.

Faulkes, C.G., D.H. Abbott, and A.L. Mellor. 1990. Investigation of genetic diversity in wild colonies of naked-mole rats (*Heterocephalus glaber*) by DNA fingerprinting. J. Zool. (London) 221: 87–97.

Figueroa, F., H. Tichy, R.J. Berry, and J. Klein. 1986. MHC polymoprhism in island populations of mice. Current Topics. Microbiol. Immunol. 127: 100–105.

Figueroa, F., E. Gunther, and J. Klein. 1988. MHC polymorphism pre-dating speciation. Nature 335: 265–267.

Fitzpatrick, J.W., G.E. Woolfenden, and M.T. Kopeny. 1991. Ecology and development-related habitat requirements of the Florida Scrub Jay (*Aphelocoma coerulescens coerulescens*). Nongame Wildlife Program Technical Report No. 8. Tallahasee, Fla.: Florida Game and Fresh Water Commission.

Germain, R.N. and D.H. Marguiles. 1993. The biochemistry and cell biology of antigen processing and presentation. Ann. Rev. Immunol. 11: 403–450.

Getz, W.M. 1981. Genetically based kin recognition systems. J. Theor. Biol. 92: 209–226.

Gibbs, H.L., P.J. Weatherhead, P.T. Boag, B.N. White, L.M. Abak, and D.J. Hoysak. 1990. Realized reproductive success of polygynous red-winged blackbirds revealed by DNA markers. Science 250: 1394–1397.

Gibbs, H., P. Boag, B. White, P. Weatherhead, L. Tabak. 1991. Detection of a hypervariable

DNA locus in birds by hybridization with a mouse MHC probe. Mol. Biol. Evol. 8: 433–446.

Gilbert, A.N., K. Yamazaki, and G.K. Beauchamp. 1986. Olfactory discrimination of mouse strains (*Mus musculus*) and major histocompatibility types by humans (*Homo sapiens*). J. Comp. Psychol. 100: 262–265.

Gilpin, M. and C. Wills. 1991. MHC and captive breeding: a rebuttal. Conserv. Biol. 5: 554–555.

Golding, B. 1993. Maximum-likelihood estimates of selection coefficients from DNA sequence data. Evolution 47: 1420–1431.

Golding, B. and J. Felsenstein. 1990. A maximum-likelihood approach to the detection of selection from a phylogeny. J. Mol. Evol. 31: 511–523.

Gyllensten, U.B. and H.A. Erlich. 1989. Ancient roots for polymorphism at the DQβ locus of primates. Proc. Natl. Acad. Sci. U.S.A. 86: 9986–9990.

Gyllensten, U.B., D. Lashkari, and H.A. Erlich. 1990. Allelic diversification at the class II *DQB* locus of the mammalian major histocompatibility complex. Proc. Natl. Acad. Sci. U.S.A. 87: 1835–1839.

Gyllensten, U., M. Sundvall, and H. Erlich, 1991. Allelic diveristy is generated by intra-exon exchange at the DRB locus of primates. Proc. Natl. Acad. Sci. U.S.A. 88: 3686–3690.

Hala, K., A.-.M. Chaussé, Y. Bourlet, O. Lassila, V. Hasler, and C. Auffrey. 1988. Attempt to detect recombination between B-F and B-L genes within the chicken B complex by serological typing, in vitro MLR and RFLP analyses. Immunogenetics 28: 433.

Harcourt, S. 1992. Endangered species. Nature 354: 10.

Hedrick, P.W. 1992. Female choice and variation in the major histocompatibility complex. Genetics 132: 575–581.

Hedrick, P.W. 1994. Evolutionary genetics of the major histocompatibility complex. Am. Nat. 143: 945–964.

Hedrick, P.W. and P.S. Miller. 1994. In: V. Loeschke, J. Tomiuk, and S.K. Jain, eds. Conservation genetics, pp. 187–204. Basel: Birkhauser.

Hedrick, P.W. and G. Thomson. 1983. Evidence for balancing selection at HLA. Genetics 104: 449–456.

Hedrick, P.W., W. Klitz, W.P. Robinson, M.K., Kuhner, and G. Thomson, G. 1991. Evolutionary genetics of HLA. In: R. Selander, A. Clark, and T. Whitman, eds. Evolution at the molecular level, pp. 248–271. Sunderland, Mass.: Sinauer.

Hill, A.V.S., C.E.M. Allsopp. D. Kwiatowski, et al. 1991. Common West African HLA antigens are associated with protection from severe malaria. Nature 352: 595–600.

Honeycutt, R.L., K. Nelson, D. Schlitter, and P. Sherman. 1991. Genetic variation within and among populations of the naked mode-rat. In: P.W. Sherman, J.U.M. Jarvis, and R.D. Alexander, eds. The biology of the naked mole-rat, pp. 195–208. Princeton, N.J.: Princeton University Press.

Hood, L. and T. Hunkapiller. 1991. The immunoglobulin superfamily. In: S. Osawa and T. Hanjo, eds. Evolution of life: fossils, molecules and culture. pp. 123–143. Tokyo: Springer-Verlag.

Horn, G.T., T.L. Bugawan, C. Long, and H.A. Erlich. 1988. Allelic variation of HLA-DQ loci: Relation to serology and to insulin-dependent diabetes susceptibility. Proc. Natl. Acad. Sci. U.S.A. 85: 6012–6016.

Howard, J.C. 1991. Disease and evolution. Nature 352: 565–566.

Hughes, A.L., 1991a. MHC polymorphisms and the design of captive breeding programs. Conserv. Biol. 5: 249–251.

Hughes, A.L. 1991b. Independent gene duplications, not concerted evolution, explain relationships among class I MHC genes of murine rodents. Immunogenetics 33: 367–373.

Hughes, A. and M. Nei. 1988. Pattern of nucleotide substitution at major histocompatibility complex loci reveals overdominant selection. Nature 335: 167–170.

Hughes, A. and M. Nei. 1989. Nucleotide substitution at major histocompatibility class II loci: evidence for overdominant selection. Proc. Natl. Acad. Sci. U.S.A. 86: 958–962.

Hughes, A.L. and M. Nei. 1992. Models of host-parasite interaction and MHC polymorphism. Genetics 132: 863–864.

Kawasaki, E.S. 1990. Amplification of RNA. In: M. Innis, D.H. Gelfand, J.J. Sninsky, and T.J. White, eds., PCR protocols: a guide to methods and applications, pp. 21–27. New York: Academic Press.

Kaufman, J. and J. Salamonsen, 1992. B-G: we know what it is, but what does it do? Immunol. Today 13: 1–3.

Kaufman, J., K. Skoedt, and J. Salmonsen. 1990. The MHC molecules of nonmammalian vertebrates. Immunol. Rev. 113: 83–117.

Kaufman, J., M. Völk, and H.-J. Wallny. 1995. A "Minimal Essential Mhc" and an "Unrecognized Mhc": two extremes in selection for polymorphism. Immunol. Rev. 143: 63–88.

Klein, J. 1986. Natural history of the major histocompatibility complex. New York: Wiley.

Klein, J. 1987. Origin of major histocompatibility complex polymorphism: the trans-species hypothesis. Hum. Immunol. 19: 155–162.

Klein, J., Y. Satta, C. O'hUigin, and N. Takahata. 1993. The molecular descent of the major histocompatibility complex. Ann. Rev. Immunol. 11: 269–295.

Klenerman, P., S. Rowland-Jones, S. McAdam, J. Edwards, S. Daenke, D. Lallo, B. Köppe, W. Rosenberg, D. Boyd, A. Edwards, P. Giangrande, R.E. Phillips, and A.J. McMichael. 1994. Cytotixic T-cell activity anatagonized by naturally occurring HIV-1 Gag variants. Nature 369: 403–407.

Klitz, W., G. Thompson, and M.P. Baur. 1986. Contrasting evolutionary histories among tightly linked HLA Loci. Am. J. Hum. Genet. 39: 340–349.

Kuhner, M.K., D.A. Lawlor, P.D. Ennis, and P. Parham. 1991. Gene conversion in the evolution of the human and chimpanzee MHC class I loci. Tissue Antigens 38: 152–164.

Lamont, S.J., C. Bolin, and N. Cheville. 1987. Genetic resistance to fowl cholera is linked to the major histocompatibility complex. Immunogenetics 25: 284–289.

Lawlor, D.A., J. Zemmour, P.D. Ennis, and P. Parham. 1988. HLA-A and B polymorphisms predate the divergence of human and chimpanzees. Nature 335: 268–271.

Marchalonis, J.J. and Schulter. 1990. Origins of immunoglobulins and immune recognition molecules. Bioscience 40: 758–768.

Manning, C.J., E.K. Wakeland, and W.K. Potts. 1992. Communal nesting patterns in mice implicate MHC genes in kin recognition. Nature 360: 581–583.

McConnell, T.J., W.S. Talbot, R.A. McIndoe, and E.K. Wakeland. 1988. The origin of MHC class II gene polymorphisms in the genus Mus. Nature 33: 651–654.

McCormack, W.T., L.W. Tjoelker, and C.B. Thompson. 1991. Avian B-cell development: generation of an immunoglobulin repertoire by gene conversion. Ann. Rev. Immunol. 9: 219–241.

Melvold, R.W. and H.I. Kohn. 1990. Spontaneous frequency of H-2 mutations. In: I.K. Egorov and C.S. David, eds. Transgenic mice and mutants in MHC research. Berlin: Springer-Verlag.

Miller, P.S. 1995. Evaluating selective breeding programs for rare alleles: examples using the Przewalski's horse and California condor pedigrees. Conserv. Biol. 9: 1262–1273.

Miller, P.S. and P.W. Hedrick. 1991. MHC polymorphism and the design of captive breeding programs: simple solutions are not the answer. Conserv. Biol. 5: 556–558.

Miller, M.M., H. Abplanalp, and R. Goto. 1988. Genotyping chickens for the B-G subregion

of the major histocompatibility complex using restriction fragment length polymorphisms. Immunogenetics 28: 374–379.

Miller, M.M., R. Goto, A. Bernot, R. Zoorob, C. Auffray, N. Bumstead, and W.W. Briles.
1994. Two *Mhc* class I and two *Mhc* class II genes map to the chicken *Rfp-Y* system
outside the *B* complex. Proc. Natl. Acad. Sci. U.S.A. 91: 4397–4401.

Morell, V. 1994. Serengeti's big cats go to the dogs. Science: 264: 1664.

Moritz, C. 1994. Defining evolutionary significant units for conservation. Trends Ecol. Evol.
9: 373–375.

Munson, L. 1993. Diseases of captive cheetahs (*Acinonyx jubatus*): results of the Cheetah
Research Council Pathology Survey, 1989–1992. Zoo Biol. 12: 105–124.

Murray, B.W., S. Malik, and B.N. White. 1995 Sequence variation of the major histocompatibility complex locus *DQB* in Beluga whales (*Delphinaptersus leucas*). Mol. Biol.
Evol. 12: 582–593.

Nei, M. and A.L. Hughes. 1991. Polymorphism and evolution of the major histocompatibility complex loci in mammals. In: R. Selander, A. Clark, and T. Whitman,
eds. Evolution at the molecular level, pp. 222–247. Sunderland, Mass.: Sinauer.

Nizetic, D., M. Stevanovic, B. Soldatovic, I. Savic, and R. Crkvenjakov. 1988. Limited
polymorphism of both classes of MHC genes in four different species of the Balkan
mole rat. Immunogenetics 28: 91–98.

Ober, C., L.R. Weitkamp, S. Elias, and D.D. Kostyu. 1993. Maternally-inherited HLA
haplotypes influence mate choice in human isolate. Hum. Genet. 53: 206 [Abstract].

O'Brien, S.J. and J.F. Evermann. 1988. Interactive influence of infectious disease and genetic
diversity in natural populations. Trends Ecol. Evol. 3: 254–259.

O'Brien, S.J., D.E. Wildt, D. Goldman, C.R., Merril, and M. Bush. 1985. The cheetah is
depauperate in genetic variation. Science 221: 459–462.

O'Brien, S.J., J.S. Martenson, C. Packer, L. Herbst, V. de Vos, P. Joslin, J. Ott-Joslin, D.E.
Wildt, and M. Bush, 1987. Biochemical and genetic variation in geographic isolates
of African and Asiatic lions. Natl. Geog. Res. 3: 114–124.

Ono, H., C. O'hUigin, H. Tichy, and J. Klein. 1993. Major histocompatibility complex
variation in two species of cichlid fishes from Lake Malawai. Mol. Biol. Evol. 10:
1060–1072.

Parham, P. 1991. The pros and cons of polymorphism: a brighter future for cheetahs? Res.
Immunol. 142: 447–448.

Parham, P. 1992. Typing for class I HLA polymorphism: past, present and future. Eur. J.
Immunogenet. 19: 347–359.

Partridge, L. 1988. The rare-male effect: what is its evolutionary significance? Phil. Trans.
R. Soc. Lond. B. 319: 525–539.

Phillips, R., S. Rowland-Jones, F.D. Nixon, et al. 1991. Human immunodeficiency virus
genetic variation that can escape cytotoxic T cell recognition. Nature 354: 453.

Potts, W.K. and P. Slev. 1995. Pathogen-based models favoring MHC genetic diversity.
Immunol. Rev. 143: 181–197.

Potts, W.K. and E.K. Wakeland, E.K. 1990. Evolution of diversity of the major histocompatibility complex. Trends Ecol. Evol. 5: 181–187.

Potts, W.K. and E.K. Wakeland. 1993. Evolution of MHC genetic diversity: a tale of incest,
pestilence and sexual preference. Trends. Genet. 9: 408–412.

Potts, W., C.J. Manning, and E.K. Wakeland, 1991. Mating patterns in seminatural
populations of mice influenced by MHC genotype. Nature 352: 619–621.

Potts, W.K., C.J. Manning, and E.K. Wakeland. 1994. The role of infectious disease,
inbreeding and mating preferences in maintaining MHC genetic diversity: an
experimental test. Phil. Trans. R. Soc. Lond. B 346: 369–378.

Rada, C., R. Lorenzi, S.J. Powis, J. van-den-Bogaerde, P. Parham, and J.C. Howard.

1990. Concerted evolution of class I genes in the major histocompatibility complex of murine rodents. Proc. Natl. Acad. Sci. U.S.A. 87: 2167–2171.

Reeve, H.K., D.F. Westneat, W.A. Noon, P.W. Sherman, and C.F. Aquadro. 1990. DNA "fingerprinting" reveals high levels of inbreeding in colonies of the eusocial naked mole-rate. Proc. Natl. Acad. Sci. U.S.A. 87: 2496–2500.

Saiki, R.K., P.S. Walsh, C.H. Levenson, and H.A. Erlich. 1989. Genetic analysis of amplified DNA with immobilized sequence-specific probes. Proc. Natl. Acad. Sci. U.S.A. 86: 6230–6234.

Sato, K., M.F. Flajnik, L. Du Pasquier, M. Katagiri, and M. Kasahara. 1993. Evolution of the MHC: isolation of class II β-chain cDNA clones from the amphibian *Xeonopus laevis*. J. Immunol. 150: 2831–2843.

She, J.X., S. Boehme, T.W. Wang, F. Bonhomme, and E.K. Wakeland. 1990. The generation of MHC class II gene polymorphism in the genus *Mus*. Biol. J. Linn. Soc. 41: 141–161.

She, J.-X., S. Boehm, T.W. Wang, F. Bonhomme, and E.K. Wakeland, 1991. Amplification of MHC class II gene polymorphism by intra-exonic recombination. Proc. Natl. Acad. Sci. U.S.A. 88: 453–457.

Singh, P.B., R.E. Brown, and B. Roser. 1987. MHC antigens in urine as olfactory recognition cues. Nature 327: 161–164.

Slade, R. 1992. Limited MHC polymorphism in the southern elephant seal: implications for MHC evolution and marine mammal population biology. Proc. R. Soc. Lond. B 249: 163–171.

Slade, R.W. and H.I. McCallum. 1992. Overdominant vs. frequency-dependent selection at MHC loci. Genetics 132: 861–862.

Slade, R.W., P.T. Hale, D.I. Francis, J.A. Marshall Graves, and R.A. Sturm. 1994. The marsupial MHC: the tammar wallaby, *Macropus eugenii*, contains an expressed DNA-like gene on chromosome 1. J. Mol. Evol. 38: 466–505.

Stott, E.J. 1991. Anti-cell antibody in macaques. Nature 353: 393.

Streilen, J.W. and W.R. Duncan. 1983. On the anomalous nature of the major histo-compatibility complex in Syrian Hamsters. Transplant. Proc. 15: 1540–1545.

Takahata, N. 1990. A simple genealogical structure of strongly balanced allelic lines and trans-specific evolution of polymorphism. Proc. Natl. Acad. Sci. U.S.A. 87: 2419–2423.

Takahata, N. 1991. Trans-species polymorphism of HLA molecules, founder principle and human evolution. In: J. Klein and D. Klein, eds. Molecular evolution of the major histocompatibility complex, pp. 29–49. Heidelberg: Springer Verlag.

Takahata, N. 1993. Allelic genealogy and human evolution. Mol. Biol. Evol. 10: 2–22.

Takahata, N. 1994. Polymorphism at *Mhc* loci and isolation by the immune system in verte-brates. In: B. Golding, ed. Non-neutral evolution. New York: Chapman and Hall.

Takahata, N. and M. Nei. 1990. Overdominant and frequency-dependent selection and polymorphism of major histocompatibility complex loci. Genetics 124: 967–978.

Takahata, N., Y. Satta, and J. Klein. 1992. Polymorphism and balancing selection at major histocompatibility complex loci. Genetics 130: 925–938.

Templeton, A.R. 1987. Inferences on natural population structure from genetic studies on captive mammalian populations. In: B.D. Chepko-Sade and Z.T. Halpin, eds. Mammalian dispersal patterns: The effects of social structure on population genetics. Chicago: University of Chicago Press.

Templeton, A.R. and B. Read. 1983. The elimination of inbreeding depression in a captive heard of Speke's gazelle, In: C.M. Shoonewald-Cox, S.M. Chambers, B. MacBryde and, L. Thomas, eds. Genetics and conservation: a reference for managing wild animal and plant populations. pp. 241–261. Reading, Mass.: Addison-Wesley.

Titus-Trachtenberg, E.A., O. Rickards, G.F. De Stefano, and H.A. Erlich. 1994. Analysis of HLA class II haplotypes in the Capaya Indians of Ecuador: a novel DRB 1 allele reveals evidence for convergent evolution and balancing selection at position 86. Am. J. Hum. Genet. 55: 160–167.

Tiwari, J.L. and P.I. Teraski. 1985. HLA and disease associations. New York: Springer-Verlag.

Troutt, A.B., M.G. McHeyzer-Williams, B. Pulendran, and G.J.V. Nossal. 1992. Ligation-anchored PCR: a simple amplification technique with single-sided specificity. Proc. Natl. Acad. Sci. U.S.A. 89: 9823–9825.

Trowsdale, J. 1993. Genomic structure and function in the MHC. Immunol. Today 9: 117–122.

Trowsdale, J., V. Groves, and A. Arnason. 1989. Limited MHC polymorphism in whales. Immunogenetics: 29: 19–24.

Trowsdale, J., J. Ragoussie, and R.D. Campbell. 1991. Map of the human MHC. Immunol. Today 12: 443–446.

Uyenoyama, M. 1988. On the evolution of genetic incompatibility systems: incompatibility as a mechanism for regulating outcrossing distance. In: R.E. Michael and B.R. Levin, eds. The evolution of sex, pp. 212–232. Sunderland, Mass.: Sinauer.

Vrijenhoek, R.C. and P.L. Leberg. 1991. Let's not throw the baby out with the bathwater: a comment on management for MHC diversity in captive populations. Conserv. Biol. 5: 252–254.

Wakeland, E.K., S. Boehme, J.X. She, C.C. Lu, R.A. McIndoe, I. Cheng, Y. Ye, and W. Potts. 1989. Ancestral polymorphisms of MHC class II genes: divergent allele advantage. Immunol. Res. 9: 123–131.

Wakelin, D. and J.M. Blackwell, eds. 1988. Genetics of resistance to bacterial and parasitic infection. London: Taylor & Francis.

Waldvogel, J.A. 1989. Olfactory orientation by birds, Curr. Ornithol. 6: 269–321.

Watkins, D.I., Z.W. Chen. A.L. Hughes, A. Lagos, A.M. Lewis, J.A. Shadduck, and N.L. Letvin. 1990a. Molecular cloning of cDNAs that encode MHC class I molecules from a New World primate (Sanguinus oedipus): Natural selection acts at positions that may affect peptide presentation to T-cells. J. Immunol. 144: 1136–1143.

Watkins, D.I., Z.W. Chen, A.L. Hughes, M.G. Evans, T.F. Tedder, and N.L. Letvin. 1990b. Evolution of the MHC class I genes of a New World primate from ancestral homologes of human non-classical genes. Nature 346: 60–63.

Watkins, D.I., Z.W. Chen, A.L. Hughes, A. Lagos, A.M. Lewis, J.A. Shadduck, and N.L.Letvin. 1990c. Syrian hamsters express diverse MHC class I gene products. J. Immunol. 145: 3483–3490.

Watkins, D.I., S.N. McAdam, L.X. Strang et al. 1992. New recombinant HLA-B alleles in a tribe of South American Amerindians indicate rapid evolution of MHC class I loci. Nature 357: 329–333.

Wildt, D.E., M. Bush, K.L. Goodrowe, C. Packer, A.E. Pusey, J.L. Brown, P. Joslin, and S.J. O'Brien. 1987. Reproductive and genetic consequences of founding isolated lion populations. Nature 329: 328–331.

Wills, C. 1991. Maintenance of multiallelic polymorphism at the MHC region. Immunol. Rev. 124: 165–220.

Wollfenden, G.E. and J.W. Fitzpatrick, J.W. 1984. The Florida Scrub Jay: Demography of a cooperative-breeding bird. Princeton, N.J.: Princeton University Press.

Xu, Y., J. Pitcovsky, L. Peterson, C. Auffrey, Y. Bourlet, B.M. Gerndt, A.W. Nordskog, S.J. Lamont, and C.M. Warner. 1988. Isolation and characterization of three class II major histocompatibility complex genomic clones from the chicken. J. Immunol. 142: 2122–2132.

Yamaguchi, M., K. Yamazaki, G.K. Beauchamp. J. Bard, L. Thomas, and E.A. Boyse. 1981. Distinctive urinary odors governed by the major histocompatibility locus of the mouse. Proc. Natl. Acad. Sci. U.S.A. 78: 5817.

Yamazaki, K., E.A. Boyse, V. Mike, et al. 1976. Control of mating preferences in mice by genes in the major histocompatibility complex. J. Exp. Med. 144: 1324–1335.

Yamazaki, K., M. Yamaguchi, P.W. Andrews, B. Peake, and E.A. Boyse. 1978. Mating preferences of F2 segregants of crosses between MHC-congenic mouse strains. Immunogenetics 6: 253–259.

Yamazaki, K., G.K. Beauchamp, I.K. Egorov, J. Bard. L. Thomas, and E.A. Boyse, 1983. Sensory distinction between H-2b and H-2bm1 mutant mice. Proc. Natl. Acad. Sci. U.S.A. 80: 5685–5688.

Yamazaki, K., G.K. Beauchamp, D. Kupniewski, J. Bard, L. Thomas, and E.A. Boyse. 1988. Familial imprinting determines H-2 selective mating preferences. Science 240: 1331–1332.

Yuhki, N. and S.J. O'Brien. 1990. DNA variation at the mammalian major histo-compatibility complex reflects genomic diversity and population history. Proc. Natl. Acad. Sci. U.S.A. 87: 836–840.

Yuhki, N. and S.J. O'Brien. 1994. Exchanges of short polymorphic DNA segments predating speciation in *Feline major* histocompatibility complex class I genens. J. Mol. Evol. 39: 22–33.

Zoorob, R., A. Bernot, D.M. Renoir, F. Choukri, and C. Auffrey. 1993. Chicken major histocompatibility complex class II B genes: analysis of interallelic and interlocus sequence variance. Eur. J. Immunol. 23: 1139–1145.

15

DNA Multilocus Fingerprinting Using Simple Repeat Motif Oligonucleotides

KORNELIA RASSMANN, HANS ZISCHLER, AND DIETHARD TAUTZ

OLIGONUCLEOTIDE FINGERPRINTING IN CONSERVATION STUDIES

The "invention" of multilocus minisatellite DNA fingerprinting in 1985 has provided conservation geneticists with a powerful new tool (Jeffreys, 1985b). Owing to the high average mutation rate of the genomic regions analysed in DNA fingerprinting, the multilocus band patterns or fingerprints normally reveal ample geneticvariation (Jeffreys et al., 1988). Thus, multilocus fingerprinting can provide information on the genetic structure of individuals (heterozygosity) or populations (allelic diversity) where less sensitive methods such as enzyme electrophoresis might fail.

Like minisatellite fingerprinting, simple sequence oligonucleotide fingerprinting detects multiple loci simultaneously in the eukaryotic genome. The technique is named after the specific hybridization probe employed: a synthetic single-stranded oligonucleotide, about 20 nucleotides in length. The sequence of this probe consists of a single, very short motif, for instance "CA" or "GTC," which is repeated several times. Such a probe can hybridize with simple repetitive DNA loci in the eukaryotic genome. Since many repetitive DNA loci in the genome consist of the same repeat motif, simple sequence oligonucleotide fingerprinting produces a multilocus fingerprint pattern (for reviews on oligonucleotide fingerprinting see: Epplen, 1988; Epplen et al., 1991, 1993). The degree of genetic variation revealed in oligonucleotide fingerprints seems to be similar to that in minisatellite fingerprints. A study analysing several loci in the human genome that hybridize with the $(CAC)_5$ oligonucleotide probe confirmed a very high average rate of germline mutations. The mutation rate of the resolvable DNA fragments containing such loci was estimated to be approximately 0.001 per locus per gamete (Nürnberg et al., 1989). The same study showed that the fingerprint patterns were somatically stable, thus providing support for the reliability of the method.

Since the advent of oligonucleotide fingerprinting in 1986 (Ali et al., 1986), a wide variety of species has been tested with this technique. Hypervariable band patterns could be obtained in all major groups, ranging from fungi and plants (Beyermann et al., 1992) to insects (Achmann et. al., 1992) and vertebrates including humans (e.g., fish, Schartl et al., 1993; amphibia, listed in Epplen et al., 1991; reptiles, Demas and Wachtel, 1991; birds, May et al., 1993; mammals, Ellegren et al., 1992).

An advantage of multilocus fingerprinting is that multiple loci can be analyzed simultaneously. Thus, like minisatellite fingerprinting, oligonucleotide fingerprinting is very effective in studies that require direct paternity assessments. This can be important in the effort to protect rare species. Mathé et al. (1993) have successfully applied oligonucleotide fingerprinting to distinguish between individuals from endangered bird species which were bred in captivity and others which were apparently caught illegally from the wild. According to Mathé et al. (1993), the public announcement of this test alone has already caused a decrease in the number of claimed "breeding successes" in these species. The German Federal Ministry of Environment, Natural Protection and Reactor Safety has encouraged the use of oligonucleotide fingerprinting for this type of application.

ADVANTAGES AND DISADVANTAGES OF THE TECHNIQUE

The strength of simple sequence oligonucleotide fingerprinting lies in its fast, reproducible, and nonhazardous technique. Since the loci that hybridize with simple sequence probes are ubiquitous in all eukaryotic genomes, establishing the fingerprinting system for a new species does not require much effort. As described below, the choice of an adequate probe/restriction enzyme combination will to some extent enable one to adjust the system to the intended study. Oligonucleotide fingerprinting is a quick method, since the shortness of the probes allows in-gel hybridization and short hybridization times. Hybridizing the DNA directly in the gel instead of transferred to a membrane is also a more sensitive means, thus reducing the amount of DNA necessary per individual (~ 1 μg is sufficient). Finally, techniques for nonisotopic oligonucleotide fingerprinting have been developed, allowing the routine application of this method in laboratories not designed for radioactive work.

However, the benefit of a multilocus fingerprinting method, namely to provide information on several loci simultaneously, can also be its major drawback. The alleles of a particular locus usually cannot be identified in a multilocus fingerprint. Therefore, oligonucleotide fingerprinting (and minisatellite fingerprinting) are less suitable for studies requiring information on allele frequency distributions, i.e. to estimate the effective population size of a population or the degree of gene flow between populations. Population studies are also often based on assumptions about the mutation rate of the loci analysed. For reasons explained below, it may be difficult to obtain a reliable hypothesis on the mutational mechanism, and thus an estimate of the mutation rate of the DNA regions responsible for the genetic variation in oligonucleotide fingerprints. Nonetheless, at the beginning of a new study, oligonucleotide fingerprinting provides a quick means to get a first estimate of the degree of genetic variability to expect. This can then focus the choice of further molecular approaches.

TECHNICAL DESCRIPTION

Methodology of Oligonucleotide Fingerprinting

The principle of oligonucleotide fingerprinting is similar to that of minisatellite multilocus fingerprinting. A hypervariable multilocus pattern is produced with bands ranging from to 3 to 20 kilobases. To achieve this, high-molecular-weight genomic DNA is digested with a frequently cutting restriction enzyme such as *Alu*I, *Hinf*I or *Rsa*I. The resulting fragments are separated on an agarose gel according to size. The DNA is then denatured in the gel and transferred to a nylon membrane or, alternatively, the DNA can be immobilized by drying the gel. The gel or the membrane is hybridized using an isotopically or nonisotopically labeled simple sequence oligonucleotide probe. The stringency of the hybridization conditions is chosen so that no DNA/probe mismatches are possible. For reuse of the gels or membranes, the hybridized probes can be easily removed by immersing them in a denaturing or in a salt-free solution. For reviews on the general methodology see Epplen, 1992; for in-gel hybridization, Schäfer et al., 1988; for nonisotopic probes, Zischler et al., 1989; for nonradioactive in-gel hybridization, Zischler et al., 1991.)

Since simple repetitive DNA is ubiquitous in all eukaryotic genomes, setting up an oligonucleotide fingerprinting system for any new species needs relatively little effort. However, certain requirements must be met. Depending on the goals of the study, one has to establish an adequate marker system that reveals the desired degree of variability. In a given species, the complexity of the fingerprint pattern can vary when using different combinations of oligonucleotide probes and restriction enzymes. When direct paternity testing or identification of individuals is intended, it is advantageous to detect as many loci simultaneously as possible. One has to be aware, however, that complex fingerprint patterns are also more difficult to interpret, since the identity of bands between gels and sometimes even between lanes of the same gel can be more ambiguous. Other studies aim to understand the genetic parameters of a certain population, or try to compare different populations in this respect. In these studies it is preferable to obtain a less complex fingerprint pattern displaying only a limited number of bands. The comparison between unrelated individuals is thus facilitated. In cases where only few bands are detected, it may even be possible to identify the alleles of a single locus (see case study later). The relative complexity of a simple sequence oligonucleotide fingerprint pattern is, of course, not only related to the probe–enzyme combination chosen, but also depends on the degree of genetic variation in a population. In populations with little allelic diversity, fewer bands can be expected in the fingerprint profile, in particular when a substantial number of the detected alleles are already fixed, as is the case in bottleneck populations.

Genomic Structure of the DNA Regions Detected in Oligonucleotide Fingerprinting

To understand the benefits as well as the problems associated with the application and interpretation of simple sequence oligonucleotide fingerprints, it is necessary to take a closer look at the structure of the DNA regions that are responsible for the variation in these fingerprints. In the following section, we will outline the main features of such loci, compare them to conventional minisatellite loci, and describe what is known about the mechanisms that cause their hypervariability.

The simple sequence oligonucleotide probes hybridize to a particular class of repetitive sequences in eukaryotic genomes, the so-called "microsatellites" or "simple sequences" (Tautz and Schlötterer, 1994). These consist of repetitions of very short sequence motifs (1–5 bp (base pairs) long) which can make up blocks of up to 100 nt (nucleotides) in length. Thus, both microsatellite loci and minisatellite loci consist of clusters of tandemly repeated DNA motifs which are ubiquitous and widely dispersed in the genome (Jeffreys and Pena, 1993; Tautz, 1993). However, minisatellites differ in several respects from microsatellites. A single repeat unit of a minisatellite is up to 100 nt in length and can be repeated several hundred times at each locus. Minisatellites at different genomic locations differ to varying extents in their basic repeat unit sequences. However, low-stringency hybridization of a genomic blot (allowing for a certain amount of probe–DNA mismatches) using a given minisatellite repeat unit as a probe will detect a whole set of loci which consist of similar repeat units. Thus, minisatellite multilocus fingerprinting yields a complex pattern of bands that originate from minisatellite loci sharing some sequence similarities, in particular with respect to certain "core" sequences (Jeffreys et al., 1985a,b). Conversely, high-stringency hybridization can be employed to detect only a specific minisatellite locus and thus may resolve a single locus fingerprint.

This is different from simple sequence oligonucleotide fingerprinting, where an increase in the stringency of the hybridization conditions will not usually result in a single locus fingerprint. The explanation for this is that the specificity of the oligonucleotide probe to a certain locus cannot be increased further. The probe consists of a short, simple repetitive DNA sequence which is exactly the same as that of many corresponding microsatellites in the genome. When these short oligonucleotides are used as probes, the hybridization conditions are already chosen such that only perfectly matching target DNA–probe hybrids are stable.

To sum up, the number of bands revealed with a given probe in simple sequence oligonucleotide fingerprinting is solely determined by the frequency and the distribution of a given microsatellite motif in the genome of the species under study. In contrast, in minisatellite fingerprinting this number depends also on the hybridization stringency and the sequence divergence of the minisatellites in relation to the probe. In both fingerprinting techniques, the intensity of a specific band will depend on the heterozygosity of the typed individual for the respective locus, on the number of repeat units per locus, and, in some cases, on the number of comigrating restriction fragments detected with the same probe. In oligo-nucleotide fingerprinting, the intensity of a band depends on yet another criterion, the number of microsatellite loci *within* one restriction fragment. Since the average microsatellite locus is very short compared to the whole restriction fragment that is detected in oligonucleotide fingerprinting, it is possible that several loci occur in the same band.

This brings us to the most decisive difference in the molecular structures forming the basis of minisatellite and oligonucleotide fingerprinting. As described above, the length of a single minisatellite can range up to several thousand bases. Therefore, the large restriction fragments detected in minisatellite fingerprinting may be mainly composed of the repeat units of a given minisatellite. In contrast, a single microsatellite locus is too short to explain the composition of the high molecular-weight bands revealed in oligonucleotide fingerprinting. The formation of such large restriction fragments is difficult to explain in the first place. The probability of obtaining large fragments with frequently cutting restriction enzymes from random DNA sequences is rather low. Thus, any fragments longer than 1 kb (kilobase) should not be composed of random DNA sequences. This was indeed found in a study in which oligonucleotide loci were cloned and subsequently sequenced. They consisted of arrays of short repeats with various different motifs as well as reiterations of higher-order repeat units composed of several of the short repeats (Zischler et al., 1992). Interestingly, only some of the motifs showed similarity with the simple sequence probe that originally was used to identify them. This implies that, in oligonucleotide fingerprinting, the same fragments, and thus the same polymorphisms, can potentially be revealed by different probes.

These molecular differences suggest that even though the multilocus patterns produced with both fingerprinting methods resemble each other, they are probably caused by different genomic mutational mechanisms. The length variability of minisatellite loci is readily explained by repeat number variation of the respective minisatellite motif. This may be caused by unequal sister chromatid exchange during mitosis or meiosis, which can lead to major alterations in allele length. A recent study has shown that polarized gene conversion mechanisms can also play a role in generating minisatellite mutants (Jeffreys et al., 1994). In contrast, the microsatellite loci detected in the high-molecular-weight bands of oligonucleotide fingerprints cannot be responsible for the extensive length variation of these bands. Slippage mechanisms are proposed to cause the gain or loss of repeat units in microsatellite loci (Tautz and Renz, 1984; Levinson and Gutman, 1987). However, these changes lie in the range of several nucleotides and cannot be resolved by the gels used for oligonucleotide fingerprinting. Such small differences can only be shown by amplifying individual microsatellite loci using PCR (polymerase chain reaction) and separating the PCR products on a polyacrylamide gel (Litt and Luty, 1989; Tautz, 1989; Weber and May, 1989). Thus the regions in which the microsatellites detected in oligonucleotide fingerprinting are embedded must be subject to extensive length variation. In these regions, several different types of mutational mechanisms might be acting. Since the compositional characteristics differ from the nature of classical mini- or microsatellites, it might even be useful to define a new class for this kind of variable region (Tautz, 1993).

CASE STUDY

Oligonucleotide Fingerprinting in Marmots

One of the oligonucleotide fingerprinting studies conducted in our laboratory dealt with the genetic structure found in an alpine marmot population (*Marmota matmota* L.) native to the Bavarian Alps (Rassmann et al., 1994). Here we will first describe how we used a two-allele locus detected with one of the probes for paternity analysis; and then we will show how we estimated the population genetic characteristics of this population on the basis of the multilocus fingerprints.

Study Population and Sampling

Most marmots analyzed in this study came from a native population in the National Park of Berchtesgaden located in the most eastern region of the Bavarian Alps. A proportion of this population (about 130 individuals per year) has been studied intensively during the past 10 years (Arnold, 1990a,b), during which blood samples were taken from almost all of the animals living in the study area (approximately 500 samples). About half of these samples were still sufficiently preserved for use in DNA fingerprinting. In addition, we sampled marmots from family groups found 13 km away from the main study population; and finally, we analysed tissue samples from several marmots from the Swiss and Austrian Alps.

DNA Fingerprinting

DNA fingerprinting methods were carried out following standard protocols (for detailed description see Rassmann et al., 1994). Initially, two oligonucleotide probes, $(CCA)_5$ and $(ATCC)_4$, were tested on the DNA of several individuals from different family groups of the main study population to detect the potential degree of polymorphism in the population. Since hybridization with these probes resulted in a very simple band pattern, involving very few and mostly invariable bands with each probe, identification of the same bands in different lanes of each gel was unequivocal (Figure 15-1a). Comparison between gels was made possible by running a standard marmot DNA digest on either side of the gel. Since no new variant bands occurred (except in one single case that was utilized for estimating a mutation rate, see below), there was no need to run internal markers in every lane to determine the length class of new alleles.

Our original intention had been to detect paternity exclusions in the studied marmot population, since field observations suggested that subordinate males of a given family group might frequently mate with their mothers. For this project we decided to use oligonucleotide fingerprinting, since it was quicker to perform than conventional minisatellite fingerprinting and because one of the probes detected two variant bands that belonged to a diallelic locus. The family study involved typing of 230 individuals from the main study population, including 170 mother–offspring pairs from 29 family groups. In this sample, the genotype frequencies of the diallelic $(CCA)_5$ locus did not deviate from Hardy–Weinberg expectations.

Paternity Testing

On the basis of these $(CCA)_5$ fingerprints, we found 3 litters out of 29 tested where the dominant male had not sired some or all of the offspring. In total, paternity of the dominant male could be excluded for 5 out of 170 offspring typed. This number seems to be rather low to support frequent polyandrous matings. However, only a certain percentage of parental exclusions can be detected in a

A   B

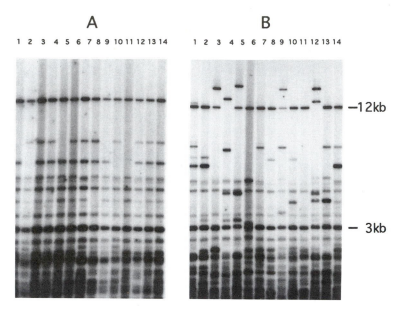

Figure 15-1 (A) Example of the hybridization pattern of the oligionucleotide probe (ATCC)₄ to *Hinfl*-digested DNA from animals of the studied alpine marmot population. Lanes 1 and 2, dominant male and female from marmot family one. Lanes 3 to 6; offspring from the dominant animals shown in lanes 1 and 2. Lanes 7 to 9, dominant male, dominant female, and one offspring from family two. Lanes 10 to 14, further dominant animals from the study population. (B) Example of the hybdridization pattern of the probe (ATCC)₄ to the *Hinfl*-digested DNA of marmots from different regions in the Alps. Lanes 1 to 7, marmots from four regions in Austria. Lanes 8 to 13, individuals from two different areas in Switzerland. Lane 14, marmot from the study area in Berchtesgaden for comparison.

Hardy–Weinberg population with a single diallelic marker system. The parental false inclusion probability can be calculated as

$$P = \sum_{i=1}^{6} \left[f_i \sum_{j=1}^{3} g_{ij} \right]$$

where P is the probability that an exclusive offspring can be detected in the population, g_{ij} is the probability that an offspring with the genotype j can be excluded for a specific parental genotype combination i, and f_i is the frequency of that parental genotype combination in the population (Table 15-1). Accordingly, even under the assumption that all parental pairs had been misassigned, only in 31% of all cases (53 offspring) could we have discovered this on the basis of the (CCA)₅ locus. Thus, parenthood of either the dominant male or the female of a family group was excluded in 9% of all offspring in which a potential mis-assignment could have been detected. However, field data on the reproductive behaviour of alpine marmots suggested that the exclusion of females was not likely. When confining the above calculation to the possibility of false paternity alone, the probability of discovering exclusions was lower (20% of all offspring),

Table 15-1 Exclusion of Offspring Genotypes that can be Detected with a Diallelic Locus for Different Parental Genotype Combinations, and the Probability of Their Occurrence in a Hardy–Weinberg Population

Parental Genotype Combination (i)	Probability of Excluding an Offspring Genotype (j) for a Parental Genotype (i), g_{ij} for $j = P, H, Q$			Frequency of Parental Genotypes in the Study Population (f_i)
	P	H	Q	
PP	0	$2pq$	q^2	0.04
QQ	p^2	$2pq$	0	0.17
PQ or QP	p^2	0	q^2	0.10
QH or HQ	p^2	0	0	0.07
PH or HP	0	0	q^2	0.34
HH	0	0	0	0.28

P, homozygous genotype of allele 1; Q, homozygous genotype of allele 2; H, heterozygous genotype; p, q, allele frequencies in the Hardy–Weinberg population (frequencies for the two alleles detected in the studied marmot population with the $(CCA)_5$ oliconucleotide probe: $p = 0.53$, $q = 0.47$). The $(CCA)_5$ genotype frequencies in the total sample size ($n = 254$) did not deviate from those expected for Hardy–Weinberg equilibrium ($G = 1.06$, df $= 1$, $p > 0.3$).

suggesting that in 15% of all detectable cases paternity of the dominant males could be excluded with the diallelic $(CCA)_5$ locus. Conventional band-sharing analysis of the multilocus fingerprint patterns achieved with both probes added some further animals to the total number of misassigned offspring. However, owing to the lack of polymorphism in the fingerprints, the sum of all exclusions did not nearly reflect the numbers suggested by the above calculations. We concluded from this project that when doing family analysis in genetically very invariant populations it might be advisable to obtain information on single loci by doing single locus fingerprinting. In this case, the advantage of getting information on the allelic state of each locus will overcome the advantage of screening multiple loci simultaneously. We therefore extended the paternity analysis in the alpine marmots using single locus PCR fingerprinting, which largely confirmed the results described here (Tautz, unpublished data).

Analysing the Genetic Structure of the Population

Further experiments were designed to explore why the resulting fingerprint patterns exhibited such a low degree of variability. First, we analyzed the reliability of the genetic markers detected by oligonucleotide fingerprinting. This involved a test of two further and presumably independent genetic marker systems (minisatellite and PCR microsatellite fingerprinting) on samples from different family groups of the main study population. However, neither of these techniques revealed a higher degree of genetic variability in the population. Finally, we analyzed individuals from geographically separated parts of the Alps with all three fingerprint techniques in order to test the potential variability of the loci studied. This analysis revealed hypervariable fingerprint patterns (Figure 15-1b), and suggested that the low genetic variability found in the Berchtesgaden marmot population was due not to a failure of the molecular tools chosen but to a reduced genetic variability in population under study.

One way in which the amount of genetic variability in a population can be measured is the observed degree of average heterozygosity (H_{obs}) in its individuals. We computed this value from the oligonucleotide fingerprints using a formula that estimates the heterozygosity values for each detected locus on the basis of the frequency of the fingerprint bands, and then averages these values over all loci. This formula is given as

$$\sum b - \frac{\sum [1 - \sqrt{(1 - b)^2}]}{L}$$

where b is the frequency of a band in the fingerprint patterns and L is the number of loci analyzed (Rassmann et al., 1994). Of course, an estimate of the number of loci is crucial for this type of analysis. However, it can easily be obtained from highly invariant multilocus fingerprints, since allelic bands are obvious and all other resolvable bands can be assumed to belong to different loci. For an alternative approach to estimating heterozygosity, see Stephens et al. (1992). Thus, the average heterozygosity detected with the two oligonucleotide probes in the Berchtesgaden marmot population yielded $H_{obs} = 0.12$.

This value suggested an exceptionally low heterozygosity usually not found in fingerprinting assays. When a low mutation rate of the analysed loci can be excluded, a low heterozygosity indicates a small genetically effective population size (the "inbreeding effective size" of population genetic theory). This can simply be the result of a small total interbreeding population size leading to inbreeding effects. It can also reflect certain events in the recent history of the studied population, such as bottlenecks or founder effects. A third explanation, that only a small fraction of all adult animals of the population reproduce, was considered unlikely on the basis of long-term demography data from the field study (W. Arnold, personal communication). To distinguish between the first two possibilities we had to estimate the current size of the interbreeding marmot population in the Berchtesgaden area. We therefore tried to estimate the extent of gene flow in this area by oligonucleotide fingerprinting marmots from another subpopulation. Indeed, we detected the same bands in the fingerprints of these animals, with the polymorphic bands in similar frequencies to those in the main study population. In parallel, a radiotelemetry study was conducted to study the migration behavior. Both studies suggested that all marmot groups throughout the Berchtesgaden area were in frequent genetic exchange. On the basis of a systematic census of all populated marmot burrows in this region, we estimated an observed effective population (census) size $Ne_{e\text{-}obs}$ of at least 1000 (500 breeding pairs).

To calculate the degree of heterozygosity expected in a population of this size we needed an estimate of the mutation rate of the loci analyzed. Since the band pattern was invariant in the fingerprints of our population, any new band found could be defined as the result of a mutation. We analysed 253 animals (i.e., 506 gametes), in which we found one mutational event with one of the oligonucleotide probes. Therefore the probablility of detecting a mutation with this probe is 0.002.

Since each locus could have contributed a new allele independently, we divided this value further by the total number of all loci that were found with this probe. A method of estimating the number of loci detected in a multilocus fingerprint has been described by Stephens et al., (1992). Thus we determined an average mutation rate of $\mu = 0.000\,17$ per gamete per locus. This mutation rate represented a minimum estimate, since it disregarded all new alleles that would have run below the separation range or that would have comigrated with other bands in the gel. Also, comparison of our result with the mutation rates for oligonucleotide fingerprinting loci found in other studies (Nürnberg et al., 1989: $\mu = 0.001$) confirmed that our calculation was very conservative.

We could now combine the values for the observed effective population size ($N_{e\text{-obs}}$) and for the mutation rate (μ) in a formula linking these variables with the expected average heterozygosity in Hardy–Weinberg populations: $H \cong 4N_e\mu/(1 + 4N_e\mu)$ (Kimura and Crow, 1964). The expected average heterozygosity for our marmot population resulted in $H_{exp} = 0.41$, which is much higher than the observed value of $H_{obs} = 0.12$. Rearranging Kimura's formula and making use of the observed average heterozygosity, we could estimate the genetically effective population size in the Berchtesgaden marmot population to be 200 at most. Thus, the population genetic analysis showed that the Berchtesgaden marmot population either represented a recent founder population or must have experienced a drastic loss of breeding pairs in its past.

CONCLUSIONS

The most likely explanation for the low degree of genetic variation in the Berchtesgaden marmot population is frequent bottlenecks due to harsh conditions during winter (Rassmann et al., 1994). Death during hibernation is the main cause for mortality in alpine marmots (Arnold, 1990b). However, the long-term behavioral study could not detect any signs of reduced fitness in this population, such as a decrease of disease resistance or an increase of the parasitic load over the 10 years of the study (W. Arnold, personal communication). Can we therefore conclude that the low degree of genetic variation has no effect on the viability of the studied population? We believe that we will not be able to answer this question entirely, since the picture we view of a single population can only represent a "snapshot" of its state in evolutionary time. Potentially, comparative studies in other populations might provide further insight into this matter. By combining molecular and behavioral/ecological approaches we may be able to relate information on the age, the viability, and the genetic variation in different populations and thus gain more information on the possible deleterious effects that low effective population sizes might cause. It is promising that for this we can now choose from a large variety of molecular tools, enabling us to adjust our methods to each individual study. Depending on the overall goals of the project, we can select a technique by considering its suitability for the samples available (kind and amount of material), the cost- and labor-intensity of the chosen method, and its sensitivity to detect genetic variation. In studies concerned with the population genetic characterization of a population or a comparison of several populations (i.e., heterozygosity within a population, gene flow between populations), we would prefer methods in which specific alleles could be characterized (such as in single locus fingerprinting or enzyme electrophoresis). For the comparison among populations, we should depend on methods that reveal a higher degree of variability among populations than within each population (in some cases single locus fingerprinting can be too sensitive a tool and may actually be less suitable than enzyme electrophoresis). Finally, projects focusing on a pedigree analysis need to use multilocus methods that promise a high level of genetic variation (i.e., oligunucleotide or minisatellite fingerprinting); however, as described here, single locus approaches may be preferable for the same purpose when the degree of polymorphism is reduced in the population under study (see case study).

ACKNOWLEDGMENTS

This work was supported by the D.F.G. and the Max-Planck Society. The field studies were conducted by Walter Arnold (MPI für Verhaltensphysiologie, Seewiesen, now University of Marburg). We thank G. Büttner and A. Türk for technical support, Terry Burke and Liz Watson for helpful comments on the manuscript, and the National Park of Berchtesgaden for sampling permission and provision of accommodation.

REFERENCES

Achmann, R., K.G. Heller, and J.T. Epplen. 1992. Last sperm male precedence in the bushcricket *Poecilimon veluchianus* (Orthoptera, Tettigonioidea) demonstrated by DNA fingerprinting. Mol. Ecol. 1: 47–54.

Ali S., C.R. Müller, and J.T. Epplen. 1986. DNA-fingerprinting by oligonucleotide probes specific for simple repeats. Hum. Genet. 74: 239–243.

Arnold, W. 1990a. The evolution of marmot society: I. Why disperse late? Behav. Ecol. Sociobiol. 27: 229–237.

Arnold, W. 1990b. The evolution of marmot sociability: II. Costs and benefits of joint hibernation. Behav. Ecol. Sociobiol. 27: 239–246.

Beyermann, B., P. Nürnberg, A. Weihe, M. Meixner, J.T. Epplen, and J. Börner. 1992. Fingerprinting plant genomes with oligonucleotide probes specific for simple repetitive DNA sequences. Theor. Appl. Genet. 83: 691–694.

Demas, S. and S. Wachtel. 1991. DNA fingerprinting in reptiles: Bkm hybridization patterns in Crocodilia and Chelonia. Genome 34: 472–476.

Ellegren, H., L. Andersson, M. Johansson, and K. Sandberg. 1992. DNA fingerprinting in horses using a simple $(TG)_n$ probe and its application to population comparisons. Anim. Genet. 23: 1–9.

Epplen, J.T. 1988. On simple repeated GATA/GACA sequences in animal genomes: a critical reappraisal. J. Hered. 79: 409–417.

Epplen, J. T. 1992. The methodology of multilocus DNA fingerprinting using radioactive or nonradioactive oligonucleotide probes specific for simple repeat motifs. In: A. Chrambach, M.J. Dunn, and B.J. Radola, eds. Advances in electrophoresis pp. 59–112. Weinheim: VCH.

Epplen, J.T., H. Ammer, C. Epplen, et al. 1991. Oligonucleotide fingerprinting using simple repeat motifs: a convenient, ubiquitously applicable method to detect hypervariability for multiple purposes. In: T. Burke, G. Dolf, A.J. Jeffreys, R. and Wolff, eds. DNA fingerprinting: Approaches and applications, pp 50–69. Basel: Birkhäuserverlag.

Epplen, C., G. Melmer, I. Siedlaczck, F.W. Schwaiger, W. Mäueler, and J.T. Epplen. 1993. On the essence of "meaningless" simple repetitive DNA in eukaryote genomes. In: S.D.J. Pena, R. Chakraborty, J.T. Epplen, and A.J. Jeffreys, eds. DNA fingerprinting: State of the science, pp. 29–45. Basel: Birkhäuserverlag.

Jeffreys, A.J. and S.D.J. Pena. 1993. Brief introduction to human DNA fingerprinting. In: S.D.J. Pena, R. Chakraborty, J.T. Epplen, and A.J. Jeffreys, eds. DNA fingerprinting: State of the science, pp. 1–20. Basel: Birkhäuserverlag.

Jeffreys, A.J., V. Wilson, and S.L. Thein. 1985a. Hypervariable "minisatellite" regions in human DNA. Nature 314: 67–73.

Jeffreys, A.J., V. Wilson, and S.L. Thein. 1985b. Individual-specific "fingerprints" of human DNA. Nature 316: 76–79.

Jeffreys, A.J., N.J. Royle, V. Wilson, and Z. Wong. 1988. Spontaneous mutation rates to new length alleles at tandem-repetitive hypervariable loci in human DNA. Nature 332: 278–281.

Jeffreys, A.J., K. Tamaki, A. MacLeod, D.G. Monckton, D.L. Neil, and J.A.L. Amour. 1994. Complex gene conversion events in germline mutation at human minisatellites. Nature Genetics 6: 136–145.

Kimura, M. and J.F. Crow. 1964. The number of alleles that can be maintained in a finite population. Genetics 49: 725–738.

Levinson, G. and G.A. Gutman. 1987. Slipped-strand mispairing: a major mechanism for DNA sequence evolution. Mol. Biol. Evol. 4: 203–221.

Litt, M. and J.A. Luty. 1989. A hypervariable microsatellite revealed by in vitro amplification of a dinucleotide repeat within the cardiac muscle actin gene. Am. J. Hum. Genet. 44: 397–401.

Mathé, J., C. Eisenmann, and A. Seitz. 1993. Paternity testing of endangered species of birds by DNA fingerprinting with non-radioactive labelled oligonucleotide probes. In: S.D.J. Pena, R. Chakraborty, J.T. Epplen, and A.J. Jeffreys, eds. DNA fingerprinting: State of the science, pp. 387–393. Basel: Birkhäuserverlag.

May, C.A., J.H. Wetton, P.E. Davis, J.F.Y. Brookfield, and D.T. Parkin. 1993. Single-locus profiling reveals loss of variation in inbred populations of the red kite (*Milvus milvus*). Proc. R. Soc. Lond. B 251: 165–170.

Nürnberg, P., L. Roewer, H. Neitzel, K. Sperling, A. Pöpperl, J. Hundrieser, H. Pöche, C. Epplen, H. Zischler, and J.T. Epplen. 1989. DNA fingerprinting with the oligonucleotide probe $(CAC)_5/(GTG)_5$: somatic stability and germline mutations. Hum. Genet. 84: 75–78.

Rassmann, K., W. Arnold, and D. Tautz. 1994. Low genetic variability in a natural alpine marmot population (*Marmota marmota*, Sciuridae) revealed by DNA fingerprinting. Mol. Ecol. 3: 347–353.

Schäfer, R., H. Zischler, U. Birsner, A. Becker, and J. T. Epplen. 1988. Optimized oligonucleotide probes for DNA fingerprinting. Electrophoresis 9: 369–374.

Schartl, M., C. Erbelding-Denk, S. Hölter, I. Nanda, M. Schmid, J.H. Schröder, and J.T. Epplen. 1993. Reproductive failure of dominant males in the poeciliid fish *Limia peruqiae* determined by DNA fingerprinting. Proc. Natl. Acad. Sci. U.S.A. 90: 7064–7068.

Stephens, J.C., D.A. Gilbert, N. Yuhki, and S.J. O'Brien. 1992. Estimation of heterozygosity for single probe multilocus DNA fingerprints. Mol. Biol. Evol. 9: 729–743.

Tautz, D. 1989. Hypervariability of simple sequences as a general source for polymorphic DNA markers. Nucleic Acids Res. 17: 6463–6471.

Tautz, D. 1993. Notes on the definition and nomenclature of tandemly repetitive DNA sequences. In: S.D.J. Pena, R. Chakraborty, J.T. Epplen, and A.J. Jeffreys, eds. DNA fingerprinting: State of the science, pp. 21–28. Basel: Birkhäuserverlag.

Tautz, D. and M. Renz. 1984. Simple sequences are ubiquitous repetitive components of eukaryotic genomes. Nucleic Acids Res. 10: 4127–4138.

Tautz, D. and C. Schlötterer. 1994. Simple sequences. Curr. Opin. Genet. Dev. 4: 832–837.

Weber, J.L. and P.E. May. 1989. Abundant class of human polymorphisms which can be typed using the polymerase chain reaction. Am. J. Hum. Gen. 44: 388–396.

Zischler, H., I. Nanda, R. Schäfer, M. Schmid, and J.T. Epplen. 1989. Deoxygenated oligonucleotide probes specific for simple repeats in DNA fingerprinting and hybridization *in situ*. Hum. Genet. 82: 227–233.

Zischler, H., A. Hinkkanen, and R. Studer. 1991. Oligonucleotide fingerprinting with $(CAC)_5$: Nonradioactive in-gel hybridization and isolation of individual hypervariable loci. Electrophoresis 12: 141–146.

Zischler, H., C. Kammerbauer, R. Studer, K.H. Grzeschik, and J.T. Epplen. 1992. A Dissecting $(CAC)_5/(GTG)_5$ multilocus fingerprints from man into individual locus-specific, hypervariable components. Genomics 13: 983–990.

Minisatellite Analysis in Conservation Genetics

TERRY BURKE, OLIVIER HANOTTE, AND IRIS VAN PIJLEN

A minisatellite locus consists of multiple tandemly repeated copies of a DNA sequence, and is usually highly polymorphic due to variation in the number of repeats (Jeffreys et al., 1985a). The repeat array typically comprises up to several hundred copies of a 10–100 bp (base pair) sequence. Minisatellites are one of the classes of loci also known as variable number tandem repeat (VNTR) loci (Nakamura et al., 1988). The other well-known class of VNTR loci is the *micro*satellites, or simple sequence repeat loci, in which the repeat comprises fewer copies (usually 10–50) of a much shorter repeat unit (1–10 bp, usually 2–5 bp) (Tautz, 1989; Weber and May, 1989). There may actually be a continuous distribution of complexity ranging from the simplest microsatellite to the better-known of the minisatellites, and beyond, and so the distinction is biologically arbitrary. Rather, it reflects the practicalities of the methods used to isolate the loci in the first place, which do indeed lead to the detection of two distinct size classes of marker loci, which in turn necessitate the use of distinct assay methods. However, there is growing evidence that the main mutational mechanism which generates the variation in the number of these different size repeat units differs between these classes of loci (Di Rienzo et al., 1994; Jeffreys et al., 1994; Goldstein et al., 1995; Slatkin, 1995).

To date, minisatellites have been detected using the traditional methods of restriction digestion and Southern blotting of agarose gels, so allowing size classes of alleles to be identified usually in the size range of 1000–20 000 bp. In contrast, microsatellites are detected through the use of the polymerase chain reaction followed, usually, by separation in acrylamide gels to give a resolution in the size range up to about 400 bp.

The first minisatellites to be described were discovered in humans and were detected as specific loci (Wyman and White, 1980; Higgs et al., 1981; Bell et al., 1982). Later, Jeffreys and colleagues found that it was possible to use a "core" probe at low stringency to detect simultaneously many minisatellite loci in which the component repeat units included the core sequence (Jeffreys et al., 1985a,b; see Burke, 1989). The bar-code-like pattern of DNA fragments so detected was found to be specific to individuals (other than identical twins) and so became known as a "DNA fingerprint." The probes initially found in humans were soon found to detect similar sequences in other organisms and have since been applied widely in many ecological studies (Burke et al., 1992; Westneat and Webster, 1994).

Multilocus DNA fingerprinting provides a powerful and relatively quick, off-the-shelf method for several applications, most especially for testing reproductive success and breeding strategies (e.g., Burke, 1989; Nybom and Schaal, 1990; Packer et al., 1991; Rico et al., 1991; Achmann et al., 1992; Westneat and Webster, 1994). However, there are circumstances in which it has proved to be highly desirable to have locus-specific data, and this has led to the development of methods for the assay of specific highly polymorphic VNTR loci (Burke et al., 1991). Single-locus profiling of either minisatellites or microsatellites may be used to satisfy this need. This chapter describes some of the recent applications of multilocus and single-locus minisatellite analysis which are of relevance to conservation genetics. The use of microsatellite loci is described by Bruford et al. (chapter 17). We will, however, attempt to compare the potential merits of the alternative approaches.

There are two main applications of minisatellite probes relevant to conservation questions: population genetics and pedigree analysis. The importance of the former is addressed elsewhere in this volume, but pedigree analysis is also important for several reasons. Understanding an organism's basic biology demands a knowledge of its mating system; this may not be easily ascertained by direct behavioral observation alone. The mating system determines the asymmetry in reproductive success among individuals and, therefore, may be important to estimates of the effective population size. The genetic management of captive populations, which may be the last hope for the survival of some species, depends on accurate pedigree data. Similarly, the legal regulation of the keeping and breeding of protected species is dependent on having available a method for verifying the registered relationships.

Minisatellites usually show the highest degree of Mendelian polymorphism of any known loci and so they are particularly valuable for confirming pedigrees. When there is only a small set of possible parents to match to their offspring, this is usually most conveniently achieved using the multilocus DNA fingerprinting technique. However, if it is necessary to check many potential parents, then it becomes desirable to establish a database of genotypes, rather then directly cross-comparing many multiple multilocus fingerprint patterns. The complexity of multilocus patterns, and the difficulty of achieving a high repeatability in them between gels and probings, makes it very difficult to store and compare the patterns in a database; this is much more easily achieved using single-locus VNTR data (Burke et al., 1991).

The high degree of polymorphism usually seen at minisatellite loci has called into question whether they are generally useful for studies of population structure. Certainly, loci having heterozygosities close to 100% will provide little resolution when one is trying to partition genetic variance within and among populations. However, lower levels of polymorphism are more characteristic of some species and there has certainly been some success in applying minisatellites to the analysis of population structure. Examples of the application of multilocus and single-locus minisatellite probes in pedigree analysis and population structure will be summarized below. Given that conservation genetics is likely to focus on populations in which genetic variability is relatively reduced, species of conservation significance are likely to be those in which other genetic markers are likely to be less informative

Choice of VNTR marker system

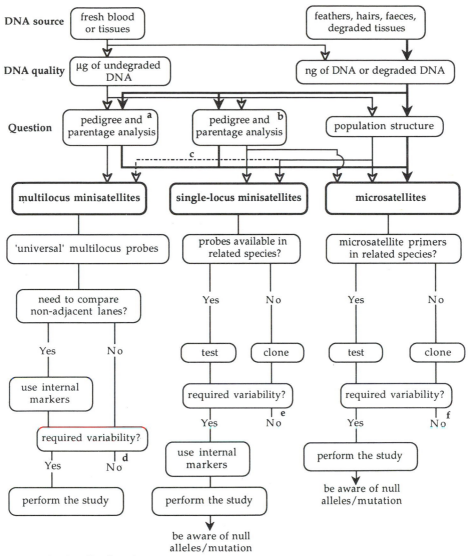

a initial and small scale studies
b where many individuals or parents have to be tested
c for less genetically variable populations
d try additional multilocus probes, new restriction enzymes
e isolate new minisatellite loci or use microsatellite loci
f isolate new microsatellite loci or use minisatellite loci

Figure 16-1 Flow chart to emphasize the key factors to be considered when deciding which VNTR locus system to use in a population study.

than usual and in which minisatellite loci are likely to become more informative.

The very significant effort required to obtain single-locus VNTR markers (either minisatellites or microsatellites) makes it important to be sure that they have a high probability of addressing the problem of concern. Once this is established, a natural first step is to check whether there are any loci already described, perhaps in related species, which might be useful. Testing available markers is relatively quick in comparison with cloning novel ones. Finally, if insufficient markers are available, it is necessary to decide whether to use minisatellites or microsatellites. The decision-making process is summarized in Figure 16-1 and the relevant issues will be discussed further below.

METHODOLOGICAL ASPECTS

Multilocus Minisatellite Fingerprinting

The most frequently used protocol for obtaining multilocus fingerprints is the one originally pioneered by Jeffreys et al. (Jeffreys et al., 1985a,b). This method uses

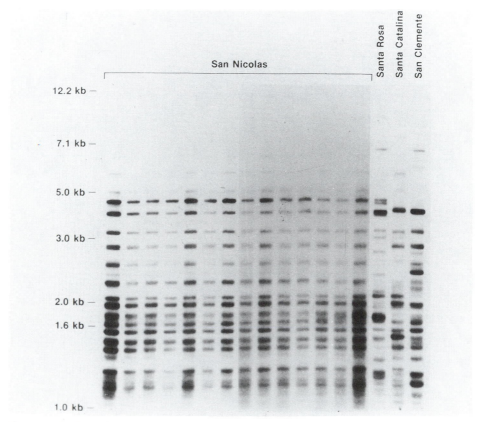

Figure 16-2 Examples of multilocus minisatellite DNA fingerprints obtained in a population study of the California Channel Island fox, *Urocyon littoralis*, using probe 33.6 (Jeffreys et al., 1985a). The foxes on San Nicolas were found to be genetically uniform, while those on the other

the hybridization of multilocus probes against Southern blots of restriction-digested genomic DNAs which have been separated in agarose gels (Figure 16-2). The original protocol has been modified and simplified since its introduction and a detailed account is provided in Bruford et al. (1992). The alternative approach uses shorter probes comprising a microsatellite repeat sequence such as $(GATA)_4$ or $(CAC)_5$ (Ali et al., 1986; Epplen et al., 1991). It seems probable that these probes detect minisatellite-containing restriction fragments through hybridization to a microsatellite repeat within the DNA flanking the minisatellite, rather than to the minisatellite DNA itself. These short probes can be hybridized directly to DNA held in dried agarose gels, avoiding the need to blot the gel onto a membrane.

Several probes have been identified which detect multiple polymorphic loci in a wide range of species (see Bruford et al., 1992; Ruth and Fain, 1993). Sometimes the use of one of the many other available cloned minisatellites (Table 16-1) as a probe at low stringency produces superior results (e.g., Gilbert et al., 1991; Scribner et al., 1994; Signer et al., 1996). Most DNA fingerprinting was originally carried out by labeling the probes with radioactive isotopes. It is now feasible to use

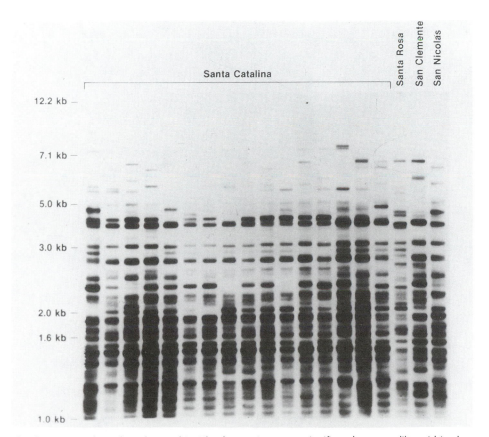

islands were moderately polymorphic. The fingerprints were significantly more alike within than between islands. (Reproduced with the permission of Macmillan Magazines Ltd. from Gilbert, D.A. et al., 1990. Nature 344: 764–766.)

Table 16-1 Available Cloned Single-locus Polymorphic Minisatellites

Group	Species	No. of probes	Reference
Arachnida	Pseudoscorpion (*Cordylochernes scorpioides*)	2[a]	Zeh et al. (1994)
Insect	Drosophila (*Drosophila mauritiana*)	1	Jacobson et al. (1992)
Fish	Atlantic salmon (*Salmo salar*)	6	Taggart and Ferguson (1990) Taggart et al. (1995)
		4	Bentzen et al. (1991), Bentzen and Wright, (1993)
		8[a]	Thomaz (1995)*
	Brown trout (*Salmo trutta*)	11	Prodöhl et al. (1994a)
	Chinook salmon (*Oncorhynchus tshawytscha*)	1	Heath et al. (1993)
	Tilapia (*Oreochromis niloticus*)	2	Harris and Wright (1995)
	Sea bass (*Dicintrarchus labrax*)	6[a]	P. Benedetti (unpublished)[b]
Amphibian	Common toad (*Bufo bufo*)	3[a]	Scribner et al. (1994)*
Reptile	Marine iguana (*Amblyrhynchus cristatus*)	3[a]	K. Rassmann (unpublished)*
Bird	Chicken (*Gallus domesticus*)	60[a]	Bruford et al. (1994), Hanotte et al. (submitted)*
	Peafowl (*Pavo cristatus*)	10[a]	Hanotte et al. (1991a,b)*
	Ruff (*Philomachus pugnax*)	13[a]	Lank et al. (1995)* E. Needham and O. Hanotte (unpublished)*
	Greylag goose (*Anser anser*)	5[a]	G. Rowe (unpublished)*
	Brent goose (*Branta bernicla*)	5[a]	G. Rowe (unpublished)*
	House sparrow (*Passer domesticus*)	16[a]	Hanotte et al. (1992b), T. Burke and E. Cairns (unpublished)*, Wetton et al. (1995)*
	Reed bunting (*Emberiza schoeniclus*)	4[a]	Dixon, (1993), Dixon et al. (1994)*
	Eastern bluebird (*Sialia sialis*)	3[a]	T. Burke and T. Robson (unpublished)*
	Pied flycatcher (*Ficedula hypoleuca*)	8[a]	D. Ross and L. Jenkins (unpublished)*
	Blue tit (*Parus caeruleus*)	9[a]	Verheyen et al. (1994)
	Willow warbler (*Phylloscopus trochilus*)	1	Gyllensten et al. (1989)
	Starling (*Sturnus vulgaris*)	4[a]	M. Double (unpublished)*
	Golden conure (*Arantinga guarouba*)	2[a]	C.Y. Miyaki (unpublished)*
	Peregrine falcon (*Falco peregrinus*)	5[a]	J. Wetton (unpublished)[c]
	Merlin (*Falco columbarius*)	3[a]	J. Wetton (unpublished)[c]
	Kestrel (*Falco tinnunculus*)	1[a]	J. Wetton (unpublished)[c]
	Red kite (*Milvus milvus*)	3[a]	May et al. (1993a,b)
	Golden eagle (*Aquila chrysaetos*)	2[a]	J. Wetton (unpublished)[c]
Mammal	Domestic dog (*Canis familiaris*)	7[a]	Joseph and Sampson (1994)
	Domestic cat (*Felis catus*)	2	Gilbert et al. (1991)
	Badger (*Meles meles*)	8[a,d]	I.A. Van Pijlen (unpublished)*
	Grey seal (*Halichoerus grypus*)	2[a]	I.A. Van Pijlen (unpublished)*

Group	Species	No. of probes	Reference
Mammal	Minke whale (*Balaenoptera acutorostrata*)	6[a]	Van Pijlen (1994), Van Pijlen et al. (1995)
	Pilot whale (*Globicephala melaena*)	1[a]	Van Pijlen (1994)*
	Red squirrel (*Sciurus vulgaris*)	2	H. Knothe (unpublished)[c]
	Mouse (*Mus musculus domesticus*)	2[e]	Kelly et al. (1991), Gibbs et al. (1993)
	Cattle (*Bos domesticus*)	36	Georges et al. (1991)
	Pig (*Sus scrofa domestica*)	3[a]	Signer et al. (1994a), Signer et al. (1996)[f]
		2	Coppieters et al. (1990, 1994)
	Human (*Homo sapiens*)	> 50	Nakamura et al. (1987, 1988), Wong et al. (1987), White (1990),
		23[a]	Armour et al. (1990)

* Available from T.B.'s laboratory at the University of Leicester.

[a] Cloned into a charomid vector.

[b] P. Benedetti Department of Biology, Universita Degli Studi di Padova, via Trieste 75, 35121 Padova, Italy.

[c] J. Wetton, Department of Genetics, Queen's Medical Centre, University of Nottingham, Nottingham NG7 2UH, U.K.

[d] All monomorphic, or nearly so, in English badgers.

[e] Unstable loci showing a high germline mutation rate (2.5–3.5%) and somatic instability.

[f] E. Signer, Department of Genetics, University of Leicester, U.K.

alkaline phosphate-conjugated oligonucleotides as probes and to detect them by chemiluminescence instead (Edman et al., 1988; Ruth and Fain, 1993).

An early concern in studies using DNA fingerprinting was that the detected DNA fragments might not be genetically independent (Jeffreys et al., 1986; Burke and Bruford, 1987; Brock and White, 1991; Hanotte et al., 1992a). This can sometimes be checked by the analysis of segregation within a family (Jeffreys et al., 1986; Bruford et al., 1992). Another concern is that minisatellites have a high mutation rate, so that nonparental bands are often found in offspring (Jeffreys et al., 1985a; Bruford et al., 1992). In practice, neither concern has proved to be a major problem when applying multilocus fingerprinting in pedigree analyses (e.g., Westneat, 1990; Amos et al., 1992; Hanotte et al., 1992a; Lifjeld et al., 1993). The implications of linkage disequilibrium among fingerprint bands for population genetic analyses have not been fully explored, but different degrees of disequilibrium within the sets of loci detected by different probes might explain occasional inconsistencies in results on the degree of genetic divergence between populations (Van Pijlen et al., 1991). The latter possibility is best minimized by using data obtained with several different probes.

Procedure for Obtaining Single-locus Probes

There are several ways in which it is possible to obtain single-locus minisatellite data. In order of decreasing likelihood of success, these are:

1. Cloning of specific minisatellite loci from the study species of interest.

2. Applying probes obtained in a previous study of a related species.
3. Testing minisatellite probes from unrelated species.
4. Identification of minisatellites belonging to a specific locus from within a multilocus DNA fingerprint.

While (1) has the highest chance of success, it also requires the most effort; (2) depends on what has been done in previous studies; (3) requires luck, but has not been tried much and may actually be quite successful; (4) is certainly dependent on luck and is unlikely to produce more than the occasional locus. We will discuss each of these approaches in turn.

Cloning Minisatellites

The keys to the successful cloning of minisatellites are (1) enriching the DNA fraction to be cloned for the presence of minisatellites, (2) selecting a cloning vector designed to accept DNA fragments in the appropriate size range, and (3) using a bacterial host designed to be most congenial to the presence of repeated DNA sequences. Most of the minisatellite probes published so far have been cloned at the University of Leicester using the protocol pioneered in humans by Armour et al. and in birds by Hanotte et al. (Armour et al., 1990; Hanotte et al., 1991b), which has since also been used at the University of Nottingham (Table 16-1). The main features of this protocol are that DNA from several different individuals is restriction-digested with the enzyme *Mbo*I; the appropriate DNA size fraction (e.g., 2–16 kb) is collected electrophoretically onto dialysis membranes; the DNA is ligated to a charomid cosmid vector designed to accept the appropriate size fraction (e.g., charomid 9–36) (Saito and Stark, 1986); the ligated DNA is then packaged using commercial packaging extracts, and an appropriate recombination-deficient bacterial host (*E.coli* NM554 or ED8767) (Raleigh et al., 1988) is used to propagate the clones. To increase the possibilities of cloning sex-specific loci, the starting DNA should be taken from the heterogametic sex. The libraries obtained in this way are so successfully enriched for minisatellites that only relatively small numbers of clones need to be screened for the presence of minisatellites and, ultimately, for polymorphism. This is most conveniently achieved by collecting clones grown initially on agar plates into an ordered array of wells containing Luria broth on a microtiter plate (Armour et al., 1990). For ecological studies, where relatively small numbers of probes are usually required (typically 3–10), it is best to screen the clones for the presence of minisatellites using the Jeffreys core probes, and then to collect only the positive colonies into microtiter plates. The protocols have been described in detail by Bruford et al. (1992).

An alternative protocol, named directed amplification of minisatellite-region DNA (DAMD) was described by Heath et al. (1993). This protocol avoids a cloning step by using the polymerase chain reaction (PCR) and minisatellite "core" repeat units as primers to obtain amplification products that can be used as minisatellite probes. One such PCR product, obtained from the chinook salmon (*Onchorynchus tshawytscha*) detected a single hypervariable locus. However, the general applicability and efficiency of this approach is as yet unproven.

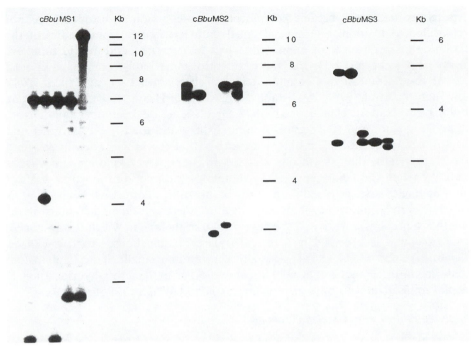

Figure 16-3 Examples of fragment patterns obtained using cloned minisatellites as probes in individuals of the species used for cloning. Hybridization patterns obtained with three common toad, *Bufo bufo*, minisatellite loci are illustrated (*Bbu*MS1, *Bbu*MS2 and *Bbu*MS3). One band (probable homozygote) or two bands (heterozygote) are present in each individual. (From Scribner, K.T., et al., 1994. Mol. Biol. Evol. 11: 737–748.) © 1994 by The University of Chicago. All rights reserved.

Several detailed accounts of the cloning of minisatellite loci have now been published (Table 16-1). Typically, 1500–3000 clones (two or three genome equivalents for the size fraction) were screened using the Jeffreys probes and 2–10% were found to be positive. Of these, about 20–30% of the probes tested were ultimately found to detect single highly polymorphic loci, though other probes also detected more complex polymorphic (usually multilocus) patterns (Figure 16-3).

A few studies have successfully used the lambda phage cloning systems which are in more general use (Wong et al., 1987; Gyllensten et al., 1989; Taggart and Ferguson, 1990; Bentzen et al., 1991; Taggart et al., 1995) or more recently the "hybrid" plasmid-bacteriophage vector called phagemid (Prodöhl et al., 1994a,b). These methods are probably less efficient and more difficult to use, but have the advantage that they may be in routine use locally.

Applying Probes from Related Species

Minisatellites are highly polymorphic because they have a high mutation rate. It therefore follows that a locus can arise and become highly polymorphic in an evolutionarily short period of time. Consequently, a locus will only be polymorphic in a related taxon if it became polymorphic in a common ancestor and has not yet drifted to fixation. The most detailed data have come from studies of the evolutionary persistence of minisatellite loci in apes and Old World monkeys (Gray and Jeffreys, 1991), and there are also some data from the attempted cross-hybridization of specific minisatellite probes in whales (Van Pijlen, 1994), fish (Bentzen et al., 1991; Harris and Wright, 1995; Heath, 1995; Taggart et al., 1995), birds (Hanotte et al., 1991a,b, 1992; Dixon et al., 1994; Sundberg and Dixon, 1996; R. Lanctot and K.T. Scribner, unpublished) and toads (K.T. Scribner and T. Burke, unpublished). The general pattern is that the probes often have utility across species in the same genus, but rarely over greater taxonomic distances (rarely beyond the family level). An example of the hybridization of a peafowl minisatellite to a series of galliform birds is shown in Figure 16-4. While there is clearly a correlation with evolutionary distance, there is little predictability about whether a probe will work in a species other than the one in which it was cloned. One exception is provided by a probe isolated in the house sparrow (*Passer domesticus*), *cPdo*MS14, which appears to detect a minisatellite locus in almost all passerine birds (Hanotte et al., 1992b), and another by the cross-hybridization of cichlid minisatellites to other fish taxa (Harris and Wright, 1995).

Applying Probes from Unrelated Species

Occasionally, a probe has been found to detect just one or two specific loci in an unrelated species. Examples include the successful use of human probes in the butterfly *Bicyclus anyana* (Saccheri and Bruford, 1993), the eastern barred bandicoot (*Perameles gunnii*) (Robinson et al., 1993), Waldrapp's ibis (*Geronticus*

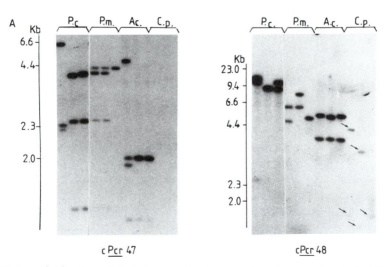

Figure 16-4 Hybridization of single-locus minisatellite probes cPcrMS47 and cPcrMS48 cloned in peafowl (*Pavo cristatus*, P.c.) against progressively more evolutionarily distant species (green peafowl, *Pavo muticus*, P.m.; Congo peacock, *Afropavo congensis*, A.c.; golden pheasant, *Chrysolophus pictus*, C.p.) (Reproduced with the permission of Birkhäuser Verlag from Hanotte, O., et al., 1991. In: Burke, T., G. Dolf, A.J. Jeffreys, and R. Wolff, eds. DNA fingerprinting: Approaches and applications, pp. 193–216.)

eremita) (Signer et al., 1994b) and the pukeko (*Porphyrio porphyrio melanotus*) (Lambert et al., 1994); the use of a chicken (*Gallus domesticus*) probe in the reed bunting (*Emberiza schoeniclus*) (Dixon et al., 1994); and the use of a mouse (*Mus domesticus*) MHC locus probe in the red-winged blackbird (*Agelaius phoeniceus*) (Gibbs et al., 1991). The evolutionary distances are such that it is extremely unlikely that the detected locus is evolutionarily homologous, but rather that a minisatellite repeat array has arisen by convergence which has the appropriate repeat unit length and sequence characteristics to allow cross-hybridization. Given that almost any minisatellite can be used as a probe at low stringency to detect many others in a multilocus fingerprint (Vergnaud, 1989; Vergnaud et al., 1991), and that the multilocus patterns include loci which exhibit widely varying degrees of intensity of hybridization (varying band intensities), occasional specific hybridization to a single locus is not surprising.

Specific Loci within Multilocus Patterns

Occasionally, it is possible to identify minisatellite fragments which are segregating as the alleles of a single locus within a specific region of a multilocus fingerprint pattern. Segregation analyses have often been used to test the frequency of allelism in multilocus profiles, but the useful, reliable instances of identifiable specific loci have involved minisatellite fragments either of high molecular weight, which have consequently exclusively occupied the top region of the multilocus profile (Amos et al., 1991; Rabenold et al., 1991b), or otherwise showing a distinctive distribution and hybridization intensity (Nybom et al., 1992; Lubjuhn et al., 1993). In several cases fragments belonging to a specific locus have been identified because of their sex-specificity (Kashi et al., 1990; Longmire et al., 1991, 1993; Rabenold et al., 1991a; Graves et al., 1992; Millar et al., 1992; Miyaki et al., 1992; Ellegren et al., 1994). The ability to detect specific loci within multilocus fingerprints is largely a matter of luck and so, while the possibility of their occurrence should be kept in mind when examining multilocus profiles, they do not provide a reliable strategy for obtaining several such loci.

Minisatellite Data

Minisatellite loci have been shown to display very high degrees of polymorphism. For example, minke whale (*Balaenoptera acutorostrata*) populations showed heterozygosities up to 0.98, with up to 50 identified alleles in 87 individuals (Van Pijlen et al., 1995). In a comparative analysis of, amongst others, minisatellite and microsatellite loci in toad populations, the minisatellite loci were found to be the most polymorphic, with an average heterozygosity of 0.68, and up to 13 identified alleles (Scribner et al., 1994). Van Pijlen (1994) also found higher levels of polymorphism at minisatellite loci than at microsatellite loci in the minke whale. In populations in which the variability has been much reduced, minisatellites are often the only markers able to detect useful polymorphism. An example of this is provided by a feral population of peafowl (*Pavo cristatus*) at Whipsnade Zoo, England, which originated from a possibly small number of founders imported into Britain. The heterozygosity at minisatellite loci in this population varied from 0.22 to 0.78, with 4–5 alleles per locus (Hanotte et al., 1991b). Given the possibility

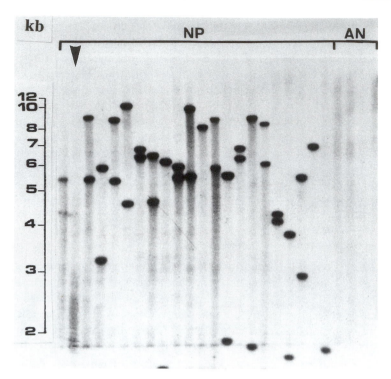

Figure 16-5 Illustration of the high numbers of alleles and heterozygosity that may be detected at a hypervariable minisatellite locus (*BacMS4* in the minke whale *Balaenoptera acutorostrata*). The vertical arrow indicates an individual in the North Pacific (NP) sample which does not show any visible alleles (i.e., homozygous "null"); none of the four Antarctic (AN) individuals showed any visible alleles either. (From Van Pijlen, I.A., et al., 1995. Mol. Biol. Evol. 12: 459–472.) © 1995 by The University of Chicago. All rights reserved.

of a very large number of alleles per minisatellite locus (Figure 16-5), it is necessary to include an internal marker fragment size ladder in each lane of the agarose gel so that the allele sizes can be estimated reasonably accurately (e.g., Dixon et al., 1994; Scribner et al., 1994; Wetton et al., 1995). The most useful size ladder is the one devised by Taggart and Ferguson (1994).

One of the problems in scoring minisatellites is that there may be many different minisatellite alleles which differ in size by increments of just one constituent repeat unit. These repeat units may be small relative to the resolution on agarose gels of DNA fragments of the size of minisatellites, and it is therefore not usually possible to distinguish every minisatellite allele. Alleles therefore have to be allocated into size classes, so that the resolution of the system is somewhat less than might be expected from the numbers of alleles that are actually present. The resolution to be used when classifying alleles can be decided objectively by measuring the error in scoring the same samples run on different gels (e.g., Scribner et al., 1994). There is also little difficulty when the number of different alleles is relatively small (say, less than 20). Once allocated to allelic classes, the data can be analysed using conventional formulae and programs, as originally developed for allozyme data. However, a scoring problem can arise when there is a quasi-continuous distribution of allele sizes; for example, using objective criteria one might determine that minisatellite A should be classed as the same as minisatellite B, and that B is the same as C, but that A is different from C. Those loci which show such large numbers of alleles are probably less useful for population genetic analyses; they are in contrast extremely useful in parentage testing, which can be achieved successfully without having to allocate minisatellites to an absolutely precise allelic class (e.g., Dixon et al., 1994; Wetton et al., 1995).

Facilities Required

The equipment required for the analysis of minisatellites is fairly standard to any normally equipped molecular laboratory, which is where one should ideally begin such studies. The only kit which is perhaps not absolutely universal is the preferred, larger style of gel tank to hold horizontal agarose gels 25 cm or 30 cm long by 20 cm wide. A rotisserie-style hybridization oven is useful, though flat hybridization chambers are adequate and may be preferable for large-scale work or multilocus fingerprinting. A fluorometer (Hoefer) is also useful for measuring the concentration of DNA digests.

The development of a series of single-locus minisatellite markers, as is also the case for microsatellites, requires a very significant investment of labor and usually takes many months to achieve in an experienced laboratory environment. Once a series of probes has been developed and tested, a skilled laboratory worker can achieve an overall average throughput (starting from tissue samples) of about three minisatellite profiles for each of 30–40 individuals per week, assuming that several hundred samples can be processed as a batch.

CASE STUDIES

As described above, the two areas most generally relevant to conservation genetics in which minisatellite data have been applied are pedigree analysis and population genetics. Additionally, it is always desirable to know the sexes of the individuals being studied and this cannot always easily be achieved morphologically. Minisatellite probes have on occasion been incidentally useful for determining the sex of individuals. Case studies which illustrate these applications will be described below.

Pedigree Analysis

Multilocus fingerprinting has been applied extensively to testing pedigrees and quantifying genetic relatedness, especially in studies of reproductive behavior, sexual selection, and sperm competition (reviewed in Burke, 1989; Burke et al., 1991; Westneat and Webster, 1994). From the perspective of conservation genetics, such studies can provide a valuable insight into the mating system and therefore allow a better estimation of the likely effective population size for a given actual population size as, without accurate data on reproductive success, behavioral observations may lead to a very poor estimate of the true variance in success among individuals. For example, males in monogamous mating systems may actually father many offspring besides those they help to raise (through extra-pair copulations (Birkhead and Møller, 1992)), and females may produce additional offspring by placing some in the broods of other females who will look after them (conspecific nest parasitism (MacWhirter, 1989)). If the mating system is unknown, it is best to begin by using the more instantly accessible multilocus fingerprinting approach, as in many cases the conclusions from behavioral observations may turn out to provide a good approximation to the true genetic mating system (e.g., Burke et al., 1989; Gyllensten et al., 1990; Hunter et al., 1992). However, when these "alternative" mating tactics are found to be common, they can be studied more efficiently using single-locus probes. This is because they allow individuals to be allocated genotypes from a set of highly informative single loci and comparisons among these genotypes can allow the parents of each offspring to be identified unequivocally.

Most of the studies in this area have so far been concerned with birds (recently reviewed in Westneat and Webster, 1994). This stems from the relative accessibility and popularity of birds for behavioral studies, as well as the ease with which blood samples can be obtained which contain adequate quantities of DNA. Among the bird species in which extra-pair paternity has been found to be common, four studies have now been published in which single-locus profiling has been used to identify the paternity of the extra-pair offspring. These concerned the blue tit (*Parus caeruleus*) (Kempenaers et al., 1992), reed bunting (Dixon et al., 1994), house sparrow (Wetton et al., 1995), and yellowhammer (*Emberiza citrinella*) (Sundberg and Dixon, 1996). Taken together, the results provide evidence that females tend to obtain extra-pair fertilizations from males with better survival prospects (blue tit) or which have survived longer than other potential partners (house sparrow and yellowhammer), consistent with the possibility that females benefit from extra-pair matings by copulating with high-quality males and so obtaining "good genes" for their offspring. An example of the use of single-locus profiles in a paternity analysis is shown in Figure 16-6.

Single-locus probes are particularly invaluable for allocating paternity in mating systems in which there are no long-term social bonds between the parents, such as in the Atlantic salmon (*Salmo salar*). Just two hypervariable minisatellite loci were sufficient to allow over 90% of redds (nests) to be assigned to their mother and her migrant mate in a population containing 97 migrant adults, and to allow the high proportion of offspring fathered by sneaking mature parr to be measured accurately at 36% (Thomaz, 1995). In the Atlantic salmon, therefore, the number of individuals contributing genetically to succeeding generations is very much greater than a count of the full-grown adults would suggest. In lekking birds, where the females visit a mating arena simply to obtain copulations with one or more of the many males present, there are again often many possible alternative fathers, making the use of single-locus probes to assign paternity worthwhile (Lank et al., 1995). Even in this kind of situation, however, good behavioral data may sometimes be sufficient to provide accurate predictions of paternity (Alatalo et al., unpublished manuscript). An example of a study of a mammal that has used VNTR data is provided by Amos et al.'s analysis of

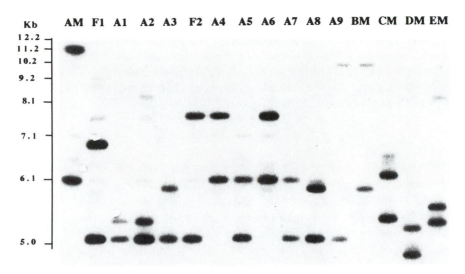

Figure 16-6 Example of the use of single-locus profiling in a paternity analysis in the reed bunting *Emberiza schoeniclus*, using probe c*Esc*MS1. Male AM was paired to both females F1 and F2; chicks A1 to A3 and A4 to A9 belonged to F1 and F2, respectively. A1 to A3, A8 and A9 were all sired by extra-pair males. BM, CM, DM and EM were neighboring males, of which BM sired chicks A3, A8 and A9, and EM sired A1 and A2 (confirmed by additional probes). (Reproduced with the permission of Macmillan Magazines Ltd. from Dixon, A., et al., 1994. Nature 371: 698–700.)

parentage in pilot whales (*Globicephala melaena*), in which both multilocus minisatellite fingerprinting and microsatellite profiling have been used to show that the fathers did not include the males present in a pod (social group) (Amos et al., 1991, 1993). The minisatellite analysis was assisted by the recognition of alleles representing a single locus within the DNA fingerprints.

It is perhaps even more important to have accurate pedigree data when managing captive populations of endangered species. This is because it is usually desirable to be able to plan matings among individuals of known relationships so as to minimize the degree of inbreeding in an attempt to maintain as much genetic variation as possible (Ralls and Ballou, 1986). However, many animals have to be maintained communally (for social behavioral or practical reasons) and it is therefore not always possible to control parentage precisely. In such situations, the parentage can be confirmed *a posteriori* using VNTR markers. A good example of this was recently provided by Signer et al. (1994b) in their study of the captive population of Waldrapp's ibis at Zurich Zoo. This analysis used a combination of multilocus and single-locus minisatellite fingerprinting. Although there were only six founders and some of these were apparently related, parentage could be assigned precisely in 29/33 cases. The exceptions were the result of a full-sib mating.

Some protected species are popular with hobbyists and their keeping is subject to legal regulation. For example, most birds of prey including the peregrine falcon (*Falco peregrinus*), goshawk (*Accipiter gentilis*) and golden eagle (*Aquila chrysaetos*) are legally protected in the United Kingdom under the Wildlife and Countryside Act 1981 and all captive individuals have to be close-ringed and registered with the Department of the Environment. It is illegal to remove birds from the wild, so all new registrations must be the declared offspring of an existing captive pair. Single-locus profiling has provided a much-needed method for parentage testing on occasions when the authorities have been concerned about the possible wild origins of newly-registered chicks of these species (Parkin et al., 1988; J.H. Wetton and D.T. Parkin, personal communication).

Population Structure

There have been only a few studies in which minisatellite markers have been used to examine the genetic relationships among populations. This is for several reasons. First, most minisatellite studies have used multilocus probes, so revealing complex multiple banding patterns which cannot be assigned to specific loci. Population genetic parameters can be estimated from such patterns (Lynch, 1991; Stephens et al., 1992; Jin and Chakraborty, 1993, 1994), but this necessitates several assumptions and there are also practical difficulties in comparing the patterns among numerous individuals (see above). Second, minisatellites are usually highly polymorphic, with heterozygosities which often approach 100%. Such loci are of limited practical use in population genetics because very large sample sizes are required to allow the genetic variation present to be partitioned into its within- and between-population components. This problem is greater with multilocus profiles and can often be avoided by using specific loci.

However, some populations have more limited levels of genetic variability and in these cases multilocus fingerprinting has been used successfully to make population genetic inferences. For example, Gilbert et al. (1990; Figure 16-2) showed that the distribution of genetic variability in populations of the California Channel Island fox (*Urocyon littoralis*) was consistent with earlier hypotheses concerning the pattern of colonization among islands, involving successive founder events during which progressively less genetic variation was conveyed to each new population. Wauters et al. (1994) found an inverse correlation between genetic variability and population size in the red squirrel (*Sciurus vulgaris*), and concluded that the loss of variability in isolated populations was due to reduced immigration. Menotti-Raymond and O'Brien (1993) concluded that the variation in multilocus minisatellite fingerprints among cheetahs, taken in conjunction with the limited polymorphism at other loci, was consistent with a hypothesis of a population bottleneck having occurred during the Pleistocene.

In their study of the eastern barred bandicoot (*Perameles gunnii*), Robinson et al. (1993) found extensive minisatellite polymorphism despite a previous failure to find any polymorphism at 27 allozyme loci (Sherwin et al., 1993). They used both multilocus and single-locus minisatellite data to conclude that there was less genetic variability present in Tasmanian populations than in the declining mainland populations. None of the alleles identified at the two specific loci was found to be present in both the mainland and Tasmanian populations. Significant genetic differences were detected between adjacent mainland populations. These authors therefore concluded that minisatellite markers were certainly of benefit for studying population differentiation and in particular for pursuing their conservation genetic goals.

A few other studies have applied single-locus minisatellite data in population genetic analyses. These have mostly been concerned with characterizing the variability of specific loci in human populations because of their importance in forensic identification (e.g., Baird et al., 1986; Wong et al., 1986, 1987; Balazs et al., 1989, 1992; Flint et al., 1989; Odelberg et al., 1989; Devlin et al., 1990; Gill et al., 1991; Deka et al., 1994). In other animals, there have been studies of salmon (Taylor et al., 1994; Heath, 1995), toads (*Bufo bufo*) (Scribner et al., 1994), and minke whales (Van Pijlen et al., 1995). Taylor et al. (1994) examined 42 populations of chum salmon (*Oncorhyncus keta*) using two minisatellite loci and were able to identify significant regional groupings. Heath et al. (1995) found a high degree of differentiation among nine populations of chinook salmon in south-western British Columbia using one or two single-locus probes. Van Pijlen et al. (1995) investigated six minisatellite loci in three distinct populations of the minke whale. Three loci showed only a small number of alleles and consequently low heterozygosities (0.00–0.47 per population), while three others were hypervariable with heterozygosities of up to 0.98. This study showed a high degree of differentiation among the three oceans, as expected, but also found that the degree of genetic variability within each ocean was correlated with the estimated population size. These results are summarized illustratively in a principal-coordinates analysis (Figure 16-7).

Scribner et al. (1994) included three cloned minisatellite loci in their study of genetic differentiation among three toad populations. This study was designed to compare the measures of genetic diversity obtained using allozyme, single-locus

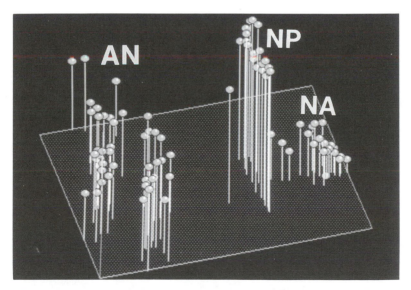

Figure 16-7 Representation of a principal-coordinates analysis of genetic similarities among minke whales from three oceans using six minisatellite loci. Each ball represents an individual minke whale and the three clusters represent the three major populations studied (NA, North Atlantic; NP, North Pacific; AN, Antarctic). (From Van Pijlen, I.A., et al. 1995. Mol. Biol. Evol. 12: 459–472.) © 1995 by The University of Chicago. All rights reserved.

minisatellite and microsatellite, and multilocus minisatellite markers. The hetero-zygosity at the six polymorphic allozyme loci was about one-half of that at the VNTR loci. Estimates of genetic distance obtained using each class of marker were generally concordant. However, estimates of the degree of spatial variation among populations, as measured using F_{st} (Wright, 1965), were greater for the allozymes than for the VNTR markers. This is probably because of the higher mutation rate at VNTR loci than at allozyme loci; consideration of the published estimates of the mutation rates at VNTR loci led to the conclusion that in this particular study migration had less influence than did mutation on the numerical estimates of interpopulational divergence.

Sex Typing

The individuals of some species, especially when immature, can be very difficult to sex morphologically or behaviorally, though it is clearly desirable to know the sexes when trying to conserve a population. It is particularly desirable to know the sexes when planning a captive breeding program. For example, parrots are notoriously difficult to sex visually and internal investigation by laparotomy is hazardous and expensive. In such cases, biochemical methods using DNA from blood samples are invaluable. Occasionally, cloned minisatellites have been found to be sex-linked (May et al., 1993b; Bruford et al., 1994; Van Pijlen, 1994; Rowe, unpublished; Hanotte et al., submitted) or sex-linked bands have been apparent in the profiles detected by minisatellite probes (Kashi et al., 1990, 1992; Longmire et al., 1991, 1993; Rabenold et al., 1991a; Graves et al., 1992; Millar et al., 1992; Miyaki et al., 1992, 1993; Ellegren et al., 1994). This approach has been tried systematically using multilocus, simple sequence repeat probes to detect sex-specific minisatellites by Longmire et al. (1993) and was successful in 6/9 bird species.

ADVANTAGES AND DISADVANTAGES OF DIFFERENT METHODS

For pedigree analysis, the use of VNTR loci is highly efficient. Random amplified polymorphic DNA (RAPD) markers have been suggested as a potential alternative (see chapter 4), but have not as yet been widely applied for this purpose, and especially not in animals. Among the VNTR systems (multilocus minisatellites, single-locus minisatellites, single-locus microsatellites), the choice among the alternatives will depend on whether enough tissue can be obtained to allow the use of Southern blotting-based methods, on the quality of the DNA that can be obtained, and on whether any single-locus markers are already available (Figure 16-1). Otherwise, single-locus methods require a substantial initial effort in setting up. If a single-locus method is definitely needed and no markers are already available, then the potential to use PCR and the possibilities for the use of automated fragment analysers to collect microsatellite genotype data from multiple loci simultaneously during electrophoretic runs (Reed et al., 1994) suggest that microsatellite-based systems will be preferred. PCR and automation potentially open the door to large-scale studies in molecular ecology in which it might be feasible to type all the individuals in a population. However, it is also possible to amplify minisatellites by PCR (Jeffreys et al., 1988b) and this approach is now being used routinely in population studies in at least one laboratory (A. Ferguson, personal communication).

For studies of population structure, the main advantage of VNTR loci over the previously widely used alternatives of mitochondrial DNA and allozymes is the much higher degree of polymorphism which characterizes them. Several authors have now argued that concerns that minisatellites might be generally *too* variable for use in estimating population genetic parameters and comparing populations were unduly pessimistic (Robinson et al., 1993; Scribner et al., 1994; Wauters et al., 1994; Van Pijlen et al., 1995), but at present microsatellites are being applied much more widely in such studies (see chapter 17).

The reason VNTR loci are so polymorphic is because they have high mutation rates (Jeffreys et al., 1988a; Dallas, 1992). This has to be kept in mind when carrying out pedigree analyses, so that either a low rate of false exclusion has to be tolerated (see Hanotte et al., 1991b) or else a low degree of mutation must be allowed for in the analysis (e.g., Burke and Bruford, 1987; Westneat, 1990). In the past, when less-mutable loci were being studied, the effect of mutation was usually ignored when quantifying population divergence using parameters such as Wright's F_{st} because the low mutation rate was probably negligible relative to the migration rate. However, this is probably not true in the case of VNTR loci (Scribner et al., 1994). The consequences for the analysis of population structure have not been fully explored, though a start has been made (Goldstein et al., 1995; Slatkin, 1995).

Another consideration with both single-locus microsatellite and minisatellite data is that "null" (undetected) alleles may occur at any locus (Armour et al., 1992; Pemberton et al., 1995; Van Pijlen et al., 1995) (see Figure 16-5). It is therefore important to avoid making parentage exclusions on the basis of incompatible apparently homozygous parental and homozygous offspring genotypes that might possibly be explained by the presence of an undetected null allele (such that the parent and offspring are actually null heterozygotes). It is never possible to rule out completely the possibility of null alleles being present by checking the Mendelian segregation of alleles in a sample of known families or by testing for the deviation of genotype frequencies from expected Hardy–Weinberg proportions. Again, it is as yet unclear what influence unidentified null alleles might have on population genetic analyses.

Finally, although there is a justifiable tendency at present for new studies requiring VNTR markers to focus on microsatellites, it is not yet clear that they will always be more advantageous than minisatellites. There are many situations in which available minisatellite probes will provide the best and most convenient solution to a problem for some time to come. There are as yet few data on both classes of marker from the same species or population. In the common toad study mentioned above, there were many more alleles detected at each of three minisatellite loci, for all the populations combined, than at one microsatellite locus (Scribner et al., 1994). A similar result was found in the minke whale (Van Pijlen, 1994; Van Pijlen et al., 1995). By contrast, in the European badger (*Meles meles*), the microsatellites were much more polymorphic than the minisatellites (Greig et al., unpublished). In general, there is also wide variation in the degree of polymorphism among loci within each class of marker. Also, when there is a reduced level of variability, which may often be the case in small and possibly endangered wild or captive populations, the increased similarity among multilocus DNA fingerprints may paradoxically make them easy to compare and therefore the most immediately accessible source of polymorphic markers (e.g., Gilbert et al., 1990). The best class of VNTR markers to use will therefore vary according to the particular question being tackled and the species concerned.

ACKNOWLEDGMENTS

We thank Lesley Barnett for preparing Figure 16-1. Our studies of minisatellites have been supported by NERC, BBSRC, and the Royal Society.

REFERENCES

Achmann, R., K.-G. Heller, and J.T. Epplen. 1992. Last-male sperm precedence in the bushcricket *Poecilimon veluchianus* (Orthoptera, Tettigonioidea) demonstrated by DNA fingerprinting. Mol. Ecol. 1: 47–54.

Alatalo, R.V., T. Burke, J. Dann, O. Hanotte, J. Höglund, A. Lundberg, R. Moss, and P.T. Rintamäki. 1996. Paternity, copulation disturbance and female choice in lekking black grouse. Anim Behav. [In press].

Ali, S., C.R. Muller, and J.T. Epplen. 1986. DNA fingerprinting by oligonucleotide probes specific for simple repeats. Hum. Genet. 74: 239–243.

Amos, B., J. Barrett, and G.A. Dover. 1991. Breeding behaviour of pilot whales revealed by DNA fingerprinting. Heredity 67: 49-55.

Amos, W., J.A. Barrett, and J.M. Pemberton. 1992. DNA fingerprinting—parentage studies in natural populations and the importance of linkage analysis. Proc. R. Soc. Lond. B. 249: 157–162.

Amos, B., C. Schlötterer, and D. Tautz. 1993. Social structure of pilot whales revealed by analytical DNA profiling. Science 260: 670–672.

Armour, J.A.L., M. Crosier, and A.J. Jeffreys. 1992. Human minisatellite alleles detectable only after PCR amplification. Genomics 12: 116–124.

Armour, J.A.L., S. Povey, S. Jeremiah, and A.J. Jeffreys. 1990. Systematic cloning of human minisatellites from ordered array charomid libraries. Genomics 8: 501–512.

Baird, M., I. Balazs, A. Giusti, L. Miyazaki, L. Nicholas, K. Wexler, E. Kanter, J. Glassberg, F. Allen, P. Rubinstein, and L. Sussman. 1986. Allele frequency distribution of two highly polymorphic DNA sequences in three ethnic groups and its application to the determination of paternity. Am. J. Hum. Genet. 39: 489–501.

Balazs, I., M. Baird, M. Clyne, and E. Meade. 1989. Human population genetic studies of five hypervariable loci. Am. J. Hum. Genet. 44: 182–190.

Balazs, T., J. Neuweiler, P. Gunn, J. Kidd, K.K. Kidd, J. Kuhl, and M.J. Liu. 1992. Human population genetic studies using hypervariable loci. 1. Analysis of Assamese, Australian, Cambodian, Caucasian, Chinese and Melanesian populations. Genetics 131: 191–198.

Bell, G.I., M.J. Selby, and W.I. Rutter. 1982. The highly polymorphic region near the human insulin gene is composed of simple tandemly repeated sequences. Nature 295: 31–35.

Bentzen, P. and J.M. Wright. 1993. Nucleotide sequence and evolutionary conservation of a minisatellite variable number of tandem repeat cloned from the Atlantic salmon, *Salmo salar*. Genome 36: 271–277.

Bentzen, P., A.S. Harris, and J.M. Wright. 1991. Cloning of hypervariable and simple sequence microsatellite repeats for DNA fingerprinting of important aquacultural species of salmonids and tilapias. pp. 243–262. In: T. Burke, G. Dolf, A.J. Jeffreys, and R. Wolff eds., DNA fingerprinting: Approaches and applications pp. 243–262. Basel: Birkhäuser Verlag.

Birkhead, T.R. and A.P. Møller. 1992. Sperm competition in birds: Evolutionary causes and consequences. London: Academic Press.

Brock, M.K. and B.N. White. 1991. Multifragment alleles in DNA fingerprints of the parrot, *Amazona ventralis*. J. Hered. 82: 209–212.

Bruford, M.W., O. Hanotte, J.F.Y. Brookfield, and T. Burke. 1992. Single-locus and multilocus DNA fingerprinting. In: A.R. Hoelzel, ed. Molecular genetic analysis of populations: A practical approach, pp. 225–269. Oxford, U.K.: IRL Press.

Bruford, M.W., O. Hanotte, and T. Burke. 1994. Minisatellite markers in the chicken genome. II. Isolation and characterization of minisatellite loci. Anim. Genet. 25: 391–399.

Burke, T. 1989. DNA fingerprinting and other methods for the study of mating success. Trends Ecol. Evol. 4: 139–144.

Burke, T. and M.W. Bruford. 1987. DNA fingerprinting in birds. Nature 327: 149–152.

Burke,. T., N.B. Davies, M.W. Bruford, and B.J. Hatchwell. 1989. Parental care and mating behaviour of polyandrous dunnocks *Prunella modularis* related to paternity by DNA fingerprinting. Nature 338: 249–251.

Burke, T., O. Hanotte, M.W. Bruford, and E. Cairns. 1991. Multilocus and single locus minisatellite analysis in population biological studies. In: T. Burke, G. Dolf, A.J. Jeffreys, and R. Wolff, eds. DNA fingerprinting: Approaches and applications, pp. 154–168. Basel: Birkhäuser Verlag.

Burke, T., W.A. Rainey, and T.J. White. 1992. Molecular variation and ecological problems. In: R.J. Berry, T.J. Crawford, and G.M. Hewitt, eds. Genes in ecology, pp. 229–254. Oxford, U.K.: Blackwell Scientific.

Coppieters, W., A. Van de Weghe, A. Depicker, Y. Bouquet, and A. Van Zeveren. 1990. A hypervariable pig DNA fragment. Anim. Genet. 21: 29–38.

Coppieters, W., C. Zijlstra, A. Van de Weghe, A.A. Bosma, L. Peelman, A. Depicker, A. Van Zeveren, and Y. Bouquet. 1994. A porcine minisatellite located on chromosome 14q29. Mammal. Genome 5: 591–593.

Dallas, J.F. 1992. Estimation of microsatellite mutation rates in recombinant inbred strains of mouse. Mammal. Genome 3: 452–456.

Deka, R., S. DeCroo, L. Jin, S.T. McGarvey, F. Rothhammer, R.E. Ferrell, and R. Chakraborty. 1994. Population genetic characteristics of the D1S80 locus in seven human populations. Hum. Genet. 94: 252–258.

Devlin, B., N. Risch, and K. Roeder. 1990. No excess of homozygosity at loci used for DNA fingerprinting. Science 249: 1416–1420.

Di Rienzo, A., A.C. Peterson, J.C. Garza, A.M. Valdes, M. Slatkin, and N.B. Freimer. 1994. Mutational processes of simple-sequence repeat loci in human populations. Proc. Natl. Acad. Sci. U.S.A. 91: 3166–3170.

Dixon, A. 1993. Parental investment and reproductive success in the reed bunting (*Emberiza schoeniclus*), investigated by DNA fingerprinting. Ph.D. thesis, University of Leicester, U.K.

Dixon, A., D. Ross, S.L.C. O'Malley, and T. Burke. 1994. Paternal investment inversely related to degree of extra-pair paternity in the reed bunting. Nature 371: 698–700.

Edman, J.C., M.E. Evans-Holm, J.E. Marich, and J.L. Ruth. 1988. Rapid DNA finger-printing using alkaline phosphatase-conjugated oligonucleotides. Nucleic Acids Res. 16: 6235.

Ellegren, H., M. Johansson, G. Hartmann, and L. Andersson. 1994. DNA fingerprinting with the human 33.6 minisatellite probe identifies sex in beavers Castor fiber. Mol. Ecol. 3: 273–274.

Epplen, J.T., H. Ammer, C. Epplen, C. Kammerbauer, R. Mitreiter, L. Roewer, W. Schwaiger, V. Steimle, H. Zischler, E. Albert, A. Andreas, B. Beyennann, W. Meyer, J. Buitkamp, I. Nanda, M. Schmid, P. Nurnberg, S.D.J. Pena, H. Poche, W. Sprecher, M. Schartl, K. Weising, and A. Yassouridis. 1991. Oligonucleotide fingerprinting using simple repeat motifs: A convenient way to detect hypervariability for multiple purposes. In: T. Burke, G. Dolf, A.J. Jeffreys, and R. Wolff, eds. DNA fingerprinting: Approaches and applications, pp. 50–69. Basel: Birkhäuser Verlag.

Flint, J., A.J. Boyce, J.J. Martinson, and J.B. Clegg. 1989. Population bottlenecks in Polynesia revealed by minisatellites. Hum. Genet. 83: 257–263.

Georges, M., A. Gunawardana, D.W. Threadgill, M. Lathrop, I. Olsaker, A. Mishra, L.L. Sargeant, A. Schoeberlein, M.R. Steele, C. Terry, X. Zhao, T. Holm, R. Fries, and J. Womack. 1991. Characterization of a set of variable number of tandem repeat markers conserved in Bovidae. Genomics 11: 24–32.

Gibbs, H.L., P.T. Boag, B.N. White, P.J. Weatherhead, and L.M. Tabak. 1991. Detection of a hypervariable DNA locus in birds by hybridization with a mouse MHC probe. Mol. Biol. Evol. 8: 433–446.

Gibbs, M., A. Collick, R.G. Kelly, and A.J. Jeffreys. 1993. A tetranucleotide repeat mouse minisatellite displaying substantial somatic instability during early preimplanta-tion development. Genomics 17: 121–128.

Gilbert, D.A., N. Lehman, S.J. O'Brien, and R.K. Wayne. 1990. Genetic fingerprinting reflects population differentiation in the California Channel Island fox. Nature 344: 764–766.

Gilbert, D.A., C. Packer, A.E. Pusey, J.C. Stephens, and S.J. O'Brien. 1991. Analytical DNA fingerprinting in lions: parentage, genetic diversity and kinship. J. Hered. 82: 378–386.

Gill, P., S. Woodroffe, J.E. Lygo, and E.S. Millican. 1991. Population genetics of four hypervariable loci. Int. J. Legal Med. 104: 221–227.

Goldstein, D.B., A.R. Linares, M.W. Feldman, and L.L. Cavalli-Sforza. 1995. An evaluation of genetic distances for use with microsatellite loci. Genetics 139: 463–471.

Graves, J., R.T. Hay, M. Scallan, and S. Rowe. 1992. Extra-pair paternity in the shag, Phalacrocorax aristotelis as determined by DNA fingerprinting. J. Zool., Lond. 226: 399–408.

Gray, I.C. and A.J. Jeffreys. 1991. Evolutionary transience of hypervariable minisatellites in man and the primates. Proc. R. Soc. Lond. B. 243: 241–253.

Gyllensten, U.B., S. Jakobsson, H. Temrin, and A.C. Wilson. 1989. Nucleotide se-quence and genomic organization of bird minisatellites. Nucleic Acids Res. 17: 2203–2214.

Gyllensten, U.B., S. Jakobson, and H. Temrin. 1990. No evidence for illegitimate young in monogamous and polygynous warblers. Nature 343: 168–170.

Hanotte, O., T. Burke, J.A.L. Armour, and A.J. Jeffreys. 1991a. Cloning, characterisation and evolution of Indian peafowl Pavo cristatus minisatellite loci. In: T. Burke, G. Dolf, A.J. Jeffreys, and R. Wolff, eds. DNA fingerprinting: Approaches and applica-tions, pp. 193–216. Basel: Birkhäuser Verlag.

Hanotte, O., T. Burke, J.A.L. Armour, and A.J. Jeffreys. 1991b. Hypervariable minisatellite DNA sequences in the Indian peafowl Pavo cristatus. Genomics 9: 587–597.

Hanotte, O., M.W. Bruford, and T. Burke. 1992a. Multilocus DNA fingerprints in gallinaceous birds: general approach and problems. Heredity 68: 481–494.

Hanotte, O., E. Cairns, T. Robson, M. Double, and T. Burke. 1992b. Cross-species hybridization of a single-locus minisatellite probe in passerine birds. Mol. Ecol. 1: 127–130.

Hanotte, O., M. Gibbs, P. Thompson, D. Dawson, A. Pugh, C. McCamley, C. Miller, L.B. Crittenden, N. Bumstead, and T. Burke. Minisatellite markers in the chicken genome. III. Minisatellite loci as genetic markers. [Submitted].

Harris, A.S. and J.M. Wright. 1995. Nucleotide-sequence and genomic organization of cichlid fish minisatellites. Genome 38: 177–184.

Heath, D. 1995. A single-locus minisatellite discriminates chinook salmon (*Onchorynchus tshawytscha*) populations. Mol. Ecol. 4: 389–393.

Heath, D.D., G.K. Iwama, and R.H. Devlin. 1993. PCR primed with VNTR core sequences yields species specific patterns and hypervariable probes. Nucleic Acids Res. 21: 5782–5785.

Higgs, D.R., S.E.Y. Goodbourn, J.S. Wainscoat, J.B. Clegg, and D.J. Weatherall. 1981. Highly variable regions of DNA flank the human alpha globin genes. Nucleic Acids Res. 9: 4213–4224.

Hunter, F.M., T. Burke, and S.E. Watts. 1992. Frequent copulation as a method of paternity assurance in the northern fulmar. Anim. Behav. 44: 149–156.

Jacobson, J.W., W. Guo, and C.R. Hughes. 1992. A *Drosophila* minisatellite contains multiple *Chi* sequences. Insect Biochem. Mol. Biol. 22: 785–792.

Jeffreys, A.J., V. Wilson, and S.L. Thein. 1985a. Hypervariable "minisatellite" regions in human DNA. Nature 314: 67–73.

Jeffreys, A.J., V. Wilson, and S.L. Thein. 1985b. Individual-specific "fingerprints" of human DNA. Nature 316: 76–79.

Jeffreys, A.J., V. Wilson, S.L. Thein, D.J. Weatherall, and B.A.J. Ponder. 1986. DNA "fingerprints" and segregation analysis of multiple markers in human pedigrees. Am. J. Hum. Genet. 39: 11–24.

Jeffreys, A.J., N.J. Royle, V. Wilson, and Z. Wong. 1988a. Spontaneous mutation rates to new length alleles at tandem-repetitive hypervariable loci in human DNA. Nature 332: 278–281.

Jeffreys, A.J., V. Wilson, R. Neumann, and J. Keyte. 1988b. Amplification of human minisatellites by the polymerase chain reaction: towards DNA fingerprinting of single cells. Nucleic Acids Res. 16: 10953–10971.

Jeffreys, A.J., K. Tamaki, A. Macleod, D.G. Monckton, D.L. Neil, and J.A.L. Armour. 1994. Complex gene conversion events in germline mutation at human minisatellites. Nature Genet. 6: 136–145.

Jin, L. and R. Chakraborty. 1993. A bias-corrected estimate of heterozygosity for single-probe multilocus DNA fingerprints. Mol. Biol. Evol. 10: 1112–1114.

Jin, L. and R. Chakraborty. 1994. Estimation of genetic distance and coefficient of gene diversity from single-probe multilocus DNA fingerprinting data. Mol. Biol. Evol. 11: 120–127.

Joseph, S.S. and J. Sampson. 1994. Identification, isolation and characterization of canine minisatellite sequences. Anim. Genet. 25: 307–312.

Kashi, Y., F. Iraqi, Y. Tikochinski, B. Ruzitsky, A. Nave, J.S. Beckmann, A. Friedmann, M. Soller, and Y. Gruenbaum. 1990. $(TG)_n$ uncovers a sex-specific hybridization pattern in cattle. Genomics 7: 31–36.

Kashi, Y., Y. Tikochinski, A. Nave, J.S. Beckmann, M. Soller, and Y. Gruenbaum. 1992. A new minisatellite probe shows highly polymorphic hybridization pattern in humans. Nucleic Acids Res. 20: 926.

Kelly, R., M. Gibbs, A. Collick, and A.J. Jeffreys. 1991. Spontaneous mutation at the hypervariable mouse minisatellite locus Ms6-hm: flanking DNA sequence and analysis of early and somatic mutation events. Proc. R. Soc. Lond. B. 245: 235–245.

Kempenaers, B., G.R. Verheyen, M. Van den Broeck, T. Burke, C. Van Broekhoven, and A.A. Dhondt. 1992. Extra-pair paternity results from female preference for high-quality males in the blue tit. Nature 357: 494–496.

Lambert, D.M., C.D. Millar, K. Jack, S. Anderson, and J.L. Craig. 1994. Single- and multilocus DNA fingerprinting of communally breeding pukeko: do copulations or dominance ensure reproductive success? Proc. Natl. Acad. Sci. U.S.A. 91: 9641–9645.

Lank, D.B., C.M. Smith, O. Hanotte, T. Burke, and F. Cooke. 1995. Genetic control of alternative mating strategies in lekking male ruff, Philomachus pugnax. Nature 378: 59–62.

Lifjeld, J.T., P.O. Dunn, R.J. Robertson, and P.T. Boag. 1993. Extra-pair paternity in monogamous tree swallows. Anim. Behav. 45: 213–229.

Longmire, J.L., R.E. Ambrose, N.C. Brown, T.J. Cade, T.L. Maechtle, W.S. Seegar, F.P. Ward, and C.M. White. 1991. Use of sex-linked minisatellite fragments to investigate genetic differentiation and migration of North American populations of the peregrine falcon (Falco peregrinus). pp. 217–229. In: T. Burke, G. Dolf, A.J. Jeffreys, and R. Wolff, eds. DNA fingerprinting: Approaches and applications, pp. 217–229. Basel: Birkhäuser Verlag.

Longmire, J.L., M. Maltbie, R.W. Pavelka, L.M. Smith, S.M. Witte, O.A. Ryder, D.L. Ellsworth, and R.J. Baker. 1993. Gender identification in birds using microsatellite DNA fingerprint analysis. Auk 110: 378–381.

Lubjuhn, T., C. Epplen, J. Brün, and J.T. Epplen. 1993. Multilocus fingerprinting using oligonucleotide probes reveals a highly polymorphic single locus system in great tits Parus major. Mol. Ecol. 2: 269–270.

Lynch, M. 1991. Analysis of population genetic structure by DNA fingerprinting. In: T. Burke, G. Dolf, A.J. Jeffreys, and R. Wolff, eds. DNA fingerprinting: Approaches and applications, pp. 113–126. Basel: Birkhäuser Verlag.

MacWhirter, R.B. 1989. On the rarity of intraspecific brood parasitism. Condor 91: 485–492.

May, C.A., J.H. Wetton, P.E. Davis, J.F.Y. Brookfield, and D.T. Parkin. 1993a. Single-locus profiling reveals loss of variation in inbred population of the red kite (Milvus milvus). Proc. R. Soc. Lond. B 251: 165–170.

May, C.A., J.H. Wetton, and D.T. Parkin. 1993b. Polymorphic sex-specific sequences in birds of prey. Proc. R. Soc. Lond. B 253: 271–276.

Menotti-Raymond, M. and S.J. O'Brien. 1993. Dating the genetic bottleneck of the African cheetah. Proc. Natl. Acad. Sci. U.S.A. 90: 3172–3176.

Millar, C.D., D.M. Lambert, A.R. Bellamy, P.M. Stapleton, and E.C. Young. 1992. Sex-specific restriction fragments and sex ratios revealed by DNA fingerprinting in the brown skua. J. Hered. 83: 350–355.

Miyaki, C.Y., O. Hanotte, A. Wajntal, and T. Burke. 1992. Sex typing of Aratinga parrots using the human minisatellite probe 33.15. Nucleic Acids Res. 20: 5235–5236.

Miyaki, C.Y., O. Hanotte, A. Wajntal, and T. Burke. 1993. Characterization and applications of multilocus DNA fingerprints in Brazilian endangered macaws. In: S.D.J. Pena, R. Chakraborty, J.T. Epplen, and A.J. Jeffreys, eds. DNA fingerprinting: State of the science, pp. 395–401. Basel: Birkhäuser Verlag.

Nakamura, Y., M. Leppert, P. O'Connell, R. Wolff, T. Holm, M. Culver, C. Martin, E. Fujimoto, M. Hoff, E. Kumlin, and R. White. 1987. Variable number of tandem repeat (VNTR) markers for human gene mapping. Science 235: 1616–1622.

Nakamura, Y., M. Lathrop, P. O'Connell, M. Leppert, D. Barker, E. Wright, M. Skolnick, S. Kondoleon, M. Litt, J.-M. Lalouel, and R. White. 1988. A mapped set of DNA markers for human chromosome 17. Genomics 2: 302–309.

Nybom, H. and B.A. Schaal. 1990. DNA "fingerprints" applied to paternity in apples (*Malus* × *domestica*). Theor. Appl. Genet. 79: 763–768.

Nybom, H., J. Ramser, D. Kaemmer, G. Kahl, and K. Weising. 1992. Oligonucleotide DNA fingerprinting detects a multiallelic locus in box elder (*Acer negundo*). Mol. Ecol. 1: 65–67.

Odelberg, S.J., R. Plaetke, J.R. Eldridge, L. Ballard, P. O'Connell, Y. Nakamura, M. Leppert, M. Lalouel, and R. White. 1989. Characterization of eight VNTR loci by agarose gel electrophoresis. Genomics 5: 915–924.

Packer, C., D.A. Gilbert, A.E. Pusey, and S.J. O'Brien. 1991. A molecular genetic analysis of kinship and cooperation in African lions. Nature 351: 562–565.

Parkin, D.T., I. Hutchinson, and J.H. Wetton. 1988. Genetic fingerprinting and its role in bird research and law enforcement. RSPB Conserv. Rev. 22–24.

Pemberton, J.M., J. Slate, D.R. Bancroft, and J.A. Barrett. 1995. Nonamplifying alleles at microsatellite loci: a caution for parentage and population studies. Mol. Ecol. 4: 249–252.

Prodöhl, P.A., J.B. Taggart, and A. Ferguson. 1994a. Cloning of highly variable minisatellite DNA single locus probes for the brown trout (*Salmo trutta*, L.) from a phagemid library. In: A.R. Beaumont, ed. Genetics and evolution of aquatic organisms. London: Chapman and Hall.

Prodöhl, P.A., J.B. Taggart, and A. Ferguson. 1994b. Single-locus inheritance and joint segregation analysis of minisatellite (VNTR) DNA loci in brown trout (*Salmo trutta* L.). Heredity 73: 556–566.

Rabenold, P.P., W.H. Piper, M.D. Decker, and D.J. Minchella. 1991a. Polymorphic minisatellite amplified on avian W chromosome. Genome 34: 489–492.

Rabenold, P.P., K.N. Rabenold, W.H. Piper, and D.J. Minchella. 1991b. Density-dependent dispersal in social wrens—genetic analysis using novel matriline markers. Anim. Behav. 42: 144–146.

Raleigh, E.A., N.E. Murray, H. Revel, R.M. Blumenthal, D. Westway, A.D. Reith, P.W.J. Rigby, J. Elhai, and D. Hanahan. 1988. McrA and McrB restriction phenotypes of some *E.coli* strains and implications for gene cloning. Nucleic Acids Res. 16: 1563–1575.

Ralls, K. and J. Ballou. 1986. Captive breeding programs for populations with a small number of founders. Trends Ecol. Evol. 1: 19–22.

Reed, P.W., J.L. Davies, J.B. Copeman, S.T. Bennett, S.M. Palmer, L.E. Pritchard, S.C.L. Gough, Y. Kawaguchi, H.J. Cordell, K.M. Balfour, S.C. Jenkins, E.E. Powell, A. Vignal, and J.A. Todd. 1994. Chromosome-specific microsatellite sets for fluorescence-based, semiautomated genome mapping. Nature Genet. 7: 390–395.

Rico, C., U. Kuhnlein, and G.J. Fitzgerald. 1991. Spawning patterns in the 3-spined stickleback (*Gasterosteus aculeatus* L.)—an evaluation by DNA fingerprinting. J. Fish Biol. 39: 151–158.

Robinson, N.A., N.D. Murray, and W.B. Sherwin. 1993. VNTR loci reveal differentiation between and structure within populations of the eastern barred bandicoot *Perameles gunnii*. Mol. Ecol. 2: 195–207.

Ruth, J.L. and S.R. Fain. 1993. The "individualization" of large North American mammals. In: S.D.J. Pena, R. Chakraborty, J.T. Epplen, and A.J. Jeffreys, eds. DNA fingerprinting: State of the science, pp. 429–436. Basel: Birkhäuser Verlag.

Saccheri, I.J. and M.W. Bruford. 1993. DNA fingerprinting in a butterfly, *Bicyclus anyana* (Satvridae). J. Hered. 84: 195–200.

Saito, I. and G.R. Stark. 1986. Charomid vectors for the efficient cloning and mapping of large or small restriction fragments. Proc. Natl. Acad. Sci. U.S.A. 83: 8664–8668.

Scribner, K.T., J.W. Arntzen, and T. Burke. 1994. Comparative analysis of intra- and interpopulation genetic diversity in *Bufo bufo*, using allozyme, single-locus micro-satellite, minisatellite, and multilocus minisatellite data. Mol. Biol. Evol. 11: 737–748.

Sherwin, W.B., N.D. Murray, J.A.M. Graves, and P.B. Brown. 1993. Measurement of genetic variation in endangered populations: bandicoots (Marsupialia: Pera-melidae). Conserv. Biol. 5: 103–108.

Signer, E.N., F. Gu, L. Gustavsson, L. Andersson, and A.J. Jeffreys. 1994a. A pseudoauto-somal minisatellite in the pig. Mammal. Genome 5: 48–51.

Signer, E.N., C.R. Schmidt, and A.J. Jeffreys. 1994b. DNA variability and parentage testing in captive Waldrapp ibises. Mol. Ecol. 3: 291–300.

Signer, E.N., F. Gu, and A.J. Jeffreys. 1996. A panel of VNTR markers in pigs. [In press].

Slatkin, M. 1995. A measure of population subdivision based on microsatellite allele fre-quencies. Genetics 139: 457–462.

Stephens, J.C., D.A. Gilbert, N. Yuhki, and S.J. O'Brien. 1992. Estimation of heterozygosity for single-probe multilocus DNA fingerprints. Mol. Biol. Evol. 9: 729–743.

Sundberg, J. and A. Dixon. 1996. Old, colourful males benefit from extra-pair copulations in the yellowhammer (*Emberiza citritiella*) revealed by single-locus DNA finger-printing. Anim. Behav. [In press].

Taggart, J.B. and A. Ferguson. 1990. Hypervariable minisatellite DNA single locus probes for the Atlantic salmon, *Salmo salar* L. J. Fish Biol. 37: 991–993.

Taggart, J.B. and A. Ferguson. 1994. A composite DNA size reference ladder suitable for routine application in DNA fingerprinting/profiling studies. Mol. Ecol. 3: 273–274.

Taggart, J.B., P.A. Prodohl, and A. Ferguson. 1995. Genetic markers for the Atlantic salmon (*Salmo salar* L.): single locus inheritance and joint segregation analysis of minisatellite (VNTR) DNA loci. Anim. Genet. 26: 13–20.

Tautz, D. 1989. Hypervariability of simple sequences as a general source of polymorphic DNA markers. Nucleic Acids Res. 17: 6463–6471.

Taylor, E.B., T.D. Beacham, and M. Kaeriyama. 1994. Population structure and identifica-tion of north Pacific Ocean chum salmon (*Oncorhynchus keta*) revealed by an analysis of minisatellite DNA variation. Can. J. Fish. Aquat. Sci. 51: 1430–1442.

Thomaz, D.M.P F. 1995. Alternative life-history strategies in male Atlantic salmon (*Salmo salar* L.). Ph.D. thesis, University of Leicester, U.K.

Van Pijlen, I.A. 1994. Hypervariable genetic markers and population differentiation in the minke whale *Balaenoptera acutorostrata*, Ph.D. thesis, University of Leicester, U.K.

Van Pijlen, I.A., B. Amos, and G.A. Dover. 1991. Multilocus DNA fingerprinting applied to population studies of the minke whale *Balaenoptera acutorostrata*. Rep. Int. Whal. Commn. (Special Issue 13): 245–254.

Van Pijlen, I.A., B. Amos, and T. Burke. 1995. Patterns of genetic variability at individual minisatellite loci in minke whale *Balaenoptera acutorostrata* populations from three different oceans. Mol. Biol. Evol. 12: 459–472.

Vergnaud, G. 1989. Polymers of random short oligonucleotides detect polymorphic loci in the human genome. Nucleic Acids Res. 17: 7623–7630.

Vergnaud, G., D. Mariat, M. Zoroastro, and V. Lauthier. 1991. Detection of single and multiple polymorphic loci by synthetic tandem repeats of short oligonucleotides. Electrophoresis 12: 134–140.

Verheyen, G.R., B. Kempenaers, T. Burke, M. Van den Broeck, C. Van Broeckhoven, and A. Dhondt. 1994. Identification of hypervariable single locus minisatellite DNA probes in the blue tit (*Parus caeruleus*). Mol. Ecol. 3: 137–143.

Wauters, L.A., Y. Hutchinson, D.T. Parkin, and A.A. Dhondt. 1994. The effects of habitat fragmentation on demography and on the loss of genetic variation in the red squirrel. Proc. R. Soc. Lond. B 255: 107–111.

Weber, J.L. and P.E. May. 1989. Abundant class of human DNA polymorphisms which can be typed using the polymerase chain reaction. Am. J. Hum. Genet. 44: 388–396.

Westneat, D.F. 1990. Genetic parentage in the indigo bunting: a study using DNA fingerprinting. Behav. Ecol. Sociobiol. 27: 67–76.

Westneat, D.F. and M.S. Webster. 1994. Molecular analysis of kinship in birds: interesting questions and useful techniques. In: B. Schierwater, B. Streit, G.P. Wagner, and R. DeSalle, eds. Molecular ecology and evolution: Approaches and applications, pp. 91–126. Basel: Birkhäuser Verlag.

Wetton, J.H., T. Burke, D.T. Parkin, and E. Cairns. 1995. Single-locus DNA fingerprinting reveals that male reproductive success increases with age through extra-pair paternity in the house sparrow (*Passer domesticus*). Proc. R. Soc. Lond. B 260: 91–98.

White, R. 1990. Detection of highly polymorphic DNA sequences in human DNA. Proceedings of the International Symposium on Human Identification 1989: Data acquisition and statistical analysis for DNA typing laboratories, pp. 1–4. Madison, Wis.: Promega Corporation.

Wong, Z., V. Wilson, A.J. Jeffreys, and S.L. Thein. 1986. Cloning a selected fragment from a human DNA "fingerprint": isolation of an extremely polymorphic minisatellite. Nucleic Acids Res. 14: 4605–4616.

Wong, Z., V. Wilson, I. Patel, S. Povey, and A.J. Jeffreys. 1987. Characterization of a panel of highly variable minisatellites cloned from human DNA. Ann. Hum. Genet. 51: 269–288.

Wright, S. 1965. The interpretation of population structure by F-statistics with special regard to systems of mating. Evolution 19: 395–420.

Wyman, A. and R. White. 1980. A highly polymorphic locus in human DNA. Proc. Natl. Acad. Sci. U.S.A. 77: 6754–6758.

Zeh, D.W., J.A. Zeh, and C.A. May. 1994. Charomid cloning vectors meet the pedipalp chelae: single-locus minisatellite DNA probes for paternity assignment in the harlequin beetle-riding pseudoscorpion. Mol. Ecol. 3: 512–522.

17

Microsatellites and Their Application to Conservation Genetics

MICHAEL W. BRUFORD, DAVID J. CHEESMAN, TREVOR COOTE, HARRIET A. A. GREEN, SUSAN A. HAINES, COLLEEN O'RYAN, AND TIMOTHY R. WILLIAMS

Microsatellites, or simple sequences, consist of tandemly repeated units, each between 1 and 10 bp (base pairs) in length, such as $(TG)_n$ or $(AAT)_n$ (Litt and Luty, 1989; Tautz, 1989; Weber and May, 1989). They are widely dispersed throughout eukaryotic genomes (e.g., Gyapay et al., 1994), and are often highly polymorphic owing to variation in the number of repeat units (e.g., Amos et al., 1993). Although the application of microsatellite markers to population genetic studies is quite recent, with a relatively small number of plant and animal studies published (still fewer of these studies have been carried out from a conservation genetics perspective), this will certainly change completely within the next few years. This is because the potential for the use of these markers in small populations, and especially in endangered species, is very great, primarily since material for microsatellite analysis can potentially be sampled noninvasively from free-living populations (e.g., Morin et al., 1994a). This fact, together with the high information content of the genetic data yielded by microsatellite loci, will serve to make these markers one of the tools of choice for many future conservation genetic studies.

In this chapter we first describe the characteristics of microsatellite loci in eukaryotic genomes; we then discuss possible patterns of microsatellite evolution and their relevance to our expectations of the characteristics of microsatellite genetic variability and allele size distributions observed in populations. Next, we describe the molecular genetic techniques and approaches required to analyse microsatellites in new species where no PCR primers for microsatellite loci are available and also where degenerate primers can be applied from species other than those under study. We then highlight examples of where these techniques have been applied to the genetics of animal populations, including those studies which have had application to conservation management. Within this context we outline the conservation questions which can be addressed using microsatellites. We take examples from our own research, focusing on "universal" primers for application to many species (in this case in apes and Old World monkeys); interspecific hybridization through the red deer and sika in

Scotland; measurement of genetic variability in fragmented populations of differing effective size, using park populations of the Cape buffalo as an example; quantification of genetic variation within and among subpopulations, with the introduced and rapidly expanding Reeves' muntjac in south-east England; and measurement of parentage and relatedness, using the examples of a wild savannah baboon social group and multiply-queened ant societies.

CHARACTERISTICS OF MICROSATELLITE LOCI

The existence of short tandem repeat, or microsatellite, loci in eukaryotic genomes has been known since the 1970s, although the large number and widespread occurrence of these sequences was not demonstrated until 1982, when Hamada et al. discovered multiple copies of the poly(dT-dG)$_n$ motif in yeast through to vertebrates. This finding was verified in 1984, by Tautz and Renz, who hybridized different microsatellite sequences to genomic DNA from a variety of organisms and found that many types of simple sequences were present. Subsequently, Tautz et al. (1986) showed that many of the simple sequences occurring in eukaryotes were 5-fold to 10-fold more frequent than equivalent-sized random motifs, and that high numbers of "cryptic" repeats, or scrambled arrangements of repetitive sequence, also occurred.

In 1985, the discovery by Jeffreys et al. of hypervariable tandem repeats in the human genome having a longer repeat unit (called minisatellites—see chapter 16) and the use of these sequences in DNA fingerprinting studies led to their widespread application in individual identification, parentage testing, and genome mapping. As with microsatellite loci, minisatellites vary in the number of tandemly repeated elements; hence a general designation for both is variable number of tandem repeat loci (VNTRs). Because the repeat units in minisatellites may be as large as 200 bp each, allele sizes can range up to 50 kb (kilobases). Consequently, conventional Southern blotting and hybridization techniques have been used to reveal minisatellite variation in many loci simultaneously (to produce the bar-code like DNA fingerprint) and specific probes have been used to reveal unit variation at single loci. Many minisatellite loci have been found to be extremely variable, with heterozygosity values greater than 90% and mutation rates sometimes exceeding 10^{-2} per generation (Bruford et al., 1992).

An advancement in the efficiency of VNTR analysis was the utilization of the polymerase chain reaction, and such systems were developed for some minisatellite loci. However, only a limited subset of variation could be analysed by PCR owing to the generally large sizes of minisatellite alleles and the limits of the efficiency of PCR in the amplification of DNA sequences much above 10 kb. A system of highly polymorphic sequences with allele sizes smaller than 500 bp and which varied over a narrow size range was desirable because variability in these loci could be assayed by PCR combined with gel electrophoresis. Moreover, variation could be assessed from minute amounts of material that might contain highly degraded DNA such as forensic or ancient samples.

Microsatellite sequences fitted these criteria well, and in 1989 three papers separately reported the isolation of microsatellites and the characterization of allelic variability at these loci using the polymerase chain reaction (Litt and Luty, 1989; Tautz, 1989; Weber and May, 1989). In these studies, microsatellites in several species were either cloned and sequenced, or were identified from sequence databases. PCR primers were designed to recognize sequences flanking the tandem repeat, and the polymorphic amplified products were separated on polyacrylamide gels. This approach allows the resolution of alleles differing by as little as one base pair, and several loci could be analyzed together on the same gel.

Since those first publications, microsatellites have been used extensively and many have been shown to be highly polymorphic, some remarkably so (Amos et al., 1993); their abundance and ubiquitous distribution making them very valuable genetic markers (Bruford and Wayne, 1993). Microsatellites have become the most important class of marker for linkage mapping in diverse organisms from humans to mosquitoes (Zheng et al., 1993; Barendse et al., 1994; Dietrich et al., 1994), and their use has enabled the identification of quantitative trait loci in a number of species (e.g., Berrettini et al., 1994). These markers have also proved valuable in forensic cases and have been used to identify the remains of humans (Jeffreys et al., 1992; Gill et al., 1994) and have been analysed in museum specimens (Ellegren, 1991). Additionally, a potentially valuable characteristic of microsatellites, and one not generally shared by minisatellites, is that primers developed in one species can be used in related taxa. PCR primers based on microsatellite loci identified in a genomic library of one whale species have been successfully applied in many other related species and a fraction of cattle-based primers are informative in sheep (Moore et al., 1991; Schlötterer et al., 1991). If many microsatellite primers prove to have a wide taxonomic range, less time and effort will be expended in the development and screening of genonic libraries.

The mechanisms underlying microsatellite allele frequency change in populations have proved difficult to determine because of the inconsistency of the patterns of allele frequency distributions that have been observed. Mutation rates at microsatellite loci have been estimated to vary between 10^{-4} and 5×10^{-6} (Dallas, 1992, Edwards et al. 1992), and the predominant means by which new length alleles are generated is thought to be intra-allelic polymerase slippage during replication (Schlötterer and Tautz, 1993; Tautz and Schlötterer, 1994). However, limitations to the direction and total number of repeat alleles may exist. Allele frequency distributions at a few loci are clearly under some functional constraints. For example a $(CGG)_n$ repeat associated with the fragile X gene (FMR-1) shows a single common allele of 29 repeat units in human populations, whereas affected individuals can have between 50 and 200 repeats (Fu et al., 1991) and a similar expansion within a protein kinase exon results in myotonic dystrophy (Brook et al., 1992).

An important question recently addressed is whether the allele frequencies seen in populations are consistent with a "stepwise" mutation model where mutational events occur as a loss or gain of single repeat units (Shriver et al., 1993; Valdes et al., 1993; Di Rienzo et al., 1994). Stepwise models were first developed in an attempt to explain allozyme allele frequency distributions but are potentially more applicable in microsatellites, where mutations might conceivably occur in a regular manner. Valdes et al. (1993) and more recently Shriver et al. (1993) have tested stepwise models for microsatellite loci. Valdes et al. analysed allele frequency data at 108 loci from families in the CEPH (Centre des Études du Polymorphisme Humaine) database from three different Caucasian populations. They found that the distributions observed were consistent with a stepwise model if the product of the effective population size and mutation rate was greater than 1. Additionally, no correlation was found between mean allele size and mutation rate.

Shriver et al. (1993) used computer simulations to estimate the expected number of alleles and their size range, heterozygosity, and frequency distribution modality given different stepwise mutation rates. They compared the simulation results with those predicted from analytical formulations of the stepwise model. Although simulated heterozygosity agreed with expectations, the average number of alleles did not, and was larger in the simulation. Shriver et al. then compared computer simulation results with real data, classifying loci into three groups: microsatellites (1–2 bp repeats), 3–5 bp microsatellites (which the authors designated short tandem repeats), and minisatellites (15–70 bp). They found that the 3–5 bp microsatellite repeat results were the closest to the stepwise simulations in number of alleles, size range, and modality, followed by microsatellite loci, with minisatellite results better explained by an infinite allele mode. A similar approach was taken by Di Rienzo et al. (1994), who examined a well-studied human population from Sardinia where both the genetic and demographic history was known. They found that a two-phase mutational process most adequately explained the frequency and distribution of different length alleles in this population, where although a stepwise simulation mutation model could explain much of the data, a second phase involving much larger mutational changes (possibly involving interallelic exchange) was required for the simulations to fit the observed distributions in a small number of cases.

One of the few direct lines of evidence of microsatellite mutations gathered to date comes from a study of mutations in recombinant strains of laboratory mice (Dallas, 1992) and in pedigrees used in human genetic mapping studies. In humans, studies have found mutations resulting in the gain of a single repeat unit, whereas mutations shortening the allele primarily occurred in two-unit steps. More data documenting the precise nature of microsatellite mutations need to be collected before a clear picture of the nature and effect of these processes emerges.

One recent and potentially very important discovery has been the occurrence of "null" alleles in microsatellites (Callen et al., 1993). Seven out of 23 loci surveyed in CEPH families showed null alleles which were recognized by the apparent noninheritance of parental alleles in some offspring. The authors demonstrated that one null allele was the result of an 8 bp deletion in the DNA flanking the microsatellite coincident with the priming site. The general lesson provided by this result is that heterozygous individuals may be being mistyped as homozygotes if null alleles are common in the population. Assignment of homozygosity in individuals on the basis of band intensity alone is a difficult task given the vagaries of PCR. Mistyping of heterozygous individuals might explain some of the heterozygote deficiencies observed in human populations and suggests caution should be used in comparing levels of heterozygosity among populations differing in the composition of alleles.

TECHNICAL DESCRIPTION

Cloning and Isolating Polymorphic Microsatellites

Several methods have been described for cloning and isolating microsatellites; however, they are all variations on a comparatively simple theme (for a good methodological example, see Rassmann et al., 1991). A schematic of this methodology is shown in Figure 17-1.

First, a *genomic library* is constructed using fragments specifically enriched for the presence of microsatellite DNA. This is done by digesting whole genomic DNA to completion with a combination of four-base-cutting restriction enzymes such as *Alu*I, *Hae*III and *Rsa*I. This yields DNA of comparatively low mean fragment size, and removes much unique and low-copy-number DNA sequence from the size fraction to be collected for cloning, leaving mostly repetitive DNA remaining, of which microsatellite DNA is one subclass. The fragment size range selected is usually 300–500 bp, which is just long enough to contain the microsatellite element plus sufficient DNA flanking the microsatellite to enable PCR primers to be designed. The cloning vectors can be either phage-type (M13, lambda) or common plasmids (e.g., pUC or similar derivatives), and ligations are normally carried out using blunt-ended fragments. Standard blunt-ended ligation protocols work adequately (Sambrook et al., 1991), though efficiency can be improved by varying the vector:insert molar ratio and using high-efficiency transformation techniques for plasmids, such as electroporation.

Genomic libraries need to yield at least 5000 recombinants since the proportion of recombinants containing microsatellites usually varies between only 0.5% and 2%. In addition, there is considerable wastage of positive recombinants in three ways: first through the detection of false positives; second, owing to the presence of clones containing insufficient or no DNA flanking the repeat sequence with which to design PCR primers; and finally, where the microsatellite locus is monomorphic or yields an uninterpretable pattern.

Positive recombinants are revealed by standard colony lifts onto nylon membrane (Buluwela et al., 1989) followed by hybridization with labeled simple sequence polymer DNA (such as poly(CA) and poly(GA), which is available commercially, or poly(AAT) or poly(GATA) which can be synthesized). As stated previously, as a rule only 0.5% to 2% of clones are usually positive for these sequences. For example, in our experience, high levels of $(CA)_n$ repeats are found in mammalian genomes, but much lower numbers are found in lower vertebrates and invertebrates. Further, owing to the wastage involved in isolating polymorphic microsatellites from positive colonies, as many as three times the number of clones will need to be sequenced, compared to the number of microsatellite loci needed for the specific application required. Standard vector preparation and sequencing methods can be used to produce the clone sequence from which microsatellite PCR primers can be designed (Sambrook et al., 1989).

Primers can be designed with a number of criteria in mind. First, the production of standard amplification protocols across all loci in a given species reduces the effort required when screening large numbers of individuals. Second,

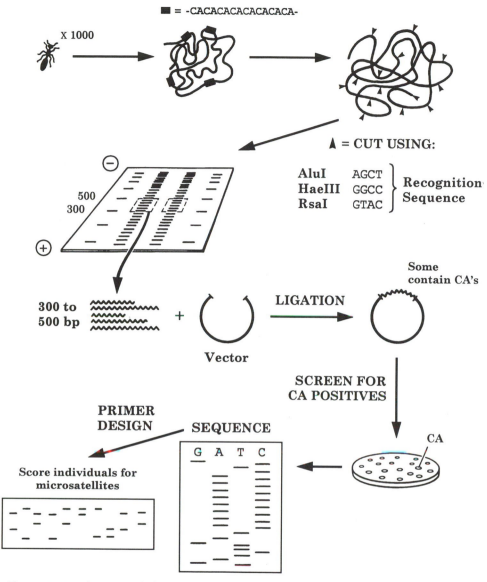

Figure 17-1 Schematic of the steps required to clone microsatellites in a new species. DNA is extracted from a number of individuals, digested with a number of four-base-cutting restriction enzymes and the restriction fragments are separated on an agarose gel. Fragments between 200 and 500 bp are isolated and ligated into a cloning vector. Following bacterial transformation, clones are plated out onto Petri dishes and colonies are lifted onto nylon membranes. Membranes are hybridized with labeled microsatellite polymers and positive clones are identified. Positive clones are isolated and insert DNA is sequenced. PCR primers are designated from DNA flanking the simple repeat sequence and the polymorphic microsatellite is amplified and subsequently separated on vertical polyacrylamide gels.

the use of standard protocols, and especially the same PCR annealing temperature, potentially allows multiplex amplification of a number of different loci in the same PCR reaction (although care must be taken to ensure that primers from different loci do not anneal or interfere with each other). Third, careful design of the allele length ranges amplified with each locus can allow several loci to be electrophoresed simultaneously, provided they have non-overlapping allele size ranges (or, if the ranges do overlap, the different microsatellite systems can be electrophoresed at different times during the run). With the advent of automated systems such as the Applied Biosystems 373 and 377 and software such as Genescan™, and Genotyper™, automated systems can now be used with high efficiency. For example, in this system primers from different loci can be labelled with markers which fluoresce at different wavelengths, allowing microsatellite systems to be multiplexed even when they have overlapping size ranges. Presently, up to 12 loci can be electrophoresed and analysed simultaneously, and the efficiency of this system may well increase in the near future. There are now many excellent primer-design software packages available which allow the design of complementary sets of primers suitable for both fluorescent labeling and radiolabeling.

Amplification protocols used are generally standard and can be carried out in a total volume of 10 µl. Radiolabeling is either by direct incorporation of labeled dNTPs or by end-labeling of one of the PCR primers. Amplifications can be carried out efficiently in 96-well microtiter plates.

Electrophoresis is normally carried out on sequencing-length gels (for radiolabeled products and automated systems), but can be carried out on smaller 20 cm vertical polyacrylamide gels. "Small" electrophoresis systems have certain advantages in that immersion-staining with silver or ethidium bromide is possible. These systems may be the most desirable in low-budget laboratories and those not equipped for radioactive procedures. The one disadvantage of small gel systems is that alleles differing by 1 to 2 bp are sometimes difficult to resolve where the total allele length exceeds 200 bp.

Scoring of microsatellite gels or autoradiograms is usually a very simple process. This is because the electrophoresis systems used usually have high resolution (to a single base pair) and because alleles usually differ in a very predictable way (multiples of the microsatellite repeat unit, e.g., 2 bp). The one major difficulty with scoring microsatellite gels is that, with mono- and dinucleotide repeat unit microsatellites, replication slippage during the amplification process can lead to the presence of sometimes confusing slippage products on the gel. These slippage products are present as less-intense bands of usually 1–5 repeat units smaller (and occasionally greater) than the actual allele. The slippage bands become relatively less intense the more they deviate in size from the native allele, and are in practice usually very easy to diagnose and ignore. However, where the second allele of a heterozygous individual overlays a slippage product from the first allele, confusion can occur, and here the difference of the relative intensity of the band is usually diagnostic (i.e., a fainter slippage product overlaid by a native allele results in a more intense band than even that of the first native allele). As with allozyme electrophoresis, the user can quickly become practised at scoring the more difficult systems. However, when there is doubt, the individual's genotype should always be confirmed by another amplification, and a reduction in the amount of *Taq* polymerase used and amplification using a smaller number of cycles will almost always resolve ambiguities. The accurate sizing of alleles is achieved by running size markers, such as a known DNA sequence, alongside the system and, in automated systems, internal size markers using a unique fluorescent label result in the sizing of alleles in each individual of even greater accuracy.

Once scored, the data are available for *analysis*. Microsatellites are codominant markers and the data generated are similar to those of allozymes, except that the number of alleles and heterozygosity revealed are almost always much higher. Population genetic, parentage, and relatedness analysis can then be carried out. There is clearly a need for generic VNTR population genetic analysis software packages to be assembled, and there is currently a dearth of such programs available. New methods and tools are needed, especially since many of the programs currently in use are not designed for hypervariable, multiallelic systems and do not take advantage of the high information content of the systems. In addition, such analytical tools need to incorporate our knowledge of how new alleles may evolve and subsequently behave in populations.

Happily, these methodologies are now being developed, and several software packages should be available shortly. Most recently, Slatkin (1995) and Goldstein et al. (1995), both take advantage of our knowledge of the predominant mode of microsatellite evolution (i.e., stepwise mutation) to derive measures of population subdivision ("*Rst*," an analogue of *Fst*; Slatkin, 1995) and genetic distance (Goldstein et al., 1995). Both measures use information on the microsatellite repeat number in the alleles themselves and utilize the mean squared difference in allele size. These have the advantage of being more reliable estimators of genetic differentiation over longer periods of time since they overcome the effects of reverse mutation and, ultimately, saturation, where fairly strict stepwise mutations have occurred during coalescence processes.

Using Degenerate Primers

As previously described, one potentially very large advantage of microsatellites in conservation genetics, especially for future studies, is the fact that primers developed for a particular species have increasingly been shown to be applicable across a wide range of related taxa (e.g., Moore et al., 1991; Schlötterer et al., 1991). With the increasing numbers of microsatellites being produced in a large range of animal and plant species, it is conceivable that in a few years the cloning of microsatellites will be unnecessary for many species, and already large numbers of cheap primers are becoming commercially available in species where extensive mapping projects are under way. However, when attempting to apply degenerate primers in new species, it is necessary to try a range of amplification protocols, and changing annealing temperature and template DNA concentration is usually sufficient to fully explore the possible applicability of the system. In our laboratory we have found that it is usually obvious after two or three experiments whether a system is going to be informative. It is possible that the frequency of "null" alleles may increase with the degeneracy of the primers used owing to a greater tendency for sensitivity to mispriming. However, there is little evidence for this phenomenon to date. In general, the use of degenerate primers offers an exciting prospect for laboratories unable to undertake the laborious and time-consuming process of cloning microsatellites in new species.

CASE STUDIES AND EXAMPLES IN POPULATION AND CONSERVATION GENETICS

Universal Primers

Perhaps the best example to date of the successful application of a set of microsatellite primers in a group of taxa is that of Schlötterer et al. (1991), who were able to show the conservation of 4 polymorphic loci in 11 cetacean species, including both toothed and baleen whales. The microsatellite loci were cloned in the long-finned pilot whale (a toothed whale), and 3 out of 4 loci were shown to be polymorphic in the fin whale (a baleen whale). The common ancestor of modern baleen and toothed whales is thought to have diverged around 35–40 million years ago, making the conservation of such sequences an extremely surprising result at the time. However, the authors also showed an extremely low rate of evolution in the unique sequence flanking the conserved microsatellites in the 11 cetaceans studied, and whether such extreme sequence conservation would be found among other taxonomic groups was open to question.

However, further evidence has continued to yield surprising levels of conservation among such loci, though perhaps nothing as extreme as Schlötterer et al.'s first observations, at least in mammals. For example Moore et al. (1991) found a conservation of only 40% of microsatellites between cattle and sheep (both even-toed ungulates). Conversely, Ammer et al. (1992) found that a microsatellite in the *MHC-DRB* intron was conserved between species as different as primates and ungulates. A large body of data is now accumulating regarding the polymorphic characteristics of human microsatellites in apes (e.g., Morin et al., 1994b) and domestic dog microsatellites in other canids (e.g., Gottelli et al. 1994; Roy et al.'s 1994), and many laboratories are now using degenerate primers routinely in population genetic analysis. Recently, microsatellite primers derived from reed buntings (*Emberiza schoenlicus*) have been shown to amplify products in a number of other avian species (Hanotte et al., 1994).

An example from our own laboratory is the development of a set of universal microsatellite markers for apes and Old World monkeys (Coote and Bruford, 1996). In a related study, we developed a set of human microsatellites which amplified polymorphic markers in savannah baboons (Altmann et al., 1996). Fourteen polymorphic microsatellites were found in baboons from a total of 85 human loci which were tested. We applied 11 of these to a panel of 24 primates, representing 14% of extant species and 6 of the 7 superfamilies. We found that polymorphic microsatellites were amplified in all apes and Old World monkeys, but that no polymorphic systems were found in any New World monkey or prosimian (see Figure 17-2). Since the primers were preselected on the basis of cross-amplification and allelic variability in the savannah baboon (a cercopithecoid monkey), it is perhaps not surprising that the primers cross-amplified

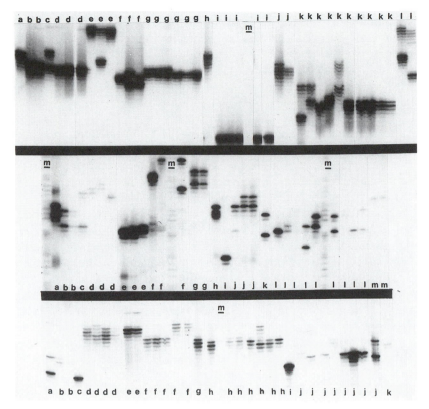

Figure 17-2 (Top) Locus D6S311: (a) de Brazza's monkey; (b) diana monkey; (c) white colobus; (d) langur sp; (e) Sulawesi macaque; (f) savannah baboon; (g) Guinea baboon; (h) mandrill; (i) orang utan; (j) gorilla; (k) chimpanzee; (l) human. (m = Molecular weight marker.) (Middle) Locus D7S503: (a) de Brazza's monkey; (b) diana monkey; (c) white colobus; (d) langur sp; (e) Sulawesi macaque; (f) savannah baboon; (g) Guinea baboon; (h) mandrill; (i) lar gibbon; (j) orang utan; (k) gorilla; (l) chimpanzee; (m) human. (m = Molecular weight marker.) (Bottom) Locus D16S420: (a) de Brazza's monkey; (b) diana monkey; (c) white colobus; (d) langur sp; (e) savannah baboon; (f) Guinea baboon; (g) mandrill; (h) orang utan; (i) gorilla; (j) chimpanzee; (k) human. (m = Molecular weight marker.)

polymorphic microsatellites in other cercopithecines and hominoid primates. The Ceboidea (New World monkeys) are thought to have diverged from the Old World higher primates some time after the separation of the North American and Eurasian land masses in the mid-Eocene, around 40–45 million years ago. However, this is a matter of some debate and it is possible that divergent evolution between the two groups started before the continents separated. These microsatellite data point tentatively to a more fundamental and ancient split than some authors have considered. The absence of cross-amplification in the prosimians and praesimians is perhaps less surprising, given that the common ancestor between these groups and the Old World higher primates is thought to be approximately 70–80 million years ago. The establishment of a set of "universal" polymorphic markers in apes and Old World monkeys potentially allows primatologists working on these species to produce data of a comparable nature within and between species, and negates the necessity for cloning microsatellites in some of the more primitive Old World monkeys, which might previously have been deemed necessary.

Hybridization

One of the best examples of the application of microsatellites to studies of intraspecific hybridization was recently published by Roy et al. (1994), who used 10 polymorphic systems derived in the domestic dog to examine perturbation of allele frequencies in gray wolf and coyote populations in zones of hybridization relative to those where hybridization was known not to occur. It was shown that hybridizing populations of gray wolves and coyotes converged in allele frequency, and more importantly that gene flow was almost unidirectional, with only the allele frequencies of hybridizing gray wolf populations being significantly perturbed. This suggested that approximately two reproducing coyotes per generation were migrating into the local wolf population. The authors also used microsatellites to reexamine the genetic status of the critically endangered red wolf. They found that red wolves shared all their microsatellite alleles with coyotes, and simulation analysis showed that the total absence of unique alleles in red wolves was inconsistent with a separate origin from other present and past sympatric canids, even allowing for the possible loss of rare red wolf-specific alleles through drift.

Another study examining the effects of hybridization in a canid involved the Ethiopian wolf, which was surveyed for variation at nine microsatellite loci (Gottelli et al., 1994). The results showed that the species had about 30–40% of the heterozygosity and allelic diversity of outbred canids. Moreover, in one area Ethiopian wolves had extensively hybridized with domestic dogs, as suggested by the presence of diagnostic dog alleles in phenotypically abnormal wolves. This study also showed a definitive case of multiple paternity, as one litter was fathered by both a dog and a wolf. Microsatellite analysis has added considerable resolution to the measurement of variation and population structure in this endangered species.

An example from our own work concerns the Scottish red deer (*Cervus elaphus scoticus*), which is a putative subspecies of the red deer (or elk) endemic to the British Isles (Cheesman et al., in press). The Scottish red deer population was virtually extirpated during the seventeenth century by hunting, with a few relic populations surviving in Kirkudbrightshire and on some of the Western Isles. Many animals were subsequently introduced from park stock elsewhere within Britain and continental Europe, and it is certain that the introduced stock has extensively hybridized with much of the indigenous population. In addition, work by Abernethy (1994) has shown extensive introgression between the population of red deer on the Argyll peninsula and an introduced population of Japanese sika deer (*Cervus nippon*), and although this is not a region thought to contain indigenous Scottish red deer, such hybridization could potentially pose a further threat.

We obtained tissue from over 300 culled red and sika deer from 13 large stock areas in Scotland. To date, five microsatellites have been applied to identify

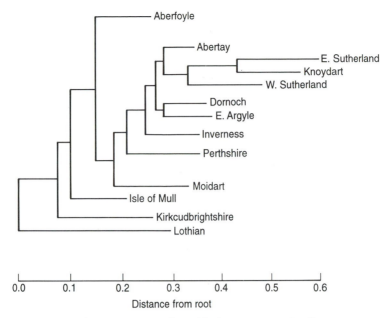

Figure 17-3 Optimized Wagner tree of modified Rogers genetic distances among Scottish red deer populations rooted using Lothian (a pure Japanese sika population) as an outgroup.

differences among individuals and populations. These markers are variously cervine, ovine, and bovine in origin. The mean heterozygosity across all five loci ranged between 0.455 (W. Sutherland) and 0.690 (Mull), and the mean number of alleles per locus ranged between 2.8 (Lothian) and 8.4 (Abertay). In general, allele frequencies within the populations studied did not depart significantly from Hardy–Weinberg equilibrium, and where departure was noted this was neither consistently in the direction of homozygote or of heterozygote excess.

Populations showed very little local inbreeding (mean $F_{IS} = 0.032$), but F_{IT} and F_{ST} were both relatively high and equivalent (0.17 and 0.22, respectively). A Wagner tree was constructed using modified Rogers distance, with the pure sika population in Lothian used as an outgroup (Figure 17-3).

Although further loci need to be genotyped, the data indicate that there is genetic substructuring both among individuals across populations and among subpopulations within Scotland. This may indicate that, athough there is some genetic differentiation among subpopulations, groups of individuals from more than one population may form contiguous breeding units. Further, the populations in Kirkcudbrightshire and Mull have relatively large genetic distances with respect to each other and to other populations in Scotland. This tree topology is also supported using unrooted UPGMA cluster analysis. This result is interesting, since historical knowledge of the Scottish red deer suggests that these populations are the two most likely to have descended directly from ancestral U.K. red deer stock.

Population Studies

Several studies describing genetic variation within and among wild populations and domestic animal breeds have recently been published. A recent example of quantification of genetic variation within and among populations concerns the black bear in three National Parks in Canada (Paetkau and Strobeck, 1994). Here, the authors found very significant differences in allele frequency distributions of four microsatellite loci among populations, and in the case of one population from Newfoundland they found surprisingly low levels of variation which may have been due to a past population bottleneck or random drift. MacHugh et al. (1994) analysed genetic variation between 6 breeds of European cattle using 12 microsatellite loci, and were able to compute genetic distances and construct dendrograms which broadly showed concordance with known breed histories. Buchanan et al. (1994) carried out a similar analysis using 8 loci in 5 different breeds of sheep, finding highly significant allele frequency differences among breeds, and were also able to construct a sensible tree using genetic distance matrices.

In a similar study to that of Paetkau and Strobeck (1994), we have been examining genetic variability and differentiation within and among park populations of Cape buffalo (*Syncerus caffer*) in South Africa (O'Ryan et al., in preparation). Here, we have used 10 microsatellites isolated in the bovine genome mapping project and these were analysed in three major populations: Kruger National Park; Natal (two populations, one at Umfolozi and one from a "daughter" population at Lake St. Lucia, which was restocked in 1974); and Cape Province (two populations, one at Addo and one from a daughter population at Hankey, which was restocked in 1979). Preliminary data show that the largest park, Krugers, had the highest levels of genetic variability, both in the number of alleles and heterozygosity. This was followed by the second largest population in Natal, which had approximately 80% of the allelic diversity found in Kruger. Over the loci tested, the allelic diversity in St. Lucia ranged between 50% and 80% of its parent population in Umfolozi. The Cape population has the lowest diversity of all, ranging between 36% and 70% of that found in Kruger. Figure 17-4 shows a histogram of allele frequency distributions where these differences are obvious simply on inspection. These results are consistent not only with known population numbers but also with park size and available suitable habitat.

To date, within-population genetic variation has been examined more extensively. For example, Amos et al. (1993) found between 3 and, remarkably at one locus, 54 alleles in long-finned pilot whales from a Faroese population. Mature males who do not disperse were found never to father offspring born within their own pod. Sixteen microsatellite loci have been surveyed in a genetically bottlenecked species, the endangered northern hairy-nosed wombat, and levels of variation were found to be severely reduced relative to an outbred closely related species, the southern hairy-nosed wombat (Taylor et al., 1994). The reduction in microsatellite variation suggested a bottleneck of just 10–20 individuals for a 120-year period.

In a study carried out in our laboratory (Williams et al., 1995, and in preparation) we have been examining the population genetic consequences of

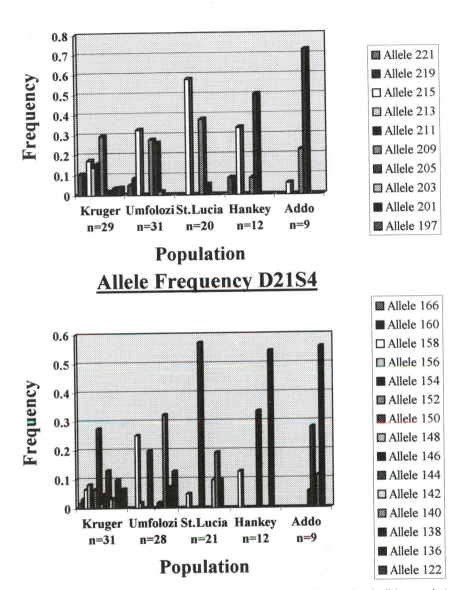

Figure 17-4   Allele frequency histograms for locus D1S6 and D21S4 in five buffalo populations. Allelic nomenclature refers to the size in base pairs.

introductions from a small founder number followed by subsequent rapid growth using the introduced Reeves' muntjac (*Muntiacus reevesi*) in south east England as a model. Reeves' muntjac is a small deer that originates in China and Taiwan. In 1894, the 11th Duke of Bedford received his first muntjac at Woburn Abbey, and in March 1901 eleven animals were taken and released into the surrounding woodland. In the late 1930s and 1940s, a number of well established feral colonies were reported, often at some distance from each other and at some distance from the then supposed center of origin—Woburn Abbey. Rapid expansion followed, and there are now several tens of thousands of these animals in southern England. The consensus opinion was that the feral populations of Reeves' muntjac originated from the stock at Woburn Abbey and investigations were designed with this in mind. Samples were taken from 18 U.K. subpopulations of muntjac representing increasing geographic distances centred on Woburn Abbey and one from Taiwan.

Genetic variability among subpopulations was assessed by an investigation of the geographic distribution of seven microsatellite loci. Most microsatellites were highly polymorphic and there was substantial variation in the number of alleles detected per subpopulation. Heterozygosity values were high, ranging from 0.48 to 0.74. Fourteen rare alleles were distributed between 10 of the 19 subpopulations. Wright's F_{IS} was calculated as a metric of the level of inbreeding within subpopulations. The mean values of this estimator ranged from -0.181 to 0.222. Nei's G_{ST} was calculated to indicate the level of genetic substructuring among subpopulations. These calculations suggested that differentiation by increasing distance was not significant over the geographic range of the sample area, possibly indicating that muntjac are unlikely to have dispersed in a natural way from any given central source. It appears that more than 11 animals have been released (for example, there are eight mitochondrial genotypes present in the population) and that dispersal may have at some stage been mediated by transportation of animals across country by man.

Microsatellite loci identified in social insects have been found to be highly polymorphic compared to allozyme loci (Choudhary et al., 1993; Evans, 1993; Hamaguchi and Ito, 1993; Hughes and Queller, 1993). Hamaguchi and Ito (1993) showed that maternity in a multiple-queen ant colony could be assigned by using five loci. Evans has shown co-dominant Mendelian inheritance in complex microsatellites in a *Myrmica* ant species with as many as 25 alleles at one locus. Choudhary et al. (1993) and Hughes and Queller (1993) have isolated highly polymorphic microsatellites in social wasps which exhibit polymodal allele size distributions. We have isolated and characterized genetic variation in microsatellites from the multiple-queened ant species *Leptothorax acervorum*, and have found one locus LXAGA2 which shows 15 alleles in 20 unrelated individuals and 80% heterozygosity, an extremely high level of genetic variation for a eusocial ant species (Figure 17-5).

Relatedness and Reproductive Success

There have been three well-publicised recent examples of the application of microsatellites to estimating relatedness and reproductive success in natural populations. In two papers, Morin et al. (1994a,b) examined relatedness, social

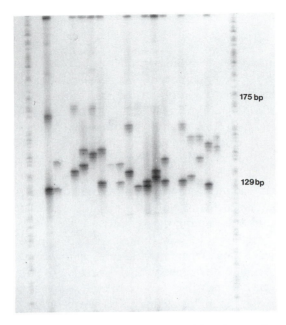

Figure 17-5   Microsatellite autoradiogram of 20 unrelated female *Leptothorax acervorum*. Alleles ranged from 129 bp to 175 bp.

structure, and parentage in a well-studied free-living chimpanzee population. Using eight human-derived microsatellite loci amplified from hair samples taken noninvasively from night nests built by different individuals, the authors were able to test certain hypotheses about relatedness within populations. Unusually for primates, the females are the dispersing sex in chimp populations, and the authors hypothesized that the philopatric males would have a higher average relatedness than within-group females. They found that the mean number of alleles shared among males per locus was significantly higher than among females, and that the results were consistent with males being on average related at the level of half-sibs (although no direct relatedness calculations were carried out). Further, significant departure from Hardy–Weinberg expectation was observed at three loci, a result suggestive of local inbreeding or a greater degree of relatedness in the social community studied (Morin et al., 1994a). In a second study, Morin et al. (1994b) analysed paternity in the same group of individuals. A large proportion of the population was typed at the eight loci, and paternity exclusions of 25 individuals were carried out (10 had the known mother also genotyped). In four cases the father was almost unambiguously identified and in a further four cases all but 2–5 fathers out of a possible 12–20 males were excluded, and in the remaining cases all living males were excluded. These studies, although perhaps not as complete as would be desirable, represent a landmark in the application of microsatellites in free-living populations and perhaps the lack of completeness of the data represents the most common situation which will confront population geneticists and behavioral ecologists studying sensitive and endangered species in the future.

THE FUTURE

Clearly, microsatellites have a very important role to play in conservation genetic studies in the near future, and there seem to be few limitations to their application to assessment of genetic variation in natural and captive populations. The fact that large numbers of polymorphic markers can be analysed from hair and fecal samples means potentially that high-precision data can be generated without invasive sampling in threatened populations. The assessment of genetic variation and its structure within and among populations and the possibility of un-ambiguous determination of parentage and precise estimation of relatedness are all now achievable in endangered species. These applications could revolutionize aspects of conservation genetics and allow previously genetically uncharacterized populations to be understood more thoroughly.

Microsatellites are, however, very limited in their applications at or above the species level since the high mean mutation rate of microsatellite loci will often obviate their application to phylogenetic analysis owing to the very large effect of reciprocal mutations. It is possible, however, that sequence analysis of unique sequence DNA flanking the microsatellite locus can be utilized in this way (e.g., Schlötterer et al., 1991). Additionally, ancient population patterns might also be obscured owing to the same multiple back-mutation effect, especially where populations have a high effective size and the microsatellite markers have a relatively high mutation rate. It is important to bear these limitations in mind when planning a conservation genetic study; however, it remains very likely that microsatellite studies of free-living populations will be extremely widely applied over the forseeable future.

ACKNOWLEDGMENTS

Much of the ongoing microsatellite research cited was carried out in the conservation genetics laboratory of the Institute of Zoology, London. Financial support was gained from the Institute of Zoology, NERC, NSF, the University of Kent, the University of Cape Town, and the EU. We acknowledge the help and support of all those who have contributed to our various microsatellite studies, including Robert Wayne, Terry Burke, Olivier Hanotte, Hugh Reed, Josephine Pemberton, Stephen Harris, Norma Chapman, Mike Cherry, Eric Harley, Andrew Bourke, Jeanne Altmann, Susan Alberts, Tim Robson, Andrew Loudon, Andrea Taylor and Helen Stanley.

REFERENCES

Abernethy, K. 1994.The establishment of a hybrid zone between red and sika deer (genus *Cervus*). Mol. Ecol. 3: 551–562.

Altmann, J., S.C. Alberts, S.A. Haines, J. Dubach, P. Muruthi, T., Coote, E. Geffen, D.J. Cheesman, R.S. Mututua, S.N. Saiyalel, R.K. Wayne, R.C. Lacy, and M.W. Bruford. 1996. Behaviorally-based predictions for genetic structure tested in a wild primate group. Proc. Natl. Acad. Sci. U.S.A., [In press].

Ammer, H., F.-W. Schwaiger, C. Kammerbauer, A. Arriens, S. Lazary, and J.T. Epplen.

1992. Exonic polymorphism versus intronic hypervariability in DRB genes: evolu-Ftionary persistence and group specific organization in simple repeat sequences. Immunogenetics 35: 330–447.

Amos, B., C. Schlötterer, and D. Tautz. 1993. Social structure of pilot whales revealed by analytical DNA profiling. Science 260: 670–672.

Barendse, W., S.M. Armitage, L.M. Kossarek, A. Shalom, B.W. Kirkpatrick, A.M. Ryan, D. Clayton, L. Li, H.L. Neibergs, N. Zhang, W.M. Grosse, J. Weiss, P. Creighton, F. McCarthy, M. Ron, A.J. Teale, R. Fries, R.A. McGraw, S.S. Moore, M. Georges, M. Soller, J.E. Womack, and D.J.S. Hetzel. 1994. A genetic linkage map of the bovine genome. Nature Genetics 6: 227–235.

Berrettini, W.H., T.N. Ferraro, R.C. Alexander, A.M. Buchberg, and W.H. Vogel. 1994. Quantitative trait loci mapping of 3 loci controlling morphine preference using inbred mouse strains. Nature Genetics 7: 54–58.

Brook, J.D., M. McCurrach, H.G. Harley et al. 1992. Molecular basis of myotonic dystrophy: Expansion of a trinucleotide (CTG) repeat at the 3′ end of a transcript encoding a protein kinase family members. Cell 68: 799–808.

Bruford, M.W., O. Hanotte, J.F.Y. Brookfield, and T. Burke. 1992. Multi- and single-locus DNA fingerprinting. In Molecular analysis of populations: A practical approach, pp. 225–269. IRL Press, Oxford.

Bruford, M.W. and R.K. Wayne. 1993b. Microsatellites and their application to population genetics. Curr. Opin. Genet. Devel. 3: 939–943.

Buchanan, F.C., L.J. Adams, R.P. Littlejohn, J.F. Maddox, and A.M. Crawford. 1994. Determination of evolutionary relationships among sheep breeds using micro-satellites. Genomics 22: 397–403.

Buluwela, L., A. Forster, T. Boehm, and T.H. Rabbits. 1989. A rapid procedure for colony screening using nylon filter. Nucleic Acids Res. 17: 452.

Callen, D.F., A.D. Thimpson, Y. Shen, H.A. Phillips, R.I. Richards, J.C. Mulley, and G.R. Sutherland. 1993. Incidence and origin of "null" alleles in the (AC)$_n$ microsatellite markers. Am. J. Hum. Genet. 52: 922–927.

Cheesman, D.J., A.S.I. Loudon, J.M. Pemberton, and M.W. Bruford. 1996. Population genetics of red deer in Scotland: evidence for a Scottish red deer enclave? Proc. 3d Int. Congr. Biol. Deer. [In press].

Choudhary, M., J.E. Strassmann, C.R. Solis, and D.C. Queller. 1993. Microsatellite variation in a social insect. Biochem. Genet. 31: 87–96.

Coote, T., M.W. Bruford. 1996. Human microsatellites applicable for analyses of genetic variation in apes and Old World monkeys. J. Hered. [In press].

Dallas, J.F. 1992. Estimation of microsatellite mutation rates in recombinant inbred strains of mouse. Mammal Genome 5: 32–38.

Dietrich, W., J.C. Miller, and R.G. Steen et al. 1994. A genetic map of the mouse with 4,006 simple sequence length polymorphisms. Nature Genet. 7: 220–245.

Di Rienzo, A., A.C. Peterson, J.C. Garza, A.M. Valdes, M. Slatkin, and N.B. Freimer. 1994. Mutational processes of simple-sequence repeat loci in human populations. Proc. Natl. Acad. Sci. USA 91: 3166–3170.

Edwards, A., H.A. Hammond, L. Jin, C.T. Caskey, and R. Chakraborty. 1992. Genetic variation at five trimeric and tetrameric tandem repeat loci in four human population groups. Genomics 12: 241–253.

Ellegren, H. 1991. DNA typing of museum birds. Nature 354: 113.

Evans, J.D. 1993. Parentage analysis in ant colonies using simple sequence loci. Mol. Ecol. 2: 393–397.

Fu, Y.-H., D.P.A. Kuhl. M. Pizzutti, M. Pieretti, J.S. Sutcliffe, S. Richards, A. Vererk, J. Holden, R. Fewick, S.T. Warren, B. Oostra, D.L. Nelson, and C.T. Caskey. 1991.

Variation of the CGG repeat at the fragile X site results in genetic instability: Resolution of the Sherman paradox. Cell 67: 1047–1058.

Gill, P., P.L. Ivanov, C. Kimpton, R. Piercy, N. Benson, G. Tully, I. Evett, E. Hagelberg, and K. Sullivan. 1994. Identification of the remains of the Romanov family by DNA analysis. Nature Genetics 6: 130–135.

Goldstein, D.B., A.R Linares, L.L. Cavalli-Sforza, M.W. Feldman, 1995. An evaluation of genetic distances for use with microsatellite loci. Genetics 139: 463–471.

Gottelli, D., C. Sillero-Zubiri, G.D. Applebaum, M.S. Roy, D.J. Girman, J. García-Moreno, E.A. Ostrander, and R.K. Wayne. 1994. Molecular genetics of the most endangered canid: The Ethiopian Wolf, Canis simensis. Mol. Ecol. 3: 301–312.

Gyapay, G., J. Morissette, A. Vignal, C. Dib, C. Fizames, P. Millassean, S. Marc, G. Bernardi, M. Lathrop, and J. Weissenbach. 1994. The 1993–94 Genethon human linkage map. Nature Genetics 7: 246–339.

Hagelberg, E., I.C. Gray, and A.J. Jeffreys. 1991. Identification of the skeletal remains of a murder victim by DNA analysis. Nature 352: 427–429.

Hamada, H., M. Petrino, and T. Kakunaga. 1982. A novel repeated element with Z-DNA-forming potential is widely found in evolutionarily diverse eukaryotic genomes. Proc. Natl. Acad. Sci. U.S.A. 79: 646–6469.

Hamaguchi, K. and Y. Ito. 1993. GT dinucleotide repeat polymorphisms in a polygynous ant, Leptothorax spinosior and their use for measurement of relatedness. Naturwissenschaften 80: 179–181.

Hanotte, O., C. Zanon, A. Pugh, C. Greig, A. Dixon, and T. Burke. 1994. Isolation and characterization of microsatellite loci in a passerine bird: the reed bunting Emberiza schoenlicus. Mol. Ecol. 3: 529–530.

Huang, T., R. Cottingham, D. Ledbetter, and H. Zoghbi. H. 1992. Genetic mapping of four dinucleotide repeat loci, DXS453, DXS458, DXS454, and DXS452 on the X chromosome using multiplex polymerase chain reaction. Genomics 13: 375–380.

Hughes, C.R. and D.C. Queller. 1993. Detection of highly polymorphic microsatellite loci in a species with little allozyme polymorphism. Mol. Ecol. 2: 131–137.

Jeffreys, A.J., V. Wilson, and L. Thein. 1985. Hypervariable "Minisatellite" regions in human DNA. Nature 314: 67–73.

Jeffreys, A.J., M.J. Allen, E. Hagelberg, and A. Sonnberg. 1992. Identification of the skeletal remains of Joseph Mengele by DNA analysis. Forensic Sci. Int. 56: 65–76.

Kwiatowski, D., E. Henske, K. Weimer, L. Ozelius, J. Gusella, and J. Haines. 1992. Construction of a GT polymorphism map of human 9q. Genomics 12: 229–240.

Litt, M. and J.A. Luty. 1989. A hypervariable microsatellite revealed by in vitro amplification of dinucleotide repeats within the cardiac muscle actin gene. Am. J. Hum. Genet. 44: 397–401.

MacHugh, D.E., R.T. Loftus, D.G. Bradley, P.M. Sharp, and P. Cunninghan. 1994. Microsatellite DNA variation within and among European cattle breeds. Proc. R. Soc. Lond. B 256: 25–31.

McDonald, D.B. and W.K. Potts. 1994. Cooperative display and relatedness among males in a lek-mating bird. Science 266: 1030–1032.

Moore, S.S., L.L. Sargeant, T.J. King, J.S. Mattick, M. Georges, and D.J.S. Hetzel. 1991. The conservation of dinucleotide microsatellites among mammalian genomes allows the use of heterologous PCR primer pairs in closely related species. Genomics 10: 654–660.

Morin, P.A., J.J. Moore, R. Chakraborty, L. Jin, J. Goodall, and D.S. Woodruff. 1994a. Kin selection, social structure, gene flow and the evolution of chimpanzees. Science 265: 1193–1201.

Morin, P.A., J. Wallis, J.J. Moore, and D.S. Woodruff. 1994b. Paternity exclusion in a community of wild chimpanzees using hypervariable simple sequence repeats. Mol. Ecol. 3: 469–478.

Paetkau, D. and C. Strobeck. 1994. Microsatellite analysis of genetic variation in black bear populations. Mol. Ecol. 3: 489–495.

Rassmann, K., C. Schlötterer, and D. Tautz. 1991. Isolation of simple sequence loci for use in polymerase chain reaction-based DNA fingerprinting. Electrophoresis 12: 113–118.

Roewer, L., M. Nagy, P. Schmidt, J.T. Epplen, and G. Herzog-Schroder. 1993. Microsatellite and HLA class II oligonucleotide typing in a population of Yanomani Indians. In: S.D.J. Pena, R. Chakraborty, J.T. Epplen, and A.J. Jeffreys, eds. DNA fingerprinting: State of the science, pp. 221–230. Basel: Birkhäuser Verlag.

Roy, M.S., E. Geffen, D. Smith, O. Ostrander, and R.K. Wayne. 1994. Patterns of differentiation and hybridization in North American wolflike canids, revealed by analysis of microsatellite loci. Mol. Biol. Evol. 11: 553–570.

Sambrook, J., E.F. Fritsch, T. and Maniatis. T. 1989. Molecular cloning. A laboratory manual, 2d edn. New York: Cold Spring Harbor Laboratory Press.

Schlötterer, C. and D. Tautz. 1993. Slippage synthesis of microsatellites. Nucleic Acids Res. 20: 211–215.

Schlötterer, C., B. Amos, and D. Tautz. 1991. Conservation of polymorphic simple sequences in cetacean species. Nature 354: 63–65.

Shriver, M.D., L. Jin, R. Chakraborty, and E. Boerwinkle. 1993. VNTR allele frequency distributions under the stepwise mutation model: A computer simulation approach. Genetics 134: 983–993.

Slatkin, M. 1995. A measure of population subdivision based on microsatellite allele frequencies. Genetics 139: 157–162.

Tachida, H. and M. Izuka. 1992. Persistence of repeated frequences that evolve by replication slippage. Genetics 131: 471–478.

Tautz, D. 1989. Hypervariability of simple sequences as a general source for polymorphic DNA markers. Nucleic Acids Res. 17: 6463–6471.

Tautz, D. and M. Renz. 1984. Simple sequences are ubiquitous repetitive components of eukaryote genomes. Nucleic Acids Res. 12: 4127–4138.

Tautz, D. and C. Schlötterer. 1994. Simple sequences. Curr. Opin. Genet. Dev. 4: 832–837.

Tautz, D., M. Trick, and G. Dover. 1986. Cryptic simplicity in DNA is a major source of genetic variation. Nature 322: 652–656.

Taylor, A.C., W.B. Sherwin, and R.K. Wayne. 1994. Genetic variation of simple sequence loci in a bottlenecked species: The decline of the northern hairy-nosed wombat (*Lasiorhinus krefftii*). Mol. Ecol. 3: 277–290.

Valdes, A.M., M. Slatkin, and N.B. Friemer. 1993. Allele frequencies at microsatellite loci: The stepwise mutation model revisited. Genetics 133: 737–749.

Weber, J.L. and P.E. May. P.E. 1989. Abundant class of human DNA polymorphisms which can be typed using polymerase chain reaction. Am. J. Hum. Genet. 44: 388–396.

Williams, T.R., M.W. Bruford, N. Chapman, R.J. Wayne, and S. Harris. 1996. Molecular genetics of an invading species: Reeves' muntjac in southern England. Proc. 3d Int. Congr. Biol. Deer. [In press].

Zheng, L., F.H. Collins, V. Kumar, and F.C. Kafatos. 1993. A detailed genetic map for the X chromosome of the malaria vector, *Anopheles gambiae*. Science 261: 605–608.

18

Noninvasive Genotyping for Vertebrate Conservation

PHILLIP A. MORIN AND DAVID S. WOODRUFF

The documentation of vertebrate diversity traditionally involved the collection of specimens for further study. Properly curated collections constitute treasure troves for biologists interested in systematics, biogeography, and evolutionary biology. The collection of specimens and their preservation was justified on the grounds that our understanding of nature depended upon it. This is no longer always the case, and in situations involving rare or threatened species it may be completely unjustified. Collections that were well-intended at the time can seem appalling in today's conservation-minded world. Four examples illustrate the conflicts between the need to conduct scientific research and the need to conserve the subjects of that research:

- The 1906 California Academy of Sciences expedition to the Galapagos discovered that the Pinzon island race of giant tortoises was, in fact, not extinct and accordingly collected the remaining 86 for study (Thornton, 1971).
- A study of reproductive cycles in African elephants was based on shooting 81 females in the course of pest-control work (Perry, 1953).
- Government zoologists shot the first Leadbeater's possum seen in Australia for a century in order to confirm its rediscovery by Eric Wilkinson, who at the time was only an amateur naturalist (Wilkinson, 1961).
- Despite the fact that all commercial whaling was banned in 1986, the International Whaling Commission continues to permit the Japanese to take whales for research purposes.

Although such collection of threatened animal species is scientifically indefensible today, their conservation management depends on the availability of both ecological and genetic data (Woodruff, 1989). Genetic data are important for the management of populations and species, especially if combined with ecological, morphological, behavioral, and demographic data when they are available. Fortunately, most of these types of data can often be gathered by simple observation, and genotyping no longer requires that individual animals be "sacrificed." Genetic data of value to conservation biologists can be obtained nonlethally from a variety of small tissue samples. Blood is the most widely used source for allozymes and other proteins and can be drawn by puncture or by tail, ear, or toe clipping (Dessauer et al., 1990). For mammals, blood has been drawn from the retroorbital sinus of mice, the femoral vein of lions, and the propatagial vein of bats, or by direct cardiac puncture. For birds, the wing vein provides a suitable source (Arctander, 1988). For reptiles, blood has been drawn from the cervical sinus of tortoises and by cardiac puncture in snakes and crocodiles (Dessauer et al., 1990). In many small mammals, however, phlebotomy provides insufficient quantities of genetic material for many analyses, and sampling has typically involved liver and muscle biopsies.

With few exceptions, collection of these tissues requires that animals be captured and restrained. Such procedures place both animals and handlers at risk. Furthermore, most tissues have to be frozen upon collection and their transport to the laboratory often requires a logistically difficult "cold-train" (Dessauer and Hafner, 1984). These shortcomings of traditional methods have made it almost impossible to study population genetic aspects of conservation of many threatened species. Such frustrations, in studies of Arabian oryx, Père David's deer, and white rhinoceros (Woodruff and Ryder, 1986, 1988; Merenlender et al., 1989), stimulated the development of methods which circumvent these problems and permit truly noninvasive genotyping. These methods permit investigators to monitor genetic variation in free-ranging populations without collecting the animals themselves. In some cases it is now possible to genotype individual animals without ever seeing them. Such methods, reviewed below, promise to revolutionize the practice of conservation genetics in the wild, in captive management, and in forensic situations.

HISTORY OF NONINVASIVE DNA GENOTYPING METHODS

Noninvasive sampling for DNA analysis became possible only after the introduction of the polymerase chain reaction (PCR) for DNA sequence amplification. The PCR method (Saiki et al., 1985; Mullis, 1990) makes use of pairs of oligonucleotide primers that bind to specific sites flanking the region of interest, and of a DNA polymerase produced by thermophilic bacteria, which functions optimally at 72–74°C and resists denaturation even at 100°C, to make up to a million copies of a specific genome region from one or a few original template copies. This allows the entire reaction to take place at temperatures sufficiently high for oligonucleotide primers to bind very specifically to a selected site in the genome. Often, primers are specific to that site in only one of a set of closely related species. Methods and applications for the PCR are covered in numerous books (e.g., Erlich, 1989; Innis et al., 1990; Herrmann and Hummel 1994).

The development of PCR-based noninvasive genotyping required the solution of several problems. First, gene sequences had to be described before site-specific primers could be made. For some mitochondrial and protein-coding nuclear genes, this problem was quickly solved through the discovery of generic or "universal" primers that amplified homologous sequences in many species (e.g., Kocher et al., 1989; Kocher and White 1989). For other loci, the synthesis of appropriate primer pairs required laborious screening of genomic libraries for specific fragments of

DNA cloned into bacteria or phages (Sambrook et al., 1989) in order to determine the sequence in the region of interest. Second, sufficient target DNA had to be consistently extractable from samples, without contamination by foreign DNA (such as human DNA introduced by handling and while performing the PCR; Orrego, 1990; Pääbo 1993). Third, the extracted DNA had to be amplifiable, that is, free of substances that might inhibit the PCR. Several tissue-specific solutions to these problems have been developed and will be reviewed below.

NONINVASIVE METHODS FOR GENOTYPING LIVE ANIMALS

Hair

One of the first reports of PCR amplification of DNA obtained noninvasively came from work related to forensics. Hair samples shed or plucked from humans were found to contain enough DNA to allow amplification of mitochondrial and nuclear loci (Higuchi et al., 1988; von Beroldingen et al., 1987; von Beroldingen, 1989; Weisburd, 1989). These methods were rapidly adapted to nonhuman primate (Morin and Woodruff, 1992; Morin et al., 1993; Woodruff, 1990, 1993) and bear (Taberlet and Bouvet, 1992) hair samples.

For hair (and feathers, see below), the primary problem is extracting the DNA from the hardened tissue of the shaft and root. This is achieved by traditional proteinase digestion followed by extraction with organic solvents (von Beroldingen, 1989; Vigilant et al., 1989; Ellergren, 1991; Taberlet and Bouvet, 1991; Morin and Woodruff, 1992, Wilson et al. 1995) and also by mechanical grinding and heat-mediated cell rupturing in the presence of a solid-state chelating agent to prevent enzymatic degradation of DNA (Walsh et al., 1991; Morin and Woodruff, 1992; Tan and Orrego, 1992; Ellergren, 1994; Morin et al., 1994a; Ohhara et al. 1994). Either method typically yields enough DNA from a single hair root to perform at least 10–20 PCR reactions (but see Gagneux and Woodruff, submitted). The former method, however, releases pigments into the DNA solution that can inhibit the PCR reaction and the multiple additional extraction steps required to overcome this problem increase the potential for contamination by nontarget DNA.

Woodruff (1993) provides a review of the first studies of primates based on hair; studies of *Pan*, *Gorilla*, *Pongo*, *Hylobates*, and *Macaca* took advantage of primer sequence homologies between human and nonhuman primates. A demonstration of the power of noninvasive genotyping is Morin's hierarchical study of genetic variation in free-ranging chimpanzees (*Pan troglodytes*) using hair samples collected from nests and orphaned individuals across Africa (Morin et al., 1994b). At the level of the social community, variation in hypervariable nuclear simple sequence repeats (SSR or microsatellite loci) were surveyed in the much-studied Gombe population to establish pedigree relationships, intrinsic variability, and population structure (Morin et al., 1993, 1994b,c). At higher levels of taxonomic organization, we used cytochrome *b* and control region sequences to demonstrate historical patterns of gene flow between populations up to about 900 km apart, and to infer genetic relationships between the three subspecies (Morin et al., 1994b). At each taxonomic level, noninvasive genotyping provided otherwise unobtainable information of importance to conservation of both wild and captive chimpanzees.

Woodruff initiated a similar study of gibbons, *Hylobates*, based on mtDNA amplified from hair plucked from or shed by captive animals (Woodruff, 1990). A preliminary phylogenetic survey of seven of the nine widely recognized species based on a short cytochrome b sequence revealed the probable existence of one additional species in the *H. concolor* species group (Garza and Woodruff, 1992). Sequence variation can be used to sort look-alike captive *H. concolor* of unknown geographic origin (Garza and Woodruff, 1994). More variable (informative) control region sequences (initially determined by Kressirer (1993) using blood) are now being used to sort the *H. lar* and *H. concolor* in Thai and North American zoos into genetically appropriate management units (Woodruff and Tilson, 1994).

There are now numerous on-going hair-based studies of other primates (Woodruff, 1993), for example, West African chimpanzees of the Thai forest (P. Gagneux, Basel and San Diego); gorilla mitochondrial control region (Garner, 1992; Garner and Ryder, 1992a,b; gorilla and orangutan SSRs (D. Field, San Diego, personal communication); gorilla control region and SSRs for paternity and population studies and *Galago* spp. cytochrome *b* and D-loop (E.J. Wickings, CIRMF, Gabon, personal communication); *Macaca* species, control region (Mubumbila et al., 1992); *Macaca sylvanicus* SSRs (F. von Segesser, Zurich, personal communication); and *Callimico goeldii* SSRs (K. Vasarhehlyi, Zurich, personal communication). Giant panda SSRs are being amplified from hair for paternity exclusion studies (Xhang et al., 1995).

Hair samples have also been used in a study important to the conservation of European brown bears. Taberlet and Bouvet (1994) collected hair samples from wild bears to determine the conservation units within the European population, and discovered not only genetically distinct eastern and western populations but distinct lineages within the western population that correspond to two different ancestral refugia. These types of data demonstrate both the usefulness of non-invasive sampling for wild, endangered species, and the application of molecular phylogeographic methods to questions critical to conservation management.

Feathers

In 1989 we introduced the potential for genotyping birds using feathers (Woodruff, 1990). A phylogenetic study of threatened Thai hornbills demonstrates the utility of the noninvasive approach (Morin et al., 1994a) and on-going studies reviewed below show that, like hair, feathers are ideal for population genetic studies. Genotyping of birds using feathers has also been demonstrated by Ellergren (1991, 1994), Leeton et al., (1993), and Cooper (1992b), but only Smith et al., (1991) have applied the method to a conservation problem. They described a new species of African shrike based on the only known specimen, which was captured, kept for one year, and released. DNA (cytochrome *b* locus, 295 bp) from a feather was amplified and sequenced, compared to homologous sequences from the other known species, and the individual was determined to be as genetically distinct from those species as they were from each other.

A hierarchical study of the endangered San Clemente Island loggerhead shrike is underway (Mundy et al., 1996). Using feathers as the DNA source, Mundy is studying variation in cytochrome *b* and control region sequences to estimate the genetic differences between the island and the mainland subspecies and SSR polymorphism to study variation, population structure, and genetic erosion in the island population.

Buccal Cells

Only one research group has reported the use of buccal cells, which may be extracted from food wadges (Takasaki and Takenaka, 1991; Inoue and Takenaka, 1993; Takenaka et al., 1993). Researchers gave sugarcane to captive and wild chimpanzees and collected the fresh fibrous wadges the animals spat out. The wadges were stored in 50% ethanol until the cells could be filtered out of a saline suspension and collected by centrifugation. They report obtaining high-quality DNA in quantities adequate for amplification of nuclear SSR loci in chimpanzees and other primates. Cells from food wadges must be preserved in a chelating buffer or ethanol to prevent enzymatic and bacterial degradation of the DNA.

Feces and Urine

Of all of the noninvasive methods developed so far, the sampling of DNA from feces has the greatest potential for field sampling of many species. Höss et al. (1992) reported the use of field-collected European brown bear feces to study these rare and widely dispersed animals. They amplified and sequenced a 141 bp segment of the mitochondrial control region and compared the variation in three Italian bears with that in other European populations. In addition, they amplified and sequenced a 356 bp portion of the chloroplast *rbcl* gene from the feces and compared it to 414 known plant *rbc*L sequences. The chloroplast sequence matched that of plants of the genus *Photinia*, and the authors concluded that this plant is a dominant component of the bears' diet. The important implications of this study are that feces can be used to study the population genetics of rare and widely dispersed species, as well as some aspects of their feeding behavior, without ever seeing the animals themselves.

Target DNA in feces suffers from degradation, but amplifiable sequences up to 600 bp have been obtained from freshly collected chimpanzee feces stored in methanol (Goldberg et al., unpublished data). One problem encountered in the extraction of DNA from some feces has been copurification of PCR-inhibiting materials, and methods have been developed to overcome the problem (Constable et al., 1995; Gerloff et al., 1995; Goldberg et al., unpublished data). Others have not reported PCR inhibition (Höss et al., 1992; Takasaki and Takenaka, 1991) including Sugiyama et al. (1993), who also studied free-ranging chimpanzees. Early reports of amplification from DNA extracted from feces have involved mitochondrial (mt)DNA, but recent studies have reported methods for amplification of single-copy nuclear loci that are of use for paternity exclusions and other studies involving nuclear DNA (Constable et al., 1995; Gerloff et al., 1995). No results have yet been reported, but results relevant to conservation are eagerly awaited.

As yet, urine has only been used as a source of DNA in human clinical studies (Gasparini et al., 1989). Both source animal DNA and DNA from infecting microorganisms such as viruses and bacteria have been successfully amplified from urine. For both captive and wild animals, if urine or feces can be collected (as in the case of hormonal studies; e.g., Czekala et al., 1990; Czekala., 1991; Pryce et al., 1994), they provide yet another source of DNA for studies of organisms and their pathogens.

Nonlethal Biopsy

Small toe, ear and tail clips have traditionally been used to mark individual animals (e.g., mice and frogs) in mark–release studies. Toe clips have been used as an initial source of nDNA for surveys of SSR variation in natural populations of small mammals in Thailand (Srikwan et al., 1996). (Once amplification protocols are optimized, hair is a satisfactory DNA source.) The purpose is to develop a noninvasive method of monitoring genetic erosion in recently fragmented populations. Theory, and some empirical data, suggests that very small populations lose variability (alleles) rapidly following isolation (range fragmentation) as a result of inbreeding and genetic drift. Characterizing a population's variability at hypervariable SSR loci may permit an assessment of the extent of genetic erosion. Our demonstration project involves monitoring several species of small rainforest mammals (*Rattus rattus, Maxomys surifer, M. whiteheadi, Chiropodomys gliroides, and Tupaia glis*) for variation at 10 loci during the 20 generations following isolation. These mammals were recently isolated on about 150 small islands when the construction of a hydroelectric dam flooded a forested valley (Lynam et al., 1992; Woodruff, 1992, 1996). When developed, the methods will be applicable to fragmented populations of species of greater concern to conservationists in Thailand and elsewhere.

Dart biopsy has been used successfully by researchers working on large mammals from which it would otherwise be nearly impossible to obtain tissue samples. In a study of whale population genetics, Baker et al. (1993) used biopsy darts to obtain tissue from free-ranging whales. Georgiadis et al., (1994) used biopsy darts to obtain several hundred samples from African elephants.

Fish Scales and Fins

Scales and nondestructively obtained fin tissue promise to provide future investigators with an adequate source of fish DNA. Although scales themselves contain little DNA, a number of cells usually adhere to their proximal surface when they are removed. C. Stepien (Case Western Reserve University, personal communication) reports successfully amplifying and sequencing mtDNA from dried tissue attached to scales. She has also successfully used 90%-ethanol preserved fin clips from tagged and released walleye (Stepien, 1995). D. Hedgecock (Bodega Marine Laboratory, U.C. Davis, personal communication) has used approximately 1-mm^2 pieces of fin tissue to amplify six nuclear loci for assignment of unpedigreed salmon to families from known crosses in the Sacramento river. Hedgecock and colleagues have also amplified mtDNA from the scales of California sardine and are attempting a retrospective study using scales found in varved marine sediments. Clearly, protocols for both scales and fin biopsy merit full development.

Eggshells

Eggshell fragments are sometimes available in and beneath bird nests. Fragments of the vitelline membrane and chorion are commonly dried on such fragments but have yet to be exploited as DNA sources (Cooper, 1994). Blood residues on eggshells of California condors have been used for sex determination utilizing a DNA hybridization-based sexing method (Longmire et al., 1993; Ryder and Chemnick, personal communication).

Deer Antlers

Annually shed antlers would appear to provide a hitherto unexploited source of DNA for genotyping of males and, in some species, both sexes for population level studies.

POSTMORTEM METHODS OF GENOTYPING

Postmortem genotyping methods are included in this review as they are also nondestructive and based on the same protocols. Studies of living organisms and populations will often have to be supplemented with recent and subfossil material from museum collections. For some of the DNA sources discussed, the distinction between noninvasive genotyping of living animals and postmortem genotyping is inappropriate. Antlers and feathers, for example, are shed annually in some species and are also available after death.

Museum Specimens

Early enthusiasm for the potential of museum specimens as genetic sources was tempered both by the difficulty in obtaining amplifiable DNA from preserved specimens and by the concern that the specimens themselves would be damaged by the removal of numerous samples (Arnheim et al., 1990; Diamond, 1990; Benford, 1992; Graves and Braun, 1992; Tan and Orrego, 1992). Fortunately, methods have been developed which require very small amounts of tissue and circumvent some of the problems associated with chemical fixation of the specimens (Cano and Poinar 1993), for example, bones (Hagelberg and Clegg, 1991; Hagelberg et al., 1989, 1991; Lee et al., 1991) and soft tissues preserved naturally or artificially (Pääbo, 1989, 1990; Pääbo et al., 1989; Smith and Patton, 1991; Cooper, 1992a, 1994). Finally, various taxidermic treatments make extraction of DNA difficult (see chapters in Herrmann and Hummel, 1994) from such tissues as skin or hide (Thomas et al., 1989, 1990) and tissues preserved in formalin, formaldehyde, or alcohol (Grody, 1994, and references therein).

Although it was long thought that there was either insufficient DNA in bone or that it was too bound up in the matrix to be accessible for amplification, high-quality DNA can now be extracted from modern bones (Lee et al., 1991) and teeth (Ginther et al., 1992). The field of ancient DNA (aDNA) has developed quickly to the point that subfossil bones up to 47 000 years old have yielded sufficient DNA for analysis (Hagelberg et al., 1994; Höss et al., 1994 provide a review). The age of the material does not seem to affect the amount or quality of DNA (though most of the aDNA sequences found in bones and teeth are very short, e.g., < 400 bp) (Brown and Brown, 1992). Nevertheless, there has been great variability in success of extraction and/or amplification of DNA from hard tissues because of variation in the conditions of preservation and of mineral or organic compounds that copurify with the DNA and inhibit amplification. This latter problem has been overcome in a number of different ways, using methods to further purify the DNA from PCR inhibitors (Cooper, 1992a; Höss and Pääbo, 1993; Goodyear, et al., 1994; Woodward et al., 1994), or by addition of bovine serum albumin (BSA) to the PCR reaction (Hagelberg et al., 1989). Once the DNA has been extracted in a way that allows amplification, analysis of short fragments (see Hagelberg et al. 1991; Hagelberg, 1994) has often been possible for historical studies of DNA sequence variation and for individual identification. One study used 1700-year-old fossil bones to elucidate the origin of modern rabbit populations on a Mediterranean island (Hardy et al., 1994).

Modern bone has not been used extensively, as it cannot usually be obtained noninvasively from living animals. Bones might be used to study some terrestrial vertebrate populations with high mortality rates, to sex dead animals (either poached or naturally deceased), or to identify animal products such as carved bone and ivory for wildlife forensic studies.

Soft tissues may retain high-molecular-weight DNA if they are frozen or dried quickly (e.g., mummified tissues of thin extremities). Dried or mummified tissue can be an excellent source of DNA from recently preserved tissues, but has yielded only highly degraded DNA in some older specimens. An early analysis by Pääbo (1989) showed that naturally and artificially dried 4000 to 13 000-year-old tissues yielded 4 amplifiable DNA, independent of the age of the material, but that the DNA was highly damaged, probably as a result of oxidation. Pääbo was able to amplify segments of DNA up to 140 bp long. Subsequently, others have successfully amplified larger fragments of DNA from naturally and artificially dried tissues. In particular, 400 bp of the mitochondrial 12S rRNA gene have been amplified and sequenced repeatedly from bones and dried soft tissues of extant and extinct ratites dated at up to 3300 years BP (Cooper et al., 1992). Similarly, mummified human and animal tissues (Cano and Poinar, 1993; van der Kuyl et al., 1992; reviewed by Pääbo, 1993, and Cooper, 1994), have yielded up to approximately 500 bp segments of DNA.

Museum specimens have been used to supplement other sample collections or to infer taxonomic relationships useful for conservation (see Roy et al., 1994). Several studies merit discussion as they illustrate the complementary use of museum specimens to extend a study of living animals and populations. Thomas et al., (1990) studied spatial and temporal continuity of kangaroo rats using museum specimens from three populations collected 55–80 years ago and modern samples from the same populations. They sequenced a 225 bp segment of two tRNAs and a portion of the control region from 43 museum specimens and 63 modern samples. Of the 23 haplotypes identified, only one was shared between historical and modern samples, but haplotypes within populations were most related to one another across time, and haplotype frequencies were not significantly different, illustrating the temporal stability of these populations. A phylogenetic reconstruction of haplotype relationships indicated that one of the populations had been genetically isolated from the others. The remaining two populations share haplotypes and the most similar haplotypes are not necessarily from the same population. These data are consistent with other evidence that suggests that the two populations were contiguous approximately 13 000 years ago and do not rule out the possibility of more recent genetic exchange.

A second study was of genetic variation among endangered and non-endangered populations of hairy-nosed wombats, and included genetic data from an extinct population that was closely related to the endangered population (Taylor et al., 1994). Inclusion of the extinct population was important for inference of genetic depletion in the endangered population, as opposed to lower levels of genetic variation due to some other evolutionary scenario or historical event.

Finally, the evolutionary history and species status of the red wolf were studied to determine the phylogenetic origins of the species and level of inter-breeding that occurred as red wolf populations declined and coyote popula-tions grew in the south-western United States (Wayne and Jenks, 1991; Roy et al., 1994). The authors compared homologous mitochondrial DNA sequences and nuclear microsatellites from extant red wolves, coyotes, and gray wolves, as well as from museum specimens of red wolves collected early in this century, before hybridization was thought to be significant. Both data sets support the theory that red wolves have a hybrid origin resulting from the interbreeding of gray wolves and coyotes, and are contrary to the theory that red wolves are ancestral to the other two species. The authors maintain, however, that this does not invalidate the specific status of the red wolf and that it should be managed to conserve it as an endangered species, as it represents an ecologically important predator and may be adapted for specific niches different from those of the progenitor species.

DISCUSSION

The noninvasive sampling methods described run the spectrum from those that require handling animals briefly to those that require only the collection of fecal wastes. Although all are simpler than traditional methods based on blood, the choice of a method will depend on the animal and biological questions being addressed. Some of the methods are now routine (hair and feathers), some are under active development (feces), and others are yet to be developed (fish scales, bird egg shells) (Eggert et al., submitted).

All the above methods are based on PCR DNA amplification. This is a relatively simple procedure once the species or genus-specific primers are available. However, the development of appropriate primer pairs and the optimization of conditions for their use remain time-consuming and expensive. For example, in a previously unstudied species it still takes several months to find 10 variable SSR loci and synthesize their primer pairs. The resolution of the genotypic patterns on electrophoretic gels involves the use of radioisotopes, or expensive nonradiometric labeling methods (silver and fluorescent dyes). Almost all conservation genetic research still involves tedious manual analyses and the results are interpreted by eye. The real costs of such genotyping are currently between $40 and $200 per sequence (300–400 bp) or SSR determination. Although automated genotyping promises to bring these costs down by an order of magnitude, the initial capital investment and maintenance costs ($150 000) are beyond the means of most investigators. Furthermore, the genotype scoring error rates of some automatic sequencing systems may be unacceptable.

The principal advantage to wildlife managers of the noninvasive methods is that DNA acquisition can be opportunistic and inexpensive. One can circumvent many of the traditional problems associated with sample acquisition. When the laboratory work becomes technically routine and less expensive, managers will be able to address questions that they cannot even consider with blood-based sampling methods. Genetic management lags behind demographic management primarily because it has been almost impossible to gather relevant data. Although this has made the wildlife manager's tasks simpler until now, it is clear that genetic aspects of management will require increased attention in the near future (Avise, 1994; Meffe and Caffoll, 1994; Tomiuk and Loeschcke, 1994). Small, isolated populations will suffer viability-threatening genetic erosion (Frankel and Soulé, 1981; Woodruff, 1994). Individual animals must be selected for translocation and reintroduction experiments. With the full development of noninvasive genotyping methods, geneticists can and should become active participants in global wildlife conservation efforts.

ACKNOWLEDGMENTS

We thank Tom Smith and Bob Wayne for making our participation in this symposium possible. Research in Woodruff's laboratory has been supported by grants from the NSF, NIH, Dept. of Navy, and the Academic Senate of the University of California. We thank Oliver Ryder, Guy Hoelzer, Bob Wayne, and one anonymous reviewer for commenting on the manuscript.

REFERENCES

Arctander, P. 1988. Comparative studies of avian DNA by restriction fragment length polymorphism analysis: Convenient procedures based on blood samples from live birds. J. Ornithol. 129: 205–216.

Arnheim, N., T. White, and W.E. Rainey. 1990. Application of PCR: organismal and population biology. BioScience 40: 174–1182.

Avise, J.C. 1994. Molecular markers, natural history and evolution. New York: Chapman and Hall.

Baker, C.S., A. Perry, J.L. Bannister, M.T. Weinrich, R.B. Abernethy, J. Calambokidis, J. Lien, R.H. Lambersten, J. Urbán Ramìrez, O. Vasquez, P.J. Clapham, A. Alling, S.J. O'Brien, and S.R. Palumbi. 1993. Abundant mitochondrial DNA variation and world-wide population structure in humpback whales. Proc. Natl. Acad. Sci. U.S.A. 90: 8239–8243.

Benford, G. 1992. Saving the "library of life." Proc. Natl. Acad. Sci. U.S.A. 89: 11098–11101.

Brown, K. and T. Brown. 1992. Amount of human DNA in old bones. Ancient DNA Newsletter 1: 18–19.

Cano, R.J. and H.N. Poinar. 1993. Rapid isolation of DNA from fossil and museum specimens suitable for PCR. BioTechniques. 15: 432–444.

Constable, J.J., C. Packer, D.A. Collins, and A.E. Pusey. 1995. Nuclear DNA from primate dung. Nature 373: 393.

Cooper, A. 1992a. Removal of colourings, inhibition of PCR, and the carrier effect of PCR contamination from ancient DNA samples. Ancient DNA Newsletter, 1: 31–32.

Cooper, A. 1992b. Seabird 12S sequences using feathers from museum specimens. Ancient DNA Newsletter. 1: 20–21.

Cooper, A. 1994. DNA from museum specimens. In: B. Hermann and S. Hummel, eds. Ancient DNA, pp. 149–165. New York: Springer-Verlag.

Cooper, A., C. Mourer-Chauviré, G.K. Chambers, A. von Haeseler, A.C. Wilson, and S. Pääbo. 1991. Independent origins of New Zealand moas and kiwis. Proc. Natl. Acad. Sci. U.S.A. 89: 8741–8744.

Czekala, N.M. 1991. Reproductive analysis of free-ranging mountain gorilla. Zoonooz 64: 14–15.

Czekala, N.M., L.H. Kasman, J. Allen, J. Oosterhuis, and B.L. Lasley. 1990. Urinary steroid evaluations to monitor ovarian function in exotic ungulates. Zoo Biol. 9: 43–48.

Dessauer, H.C. and M. S. Hafner, eds. 1984. Collections of frozen tissues: Value, management, field and laboratory procedures, and Directory of existing collections. Association of Systematic Collections. Lawrence, Kans.: University of Kansas Press.

Dessauer, H.C., C.J. Cole, and M.S. Hafner. 1990. Collection and storage of tissues. In: D.M. Mills and C. Moritz, eds. Molecular systematics, pp. 25–41. Sunderland, Mass.: Sinauer.

Diamond, J. 1990. Old dead rats are valuable. Nature 347: 334–335.

Eggert, L.S., L.G. Chemnick, E.A. Oakenfeld, D. Irwin, and O.A. Ryder. 1996. Genetic studies for conservation: non-invasive sampling and simplified DNA preparation. Mol. Ecol. [Submitted].

Ellergren, H. 1991. DNA typing of museum birds. Nature 354: 113.

Ellergren, H. 1994. Genomic DNA from museum bird feathers. In: B. Herrmann and S. Hummel, eds. Ancient DNA, pp. 211–217. New York: Springer-Verlag.

Erlich, H.A., ed. 1989. PCR technology: Principles and applications for DNA amplification. New York: Stockton Press.

Frankel, O.H. and M.E. Soulé. 1981. Conservation and evolution. Cambridge, U.K.: Cambridge University Press.

Gagneux, P. and D.S. Woodruff. 1996. Microsatellite scoring errors associated with non-invasive geotyping based on nDNA amplified from hair. Mol. Ecol. [Submitted].

Garner, K.J. 1992. The world's largest collection of gorilla hair. Zoonooz 65: 12–13.

Garner, K.J. and O.A. Ryder. 1992a. Mitochondrial DNA D-loop gorillas. In: *Abstracts of XIVth Congress International Primatological Society*, Strasbourg, France, August 16–21, pp. 365.

Garner, K.J. and O.A. Ryder. 1992b. Some applications of PCR to studies of wildlife genetics. Symp. Zool. Soc. Lond. 64: 167–181.

Garza, J.C. and D.S. Woodruff. 1992. A phylogenetic study of the gibbons (*Hylobates*) using DNA obtained non-invasively from hair. Mol. Phylogeny and Evolut. 1: 202–210.

Garza, J.C. and D.S. Woodruff. 1994. Crested gibbon (*Hylobates* (*Nomascus*)) identification using noninvasively obtained DNA. Zoo Biol. 13: 383–387.

Gasparini, P., A. Savoia, P.F. Pignatti, B. Dalapiccola, and G. Novelli, 1989. Amplification of DNA from epithelial cells in urine. N. Engl. J. Med. 320: 809.

Georgiadis, N., L. Bischof, A.R. Templeton, J. Patton, W. Karesh and D. Western. 1994. Structure and history of African elephant populations: I. eastern and southern Africa. J. Hered. 85: 100–104.

Gerloff, U., C. Schlötterer, K. Rassmann, I. Rambold, G. Hohmann, B. Fruth, and D. Tautz. 1995. Amplification of hypervariable simple sequence repeats (microsatellites) from excremental DNA of wild living bonobos (*Pan paniscus*). Mol. Ecol. 4: 515–518.

Ginther, C., L. Issel-Tarver, and M.-C. King. 1992. Identifying individuals by sequencing mitochondrial DNA from teeth. Nature Genetics 2: 135–138.

Goldberg, T.L., A.D. Yoder, and M. Ruvolo. 1995. Improved method for isolation and PCR-amplification of DNA from feces, unpublished data.

Goodyear, P.D., S. MacLaughlin-Black, and I.J. Mason. 1994. A reliable method for the removal of co-purifying PCR inhibitors from ancient DNA. BioTechniques 16: 232–234.

Graves, G.R. and M.J. Braun. 1992. Museums: Storehouses of DNA? Science 255: 1335–1336.

Grody, W.W. 1994. Screening for pathogenic DNA sequences in clinically collected human tissues. In: B. Hermann and S. Hummel, eds. Ancient DNA, pp. 69–91. New York: Springer-Verlag.

Hagelberg, E. 1994. Mitochondrial DNA from ancient bones. In: B. Herman and S. Hummel, eds. Ancient DNA, pp. 195–204. New York: Springer-Verlag.

Hagelberg, E. and J.B. Clegg. 1991. Isolation and characterization of DNA from archaeological bone. Proc. R. Soc. Lond., B 244: 45–50.

Hagelberg, E., B. Sykes, and R. Hedges. 1989. Ancient bone DNA amplified. Nature 342: 485.

Hagelberg, E., L.S. Bell, T. Allen, A. Boyde, S.J. Jones, and J.B. Clegg. 1991. Analysis of ancient bone DNA: techniques and applications. Phil. Trans. R. Soc. Lond. B 333: 399–407.

Hagelberg, E., M.G. Thomas, C.E. Cook Jr., A.V. Sher, G.F. Baryshnikov, and A.M. Lister. 1994. DNA from ancient mammoth bones. Nature 370: 333–334.

Hardy, C., J.-D. Vigne, D. Casñe, N. Dennebouy, J.-C. Mounolou, and M. Monnerot. 1994. Origin of European rabbit (*Oryctolagus cuniculus*) in a Mediterranean island: Zooarchaeology and ancient DNA examination. J. Evol. Biol. 7: 217–226.

Herrmann, B. and S. Hummel, eds. 1994. Ancient DNA. New York: Springer-Verag.

Higuchi, R., C.H. von Beroldingen, G.F. Sensabough and H. Erlich. 1988. DNA typing from single hairs. Nature. 332: 543–546.

Höss, M. and S. Pääbo. 1993. DNA extraction from Pleistocene bones by a silica-based purification method. Nucleic Acids Res. 21: 3913–3914.

Höss M., M. Kohn, S. Pääbo, F. Knauer, and W. Schröder. 1992. Excrement analysis by PCR. Nature 359: 199.

Höss, M., O. Handt, and S. Pääbo. l994. Recreating the past by PCR. In: K.B. Mullis, F. Ferre, and R.A. Gibbs, eds. The polymerase chain reaction, pp. 257–264. Boston: Birkhauser.

Innis, M.A., D.H. Gelfand, J.J. Sninsky, and T.J. White, eds. 1990. PCR protocols. A Guide to methods and applications. San Diego: Academic Press.

Inoue, M. and O. Takenaka. 1993. Japanese macaque microsatellite PCR primers for paternity testing. Primates 34: 37–45.

Kocher, T.D. and T.J. White. 1989. Evolutionary analysis via PCR. In: H.A. Erlich, ed. PCR technology: Principles and applications for DNA amplification, pp. 137–147. New York: Stockton Press.

Kocher, T.D., W.K. Thomas, A. Meyer, S.V. Edwards, S. Pääbo, F.X. Villablanca, and A.C. Wilson. 1989. Dynamics of mitochondrial DNA evolution in animals: Amplification and sequencing with conserved primers. Proc Natl. Acad. Sci. U.S.A. 86: 6196–6200.

Kressirer, P. 1993. Eine Molekulare Phylogenie Der Gibbons (Hylobatidae) (Hinweisse auf Vererbte Gesangsmerkmale). Master's thesis, Ludwig-Maximillian-Universitat, Munich.

Lee, H.C., E.M. Pagliaro, K.M. Berka, N.L. Folk, D.T. Anderson, G. Ruano, T.P. Keith, P. Phipps, G.L. Heffin Jr, D.D. Garner, and R.E. Gaensslen. 1991. Genetic markers in human bone: I. Deoxyribonucleic acid (DNA) analysis. J. Forensic Sci. 36: 320–330.

Leeton, P., L. Christidis, and M. Westerman. 1993. Feathers from museum bird skins—a good source of DNA for phylogenetic studies. Condor 95: 465–466.

Longmire, J.L., M. Maltbie, R.W. Pavelka, L.M. Smith, S. Witte, O.A. Ryder, and R.J. Baker. 1993. Sex identification in birds using DNA fingerprint analysis. Auk 110: 378–381.

Lynam, A.J., S. Srikwan, and D.S. Woodruff. 1992. Species persistence and extinction following rainforest fragmentation at Chiew Lam, Surat Thani Province, Thailand. Science Society of Thailand, 18th Congress on Science and Technology of Thailand. Bangkok, October 27–29. Abstracts, pp. 570–571.

Meffe, G.K. and C.R. Carroll. 1994. Principles of conservation biology. Sunderland, Mass.: Sinauer.

Merenlender, A.M., D.S. Woodruff, O.A. Ryder, R. Koch, and J. Vahala. 1989. Allozyme variation and differentiation in Africa and Indian rhinoceroses. J. Hered. 80: 377–382.

Morin, P.A. and D.S. Woodruff. 1992. Paternity exclusion using multiple hypervariable microsatellite loci amplified from nuclear DNA of hair cells. In: R.D. Martin, A.F. Dixson, and E.J. Wickings, eds. Paternity in primates: Genetic tests and theories. pp. 63–81. Basel: Karger.

Morin, P.A., J. Wallis, J.J. Moore, R. Chakraborty, and D.S. Woodruff. 1993. Non-invasive sampling and DNA amplification for paternity exclusion, community structure, and phylogeography in wild chimpanzees. Primates 34: 347–356.

Morin, P.A., J. Messier, and D.S. Woodruff. 1994a. DNA extraction, amplification, and direct sequencing from hornbill feathers. J. Sci. Soc. Thailand 20: 31–41.

Morin, P.A., J.J. Moore, R. Chakraborty, L. Jin, J. Goodall, and D.S. Woodruff. 1994b. Kin selection, social structure, gene flow, and the evolution of chimpanzees. Science, 265: 1193–1201.

Morin, P.A., J.J. Moore, J. Wallis, and D.S. Woodruff. 1994c. Paternity exclusion in a community of wild chimpanzees using hypervariable simple sequence repeats. Mol. Ecol. 3: 469–478.

Mubumbila, M.V., S. Ducelaud, J.L., Toussaint, and J. Kempf. 1992. PCR amplification and identification of D-loop region in primate mtDNA. In: *Abstracts of XIVth Congress International Primatological Society*, Strasbourg, France, August 16–21, pp. 370.

Mullis, K. 1990. The unusual origin of the polymerase chain reaction. Sci. Am. 262: 56–65.

Mundy, N.I., C. Winchell, and D.S. Woodruff. 1996. Tandem repeats and heteroplasmy in the mitochondrial DNA control region of the loggerhead shrike. (*Lanius ludovicianus*). J. Heredity 87: 21–26.

Ohhara, M., Y. Kurosu, and M. Esumi. 1994. Direct PCR of whole blood and hair shafts by microwave treatment. BioTechniques 17: 726–728.

Orrego, C. 1990. Organizing a laboratory for PCR work. In: M.A. Innis, D.H. Gelfand, J.J. Sninsky, and T.J. White, eds. PCR protocols: A guide to methods and applications, pp. 447–454. San Diego: Academic Press.

Pääbo, S. 1989. Ancient DNA: extraction, characterization, molecular cloning, and enzymatic amplification. Proc. Natl. Acad. Sci. U.S.A. 86: 1939–1943.

Pääbo, S. 1990. Amplifying ancient DNA. In: M.A. Innis, D.H. Gelfand, J.J. Sninsky, and T.J. White, eds. PCR protocols. A guide to methods and applications, pp. 159–166. San Diego: Academic Press.

Pääbo, S. 1993. Ancient DNA. Sci. Am. 269: 86–92.

Pääbo, S., R.G. Higuchi, and A.C. Wilson. 1989. Ancient DNA and the polymerase chain reaction. J. Biol. Chem. 264: 9709–9712.

Perry, J.S. 1953. The reproduction of the African elephant, *Loxodonta afriana*. Phil. Trans. R. Soc. Lond. B 237: 93–149.

Pryce, C. R., F. Schwarzenberger, and M. Dobeli. 1994. Monitoring fecal samples for estrogen excretion across the ovarian cycle in Goeldi's monkey (*Callimico goeldii*). Zoo Biol. 13: 219–230.

Roy, M. S., D.J. Girman, A.C. Taylor, R.K. Wayne. 1994. The use of museum specimens to reconstruct the genetic variability and relationships of extinct populations. Experientia 50: 551–557.

Saiki, R.K., S. Scharf, F. Faloona, K.B. Mullis, B.T. Horn, H.A. Erlich, and N. Arnheim. 1985. Enzymatic amplification of B-globin genomic sequences and restriction site analysis for diagnosis of sickle cell anemia. Science 230: 1350–1354.

Sambrook, J., E.F. Fritsch, and T. Maniatis, eds. 1989. Molecular cloning: a laboratory manual. Cold Spring Harbor, N.Y.: Cold Spring Harbor Press.

Smith, M.F. and J.L. Patton. 1991. PCR on dried skin and liver extracts from the same individual gives identical products. Trends Genet. 7: 4.

Smith, E.F.G., P. Arctander, J. Fjeldsa, and O.G. Amir. 1991. A new species of shrike (*Laniidae: Laniarius*) from Somalia, verified by DNA sequence data from the only known individual. Ibis. 133: 227–235.

Srikwan, S., D. Field, and D.S. Woodruff. 1996. Noninvasive genotyping of free-ranging rodents with heterologous PCR primer pairs for hypervariable nuclear microsatellite loci. J. Sci. Soc. Thailand, [In press].

Stepien, C.A. 1995. Population genetic divergence and geographic patterns from DNA sequences: Examples from marine and freshwater fishes. In: J.L. Nielsen, ed. Evolution and the aquatic ecosystem: defining unique units in population conservation, pp. 263–287. Am. Fisheries Soc. Symposium 17, Bethesda, Maryland.

Sugiyama, Y., S. Kawamoto, O. Takenaka, K. Kumazaki, and N. Miwa. 1993. Paternity discriminated and inter-group relationships of chimpanzees at Bossou. Primates 34: 545–552.

Taberlet, P. and J. Bouvet. 1991. A single plucked feather as a source of DNA for bird genetic studies. Auk 108: 959–960.

Taberlet, P. and J. Bouvet. 1992. Bear conservation genetics. Nature 358: 197.

Taberlet, P. and J. Bouvet. 1994. Mitochondrial DNA polymorphisms, phylogeography, and conservation genetics of the brown bear *Ursus arctos* in Europe. Proc. R. Soc. Lond., Biology 255: 195–200.

Takasaki, H. and O. Takenaka. 1991. Paternity testing in chimpanzees with DNA amplification from hairs and buccal cells in wadges: A preliminary note. In: A. Ehara, T. Kimura, O. Takenaka, and M. Iwamoto, eds. Primatology Today (Proc. 13th Congr. Int. Primatol. Soc.) pp. 613–616. Amsterdam: Elsevier.

Takenaka, O., H. Takasaki, S. Kawamoto, M. Arakawa, and A. Takenaka. 1993. Polymorphic microsatellite DNA amplification customized for chimpanzee paternity testing. Primates, 34: 27–35.

Tan, A.-M. and C. Orrego. 1992. DNA stabilization and amplification from museum collections of extracts originally intended for allozyme analysis. Mol. Ecol. 1: 195–197.

Taylor, A.C., W.B. Sherwin, and R.K. Wayne. 1994. Genetic variation of microsatellite loci in a bottlenecked species: the northern hairy-nosed wombat *Lasiorhinus krefftii*. Mol. Ecol. 3: 277–290.

Thomas, R.H., W. Schaffner, A.C. Wilson, and S. Päabo. 1989. DNA phylogeny of the extinct marsupial wolf. Nature, 340: 465–467.

Thomas, W., S. Päabo, F.X. Villablanca, and A.C. Wilson. 1990. Spatial and temporal continuity of kangaroo rat populations shown by sequencing mitochondrial DNA from museum specimens. J. Mol. Evol. 31: 101–112.

Thornton, I. 1971. Darwin's islands: A natural history of the Galapagos. New York: Natural History Press.

Tomiuk, J. and V. Loeschcke. 1994. The genetic monitoring of primate populations for their conservation. In: V. Loeschcke, J. Tomiuk, and S.K. Jain, eds., Conservation genetics, pp. 401–406. Basel: Birkhauser.

van der Kuyl, A.C., J. Dekker, J. Clutton-Brock, R. Perozonius, and J. Goudsmit. 1992. Sequence analysis of mitochondrial DNA fragments from mummified Egyptian monkey tissue. Ancient DNA Newsletter. 1: 17–18.

Vigilant, L., R. Pennington, H. Harpending, T.D. Kocher, and A.C. Wilson. 1989. Mitochondrial DNA sequences in single hairs from a southern African population. Proc. Natl. Acad. Sci U.S.A. 86: 9350–9354.

von Beroldingen, C. 1989. Applications of PCR to the analysis of biological evidence. In: H.A. Erlich, ed. PCR technology, pp. 209–224. New York: Stockton Press.

von Beroldingen, C.H., R.G. Higuchi, G.F. Sensabaugh, and H.A. Erlich. 1987. Analysis of enzymatically amplified HLA-DQ α DNA from single human hairs. Am. J. Hum. Genet. 41: 725.

Walsh, P.S., D.A. Metzger, and R. Higuchi. 1991. Chelex® 100 as a medium for simple extraction of DNA for PCR-based typing from forensic material. BioTechniques. 10: 506–513.

Wayne, R.K. and S.M. Jenks. 1991. Mitochondrial DNA analysis implying extensive hybridization of the endangered red wolf *Canis rufus*. Nature 351: 565–568.

Weisburd, S. 1989. Fingerprinting DNA from a single hair. Science. 133: 262.

Wilkinson, H. E. 1961. The rediscovery of Leadbeater's possum, *Gymnobelideus leadbeateri* McCoy. Victorian Naturalist 74: 97–102.

Wilson, M.R., D. Polanskey, J. Butler, J.A. DeZinno, J. Replogle, and B. Budwole. 1995. Extraction, PCR amplification and sequencing of mitochondrial DNA from human hair shafts. BioTechniques 18: 662–669.

Woodruff, D.S. 1989. The problems of conserving genes and species. In: D. Western and M. Pearl, eds. Conservation for the twenty-first century, pp. 76–88. New York: Oxford University Press.

Woodruff, D.S. 1990. Genetics and demography in the conservation of biodiversity. J. Sci. Soc. Thailand 16: 117–132.

Woodruff, D.S. 1992. Genetics and the conservation of animals in fragmented habitats. In: In harmony with nature (Proc. Int. Conf. Tropical Biodiversity), pp. 258–272. Kuala Lumpur: Malay Nature Society.

Woodruff, D.S. 1993. Non-invasive genotyping of primates. Primates 34: 333–346.

Woodruff, D.S. 1996. Biodiversity: conservation and genetics. Proceedings of the 2d Princesss Chulabhorn Science Congress, Bangkok, Thailand, November 2–6, November 2–6, [In press].

Woodruff, D.S. and O.A. Ryder. 1986. Genetic characterization and conservation of endangered species: Arabian oryx and Père David's Deer. Isozyme Bull. 19: 33.

Woodruff, D.S. and O.A. Ryder. 1988. Genetic characterization and conservation of endangered species. Symposium on Asian Pacific Mammalogy, Huirou, P.R.C., July 26–30, p. 60.

Woodruff, D.S. and R.L. Tilson. 1994. Genetic aspects of gibbon management in Thailand. Proc. Thai Gibbon PHVA. Khao Yai, Thailand. Apple Valley, Minn. CBSG/SSC/IUCN.

Woodward, S.R., M.J. King, N.M. Chiu, M.J. Kuchar, and C.W. Griggs. 1994. Amplification of ancient nuclear DNA from teeth and soft tissues. PCR Methods Applic. 3: 244–247.

Xhang, Y.-P., O.A. Ryder, Q.-g. Zhao, Z.-y. Fan, G.-z. He, A.-j. Zhang, H.-m. Zhang, T.-m. He, and C. Yucun. 1995. Non-invasive giant panda paternity exclusion. Zoo Biol. 13: 569–574.

19

Future Applications of PCR to Conservation Biology

DEIDRE CARTER, REBECCA REYNOLDS, NICOLA FILDES, AND THOMAS J. WHITE

Since the first article on the polymerase chain reaction was published (Saiki et al., 1985), it has become possible to do experiments that could not be considered previously. Further advances in the method and its applications over the past nine years have extended PCRs utility to many new fields such as molecular archeology and conservation biology (White et al., 1989; Arnheim et al., 1990). In this article we review several recent modifications to the PCR technique that overcome some of its limitations and promise to make its use routine among conservation biologists by the end of the decade.

LONG PCR

Although PCR has been used extensively to amplify relatively short (less than 1000 base pairs (bp)) fragments of DNA, it has been difficult to obtain efficient yields of larger targets, especially those over 5 kb (kilobase). However, this limitation has recently been overcome (Barnes, 1994, Cheng et al., 1994a) by using a combination of thermostable enzymes and a specific set of buffer conditions. This has permitted DNA fragments up to 22 kb and 42 kb to be amplified from total human DNA and from phage DNA, respectively.

The implications of this technical advance for the conservation biologist are manyfold. For years, the mitochondrial genome has been a favored molecule for analysis of intraspecific variation and has been used for studies of population structure, hybridization, and introgression and for systematics (Burke et al., 1991). Most genetic experiments on mitochondrial DNA (mtDNA) have required its purification prior to restriction fragment analysis, or have used radioactive Southern blots to find variation. Mitochondrial DNA sequence analysis has mainly targeted the variable portion of the control region or the cytochrome *b* gene. If the long PCR conditions could specifically amplify the entire mitochondrial genome from preparations of total DNA, it would be possible to rapidly obtain more information on its variability. This task has now been accomplished (Cheng

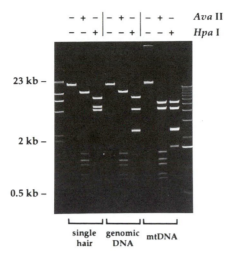

Figure 19-1 Human mitochondrial DNA was amplified from three sample types: total genomic DNA from a hair; total DNA from a cell line; and purified mtDNA from placental tissue. The first lane in each set of three is undigested amplified DNA; the second and third lanes were digested with the indicated restriction enzyme. (Reproduced with the permission of Cheng et al., 1994c.)

et al., 1994c) as shown in Figure 19-1. The complements of primers from a conserved portion of the cytochrome *b* gene (Kocher et al., 1989) were used to amplify a 16.3 kb fragment with high yield and specificity. Thus, in the future, conservation biologists may be able to use these primers on a wide variety of vertebrates and avoid the need to purify mtDNA or use radioactive Southern blots.

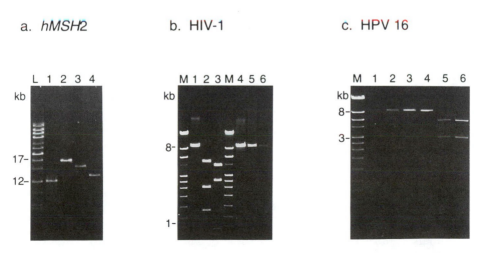

Figure 19-2 (a) Amplification of segments within the *hMSH2* gene using primers from adjacent or remote exons. (b) Amplification of 8.3 kb from a proviral HIV-1 plasmid (lane 1) and verification by restriction digestion (lanes 2 and 3). (c) Amplification of the HPV16 genome from plasmid (lanes 1–3) and cervical lavage specimens (lane 4). Verification of plasmid (lane 5) and clinical (lane 6) products. (Reproduced with the permission of Cheng et al., 1994b.)

Other demonstrations of the breadth of applications of long PCR were reported by Cheng and coworkers (1994b). These include the ability to characterize large introns in a genomic clone of the human *hMSH2* gene (Figure 19-2a), to amplify nearly the entire proviral HIV-1 genome (Figure 19-2b), and to amplifly the essentially uncultureable human papilloma virus type 16 genome directly from a clinical specimen (Figure 19-2c). The sizes of these targets ranged from 7.9 to 17 kb. Since introns are often regions of high variability, the ability to analyze them with primers based on more conserved flanking exon sequences may be useful for studies of variation. The ability to detect entire genomes of viral pathogens from unpurified samples may assist studies of host–parasite interactions and symbioses. Furthermore, it should be possible to examine the significance of linked mutations in different parts of a viral genome for coevolution, tissue tropisms, or competence for replication or transcription.

APPLICATION OF TESTS FOR HUMAN IDENTIFICATION AND FORENSIC ANALYSIS TO ENDANGERED SPECIES

One of the most important developments in the field of human identity testing is the use of DNA typing to analyze biological samples. PCR has found an increasing role in such testing because of its ability to use minute amounts of old or degraded DNA as a suitable template for amplification and subsequent typing. As a result, human identity tests have been produced that target single-copy genes with multiple alleles, e.g., HLA DQA1, or variable number of tandem repeat loci such as D1S80, or that coamplify multiple loci containing two or three alleles. Such tests have been used for forensic casework and the coamplified markers (with two or three alleles) alone provide a high discrimination power (0.9997) or paternity index (95%) with human samples (unpublished).

The validation studies for human forensic DNA typing systems include testing of samples from other primates, birds, and so on, as shown in Figure 19-3 for the D1S80 locus. The gorilla sample shows two allelic bands, though these are of slightly different mobility from the human allelic standards which differ by 16 bp. Nonetheless, if this preliminary result is confirmed following tests of additional gorillas, the D1S80 PCR system may be useful in studies of kinship in these primates.

A similar opportunity arises with a test designed to coamplify six human genetic loci. Figure 19-4 shows the amplification products of the HLA DQA1, low-density lipoprotein receptor, glycophorin A, hemoglobin G gammaglobin, D7S8, and group-specific component loci from a single PCR reaction. Several of the primate samples were amplified at some or all loci and could also be typed for some of the alleles as shown in Figure 19-5. In this figure, the products from a single amplification reaction are simultaneously hybridized to allele-specific oligonucleotide probes bound to a nylon strip in a "reverse dot blot" format (Saiki et al., 1989). If these results are confirmed by testing more individuals from the other species, and by DNA sequencing to confirm the alleles, this approach offers

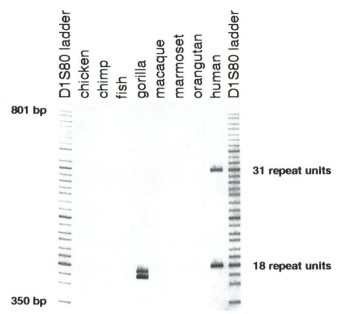

Figure 19-3   PCR amplification of the D1S80 locus from multiple species.

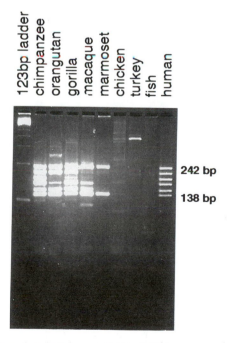

Figure 19-4   Coamplification of six loci from primate samples.

**human
chimpanzee
orangutan
gorilla
macaque
marmoset
chicken
turkey**

Figure 19-5   Simultaneous genotyping of five loci from primate samples.

potential for individual identification, paternity analyses, and tracking of individuals from noninvasive specimens.

MULTILOCUS GENOTYPING

Although the motivation for studying populations of human pathogenic microorganisms is quite different from that for studying endangered species, certain similar problems are posed by both groups of organisms. Cultivation may be difficult, impossible, or inappropriate, making the acquisition of adequate quantities of material problematic. It is also usually desirable to keep handling of the organisms to a minimum (albeit for different reasons) which can cause additional difficulties with obtaining material of high quantity and quality. And as with most population biology studies, it is often necessary to analyze large numbers of different organisms or isolates. Techniques that have been developed to overcome these problems in the study of one group of organisms may therefore be useful in the study of the other. Ideally, such techniques should be robust and easy to perform, require minimal amounts of starting material, and be relatively insensitive to the quality of the material under analysis.

We are currently interested in the development and application of techniques to study *Histoplasma capsulatum*, a class 3 human mycopathogen. In order to minimize exposure, the fungus is killed by heat prior to DNA extraction. This causes the DNA to fragment, making it unsuitable for restriction fragment length polymorphism (RFLP) or random amplified polymorphic DNA (RAPD) analysis. PCR using arbitrary primers can be used, however, to generate allelic bands from different individuals, which may then be analyzed for variation at the sequence level. Two types of sequence polymorphisms are being screened for: (1) minor, biallelic polymorphisms such as nucleotide substitutions, small insertions and small deletions, and (2) multiallelic microsatellite and repetitive DNA polymorphisms (Carter et al., 1995).

SCREENING FOR MINOR SEQUENCE POLYMORPHISMS

Initial DNA amplification is based on RAPD or arbitrarity primed PCR amplification (AP-PCR) (Welsh and McClelland, 1990; Williams et al., 1990; and see chapter 4 in this book), with the modification that a pair of arbitrary primers is used, with one primer held uniform in every reaction. By modifying this uniform primer, specific modifications can be introduced into the amplified product. We have chosen the Ml3-40 sequencing primer as the uniform primer, which is then paired with different 10-mer RAPD primers.

The following steps are involved in the screening of arbitrarily primed PCR products for sequence variation. This is illustrated in Figure 19-6.

1. Amplification of DNA from a single isolate using various RAPD primers with and without the M13-40 primer. RAPD/M13-40 primer combinations producing bands suitable for single-strand conformational polymorphism (SSCP) analysis (Orita et al., 1989) (i.e., 100–500 bp) are selected (Figure 19-6a).
2. Amplification of DNA from six "tester" strains with the primers selected in step 1. (Figure 19-6b). If the suitable band is amplified consistently from all six isolates, it is excised from the gel for reamplification.
3. Reamplification of the excised band for fluorescent SSCP, using a fluorescently labeled M13-40 primer plus the appropriate RAPD primer.
4. Fluorescent SSCP (Figure 19-6c). This uses the Applied Biosystems automated sequencing apparatus with Genescan software. As the apparatus is capable of detecting four different florophores, three PCR products that have been amplified using different labels may be run in a single lane, along with size standards that have been labeled with the fourth fluorophor. Many different samples can therefore be analyzed on a single gel, and the simultaneous electrophoresis of size standards allows mobility differences to be assessed very accurately.
5. Reamplification of DNA showing mobility differences with 5'-biotin–M13-40. Introduction of biotin into one strand of the PCR product allows it to be adsorbed onto streptavidin-coated magnetic beads. Alkaline denaturation separates the bound strand from its complement, and either strand can then be purified and used as a template in single-stranded sequencing (Bowman and Palumbi, 1993). Sequencing reveals the position and nature of the polymorphism causing the mobility shift seen in SSCP analysis. Sequencing information also allows specific primers to be designed that flank the polymorphic sequence and can directly amplify it from the DNA of *H. capsulatum*.
6. Analysis of the polymorphism in additional isolates. If an RFLP is associated with the sequence change, restriction digestion of the amplified DNA will allow the two alleles to be easily resolved. Otherwise, presence or absence of the polymorphism can be assessed by SSCP or by allele-specific hybridization (Saiki et al., 1989).

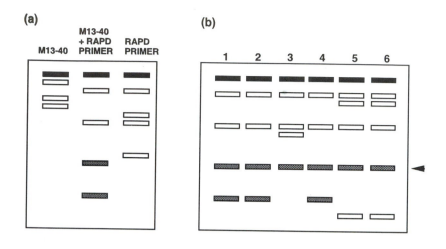

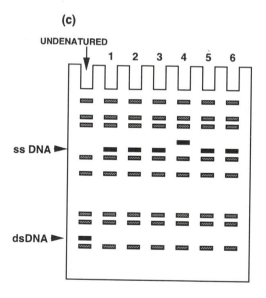

Figure 19-6 Diagram showing the analysis of arbitrarily primed PCR product for sequence variation. (a) Agarose gel of arbitrarily primed PCR product using primers independently and in combination. Bands produced by both primers that are suitable for SSCP are shadowed. (b) Amplification of DNA from additional isolates. One of the bands seen in (a) is consistently amplified and is selected for reamplification using fluorescently labeled M13-40. (c) SSCP analysis of the reamplified bands. On this diagram, size standards are shown as shaded bands, and the samples as black bands. Denatured, single-stranded DNA migrates more slowly than the undenatured, double-stranded control. Only one of the two single-stranded bands is seen as the fluorophor is introduced through the M13-40 primer alone. DNA from isolate 4 shows altered mobility and can be reamplified with 5′-biotin–M13-40 for direct sequencing.

RETRIEVAL OF MICROSATELLITE AND REPETITIVE DNA SEQUENCES

Microsatellite screening uses the same amplified DNA from step 1 above. This is electrophoresed in an agarose gel and blotted onto a nylon membrane (Figure 19-7a). The membrane is then hybridized with Quicklight microsatellite sequences (FMC), sprayed with the Lumi-Phos Dioxetane developing agent supplied by the manufacturer, and exposed to autoradiograph film (Figure 19-7b). Bands found to hybridize with the microsatellite probe are identified in the amplified DNA profile, and are excised and reamplified with 5'-biotin–M13-40. Direct sequencing

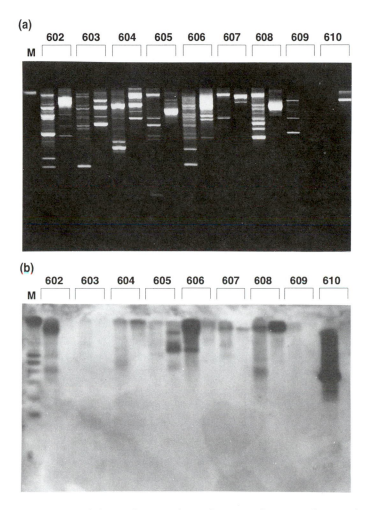

Figure 19-7 (a) Agarose gel electrophoresis of DNA from a single *H. capsulatum* isolate amplified by RAPD primers 602–610. Each pair of lanes uses the RAPD primer respectively with and without the M13-40 primer. (b) Blot of gel shown in (a) hybridized with chemiluminescently labeled (CA)$_n$. A strongly hybridizing band is seen in the band containing DNA amplified by M13-40 + 610. Direct sequencing confirmed the presence of a polymorphic (CA)$_n$ microsatellite.

Table 19-1 Molecular Genotypes for a Small Population of *H. capsulatum* Isolates Based on One $(CA)_n$ Microsatellite (HSP) and Two Repetitive DNA Sequences (610.1 and 613.2). Most isolates have a unique overall molecular genotype.

Band																	Isolate	
	H1	H2	H3	H4	H5	H6	H7	H8	H9	H10	H11	H12	H13	H14	H15	H17	H18	
HSP	1	2	1	1	3	4	5	1	6	6	1	1	1	7	3	5	1	
610.1	1	2	1	1	1	3	3	2	4	4	5	1	3	3	3	3	3	
613.2	1	2	3	3	3	4	5	5	–	–	6	7	8	7	8	6	9	
Genotype	A	B	C	C	D	E	F	G	H	H	I	J	K	L	M	N	O	

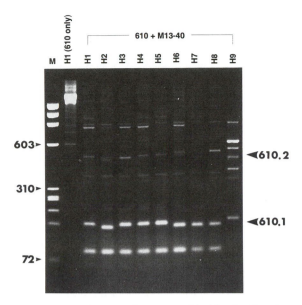

Figure 19-8 Agarose gel electrophoresis of DNA amplified by RAPD primer 610 + M13-40. In addition to the band containing the $(CA)_n$ microsatellite found above (band 610.2), a smaller band also appears to be variable in length between the different isolates (band 610.1). Sequencing revealed the presence of a polymorphic repetitive sequence in this band.

is then performed as in step 5 above to identify the hybridizing sequence and to develop specific flanking primers for direct amplification of the microsatellite. These may incorporate a fluorophor if the microsatellite is to be analyzed using the ABI automated gene sequencer. Analysis may be done by SSCP or by assessing length differences on a sequencing gel. The former is preferable when the micro-satellite incorporates base changes as well as differences in repeat unit number.

Examination of RAPD profiles also reveals occasional bands that differ slightly in mobility between different isolates (Figure 19-8). These have also been found to contain polymorphic repetitive sequences. SSCP analysis of these sequences in a number of different *H. capsulatum* isolates revealed considerable polymorphism in their base composition and/or repeat unit number. Scoring each allele for one microsatellite (HSP) and two repeat sequence polymorphisms (610.1 and 613.2) gives an overall genotype for each isolate (Table 19-1). These three multiallelic polymorphisms give unique genotype profiles for most members of this population and will be useful for the typing of closely related isolates.

ADVANCES IN PCR INSTRUMENTATION

Although it has not yet been widely appreciated, there is now no fundamental barrier to producing a small, portable thermocycler that could simultaneously perform PCR and detection. Work by Higuchi and Watson has shown that the DNA products of a PCR reaction can be detected by measuring the increase in fluorescence that occurs as a dye binds to amplified DNA (Higuchi et al., 1992, 1993). Such approaches allow us to anticipate devices that have extremely high throughput for samples and are either qualitative or quantitative in their readout. Alternatively, a single device using microinstrumentation could be handheld or even disposable. Allowing some latitude for speculation in an article with this title, we predict the availability of small PCR devices for use in field experiments by the end of the decade.

CONCLUSIONS

The methods presented in this chapter represent modifications or refinements to existing PCR techniques. These were developed to overcome limitations encountered in the application of existing techniques to certain fields of research, or to extend these techniques to allow a higher level of resolution to be obtained. They may be of benefit to the field of conservation molecular genetics when similar limitations, or the desire for similar high resolution studies, are encountered.

We predict that the long PCR method will reinvigorate the use of the RFLP technique as a means of detecting variation since the combination of long PCR and RFLP will allow sampling of larger genomic fragments of up to 20–30 kb. Yet the digests will be visible by fluorescent staining of gels and thereby eliminate the need for Southern blotting and the use of radioactive probes (see chapter 12). The long PCR method will also extend the usefulness of the ribosomal DNA repeat unit as a tool for the study of mating systems and population structure (Burke et al., 1991). In recent work (M. Lopez and R. Garcia, personal communication) the entire 13.2 kb repeat unit from *Leishmania* kinetoplasts could be amplified and the variation in the internal transcribed spacers and intergenic spacers analyzed by RFLP.

The carryover of methods and genetic markers originally developed for human forensics has already occurred in a study of chimpanzee paternity (Morin et al., 1994). The rapid method for identifying short tandem repeat (STR) markers described above should only accelerate such conservation studies since it eliminates the need for cloning total genomic libraries from new species being studied.

Techniques optimized to produce robust results using DNA that is of low quantity and/or quality should benefit studies of conservation genetics as the organism under study is often so rare that single hairs or feathers found in the field must be relied on as sample material (for example, see chapter 8). The relative insensitivity of the methods to DNA degradation also allows greater flexibility in sample storage and transport from the field to the laboratory. Last, the analysis is rapid and relatively inexpensive, so it can be performed on large numbers of different samples.

The type of polymorphism to be screened for using these techniques will depend on the application. Microsatellites, which are highly polymorphic, allow discrimination at the individual level. They can be used for estimating diversity within populations and for examining intrapopulation relationships such as paternity/maternity or examining the extent of inbreeding. Single base substitutions evolve more slowly and are therefore suited to the study of more distant relationships, such as between populations. They can be used to analyze the occurrence of hybridization between different populations or different species, or to establish conservation units within established species groups.

Finally, miniaturization and automation of amplification detection systems should permit rapid identification of endangered or illegally imported species. The recent example of "field" analysis of whale meat specimens performed with a portable thermocycler and gel apparatus in a hotel room is illustrative of the many purposes to which small handheld devices may be put by conservation biologists (Baker and Palumbi, 1994). Combined with noninvasive sampling of hair or excreta, these rapid methods may ultimately be as commonplace as binoculars in the field.

ACKNOWLEDGMENTS

We thank our colleagues S. Cheng, R. Higuchi, S.-Y. Chang, P. Gravitt, R. Respess, M. Stoneking, and S. Cosso for permission to use results prior to publication, and J. Taylor and A. Burt for collaborating on the multilocus genotyping techniques for human fungal pathogens.

REFERENCES

Arnheim, N., T. White, and W.E. Rainey. 1990. Application of PCR: Organismal and population biology. BioScience 40: 174–182.

Baker, C.S. and S.R. Palumbi. 1994. Which whales are hunted? A molecular genetic approach to monitoring whaling. Science 265: 1538–1539.

Barnes, W.M. 1994. PCR amplification of up to 35 kb DNA with high fidelity and high yield from λ bacteriophage templates. Proc. Nat. Acad. Sci. U.S.A. 91: 2210–2220.

Bowman, B.H. and S.R. Palumbi. 1993. Rapid production of single stranded sequencing template from amplified DNA using magnetic beads. Methods Enzymol. 224: 399–406.

Burke, T., W.E. Rainey, and T.J. White. 1991. Molecular variation and ecological problems. In: R.J. Berry, T.J. Crawford and G.M. Hewitt, eds. Genes in ecology, pp. 229–254. Oxford, U.K.: Blackwell Scientific Publications.

Carter, D.A., A. Burt, and J.W. Taylor. 1995. Direct analysis of specific bands from arbitrarily primed PCR reactions. In: M.A. Innis, D.H. Gelfand, and J.J. Sninsky, eds. PCR strategies, pp. 325–332. New York: Academic Press.

Cheng, S., C. Fockler, W.M. Barnes, and R. Higuchi. 1994a. Effective amplification of long targets from cloned inserts and human genomic DNA. Proc. Nat. Acad. Sci. U.S.A. 91: 5695–5699.

Cheng, S., S.-Y. Chang, P. Gravitt, and R. Respess. 1994b. Long PCR. Nature 369: 684–685.

Cheng, S., R. Higuchi, and M. Stoneking. 1994c. Complete mitochondrial genome amplification. Nature Genetics 7: 350–351.

Higuchi, R., G. Dollinger, P.S. Walsh, and R. Griffith. 1992. Simultaneous amplification and detection of specific DNA sequences. Bio-Technology 10: 413–417.

Higuchi, R., C. Fockler, G. Dollinger, and R. Watson 1993. Kinetic PCR analysis—real-time monitoring of DNA amplification reactions. Bio-Technology 11: 1026–1030.

Kocher, T.D., W.K. Thomas, A. Meyer, S.V. Edwards, S. Pääbo, F.X. Villablanca, and A.C. Wilson. 1989. Dynamics of mitochondrial evolution in animals: amplification and sequencing with conserved primers. Proc. Natl. Acad. Sci. U.S.A. 86: 6196–6200.

Morin, P.A., J.J. Moore, R. Chakraborty, L. Jin, J. Goodall, and D.S. Woodruff. 1994. Kin selection, social structure, gene flow, and the evolution of chimpanzees. Science 265: 1193–1201.

Orita, M., H. Iwahana, H. Kanazawa, K. Hayashi, and T. Sehiya, 1989. Detection of polymorphisms of human DNA by gel electrophoresis as single-stranded conformation polymorphisms. Proc. Natl. Acad. Sci. U.S.A. 86: 2766–2770.

Saiki, R.K., S. Scharf, F. Faloona, K.B. Mullis, G.T. Horn, H.A. Erlich, and N. Arnheim. 1985. Enzymatic amplification of β-globin genomic sequences and restriction site analysis for diagnosis of sickle cell anemia. Science 230: 1350–1354.

Saiki, R.K., D.S. Walsh, and H.A. Erlich, 1989. Genetic analysis of amplified DNA with immobilized sequence-specific oligonucleotide probes. Proc. Natl. Acad. Sci. U.S.A. 86: 6230–6234.

Welsh, J. and M. McClelland, 1990. Fingerprinting genomes using PCR with arbitrary primers. Nucleic Acids Res. 18: 7213–7218.

White, T.J., N. Arnheim, and H.A. Erlich. 1989. The polymerase chain reaction. Trends Genet. 5: 185–189.

Williams, J.G., A.R. Kubelik, K.J. Livak, J.A. Rafalski, and S.V. Tingey, 1990. DNA polymorphisms amplified by arbitrary primers are useful as genetic markers. Nucleic Acids Res. 18: 6531–6535.

II

ANALYSIS

20

Estimation of Effective Population Size and Migration Parameters from Genetic Data

JOSEPH E. NEIGEL

In recent years, the arrival of molecular genetic data has brought forth a renaissance in the field of population genetics. Classical population genetic models defined the allele as the unit of variation, and were formulated in terms of the frequencies of alleles within populations. Now DNA sequence data has redefined the description of genetic variation in terms of the nucleotide sequence differences that distinguish alleles. In those cases in which these differences are the result of an orderly stepwise process of change, they can be used to infer relationships of descent among sequences. New theoretical models, often referred to as "coalescent models" (Hudson, 1990) have been developed to incorporate this richer, more historical view of genetic variation. Other developments include methods that can detect highly variable portions of the genome. While these methods have gained most attention in applications to forensics and paternity analysis, they are now being applied to population genetics as well (see other chapters in this volume, and references therein).

Conservation biology is also a rapidly developing field, with a new emphasis on genetic variation in both natural and managed populations (see, for example, the volume edited by Soulé, 1987). The combination of molecular data, new theoretical models, and computerized methods of analysis holds the promise of new and powerful tools that can be applied to conservation genetics. There are essentially two ways that molecular genetic markers can be used in conservation biology. The first is motivated by the concern that genetic variation is an important determinant of population viability and adaptability. The aim is simply to use genetic markers to indicate current levels and distributions of genetic variation. The main assumption is that a given set of genetic markers is representative of the variation that is relevant to conservation. The second use is the measurement of processes, such as migration, which may be important for ecological as well as genetic reasons, but are expected to produce measurable effects on patterns of genetic variation. Genetic marker data can thus provide an "indirect" view of these processes, but its analysis and interpretation must be based upon theoretical population genetic models. A major assumption of indirect uses of genetic markers is the suitability of the models upon which these analyses are based.

The purpose of this chapter is to consider how molecular population genetic data can be analyzed to provide inferences about two parameters that are regarded as important to conservation genetics: effective population size and migration. The term effective population size may refer to any of several quantities, but all are concerned with the rate at which some measure of genetic variation changes in a finite population (Ewens, 1982). Migration may also be defined in several ways, with a principal distinction between the migration *rate*, which is the proportion of individuals that move between populations, and the migration *distance*, which is how far a typical individual moves in one generation (Slatkin, 1985). Because these parameters represent the rates of processes, they must be estimated "indirectly" from genetic marker data. This chapter is largely concerned with the theoretical models that relate these parameters to molecular population genetic data.

There are many ways to approach the question of what is the "best way" to estimate either effective population size or migration parameters. The question that must be answered first is: What use will be made of these estimates? The method of choice from an evolutionary perspective may not be the best for conservation biology. In particular, the time-scales appropriate for evolutionary studies may be inappropriate for addressing the more immediate concerns of conservation biology. Therefore, along with offering some definitions of the quantities to be estimated, I will attempt to summarize some of the major reasons why they are of importance in conservation biology. These will hopefully serve as criteria for evaluating alternative approaches to estimating these parameters.

EFFECTIVE POPULATION SIZE

Much of the early development of population genetics theory incorporated a set of simplifying assumptions that together defined an "ideal population" (see, for example, Wright 1969). In an ideal population of N diploid individuals, each generation is constituted from a sample of $2N$ gametes drawn randomly from the individuals of the preceding generation. Thus every individual in the population is considered equal in reproductive capacity; there is no overlap in the individuals from one generation to the next; mating occurs at random; and the population remains at a constant size over time. Because gametes are sampled randomly, some alleles carried by gametes will, by chance, be represented more than other alleles. These sampling accidents are expected to produce a range of related effects that accumulate over time (Figure 20-1). These include random fluctuations of allele frequencies (a process referred to as genetic drift), loss of alleles from the population (allele extinction), and a decrease in heterozygosity (considered a form of "inbreeding"). Application of probability theory to the ideal population model allows predictions to be made about the expected rates of these processes that occur as consequences of gamete sampling.

There is variety of ways in which real populations can depart from the ideal population described above. These include unequal sex ratios, overlapping generations, differential reproductive success, and changes in population size. These deviations cause departures from predictions based on the ideal population model (a good introduction to this topic is provided by Hartl and Clark, 1989).

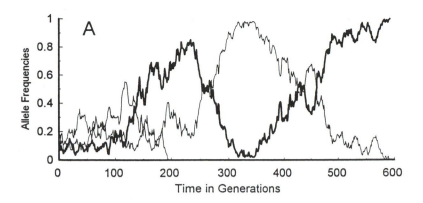

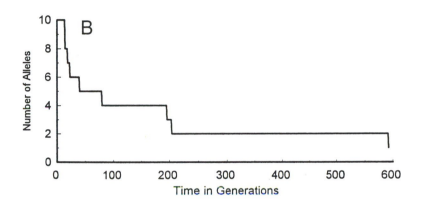

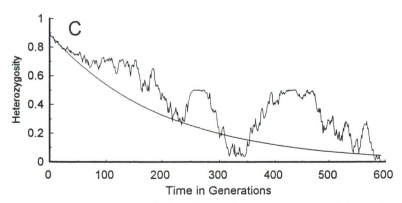

Figure 20-1 Computer simulation of the effects of genetic drift. An ideal population was initialized with 10 alleles at one locus. In panel (A) the frequencies of three alleles are monitored, with the trajectory of the allele that eventually reaches fixation in bold. In panel (B) the decline in the number of alleles over time is shown. In panel (C) the smooth curve represents the expected decline in heterozygosity, which can be contrasted with the more erratic trajectory of the actual changes in heterozygosity at a single locus.

Depending on the species, effective population sizes may be nearly the same as the "census size" (e.g., vertebrate examples in Nunney and Elam, 1994) or orders of magnitude lower (e.g., plant and invertebrate examples in Orive, 1993). For any nonideal population, its effective size, represented as N_e, is the size that an ideal population would be that behaved in the same way. More precise definitions of effective population size must be made specifically with respect to one of the process that is dependent on population size (Ewens, 1982). Thus, the inbreeding effective population size is defined with respect to the probability of homozygosity due to common ancestry; the variance effective population size is defined with respect to the variance of gene frequencies over time; and the extinction effective population size is defined with respect to the loss of alleles from the population. The distinctions among these definitions of N_e, which are in some cases numerically equivalent, are often considered trivial. However, these distinctions may assume considerable importance in conservation biology, where a specific effect of finite population size is considered detrimental. Management plans or breeding programs may affect each of these population size effects differently (see review by Simberloff, 1988).

Inbreeding Effective Population Size

Inbreeding depression is a reduction in fitness that occurs in the progeny of matings between related individuals. At least two mechanisms may account for inbreeding depression (reviewed by Charlesworth and Charlesworth, 1987). The one that is most easily demonstrated is an increase in the proportion of genetic loci that are homozygous for rare, recessive alleles with deleterious effects. These alleles are generated by mutation, and can persist as rare alleles in large outbreeding populations because they will generally occur as heterozygotes, which are not subject to selection. In small populations, there is a higher probability that matings will occur between related individuals that carry the same alleles. The increase in the occurrence of homozygotes for recessive deleterious alleles will result in inbreeding depression. Over time, selection may eliminate these rare alleles in small populations, and so inbreeding depression may be reduced in populations that have been small for many generations. In effect, the combination of in-breeding and selection can "purge" these deleterious alleles. A second potential cause of inbreeding depression is a decrease in the proportion of genetic loci that are heterozygous for pairs of alleles that confer greater fitness as hetero-zygotes than they would as either of the alternative homozygotes. The classic example of heterozygote superiority is the human hemoglobin locus, for which there is an allele that causes sickle cell anemia as a homozygote, but resistance to malaria (without severe anemia) as a heterozygote (Allison, 1955). Few other examples of heterozygote advantage are so well documented but, if it were a general phenomenon a reduction in the proportion of heterozygous loci would reduce fitness. Unlike the effect of deleterious recessive alleles, selection could not temper a population against the effects of loss of heterozygosity. Furthermore, the potential to restore heterozygosity will be reduced if alleles are lost from a population (mutation can replenish alleles, but only over relatively long time-scales). Additional mechanisms of inbreeding depression may involve multilocus interactions, such as epistasis, although these complicated mechanisms are seldom considered.

There are two occasions when it would be desirable to obtain an estimate of inbreeding effective population size. First, an estimate of inbreeding N_e from an undisturbed population could serve as a reference point. Inbreeding depression would be expected to occur if N_e suddenly became much smaller. The ideal estimate of N_e for this purpose would represent a long-term average. Thus, a measure of N_e that was relatively insensitive to short-term changes would be desirable. The second occasion would be in a threatened or managed population. An estimate of N_e could be used to evaluate the effectiveness of a management strategy in reducing inbreeding, especially if N_e estimates from undisturbed populations were available for comparison. In this case, the ideal estimate of N_e should reflect the current status of the population, and not its history prior to disturbance.

Variance Effective Population Size

Genetic drift has been considered to be an important force in evolution because it provides opportunities for evolutionary changes beyond those that would occur by natural selection acting alone, including genetic divergence between isolated populations. In small populations, random changes in gene frequency that occur from genetic drift may reduce the effects of selection. In the "shifting-balance theory of evolution" proposed by Sewall Wright (1932), evolution within populations can become highly constrained when genetic variation becomes caught in a balance of selective forces. This balance of selective forces may prevent the formation of new, and potentially advantageous, combinations of genes. Genetic drift within small populations can loosen this constraint, and thereby allow new combinations of genes to form, which may in turn precipitate sudden shifts to a new balance of selective forces. In a species divided into many separate populations, a sudden shift to a new and better adapted combination of genes in one population could be propagated to others. In some cases, this may involve extinction and recolonization of local populations, In somewhat anthropomorphic terms, each population is an "experiment" in finding new adaptations, and the results of a successful experiment can be shared by the entire species. The validity of Wright's theory has been extensively debated, and is still an open question. It is certainly reasonable to expect that the theory will apply more to some species than to others. An important implication for conservation biology is that processes that may be viewed as detrimental to an individual population may enhance the adaptive potential of the species as a whole.

Extinction Effective Size

The extinction of alleles from a population is irreversible if there are effectively no sources to replenish them. Thus, for a species that consists of essentially a single small population, loss of alleles is a special concern. The upper limit for heterozygosity is set by the number of alleles at each locus, as is the number of possible combinations of genes that can be formed. However, it is not clear at which point the loss of alleles begins to impact the viability or adaptive potential of a population. As indicated above, a general benefit from heterozygosity per se has been difficult to demonstrate. Furthermore, variation in quantitative traits may be rather resistant to the loss of alleles (Robertson, 1966). These concerns become important in the design of management or captive breeding programs, because the best strategies for maintaining a high extinction N_e will reduce the inbreeding and variance N_e values. For example, allelic diversity can be preserved by maintaining a large number of small populations. Within each population, high levels of genetic drift and inbreeding will occur, and alleles will be lost. But because different populations will not lose the same alleles, a large number of alleles can be maintained indefinitely among all the populations as a whole. Furthermore, remixing separately maintained populations will restore much of the original heterozygosity. Thus, if the goal is to eventually reestablish a species into its former range, this strategy has merit.

METHODS FOR THE ESTIMATION OF EFFECTIVE POPULATION SIZE

Temporal Method

It would seem relatively straightforward to estimate variance effective population size by determining allele frequencies from a single population in successive generations, and calculating the variance, F, directly from these frequencies. Development of this approach has been based on a theoretical population in which mating is completely random, there is no migration from other populations, and generations are discrete and nonoverlapping. Waples (1989) has provided a general synthesis of the temporal method, and tested its performance with simulations. The expected variance in allele frequency, $E(F)$ over t generations in a population with variance effective size N_e is

$$E(F) \approx 1 - \left(1 - \frac{1}{2N_e}\right)^t$$

Note that this equation relates N_e to the *expected* variance in allele frequency. Because genetic drift is random, a specific prediction cannot be made about the net change in frequency of a particular allele over a period of time (see Figure 20-1). Thus, observations for multiple time points and/or multiple independent alleles are essential for the estimation of N_e. In addition, differences in allele frequencies between samples drawn from a population will occur from random sampling error as well as any actual temporal changes in the population. Estimation of N_e from temporal samples thus requires both a method for combining data for multiple alleles and/or multiple time points, and a correction for sampling error. Several formulas have been proposed for combining data from several alleles at a single locus. Although no one method appears to work best in all situations, Nei and Tajima's (1981) estimate, $\hat{F}_c$ appears to be generally reliable

$$\hat{F}_c = \frac{1}{K} \sum_{i=1}^{K} \frac{(x_i - y_i)^2}{\frac{1}{2}(x_i + y_i) - x_i y_i}$$

where K is the number of alleles at the locus, x_i is the frequency of the ith allele in the sample collected at generation 0, and y_i is the frequency of the ith allele in the sample collected at generation t. A weighted average, $\hat{F}_c$, for multiple loci can be calculated as

$$\hat{F}_c = \frac{\sum K_j F_{cj}}{\sum K_j}$$

where the index j specifies the locus.

If the effective population size N_e is not too small and the number of generations t is not too large, the exponential increase in the expected population allele frequency variance, $E(F)$ can be approximated as a linear function of t

$$E(F) \approx \frac{t}{2N_e}$$

With the addition of a correction for sampling variance (suggested by Krimbas and Tsakas, 1971), the expected variance in the estimator, $\hat{F}_c$, can be expressed in terms of effective population size, number of generations, and sample sizes

$$E(\hat{F}_c) \approx \frac{1}{2S_0} + \frac{1}{2S_t} + \frac{t}{2N_e}$$

where S_0 is the size of the sample taken at generation 0, and S_t is the size of the sample taken at generation t. This has led to the following as an estimate of variance effective population size

$$\hat{N}_e = \frac{t-2}{2\left[\hat{F}_c - \dfrac{1}{2S_0} - \dfrac{1}{2S_t}\right]}$$

The distribution of this estimator is approximately χ^2 with n degrees of freedom equal to one less than the number of alleles used to estimate $\hat{F}_c$

$$\chi^2 = \frac{n\hat{F}}{E(\hat{F})}$$

This estimator involves several mathematical approximations that appear valid in most situations but need not be made for those situations in which they are questionable (see Waples, 1989). In general, the method is most effective for populations with relatively small values of N_e. A typical application might involve 10 independent genetic markers, with samples of 100 individuals taken five generations apart. The theoretical accuracy of the estimate can be increased by increasing the number of genetic markers used, the length of the time interval (measured in generations) between samples, or the number of individuals sampled. However, the practical limitations of this approach may depend on how well the population under study fits the simple model upon which this estimator is based. Migration, in particular, has the potential to profoundly impact allele frequencies. If migrants are continuously received from a larger population or metapopulation with relatively stable allele frequencies, temporal variation in the recipient population will be reduced, and temporal estimates of N_e will be biased upward (Nei and Tajima, 1981). On the other hand, episodic migration from sources with different allele frequencies may cause rapid, although transient, shifts in allele frequencies, which would bias temporal estimates of N_e downward. Selection on the genetic marker, or at loci linked to the markers may also introduce inaccuracies. This is likely to occur precisely under the conditions where estimates of N_e would be sought: small populations subject to inbreeding depression.

Estimates of N_e from DNA Sequences

The description of genetic variation in terms of DNA sequence differences rather than allele frequencies has led to the development of corresponding "gene genealogical" models in population genetics (reviewed by Hudson, 1990). Within a pedigree that traces the descent of individuals from their ancestors, the pedigrees of individual genes can also be traced. These genealogies are subject to the effects of finite population sizes. In particular there is an expected relationship between inbreeding effective population size and parameters that can be determined from gene genealogies.

At present, an estimate of N_e based on DNA sequence data must begin with some rather restrictive assumptions about DNA sequence evolution. The approximate mutation rate must be known, so that it is possible to convert measures based on DNA sequence differences into estimates of time. This mutation rate must also be high enough to provide sufficient variation for statistical analyses. The majority of these mutations should be simple single nucleotide substitutions, rather than length changes or rearrangements. Otherwise, it would be difficult to determine genealogical relationships among the sequences. Finally, there can be no recombination among DNA sequences. Recombination would create sequences with ambiguous genealogical relationships. In practice, these restrictions have limited most attempts at estimates of N_e from DNA sequence data to the use of animal mitochondrial DNA (mtDNA). In addition to coming closest to meeting the requirements listed above, animal mtDNA is relatively easy to isolate and analyze. However an important caveat is that, as a general rule, mtDNA appears to be inherited maternally in animals (Avise et al., 1987). Thus, only the effective number of females in a population can be estimated.

There are basically two steps to converting DNA sequence data to an estimate of effective population size. First, the DNA sequence data is used either to infer an mtDNA genealogy (also referred to as a "phylogeny") from which statistics are derived, or statistics may be estimated directly from the sequence data. The second step is to convert these statistics into an estimate of effective population size. In general, these methods do not estimate N_e directly, but rather the product $4N_e\mu$, often represented simply as θ, where μ is the mutation rate. An estimate of μ is therefore needed to convert an estimate of θ into an estimate of N_e.

A simple approach is suggested by Watterson's (1975) analysis of the number of segregating (variable) nucleotide sites expected for a sample of sequences. In a sample of n sequences that have a mutation rate μ for the entire sequence and are drawn from a population of effective size N_e, the expected number of segregating sites $E(K)$ is approximated by

$$E(K) \approx \theta \sum_{i=1}^{n-1} \frac{1}{i}$$

Rearrangement of this expression provides an estimate of N_e that is based on a simple count of the number of segregating sites in a sample of DNA sequences and an estimate of μ. This approach avoids the problem of converting sequence data to estimates of divergence times. However, it still assumes that the occurrence of mutations follows a Poisson distribution, and that the parameter μ is known.

Other approaches have been based on the expected average divergence time for a pair of randomly chosen sequences, which is $2N_e$. Nei and Tajima (1981) suggested first calculating the nucleotide diversity index, π, defined as the average number of differences between two DNA sequences *at each nucleotide site* within a sequence (Nei and Li, 1979). Assuming the probability of multiple mutations at a single site is low enough to ignore, and the mutation rate μ (in this case, per nucleotide site) is known, effective population size can be estimated as

$$\hat{N}_e = \frac{\pi}{4\mu}$$

A constant of 4 rather than 2 appears in this expression because mutations can occur in both lines leading from two sequences back to a common ancestor. Avise et al. (1998) proceeded in a slightly different way, using the average estimated divergence time between mtDNA sequences as a direct estimate of $2N_e$. In place of a single mutation rate parameter, this approach can incorporate any method to convert sequence data to estimates of divergence time, and does not necessarily involve an explicit estimate of the mutation rate.

Felsenstein (1992) criticized the above approaches because they are based on statistics that represent only a fraction of the information potentially available in sequence data. This criticism is justified because the variance of these estimates tends to be rather high. To demonstrate this point, he developed a method that could be used if the gene genealogy for a random sample of sequences were known with complete accuracy. In this case, the distribution of time intervals between successive branch events in the genealogy contains all of the information that can be used to estimate N_e. A maximum-likelihood estimation procedure based on this distribution proved to have a much lower variance than methods based either on the number of segregating sites or on pairwise sequence comparisons. Unfortunately, gene genealogies cannot be determined with complete accuracy, and so this method could not be directly applied to real data sets.

More recently, Fu (1994) developed a method for estimating N_e from DNA sequence data that uses all of the information in an estimated gene genealogy. This method is illustrated for the hypothetical gene genealogy in Figure 20-2. Each branching event is numbered starting from the root of the genealogical tree. Coalescent theory predicts that as a consequence of the extinction of individual gene lineages over time, the time intervals between the deeper branch points in a

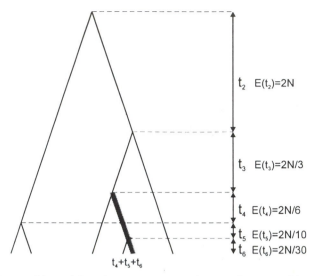

Figure 20-2 Expected branch lengths in a gene genealogy. Each expected branch length is the sum of expected time intervals between ordered branch points, and is proportional to effective population size.

genealogy will tend to be greater than those between the more shallow branches. More precisely, the expected time between the $(k-1)$th and kth branch points is

$$E(t_k) = \frac{4N_e}{k(k-1)}$$

From Figure 20-2 it can be seen that the length of each branch in the genealogy is the sum of one or more of these time intervals between branch points. Thus, expected time intervals between selected branch points can be summed to provide the expected time traversed by any given branch. For example, the expected length of the branch indicated by shading is

$$t_4 + t_5 + t_6$$

The number of mutations expected to occur in the branch is the product of its length and the mutation rate

$$E(n_i) = \mu l_i$$

By comparing the number of mutations observed on each branch with its expected length, each branch can furnish an estimate of θ. An overall estimate of θ can be based on a weighted average of estimates from each branch. The problem is to assign values to weighting factors to yield an estimate that is unbiased and has the minimum variance. This **best linear unbiased estimator (BLUE)** can be determined by solving a system of equations that include the weighting factors and the covariances of the estimates. A complication arises because, although the covariances can be determined from a coalescent model, they are functions of the parameter θ, which is itself being estimated. However, an iterative procedure can be applied, starting with an initial estimate of θ (i.e., based on some other method) to yield a new estimate of θ, which is then fed back into the procedure.

A potential weakness of this method is that the development of the BLUE estimator is based on the topology of the gene genealogy, which itself may be inaccurate. In a series of simulations, it was shown that the combination of genealogies inferred by the UPGMA procedure (one of the simplest methods of inferring a genealogy) together with the BLUE procedure should provide good estimates of θ. Estimates with this combination, UPBLUE, had low variance and a small bias that can be easily corrected (Fu, 1994). A FORTRAN program that takes a distance matrix as its input and outputs the UPBLUE estimate of θ, along with other estimates of θ is available from Fu.

Unlike an estimate of N_e that has been based on temporal variation in allele frequencies observed over a set period of time, an estimate from a DNA sequence genealogy reflects the end results of a process that may have occurred over a very long period of time. For this reason, such methods are sometimes said to estimate "evolutionary effective population size" (Ball et al., 1990). The expected time for two DNA sequences to trace back to a single point in their genealogy is $2N_e$ generations. For species with large effective population sizes, this could be many thousands of generations. For many species, changes in population size that occurred during the Pleistocene (12 000 to 1.8 million years ago) could easily be reflected in sequence-based estimates of N_e.

As with temporal-based estimates of N_e, a basic assumption of DNA sequence-based estimates is that the population being sampled has been closed with respect to migration. Whether or not this assumption holds may be difficult to determine directly, because, even if migration is not presently occurring, it may have occurred recently enough in the past to influence the present genealogical structure of the population. At present, there are no methods available for the estimation of N_e from genetic data for populations with migration. However, the problem of migration has been studied in its own right, and there are methods for the estimation of migration parameters that are analogous to those used for estimation of N_e. A rational approach to the estimation of N_e should probably begin by defining the population to be studied with respect to its history and possible sources of migration. Methods appropriate for this are discussed in the remainder of this chapter.

MIGRATION

Migration is important as both an ecological and as a population genetic process. It has the immediate ecological effect of altering population size or density. Colonization of new habitats and recolonization of formerly occupied habitats are also direct consequences of migration. A major population genetic effect of migration is to reduce genetic divergence between populations, which might otherwise occur as a result of either genetic drift or natural selection. Theoretical results indicate that surprisingly small levels of migration, a few individuals per generation, can reduce divergence due to genetic drift. Thus migration is sometimes viewed as a constraining force in evolution that prevents both adaptation to local conditions and the formation of new species (see for example, Mayr, 1963). However, migration also reduces inbreeding, and as proposed by Wright (1932), may allow the spread of favorable adaptations among populations.

The terms migration and gene flow are often used interchangeably, although the latter refers strictly to movements of genes within or between populations. Equating these terms assumes that migrants are as successful at reproduction as residents. A distinction also arises for species in which dispersal of gametes may occur independently of the movement of individuals. Thus it should be kept in mind that genetic measures of migration are actually measures of gene flow. In population genetics, the term migration is generally applied to any movement of individuals that affects the mating structure of populations. Most theoretical treatments assume that populations consist of discrete units, within which mating is random. In this case, movements of individuals between these populations constitute migration, and can be quantified by the migration rate m, the proportion of individuals that enter a population by migration. A different case is presented by individuals that are distributed continuously (rather than in discrete units) but tend to mate more often with nearby individuals. Here any movement may alter mating interactions, and so can be considered migration. A useful way of quantifying these movements in continuous populations is with the standard dispersal distance, σ_d, the standard deviation of the distances moved from the site of birth to the site of reproduction.

In the management of natural populations, it would be useful to use migration parameters to predict how demographic changes in one population would affect others. For example, through migration, one population (a source) could prevent a decline in size of another population (a sink). If this were known, it could be predicted that loss of the source population would also cause the loss of the sink population (Pulliam, 1988). However genetically based measurements of migration generally do not specify directionality. A single parameter, the migration rate, is estimated, and assumed to represent migration in both directions. Furthermore, there is no reason to expect that migration rates will remain constant as populations undergo demographic changes. Thus, it is important to consider the time-frame over which a migration estimate is relevant.

F_{ST} Based Methods of Estimating Migration Rate

Genetic drift and migration have opposite effects on the distribution of genetic variation among populations. Over time, genetic drift will result in the divergence of allele and genotype frequencies between isolated populations. Because migrants introduce alleles at frequencies that reflect their source populations, they reduce any genetic differences between populations. The basic theory that relates migration to the distribution of genetic variation was developed by Wright (1951, 1965). He defined a set of correlation coefficients ("F-statistics") in terms of the correlations between gametes. If gamete are represented as random variables that reflect their ancestry, then a correlation coefficient can be defined for the pairs of gametes that combine to form zygotes. Positive correlation coefficients result when gametes of common ancestry combine more frequently than expected. F_{ST} was defined as "the correlation between random gametes within a population, relative to gametes of the total population" (Wright, 1965). If gametes drawn from the same population are more likely to have a common ancestor than gametes drawn from different populations, F_{ST} is positive. When allele frequencies vary among populations, this implies that gametes within an individual population are correlated, and F_{ST} has a positive value. Considering just one allele, F_{ST} can be equated with the standardized variance of the allele's frequency

$$F_{ST} = \frac{V(q)}{\bar{q}(1 - \bar{q})}$$

where $V(q)$ is the variance in frequency among populations, and $\bar{q}$ is the average frequency over populations.

F_{ST} is often used to estimate migration rate because a variety of theoretical models indicate a robust relationship between F_{ST} and the product of the migration rate and effective population size: $N_e m$. For example, for a model in which an infinite number of subpopulations exchange migrants at rate m, this relationship is

$$F_{ST} = \frac{1}{4Nm + 1}$$

Figure 20-3 shows diagrammatic representations of several other models, and

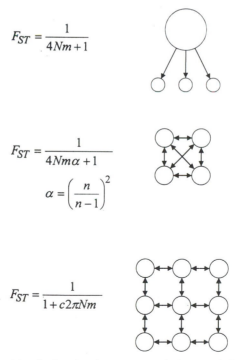

$$F_{ST} = \frac{1}{4Nm + 1}$$

$$F_{ST} = \frac{1}{4Nm\alpha + 1}$$

$$\alpha = \left(\frac{n}{n-1}\right)^2$$

$$F_{ST} = \frac{1}{1 + c2\pi Nm}$$

Figure 20-3   Several models of migration between populations, and the expected relationship between F_{ST} and Nm under each. (See Slatkin and Barton, 1989.)

the expected relationship between $N_e m$ and F_{ST} (see Slatkin and Barton, 1989). These models suggest that $N_e m$ can be determined by applying an equation of the form

$$N_e m = \frac{1 - F_{ST}}{\alpha F_{ST}}$$

where α is a constant that depends on the model. However, there are some limitations to this approach. First, the expected value of F_{ST} is inconveniently small for values of $N_e m$ much over 10. Thus, in general, only relatively low rates of migration can be estimated with much precision. A second potential problem with the interpretation of F_{ST} is the possibility that the populations have not reached an equilibrium between genetic drift and migration. Crow and Aoki (1984) showed that the number of generations, t, for F_{ST} to be near an equilibrium value is

$$t \approx \frac{1}{2m + \dfrac{1}{2N_e}}$$

If the migration rate m is relatively large, then an equilibrium will rapidly be approached. However, if both m and $1/2N_e$ are small, the approach to equilibrium will occur slowly, on the order of $1/2N_e$ generations. Thus, while F_{ST} can be rapidly lowered by a high migration rate, the elevation of F_{ST} by a reduction in migration may be much slower. As a result, observations based on F_{ST} cannot distinguish between the immediate effects of ongoing migration and the residual effects of past migration.

There has also been some confusion over how F_{ST} should be estimated from actual data, and how it should be interpreted. Part of this confusion is due to the distinction between F_{ST} as a demographic parameter which can be defined without reference to genetic variation, and F_{ST} as a statistic which is calculated from genetic data (see Weir and Cockerham, 1984). This distinction becomes important when considering multiple alleles or multiple loci. A demographic definition of F_{ST} implies that there is only a single value, although F_{ST} statistics are expected to vary among alleles and loci. Weir and Cockerham (1984) developed a method to estimate F_{ST} as a single parametric value, which they defined as θ, to distinguish it from other definitions of F_{ST}. (Note that this use of the symbol θ differs from its use earlier in this chapter, where it represented $4N_e\mu$). Nei (1973) introduced the statistic G_{ST}, which is widely used because it can be calculated for multiple alleles and multiple loci. However, G_{ST} cannot be considered an estimate of a single parameter because it its expected to vary with the number of populations observed.

It is important to consider two components of variance in allele frequencies among population samples. One is the actual variance in allele frequencies among the populations, the other is the variance due to sampling a limited number of individuals from each population. If corrections are not applied for the latter component of variance, the estimate of F_{ST} will be upwardly biased. Methods for obtaining unbiased estimates of F_{ST} and related quantities have been developed by Nei and Chesser (1983) and Weir and Cockerham (1984). The latter method also corrects for the sampling error associated with a small number of populations.

Estimating Migration Parameters from DNA Data

As with estimates of effective population size, there are two basic kinds of genetic data that can be used to estimate migration parameters: allele frequencies in population samples and sets of DNA sequences sampled from populations. Some forms of DNA sequence variation do not reflect orderly changes of character states (e.g., microsatellite length variation), and so cannot be used to infer genealogical relationships among sequences. Such variation can be analyzed in essentially the same ways that allozyme variation is analyzed. However, estimates of F_{ST} that are based upon the variance in allele frequencies among populations are sensitive to mutation rates. For example, G_{ST} is generally expressed as the ratio of two quantities, D_{ST}/H_T (Nei, 1973). D_{ST} represents the covariance of genes within subpopulations, and is called the average gene diversity between subpopulations. H_T represents the variance of genes in the total population, and is called the gene diversity in the total population. Higher mutation rates increase both D_{ST} and H_T. For low mutation rates, the relationships between these gene diversity indices and mutation rates are approximately linear, so that the proportionate effects on both D_{ST} and H_T are nearly the same, and their ratio is nearly independent of mutation rate. However, as gene diversity approaches its maximum value of one, the rate at which gene diversity increases with mutation rate declines. In effect, gene diversity becomes "saturated" as a higher proportion of mutations occur in alleles that have already been differentiated by previous mutations. Because D_{ST} is generally higher than H_T, the saturation effect is greater for D_{ST}, and so D_{ST}/H_T becomes smaller. In practice, this effect is negligible if mutation rates are at least several orders of magnitude lower than migration rates, as is generally assumed for allozyme loci. However hypervariable makers, which may have mutation rates that approach or exceed migration rates, may be expected to exhibit lower values of G_{ST}. This suggests that extremely variable markers may be of limited usefulness in detecting and measuring genetic divergence between populations.

The basic theory represented by the use of F_{ST} as a population structure parameter has now been extended to include maternally transmitted mtDNA (Takahata and Palumbi, 1985) as well as genealogical relationships among DNA sequences (Slatkin, 1991). However, this approach does not represent a very effective use of mtDNA sequence data, since the entire mitochondrial genome is analyzed as if it were a single polymorphic locus. The information represented by the genealogical relationships among mtDNA sequences is essentially discarded. For an alternative, more appropriate use of mtDNA data, Slatkin and Maddison (1989) developed a method for estimating the product of effective population and migration rate ($N_e m$) from gene genealogies. This method treats the geographic location of each individual as a character, and uses a parsimony analysis to determine the minimum number of migration events that could reconcile this location "character" with the mtDNA phylogeny. Computer simulations provide a simple function to convert the minimum number of migration events to an estimate of $N_e m$.

Neigel et al. (1991) introduced a method for estimating σ_d, the standard deviation of the distances moved from the site of birth to the site of reproduction, from mitochondrial DNA data. The premise of the method is that for many species the mutation rate of mtDNA is high relative to the rate at which new mutations are dispersed geographically. As a result, mtDNA lineages (the branches on a genealogical tree) should develop a hierarchical geographic structure that reflects their genealogical structure. This process can be modeled as a random walk, in which each mtDNA lineage originates at a unique geographic location and spreads outward at a rate dependent on σ_d. A prediction of this model is that the variance of the geographic locations of the individuals within a lineage should be roughly proportional to both the age of the lineage and σ_d. Geographical surveys of mtDNA variation, along with an estimate of the rate at which mtDNA sequences mutate, are needed for this method. A computer program, PHYFORM, is available to convert these data into estimates of σ_d. On the basis of simulation studies, Neigel and Avise (1993) found this approach to be fairly robust under various types of population density regulation and dynamic change. One advantage of this approach is that it allows separate estimates of σ_d to be made from lineages of different ages. It is thus possible to examine the possibility of historical changes in this parameter.

CONCLUSIONS

Genetic drift and migration operate over a range of time-scales, and are important determinants of both the immediate viability and the long-term adaptive potential of species. Genetic drift in small populations may result in inbreeding depression and loss of adaptive potential. However, genetic drift may also play a positive role in creating novel opportunities for evolutionary change. Migration can relieve inbreeding depression, but may also act as a constraining force in evolution by preventing local adaptation. Management decisions should be based on an awareness of both the positive and negative roles that genetic drift and migration may play.

Rates of genetic drift and migration can be estimated by the analysis of molecular genetic markers, but these are indirect estimates, based on the predictions of theoretical models. Methodological choices should be based on appropriate models and a clear understanding of the parameters they embody. Recent developments in both molecular genetics technology and population genetics theory have provided new methods for the study of genetic drift and migration in natural populations. In particular, methods based on DNA sequences and coalescence models have the potential to avoid the equilibrium assumptions of classical models, and to resolve historical patterns of genetic drift and migration.

ACKNOWLEDGMENTS

This chapter has benefited from collaborative work with J.C. Avise and R.M. Ball, and discussions with M. Slatkin. Supported by NSF/LEQSF grant (1992–1996)-ADP-02.

REFERENCES

Allison, A.C. 1955. Aspects of polymorphism in man. Cold Spring Harbor Symp. Quant. Biol. 20: 239–255.

Avise, J.C., J. Arnold, R.M. Ball, E. Bermingham, T. Lamb, J.E. Neigel, C.A. Reed, and N.C. Saunders, 1987. Intraspecific phylogeography: The mitochondrial DNA bridge between population genetics and systematics. Annu. Rev. Ecol. Syst. 18: 489–522.

Avise, J.C., R.M. Ball, and J. Arnold. 1988. Current versus historical population sizes in vertebrate species with high gene flow: A comparison based on mitochondrial DNA lineages and inbreeding theory for neutral mutations. Mol. Biol. Evol. 5: 331–344.

Ball, R.M., J.E. Neigel, and J.C. Avise. 1990. Gene genealogies within the organismal pedigrees of random mating populations. Evolution 44: 360–370.

Charlesworth, D. and B. Charlesworth. 1987. Inbreeding depression and its evolutionary consequences. Ann. Rev. Ecol. Syst. 18: 237–268.

Crow, J.F. and K. Aoki. 1984. Group selection for a polygenic behavioral trait: Estimating the degree of population subdivision. Proc. Natl. Acad. Sci. U.S.A. 81: 6073–6077.

Ewens, W.J. 1982. On the concept of effective population size. Theor. Pop. Biol. 21: 373–378.

Felsenstein, J. 1992. Estimating effective population size from samples of sequences: inefficiency of pairwise and segregation sites as compared to phylogenetic estimates. Genet. Res. 59: 139–147.

Fu, Y. 1994. A phylogenetic estimator of effective population size or mutation rate. Genetics 136: 685–692.

Hartl, D.L. and A.G. Clark. 1989. Principles of population genetics, 2d ed. Sunderland, Mass.: Sinauer Associates.

Hudson, R.R. 1990. Gene genealogies and the coalescent process. In: D. Futuyma and J. Antonovics, eds. Oxford Surveys in Evolutionary Biology, Vol. 7, pp. 1–44. New York: Oxford University Press.

Krimbas, C.B. and S. Tsakas. 1971. The genetics of Dacus oleae. V. Changes of esterase polymorphism in a natural population following insecticide control—selection or drift? Evolution 25: 454–460.

Mayr, E. 1963. Animal species and evolution. Cambridge, Mass.: Harvard University Press.

Nei, M. 1973. Analysis of gene diversity in subdivided populations. Proc. Natl. Acad. Sci. U.S.A. 70: 3321–3323.

Nei, M. and R.K. Chesser. 1983. Estimation of fixation indices and gene diversities. Ann. Hum. Genet. 47: 253–259.

Nei, M. and W.H. Li. 1979. Mathematical model for studying genetic variation in terms of restriction endonuclease. Proc. Natl. Acad. Sci. U.S.A. 76: 5269–5273.

Nei, M. and F. Tajima. 1981. Genetic drift and estimation of effective population size. Genetics 98: 625–640.

Neigel, J.E. and J.C. Avise. 1993. Application of a random walk model to geographic distributions of animal mitochondrial DNA variation. Genetics 135: 1209–1220.

Neigel, J.E., R.M. Ball, and J.C. Avise. 1991. Estimation of single generation migration distances from geographic variation in animal mitochondrial DNA. Evolution 45: 423–432.

Nunney, L. and D.R. Elam. 1994. Estimating the effective population size of conserved populations. Conserv. Biol. 8: 175–184.

Orive, M.E. 1993. Effective population size in organisms with complex life histories. Theor. Pop. Biol. 44: 316–340.

Pulliam, H.R. 1998. Sources, sinks and population regulation. Am. Nat. 132: 652.

Robertson, A. 1966. Artificial selection in plants and animals. Proc. R. Soc. Lond., Ser. B 164: 341–349.

Simberloff, D. 1988. The contribution of population and community biology to conservation science. Ann. Rev. Ecol. Syst. 19: 473–511.

Slatkin, W.M. 1985. Gene flow in natural populations. Ann. Rev. Ecol. Syst. 16: 393–430.

Slatkin, M. 1991. Inbreeding coefficients and coalescence times. Genet. Res. Camb. 58: 167–175.

Slatkin, M. and N.H. Barton. 1989. A comparison of three methods for estimating average levels of gene flow. Evolution 43: 1349–1368.

Slatkin, M. and W.P. Maddison. 1989. A cladistic measure of gene flow inferred from the phylogenies of alleles. Genetics 123: 603–613.

Soulé, M.E., ed. 1987. Viable populations for conservation. Cambridge, U.K.: Cambridge University Press.

Takahata, N. and S.R. Palumbi. 1985. Extranuclear differentiation and gene flow in the finite island model. Genetics 109: 441–457.

Waples, R.S. 1989. A generalized approach for estimating effective population size from temporal changes in allele frequency. Genetics 121: 379–391.

Watterson, G. 1975. On the number of segregating sites in genetical models without recombination. Theor. Pop. Biol. 7: 256–276.

Weir, B.S. and C.C. Cockerham. 1984. Estimating F-statistics for the analysis of population structure. Evolution 38: 1358–1370.

Wright, S. 1932. The roles of mutation, inbreeding, crossbreeding and selection in evolution. Proc. 6th Int. Cong. Genet. 1: 356–366.

Wright, S. 1951. The genetical structure of populations. Ann. Eugen. 15: 323–354.

Wright, S. 1965. The interpretation of population structure by F-statistics with special regard to systems of mating. Evolution 19: 395–420.

Wright, S. 1969. Evolution and the genetics of populations, vol. 2. The theory of gene frequencies. Chicago: University of Chicago Press.

21

Simulation Models of Bottleneck Events in Natural Populations

JOHN HALLEY AND A. RUS HOELZEL

In this chapter we discuss how simulation models can be used to study the development of natural populations after bottleneck events. We will show how the consequent genetic effects can be used to infer the characteristics of a bottleneck, and compare models based on mitochondrial DNA (mtDNA) with those based on nuclear DNA. We will illustrate these with respect to three examples: the Northern elephant seal following the bottleneck of 1884, the Ngorongoro crater lions following the *Stomoxys* epidemic in 1962, and the Asian lions in the Gir forest.

For the vertebrate species we consider in our examples, the size of a bottleneck (N_b) and its imposed duration (ΔT) have different effects on the diversity of mitochondrial DNA (mtDNA) and nuclear DNA. This is because animal mtDNA is passed as a haplotype ($1N$), and only through matrilines, while nuclear DNA is diploid ($2N$) and is passed on from both parents. A single breeding pair will pass on only one mtDNA genotype, but up to four nuclear genotypes. Different regions in either genome may be evolving at differing rates, and therefore the choice of genetic marker can affect the resolution of the analysis. Typically, mtDNA evolves more rapidly than nuclear DNA.

The quality of the information provided by retrospective studies may not be very good, but sometimes it is the only information available to us. In other cases, when the bottleneck is relatively recent, genetic information may be combined with demographic information to give a better estimate. Different types of DNA may also provide partially independent estimates of the same quantity.

Although simulation experiments have neither the explanatory power of analytic models nor the realism of experiments, they provide useful tools of investigation where theoretical analysis may be difficult and experiments impossible.

This chapter has two main sections: one about mitochondrial DNA and one about nuclear DNA (from allozymes). In each section we will begin with a qualitative discussion about how each type of DNA is affected by a bottleneck, drawing on analytic results from Appendix B. After this, we discuss the simulation model. Then we discuss the results of the model when it is applied to some real examples. We include a final section on the sensitivity of our model to uncertainty in the parameters.

MITOCHONDRIAL DNA

The Effect of a Population Bottleneck on Mitochondrial DNA Diversity

Mitochondrial DNA is haploid and inherited primarily along matrilines in vertebrate species. If a population is subjected to an intense bottleneck, much of the natural variation among haplotypes will be lost: first, because of the simple loss of individuals with rare haplotypes and, second, because of continued loss of individuals after the bottleneck. Thus the effective length of the bottleneck (T_b) is the sum of the "imposed bottleneck time" ΔT, during which the population is prevented from recovering naturally, and the natural recovery time, which takes about $1/r$ years, where r is the growth rate, defined in the list of symbols (see Appendix A).

$$T_b = \Delta T + \frac{1}{r} \tag{1}$$

During this time the genetic variation decays approximately exponentially. The rate of loss of mitochondrial haplotypes (matrilines) falls off as population growth gathers pace. This effect is shown in Figure 21-1 where a model healthy population is subjected to a 1-year bottleneck, and then allowed to recover. This scenario assumes female-only transmission and neutrality of matrilines. The time-scale is short enough for mutation not to be significant, and the model does not include immigration.

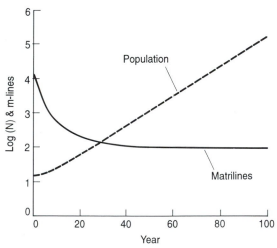

Figure 21-1 Loss of genetic variation continues even while the population recovers from a bottleneck. Here the genetic material is mtDNA and the bottleneck of 15 individuals lasts for a single year. The data are for elephant seals (Hoelzel et al., 1993) and the two curves are averaged from the output of 100 runs. Note that the intercept on the y-axis is the number of matrilines immediately after the bottleneck (i.e., when $N_t = 15$), not before it.

In the special case of a 1-year bottleneck, the final number of matrilines in a population which has completely recovered (in numbers) from a bottleneck is given (see Appendix B) by

$$m_\infty \approx \frac{vN_b}{2}\left[1 - \left(\frac{\mu}{b}\right)^{1/v}\right] \tag{2}$$

Thus, the final mitochondrial genetic diversity is dependent on v, the initial mitochondrial diversity; on N_b, the number of reproductive adults that survive the bottleneck; on μ, the yearly death rate per reproductive adult; and on b, the yearly birth rate per reproductive adult.

The characteristics that make mtDNA especially susceptible to reduced variability following a population bottleneck make it a very useful marker for these simulation studies. This is because the level of haplotype diversity will be directly related to the number of matrilines that remain in the populations. The models track the growth of the population and can measure the demographic fluctuation of matriline number directly.

A Simulation Model of mtDNA

In an earlier paper (Hoelzel et al., 1993), we described how to use mtDNA, in conjunction with population data, to back-predict population parameters for the bottleneck by using a Monte-Carlo simulation model.

This model was realized as a computer program. This model included both the demographic and genetic processes, using previously measured biological data (see next section). The seed of the random number generator was changed each run to ensure an independent trial.

The population dynamics in the model was fully age-structured with overlapping generations. A complete population matrix was generated for the females. The columns corresponded to matrilines while the rows correspond to cohorts. Only a simple vector of cohort size was needed to describe the male population. The model assumed that reproduction occurs seasonally.

Each run of the simulation required the setting up of an initial population with a stable age-structure. This could either be done by solving the von Foerster equation for the age structures (Wood, 1994), or by starting several years before the bottleneck, so that the population has settled down before the bottleneck is applied. We used the former method.

Initially the matrilines were filled at random according to the coefficient of polymorphism v for the population, using a multinomial random variable with all outcomes equally likely, so that the initial population had the required polymorphism. The bottleneck was simulated by choosing N_b animals at random from the population, according to a multinomial trial. The multinomial random variable was individual-based up to a certain number of trials, above which the law of large numbers was assumed to apply. A bottleneck could be simulated at any time during the run. If the population happened to be lower than N_b at that time, the bottleneck routine was ignored. Thus, the bottleneck was assumed to be the same as a population ceiling of size N_b.

The growth in the population of each matriline was found by calculating the numbers of recruits into that matriline, based on female reproductive success. This process was repeated every year, for the duration of the run. To apply density dependence, these were pooled and then selected according to the gene frequencies of the population matrix. Provided the species was polygynous, the existence of at least a single male is sufficient to prevent reproductive failure. The only limit on the number of recruits (apart from the reproductive capacity of the population) was the difference between the carrying capacity and the current population size, unless the population was actually above carrying capacity, in which case complete reproductive failure was assumed to occur.

The simulation model required the following inputs.

Number of years to run. This is the number of years that the model runs for in each trial.

Number of age classes. This is the number of years for which the organism lives.

Initial mitochondrial variation. If there are 20 different matrilines per 100 females, this parameter is 0.2.

Male–female birth ratio. Sex ratio at birth.

Mortality by age. This must be specified for both males and females. This is the average probability that an individual of a given age and sex died in the course of the season.

Reproductive success by age. This is also specified for both sexes. For females of a given age it means the average number of children she has. For a male of a given age, it is the probability that a given mating event is attributed to a male of that age.

Carrying capacity. The simulations assume that the limiting resource is constant. Our simulations all assume a ceiling model, where the population can grow up to a ceiling level but not above it.

Number of runs. The random number generator is given a different seed for each run.

Bottleneck information. The size of the bottlenecks, their duration, and the times at which they occur can be specified.

An Example: The Northern Elephant Seal

Around the turn of this century, the Northern elephant seal (*Mirounga angustrostris*) was hunted to the edge of extinction, not only for blubber but also by collectors. The last major hunt occurred in 1884, when 153 were killed. Thereafter, the population was mostly left alone and managed to recover so that by 1922 the population was estimated at 350. In 1960 the population was censused at about 15 000. The current population is well in excess of 100 000. In our earlier paper (Hoelzel et al., 1993), we used the above method to estimate the limits of the size and duration of the population bottleneck of the late 1880s. We also measured mtDNA diversity and found 2 haplotypes among 40 seals, compared to 23 among 27 Southern elephant seals (*M. leonina*).

Since the population was in the recovery phase, we could use another piece

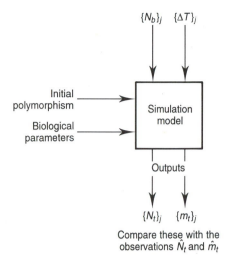

Figure 21-2 Strategy for using a simulation model to back-calculate features of a historical bottleneck. The biological parameters and the initial polymorphism are estimated and put into the model. The simulation model is then run for a range of different bottleneck sizes N_b and imposed bottleneck duration Δt. Each pair of values is repeated 500 times, giving a range of different values for the output parameters N_t and m_t. We then compare these with the measured values of N_t and m_t, looking for the best match.

of information, namely our estimate of the population size, which in the recovery phase will be given by equation (B2) on average.

The general strategy is shown in Figure 21-2, the basis being the Monte Carlo simulation model of the population. The aim was to estimate the likely values of the bottleneck size (N_b) and imposed bottleneck duration (ΔT), both of which were inputs to the simulation model. To solve this inverse problem, we performed a large number of replicates of the simulation for each of a range of possible values of the pair of bottleneck parameters $\{N_b, \Delta T\}$. Corresponding to each such pair of inputs there were 500 pairs of outputs. The pair of bottleneck parameters $\{N_b, \Delta T\}$ leading to the set of outputs $\{\bar{m}_t, \bar{N}_t\}$ which is most compatible with the observed data $\{m_t, N_t\}$, is what we are looking for.

Several different definitions of compatibility are available. The method used in Hoelzel et al. (1993) is the simplest. For each pair of values $\{N_b, \Delta T\}$ we counted the number of *hits* out of a total of 500 runs. A *hit* was defined as follows: $m_{76} = 2$ and $14\,000 \leq N_{76} \leq 16\,000$. $\hat{N}_t$ was the censused figure of $15\,000$ in 1960, 76 years after the last major hunt, subject to expected measurement error of ± 1000 animals. The relative proportions of successful trials gave the relative likelihood of combinations of $\{N_b, \Delta T\}$. The result was that 95% of the runs compatible with the observed data fell in the range where $N_b \leq 30$ and $\Delta T \leq 20$ years. Specifically, for $\Delta T = 1$ the most hits (8/500) occurred when $N_b = 10$; for $\Delta T = 5$ the most hits (8/500) were at $N_b = 15$; for 10 years it was $N_b = 20$ with 15 hits; for 15 years it was 25 seals with 11 hits (Hoelzel et al., 1993).

The choice of an a priori probability distribution had to be based on a sensible assumption for the size and duration of the bottleneck in the first place. In our

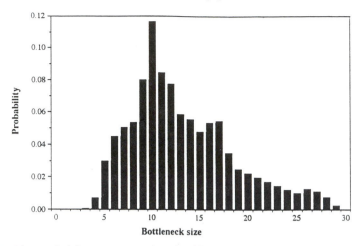

Figure 21-3 The probabilities associated with different bottleneck sizes assuming that the Northern elephant seal suffered a single one-year bottleneck.

calculations we assumed that the bottleneck was equally likely for all bottleneck sizes less than 100. Furthermore, the length of the bottleneck with $N_b \leq 100$ could not have been more than 25 years, since the population had reached 350 by 1922.

Historical evidence, however, suggests that there was only a 1-year bottleneck. Figure 21-3 shows the results when we assume a 1-year bottleneck on the basis of historical records (Hoelzel et al., 1993). We assumed that $1 \leq N_b \leq 30$ and a uniform a priori distribution for N_b in this range. We also used a slightly more sophisticated measure of the relative goodness of fit, having noted the large number of rejected runs in our calculations in Hoelzel et al. (1993). For each bottleneck size N_b, we ran 500 simulations and scored the number of hits on $m_{76} = 2$. Then the corresponding distribution of populations $\{N_{76}\}$ (given $m_{76} = 2$) was fitted to a log-normal density function with an area scaled by the number of hits. The value of this density at $N_{76} = 15\,000$ the provided an a postiori probability for a bottleneck size N_b. Figure 21-3 shows the distribution of these probabilities. The mode of this distribution is at $N_b = 10$, which agrees with our earlier result. Note, however, that the mean is nearer $N_b = 12$.

NUCLEAR DNA: ALLOZYMES

Genes at a nuclear locus in a diploid species have two alleles, and these are usually passed on to the next generation according to the Mendelian rules of independent assortment. Thus, parents with genotypes AB and ab will produce offspring with genotypes Aa, Ab, Ba, or Bb with equal probability. Factors such as natural selection and gene conversion can cause observed genotype frequencies to differ from these theoretical expectations, but we assume Mendelian segregation for the purpose of our model.

The Loss of Heterozygosity Following a Bottleneck

The effect of a population bottleneck on the average heterozygosity of nuclear DNA is more complicated than that on mitochondrial haplotype diversity. It has been discussed by Nei (1975) and McCommas and Bryant (1989). The exact way in which a population bottleneck affects heterozygosity may be affected by factors such as natural selection and epistatic complexes. However, in the absence of mutation or selection, for a constant population size with random mating and equal sex ratios, average heterozygosity is believed to approximately obey the following equation (see Nei et al., 1975; Maynard-Smith, 1989)

$$H_{\tau+1} \approx \left(1 - \frac{1}{2N_e}\right)H_\tau \tag{3}$$

Here H_τ is average heterozygosity in generation τ. When the numbers of reproductivity active males and females in the population are unequal, the effective population size, N_e is given by

$$N_e = \frac{4MF}{M + F} \tag{4}$$

The result assumes discrete generations. Many of the species being studied have overlapping generations, and each individual may produce offspring in different years (iteroparity). This results in a fragmentation of the population such that only some reproductively active individuals are reproducing in a given year. In simulations, the effect of overlapping generations and iteroparity can be approximated by calculating the effective number of breeding individuals per year. An analytical approach to the problem, using equivalent generation times, is dicussed by Charlesworth (1980).

When a population is suffering from an imposed bottleneck (or is kept constant in a limited environment), it will lose heterozygosity approximately at a rate of $1/2N_e$ per generation. Over several generations, the heterozygosity thus falls off exponentially (Appendix B), so that in the limit, if the bottleneck is imposed indefinitely, the population loses all variation due to drift and becomes homozygous. The time-scale on which this occurs is $2N_{eb}$ generations, where N_{eb} is the effective population in the bottleneck.

For a population recovering from a bottleneck, however, loss occurs primarily in the early stages of recovery, when the population is still small (Appendix B). Here, the loss of variation is interrupted by the recovery of the population's numbers, and the average heterozygosity stabilizes at a value of

$$H_\infty \approx H_0 \exp\left(-\frac{1}{2rT_g N_{eb}}\right) \tag{5}$$

H_∞ is the eventual equilibrium value of heterozygosity and T_g is the generation time. Thus, the eventual heterozygosity may be used to infer something about the bottleneck. Note, however, that, in contrast to equation (2) for mtDNA, the final heterozygosity here is directly proportional to the initial heterozygosity H_0.

For isolated populations, heterozygosity eventually increases again owing to the accumulation of mutations. On human time-scales, this is a very slow process. However, if the population is in contact with another population of greater diversity, heterozygosity will recover much faster by the immigration of new stock. The corresponding heterozygosity for an island population with immigration is (see Appendix B)

$$H_\infty \approx \frac{\xi H_s}{\xi + (1 - \xi)/(2N_e)} \tag{6}$$

ξ is the average proportion of new immigrants in the island population each generation, and H_s is the heterozygosity of the mainland population. This process is dominated by the time constants of $1/\xi$ and $2N_e$. Note that the value of H_∞ in equation (6) is no longer dependent on the size of the bottleneck. This is in contrast to equation (5), which describes a truly isolated population. Immigration masks the effect of population bottlenecks, and limits the usefulness of our back-prediction techniques. The time constant $1/\xi$ is the time it takes for the population to lose its *genetic memory* of the bottleneck. Thus, if we wish to use our model to make back-predictions, we must do so on time-scales less than $1/\xi$.

Simulation Model

The structure and operation of the simulation model are similar to those of the mitochondrial DNA model. The model considers heterozygosity and poly-morphism at a single locus. The maximum number of alleles is limited. In this model, mutation or immigration events transform genes between allelic states. The model also calculates N_e for each model year, allowing us to use equation (3) to give an estimate of average heterozygosity. However, this estimate takes into account only N_e, and is dependent on an initial estimate of pre-bottleneck heterozygosity. A better approximation for average heterozygosity can be found by running the model for each locus.

The heterozygosity at a specified locus is given by the formula (Nei, 1986)

$$h = 1 - \sum_{i=1}^{N_g} x_i^2 \tag{7}$$

Here N_g is the number of alleles in the population at this locus, and the x_i are the allele frequencies. The average heterozygosity is the average of h over all loci. Note that using this formula in the simulation allows a direct calculation of the heterozygosity from gene diversity each year.

The inputs are the same as for the mtDNA model except that now, instead of initial matriline polymorphism, we require

- Initial average heterozygosity.
- Number of alleles at a specified locus.
- Initial allele frequencies at the locus.
- Immigration (or mutation) rate.

The output now includes the heterozygosity at the specified locus, average heterozygosity, and a list of gene frequencies as well as population number.

This model is only partially individual-based. The genetic material is described by means of two population matrices, one for male genetic material and one for female genetic material. The pairing inside individuals is ignored. For example, the death of individuals in each age class is accomplished by removing a weighted number of pairs of individual units at random from each cohort.

The birth process is modeled in a similar way, with recruits either male or female according to a specified birth-ratio. Each new individual receives two new alleles from the male and female "pools" according to the probabilities of those alleles occurring in the reproductively successful age classes.

The advantage of this kind of model is to allow the investigation of more complicated reproductive strategies over a wide range of population sizes.

Examples: Isolated Lion Populations in Africa and Asia

About 3000 African lions (*Panthera leo leo*) live in Serengeti National Park, Tanzania. This is quite a large population, and has always been so in recorded history. There is a much smaller population of about 100 lions nearby, in the Ngorongoro Crater. The two populations are connected by occasional dispersal. Following an epizootic of biting flies (*Stomoxys calcitrans*), the crater population fell to about 10 lions in 1962. The population numbers had recovered by the mid 1970s but the average heterozygosity remains lower than that of the Serengeti lions. Is this attributable to the crash of 1962? These populations, together with the Asian lions (*P.l.persica*), were studied by Wildt et al. (1987).

The Asian lions are located in Gir forest Nature reserve in India (O'Brien et al., 1987). In the early 1900s their population was driven down to about 20 (Caldwell, 1938). Since then, under protection from the state, their numbers have increased to about 250. However their heterozygosity remains extremely low. A sample of 28 animals was found to be completely monomorphic at 46 different loci. Unlike the Ngorongoro lions, this population is effectively isolated.

The Lions of Ngorongoro Crater

Tables 21-1 and 21-2 shows the expected levels of heterozygosity for the crater lions on the basis of a series of 500 runs of the model. We assumed that the initial 10 lions in the relict population in 1962 were taken from a stock with an average heterozygosity of 3.3%, that of the Serengeti population. Following Packer et al. (1990) we assumed that loci assort independently. Since our model is only a single-locus model, we ran it separately (500 times) for each of the seven polymorphic loci found in the Serengeti population.

We adjusted the observed birth rates upward (multiplying by 2.3) to give the population growth rate observed (Packer et al., 1990) during the recovery phase. This is because Packer et al. (1988) measured their demographic parameters at a time when density dependence was limiting population growth. We could alternatively have adjusted down the death rate by the same factor.

The predicted average heterozygosity from Table 21-1 is 2.3%; this is calculated by using equation (7) for each locus and taking the average of this for all 47 loci. This is in reasonable agreement with the measured value of 2.2%. However, caution must accompany this prediction because of the large error

Table 21-1 Observed and Predicted Final Heterozygosity of the Crater Lions, for Each of the Loci Polymorphic in Serengeti Park. We assume that the initial population is genetically identical to the Serengeti park lions. The heterozygosity was predicted on the basis of 500 runs of the model for each locus. The carrying capacity was set at 100. The demographic parameters were calculated from Packer et al. (1988)

Locus	Heterozygosity in Serengeti	Observed Current H_t	Predicted Current H_t	Variance of Prediction
ADA	0.34	0.26	0.22	0.2
IDH1	0.39	0.11	0.29	0.18
GPT	0.26	0.17	0.20	0.2
TF	0.50	0.49	0.35	0.15
GOT, MPI, MDH2	0.02	0	0	0.055
40 others	0	0	0	0
Average	3.3%	2.2%	2.3%	1.9%

margin and for the following reasons. First, we have ignored immigration after the bottleneck and, second, in assuming the initial population was genetically identical to Serengeti park we have ignored any degree of isolation of the crater lions prior to the bottleneck. Note that according to equation (3), the heterozygosity of an isolated population declines exponentially. In the long term, it is likely that diversity has been maintained in the crater by immigration. Thus, in deciding a reasonable value of initial heterozygosity, we must consider immigration into the crater. Although no successful immigration has been recorded in the crater since 1975, there were several males who immigrated before then (Packer et al., 1990). The good agreement between the predicted and observed final heterozygosity may be due to the cancellation of two opposing errors: ignoring immigration versus starting with an artificially high heterozygosity.

Table 21-2 Observed and Predicted Allele Frequencies in Crater Population for Each of the Loci Polymorphic in Serengeti Park

Locus	Serengeti Allele Frequencies (Assumed starting values for crater population)			Most Common Allele Frequency		
	a	b	c	Observed	Predicted	Prediction variance
ADA	0.79	0.19	0.02	0.85	0.84	0.17
IDH1	0.74	0.26	–	0.94	0.78	0.16
GPT	0.85	0.13	0.03	0.91	0.78	0.16
TF	0.49	0.51	–	0.91	0.85	0.16
GOT, MPI, MDH2	0.99	0.01	–	0.56	0.73	0.14
40 others	1	–	–	1	1	0

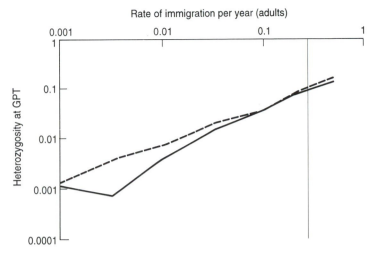

Figure 21-4   The heterozygosity expected in a population of lions with a ceiling population of $N = 100$ for various levels of immigration, with and without a bottleneck ($N_b = 10$) 25 years before the end of the run. The graphs are the average results of 400 runs (except for very small immigration values for which it is 1600 runs). The vertical line is the observed number of immigrants into Ngorongoro Crater interpreted as a rate (7/25 per year). Solid line corresponds to $N_b = 10$ and broken line corresponds to the bottleneck-free trials.

We tested the importance of immigration by simulating the population dynamics and genetics over 1000 years, for various rates of immigration into the crater. Immigrants were assumed to come from Serengeti and were counted individually. Our simulation model is less detailed than the one used by Packer et al. (1990), which made use of the known genealogies of the crater population. In that model, the demographics were known and only the genetics, in particular the initial heterozygosity, were subject to estimation. Packer et al., concluded, on the basis of their simulations, that the initial heterozygosity was lower than that of the Serengeti population.

Figure 21-4 shows the expected heterozygosity at the GPT locus. The vertical line is the overall rate of immigration (of lions) measured since the bottleneck (7 lions per 25 years). The two curves represent the heterozygosity with and without a bottleneck 25 years before the end of the run. Thus, at low immigration rates, many replicates were necessary (up to 1600), and even then the variance in the output was very large. This explains the lack of smoothness in the graphs in regions of low immigration. Note that, in this figure, the significance of a single bottleneck is not great. The long-term heterozygosity is dominated by the average immigration rate. Thus, the reduced average heterozygosity of the crater lions may reflect their long-term isolation rather than the effects of the 1962 bottleneck.

Current observations (Packer et al., 1990) suggest that native coalitions of males effectively prevent immigration of other lions into the crater. Such a *closed shop* effect would serve to increase the genetic isolation of the crater population, which would increase the rate of loss of heterozygosity. It would break down at times when the crater population was seriously depleted. Thus bottlenecks such as occurred in 1962 might even serve to *sustain* heterozygosity in the long term, by permitting more individuals to immigrate.

The Asian Gir Forest Lions

In contrast to the Crater lions, the Asian population showed a remarkable lack of heterozygosity (O'Brien et al., 1987; Wildt et al., 1987). We repeated the calculations performed in the previous section for the Indian lions. We assumed a single-year bottleneck of 20 lions followed by a recovery to a carrying capacity of 250, and ran the simulation for 70 years. Although the date of the bottleneck is not known, it makes little difference since the population (with the growth rate we assume) will reach carrying capacity quickly and thereafter suffer little change. Otherwise we assumed the same demographic data as for the African lions, and the same initial heterozygosity. This time the predicted heterozygosity stabilized at 2.6%. This is much too high to account for the homozygosity of the 28 Gir lions tested by O'Brien et al. (1987). The period to which the Gir lions were subject to the bottleneck must have been much longer or the bottleneck size much smaller.

Figure 21-5 shows the various bottleneck sizes which could account for the Gir lion homozygosity on the basis of simulations carried out by us. Because of constraints of computing time, we used only the GPT locus and assumed that all initially polymorphic loci started with a heterozygosity of 26% (an average value for polymorphic loci, see Table 21-1). Thus if the average heterozygosity was initially 3.3%, then 6 out of the 47 were polymorphic with 26% heterozygosity. To measure zero heterozygosity over the same 47 loci (ignoring the finite sample size of animals) after 70 years (with 90% confidence) would require that the number of polymorphic loci had fallen to 1 in 446 or less. Figure 21-5 shows the trials that were compatible with this requirement.

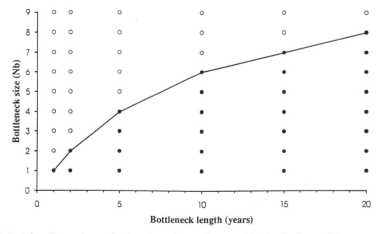

Figure 21-5   The dimensions of a bottleneck consistent with the findings of heterozygosity for the Gir Forest lion population. Dark circles are averages of runs (out of 3500 with at least 400 to survive) where the final heterozygosity is compatible with the observation of zero heterozygosity for the Gir lions.

The solid circles in Figure 21-5 correspond to simulation trials where bottleneck sizes and durations gave results compatible with the observation that the population of 250 was completely homozygous over a sample of 47 loci. Conversely, open circles represent results incompatible with such an observation. The border between the two zones is approximately logarithmic in shape as a function of bottleneck length. So, for a single-year bottleneck, any number greater than 1 individual is likely to yield a greater heterozygosity; this number is 2 for a 2-year bottleneck, 6 for a 10-year bottleneck, and 8 for a 20-year bottleneck. The required length of bottlenecks explaining the current Gir lion homozygosity seem unrealistically high.

One factor which leads to a higher expectation of homozygosity is a lower growth rate in the recovery phase. Growth rate is very high in Ngorongoro crater, perhaps owing to the abundance of prey. Such suitable conditions are unlikely to have prevailed in Gir forest. Another likely factor is the possibility of all reproductive success being confined to a very few males at the time of the bottleneck. This would reduce N_{eb} in equation (5), leading to lower eventual heterozygosity.

SENSITIVITY TO PARAMETER UNCERTAINTY

In a scheme like this, it is important that our answer be reasonably robust to variations in the parameters.

In the elephant seal calculations, one of the most important unknown quantities was the initial mitochondrial polymorphism. This had to be estimated using the Southern elephant seal as a model. Tests of this conducted by us (Hoelzel et al., 1993) showed that the final answer was relatively insensitive to initial mitochondrial variation. This is illustrated by the analytic model of equation (2). Over regions of high initial polymorphism v, the output is remarkably insensitive to v. For example, after a bottleneck size of $N_b = 10$, equation (2) predicts the final numbers of matrilines to be 1.65, 2.05, 2.32, and 2.50, given intitial polymorphisms of 0.4, 0.6, 0.8, and 1.0, respectively. Thus, a poor knowledge of initial polymorphism is not likely to be a serious problem in back-predictions using mitochondrial DNA. However, this is not so for allozyme back-predictions, since the final heterozygosity is proportional to the initial heterozygosity. Here much better estimates of the initial genetics will be necessary.

There are time-scales for the applicability of back-predictive models. Density dependence will limit the power of demographic information in any back-prediction scheme. As the contribution of density dependence becomes more important, the memory of the bottleneck signal in the demographics will be lost. The time-scale for the loss of genetic memory will be of the order of $1/\xi$, where ξ is the probability that one of two genes, taken at random in the island population, is an immigrant. Therefore, the number of generations over which the island population loses its genetic memory is approximately N_e/i_g, where i_g is the total effective number of immigrants per generation.

In general, it is difficult to distinguish between short bottlenecks with small

N_b, and long bottlenecks with large N_b. In Figure 21-5, it is possible to see how the current parameters of the Gir lion population might be satisfied by a range of different kinds of bottlenecks (dark circles). When we have more than one piece of information (as was the case with the elephant seals), then it is possible to narrow this range further.

Variable polygyny is another problem. Normally observed mating patterns may not apply around the time of a population bottleneck. If extreme polygyny prevailed, for example, N_{eb} would be lower and, by equation (5), the final heterozygosity would be correspondingly reduced. We ran the model, repeating the numerical experiment for the Gir Forest lions for a range of different degrees of polygyny. We assumed that $N_b = 8$. When $N_b/N_{eb} \approx 12.3$, heterozygosity at the GPT locus fell from 26% to 10%. At the other extreme, for $N_b = N_{eb}$ the heterozygosity fell to 15%. This suggests that the effects of changes in family size and polygyny, while significant, are not enormous. Nevertheless, more work needs to be done in this area.

CONCLUSIONS

With reference to specific organisms we have shown how simulation models, assisted by analytic models, can be used to study the population genetics of organisms recovering from bottlenecks. In the case of the Northern elephant seal we used mtDNA to infer the size of the bottleneck imposed on the population in the late nineteenth century. Nuclear DNA was used to study the bottlenecks suffered by the Ngorongoro crater and Asian lion populations.

The heterozygosity observed for the crater lions is commensurate with what we would expect from a bottleneck, assuming the survivors to be genetically the same as the Serengeti population. However, if immigration is included in the model, the results are profoundly altered. It would then seem more likely that the reduced heterozygosity of the Ngoronogoro population is the result of the long-term isolation of those lions from the main Serengeti population, rather than as a response to the bottleneck of 1962.

The genetic impoverishment of the Asian lion population is greater than we would normally expect from a single bottleneck of 20 animals followed by recovery to a carrying capacity of 250 animals. Assuming the same growth rates for Asian lions as those from Ngorongoro, we can assess the types of bottlenecks that are responsible for their current homozygosity. This prediction is likely to overestimate the severity of the bottleneck since the growth rates in the Gir forest are unlikely to have been as great as those in Ngorongoro. Lower recovery rates for these lions, combined with extreme polygyny in the early stages of recovery, might explain some of this spectacular loss.

Even under relatively ideal conditions, such as allowed the back-prediction of the population for the elephant-seal bottleneck, the uncertainty is very great owing to the inherent stochasticity of the system and the small numbers of items of data used in the inference. However, recognizing this inherent limitation, we believe that simulation models provide an extremely powerful tool for analyzing the past history of populations as well as for predicting their future development.

REFERENCES

Caldwell, K. 1938. The Gir lions. J. S. Preservation of the Fauna of the Empire 34: 62–68.

Charlesworth, B. 1980. Evolution in age-structured populations. Cambridge, U.K.: Cambridge University Press.

Hoelzel, A.R., J. Halley, C. Campagna, T. Arnbom, B. Le Boeuf, S.J. O'Brien, K. Ralls, and G.A. Dover. 1993. Elephant seal genetic variation and the use of simulation models to investigate historical population bottlenecks. J. Hered. 84: 443–449.

Maynard-Smith, J. 1989. Evolutionary genetics. Oxford, U.K.: Oxford University Press.

McCommas, S.A. and E.H. Bryant. 1990. Loss of electrophoretic variation in serially bottle-necked populations. Heredity 64: 315–321.

Nei, M., 1986. Molecular evolutionary genetics. New York: Columbia University Press.

Nei, M.T. Maruyama, and R. Chakraborty. 1975. The bottleneck effect and genetic variability in populations. Evolution 29: 1–9.

O'Brien, S.J., P. Joslin, L.L. Smith III, R. Wolfe, N. Schaffer, E. Heath, J. Ott-Joslin, P.P. Rawal, K.K. Bhattacharjee, and J.S. Martenson. 1987. Evidence for African origins of founders of the Asiatic lion species survival plan. Zoo Biol. 9: 99–116.

Packer, C., L. Herbst, A.E. Pusey, J.D. Bygott, J.P. Hanby, S.J. Cairns, and M. Borgerhoff Mulder. 1988. Reproductive success of lions. In: T.H. Clutton-Broch, ed. Reproductive Success, pp. 363–383. Chicago: University of Chicago Press.

Packer, C. A.E. Pusey, H. Rowley, D.A. Gilbert, J.S. Martenson, and S.J. O'Brien. 1990. Case study of a population bottleneck: Lions of the Ngorongoro Crater. Conserv. Biol. 5: 219–230.

Richter-Dyn, N. and N.S. Goel. 1972. On the extinction of a colonising species. Theor. Pop. Biol. 3: 406–433.

Wildt, D.E., M. Bush, K.L. Goodrowe, C. Packer, A.E. Pusey, J.L. Brown, P. Joslin, and S.J. O'Brien. 1987. Reproductive and genetic consequences of founding isolated lion populations. Nature 329: 328–331.

Wood, S.N. 1994. Obtaining birth and mortality patterns from structured population trajectories. Ecol. Monogr. 64: 23–44.

APPENDIX A. A LIST OF SYMBOLS

Symbol	Meaning
b	The yearly birthrate per individual.
h	The heterozygosity at a specific locus.
H_τ	The average heterozygosity in generation τ.
m_0	Number of matrilines at the start of the bottleneck.
m_∞	Number of matrilines left after the population has recovered.
m_t	The number of matrilines in the population at year t.
$\hat{m}_t$	The observed number of matrilines in the population at year t.
n_m	The average initial number of individuals in a single matriline.
N_b	The number of individuals in the bottleneck.
N_{eb}	The effective number of individuals in the bottleneck.
N_g	The number of alleles on a specific locus.
N_t	The population size at time t.

$\hat{N}_t$ The observed population size at time t.

Q_τ The average probability that two genes taken randomly, one from an "island" population and the other from the "mainland," are different.

r Yearly growth rate as per-generation growth rate.

t Time (in years).

T Time (in years) between the start of the bottleneck and the time of observations.

T_b The effective length of the bottleneck in years: $T_b = \Delta T + 1/r$.

T_g Generation time (years).

ΔT The length for which the bottleneck is imposed; that is, the time for which the population is externally prevented from recovering.

x_i The frequency of allele i at a specific locus.

μ The yearly death rate per individual.

v Mitochondrial polymorphism. If $v = 0.8$, for example, we expect 80 haplotypes per 100 females.

ζ The probability that two genes taken at random from an island population contain an immigrant gene.

τ Time (in generations).

APPENDIX B. AN ANALYTIC MODEL FOR THE LOSS OF GENETIC VARIATION AFTER A BOTTLENECK

Mitochondrial DNA

Since matrilines are assumed neutral, each may change in relative value within the bottleneck period, more or less as a random walk on $(0, N_b]$, until it goes extinct. Let v be the polymorphism of matrilines in the population. Assuming equal proportions of each matriline to be present in the population, the number of such matrilines is given by

$$m_0 = v\,\frac{N_b}{2} \tag{B1}$$

Assuming equal proportions of each matriline, each of these contains approximately $n_m \approx 1/v$ females.

When the population is in the recovery phase, it grows (on average) exponentially, according to

$$N_t = N_b e^{rt} \tag{B2}$$

The population associated with each matriline will grow independently according to the same equation. The probability that a population, with a positive average growth rate, is lost by demographic stochasticity by a time t, having started with n_m individuals, is (Richter-Dyn and Goel, 1972, equation 3.5)

$$E_{n_m}^{(t)} = \left(\frac{e^{rt} - 1}{(b/\mu)e^{rt} - 1}\right)^{n_m} \tag{B3}$$

Here b and μ are respectively the birth and death rates per individual per year. Because of the exponential factor, at times $t \gg 1/r$ the probability of loss of one matriline will stabilize at $(b/\mu)^{n_m}$. Thus, the average number of matrilines surviving after the population has effectively recovered is m_∞, which is given by

$$m_\infty \approx m_0\left[1 - \left(\frac{\mu}{b}\right)^{n_m}\right] = \frac{vN_b}{2}\left[1 - \left(\frac{\mu}{b}\right)^{1/v}\right] \tag{B4}$$

The average number of matrilines becomes "frozen" after the population begins to grow rapidly towards restoration. This number is proportional to N_b, and depends also on μ, b, and v. Note that $b - \mu = r$; then

$$m_\infty \approx m_0\left[1 - \left(\frac{1-r}{b}\right)^{n_m}\right] \tag{B5}$$

In many situations, the dependence on v is weak for intermediate values. This is because for high v, the initial number of matrilines m_0 is high. However, n_m will be very small on average and hence a given matriline is vulnerable to extinction. When v is smaller we expect fewer, more secure, matrilines.

The number of matrilines lost, when the bottleneck lasts for more than 1 year, is more problematic. A crude estimate of the total number of matrilines lost in a multi-year bottleneck can be found, provided equation (B5) holds, by replacing r by $1/T_b$ in (B5), using equation (1).

Nuclear (Allozyme) DNA

If the loss of heterozygosity obeys the equation

$$H_{\tau+1} = \left(1 - \frac{1}{2N_e}\right)H_\tau \tag{B6}$$

when a population is kept constant it will lose heterozygosity at a rate of $1/2N_e$ per year. Thus, if we approximate $(H_{\tau+1} - H_\tau) \to dH/d\tau$ and integrate the resulting differential equation, we have

$$H_\tau = H_0\,\exp\left(-\frac{\tau}{2N_{eb}}\right) \tag{B7}$$

If the bottleneck is imposed indefinitely, the population eventually loses all variation due to drift, becoming homozygous. The time-scale on which this occurs is $2N_{eb}$ generations.

For a population recovering from a bottleneck, according to equation (B2), equation (B6) can be integrated approximately, using the same approximation for $dH/d\tau$

$$H_\tau = H_0\,\exp\left(-\frac{(1 - e^{-rt})/(rT_g)}{2N_{eb}}\right) \tag{B8}$$

Here, the loss of variation is interrupted by the recovery of the population's numbers, on a time-scale of $1/r$. The heterozygosity stabilizes at a value of

$$H_\infty \approx H_0 \exp\left(-\frac{1}{2rT_g N_{eb}}\right) \tag{B9}$$

If the population is in contact with another population, H_τ will increase owing to immigration. If the source of immigrants is a large "mainland" population, which can be regarded as independent of the size and homozygosity of the recovering population, then the system obeys the pair of equations

$$H_{\tau+1} = (1 - \xi)\left(1 - \frac{1}{2N_{eb}}\right)H_\tau + \xi Q_\tau \tag{B10}$$

$$Q_{\tau+1} = (1 - \xi)Q_\tau + \xi H_S \tag{B11}$$

Here Q_τ is the probability that two genes taken at random, one from the "island" population and one from the "mainland" population, are different; ξ is the probability that two random genes in the island population contain an immigrant gene; H_s is the heterozygosity of the "mainland" population. The derivation follows Maynard-Smith (1989, p. 158) but assumes a constant "mainland" heterozygosity.

The expected equilibrium average heterozygosity for such an island population is found from these equations (letting $H_{\tau+1} = H_\tau$ and $Q_{\tau+1} = Q_\tau$) to be

$$H_\infty = \frac{\xi H_s}{\xi + (1 - \xi)/(2N_e)} \tag{B12}$$

22

Assessing Relatedness and Evolutionary Divergence: Why the Genetic Evidence Alone Is Insufficient

RICHARD A. NICHOLS

Modern molecular genetics provides such large quantities of precise data that there is a tendency to overlook its limitations. Certainly genetic studies can provide much evidence that is pertinent to conservation biology, including indications of past fluctuations in population size, the extent of gene flow between populations, and relatedness within populations. Part of my purpose here is to show that this genetic evidence can only be interpreted in conjunction with an understanding of a species' biology, and hence to exhort conservation geneticists to make full use of their knowledge rather than to rely on the genetic information alone.

These limitations on inference from genetic data are seen particularly clearly when we consider the effects of evolutionary history on the ancestry of genes in a population. The ancestry of a group of genes can be visualized as a tree: a gene genealogy in which the open branches represent a sample of alleles and the nodes their common ancestors (see Figure 22-1).

Molecular genetic data is especially amenable to this type of analysis, and both the applied and theoretical fields are flourishing. Recent advances are reviewed by Hudson (1990) and Ewens (1990). In the case of segments of DNA that are unrecombined, it may actually be possible to attempt to reconstruct a genealogy (Avise, 1994). Recombination introduces complications, but can be encompassed (Kaplan and Hudson, 1985).

Tracing the ancestry of genes tends to highlight the importance of the *time* to a common ancestor. Differences between alleles have accumulated since their common ancestor; hence those differences can be used to reconstruct their history. I will argue, however, that the variation in timings for a particular evolutionary history is disappointingly large. In addition, the same genetic pattern can arise from different causes. Consequently, we must draw on other sources of information about a species' past, including biogeographical, historical, and geological evidence to date the origin of genetic diversity. Particularly clear examples include the timing of large-scale colonizations that followed the last ice age (Hewitt, 1989). This knowledge of timing can be used to support some interpretations of the genetic data, and to rule out others.

In the following section the basic principles will be developed in the setting of a single randomly mating population. Next, modifications will be introduced to more closely encompass the complexities of natural populations. Finally, the implications for interpreting genetic data will be illustrated by considering its use in calculations to determine parentage.

WITHIN-POPULATION GENETIC DIVERSITY

Figure 22-1 shows two copies of a gene genealogy characteristic of a random breeding population of constant size. Other individuals alive at the time of the common ancestors can be omitted because they left no offspring in the sample; the picture is therefore much simplified. Such methods are as old as the Neo-Darwinian synthesis. Wright's work on inbreeding and co-ancestry (see Wright, 1969) and Malecot's (1948) analysis of isolation by distance drew heavily on the same principles. The shape is typical of a neutral marker locus in a population with an effective size of N alleles, such as an autosomal locus in an idealized population of $N/2$ diploid individuals. A pair of alleles drawn from the population have probability $1/N$ of sharing a common ancestor in the previous generation. This is simply a consequence of there being N possible parent alleles in the previous generation. If two lineages converge at a common ancestor they are said to coalesce.

The genealogy tends to branch much more profusely near the terminal twigs, whereas the basal boughs are longer. This pattern is a consequence of the way in

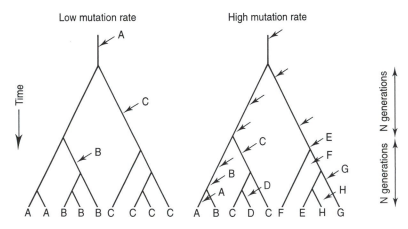

Figure 22-1   An example of a gene genealogy in a single population. The open branches of the tree represent genes in a present-day population. The nodes represent common ancestors. The scale shows the expected number of generations to a common ancestor for the whole population: twice the population size (N). For a period averaging N generations there were only two common ancestors. Arrows on a branch represent a mutation. Note that when the mutation rate is high, there is a mutation along the route connecting any one allele with most of the others. This indicates that there are many different alleles in the population. Another consequence of high mutation rate is that most branches have mutations in the recent past. This may tend to obliterate information about the previous history of the population.

which opportunities for coalescence change with the number of ancestral lineages. At the time when there were only two ancestral lineages, there was only one pair that could coalesce. In the recent past, when there were N_a lineages, there were $\binom{N_a}{2}$ opportunities for coalescence, hence the rate was much higher [$\binom{N_a}{2}$ represents the number of ways of choosing two lineages from N_a].

We have seen that the rate of coalescence for a pair of lineages is $1/N$ per generation, hence the expected time from two ancestors to one is N generations. The rate of coalescence in recent generations is so much faster that the expected time back to two common ancestors for the whole population is only $\approx N$ generations as well.

This typical geometry shows that the population will have spent a large proportion of its history as two or three lineages. Slatkin and Hudson (1991) demonstrate that one implication is that mitochondrial DNA from a single population can typically be divided into two or three distinct categories. The naive view might be that these categories had a history of reproductive isolation, whereas in fact they differ because of the mutations that accumulated in each lineage during this early history.

The expected number of substitutions distinguishing two mitochondrial DNA sequences is a function of the time since their common ancestor. Other loci provide less direct information about time. At minisatellite and microsatellite loci, mutation seems to involve both large and small changes in length (Jeffreys et al., 1994; DiRienzo et al., 1994), so the difference in allele length is not directly interpretable. Slatkin (1995) has found, however, that it is possible to make efficient use of the allele length data if the variance in mutational changes is constant. For many other loci it may only possible to tell that there has been either at least one mutation or none since the common ancestor. That being the case, at loci with high mutation rates recent mutations obscure the earlier history of the population, whereas lower rates provide only low resolution (Figure 22-1). At first sight this would seem a convincing argument for expending the time and effort required to obtain DNA sequences. However, we will see later that the apparent, informativeness of sequence data can be illusory.

The relationship between heterozygosity, mutation rate, and population size, derived by Kimura and Crow (1964), is a description of the relative rates of mutation and coalescence in Figure 22-1. As we trace back the ancestry of two lineages, mutation occurs in one or the other at a rate of 2μ per generation. The rate of coalescence is $1/N$; hence the probability that the first event is a mutation is given by

$$\frac{2\mu}{(1/N) + 2\mu} = \frac{2N\mu}{1 + 2N\mu} \tag{1}$$

(ignoring multiple events in the same generation). If all mutations generate unique alleles, then this is the heterozygosity. This equilibrium is reached after $\approx 2N$ generations when the full genealogy has formed.

We can use these rules of thumb to consider the recovery and loss of genetic vatiation due to fluctuations in population size. Menotti-Raymond and O'Brien (1993) showed that while cheetah are almost monomorphic at allozyme loci, there is considerable variation for minisatellite VNTRs (variable number tandem repeat loci). The VNTR loci have mutation rates around three orders of magnitude higher (10^{-3} rather than 10^{-6}). It may be that the VNTR loci have accumulated genetic variation much faster than the allozyme loci since the putative population bottleneck 10 000 years ago. The cheetah population size, however, is estimated at only around 20 000. If they were to continue at their present numbers, the allozyme loci would only recover a small amount of variation because of the low mutation rate. If the effective population size were 1/10 of the 20 000, then the equilibrium heterozygosity of 0.4% would be reached after around 8000 generations, or 48 000 years. This may not be detectably different from the present-day situation.

SUBDIVIDED POPULATIONS

The preceding derivations applied to a single panmictic population. Figure 22-2 represents a slightly more realistic representation of some natural populations. The population is divided into subpopulations with reduced gene flow between them. Within each subpopulation the rate of coalescence is of the order of $1/N_s$, (where N_s is the subpopulation size). In the central subpopulation of Figure 22-2, two branches of the genealogy are derived from immigrants. The coalescence time of these branches is much greater. To find the common ancestor, we have to look back to the time when the lineages were previously in the same subpopulation. This ancestry of migrants is like a genealogy similar to that seen in a single population, but with much longer branches. As we trace back lineages from separate subpopulations, the rate of coalescence depends on the probability of one or other migrating, the probability that the migrant came from the same subpopulation as the other, and the probability of (and time to) coalescence given that the two lineages are in the same subpopulation.

Figure 22-2 A gene genealogy in a subdivided population. The three basal circles represent different subpopulations in the present-day population. Lineages that can be traced back within the subpopulation are represented by those that stay within the area above the circles. Migrant lineages cross into the realm of other subpopulations.

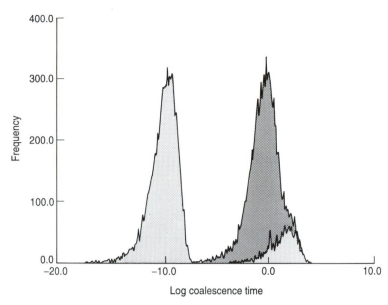

Figure 22-3 A comparison of coalescence times in a single panmictic and a subdivided population. The distribution of coalescence times for pairs of lineages drawn from a simulated population of 1 000 000 individuals. Black distribution: a single panmictic population. Grey distribution: population subdivided into 10 000 demes with migration equiprobable to other islands (an island model) at a rate of 0.1 per island, per generation. The time-scale is ln(number of generations/1 000 000).

Rather startlingly, the expected time to a common ancestor of two alleles drawn from a subpopulation can be unaffected by the extent of population subdivision (Strobeck, 1987). In simple models, the reduced coalescence time within sub-populations is exactly balanced by the increased time for migrants. It is the *distribution* of coalescence times that changes (Nichols and Beaumont 1996; Figure 22-3). The distribution for this subdivided population is bimodal. The left-hand peak represents coalescences within a subpopulation, the right-hand peak coalescences of migrants. Note that time-scale is logarithmic, so that the difference between the shorter and longer times for the migrants is three orders of magnitude. When a subpopulation has migrant ancestors, the time interval from two to one common ancestors will be drawn from this distribution. As we have already seen, this interval makes up a large proportion of the total for the genealogy.

This huge variability means that the genealogy of a single locus can only give very imprecise information about the evolutionary history of the population. This observation is pertinent to the interpretation of sequence data; Karl (chapter 3) has shown that inferences about migration drawn from single loci could be deceptive. When a locus shows very little variation, it may imply a history as a discrete population, but Figure 22-3 shows that some loci will have recent common ancestry even in a large subdivided population. Hence, collecting more individuals to type the same locus would only reveal more descendants of the same common ancestor. Instead, scoring extra loci proves far more productive. To understand why extra loci are more informative, consider tracing back the lineages at two unlinked loci from the same individual. We could have chosen the maternal allele from one locus and the paternal allele from the other. Hence, with a 50% probability, the lineages will diverge. The same applies to each of the preceding generations, so that after we have traced back a handful of generations, the genealogies of distinct loci are essentially independent.

This principle is important because it can show that apparently reasonable sampling strategies are ill-advised. Standard sampling theory indicates that allele frequency estimates will be improved if larger samples are taken from a sub-population: the variance of the estimate decreases inversely with the sample size. It might therefore seem rational to sample 50 or more individuals from each subpopulation. In most cases, however, biologists are not interested in the allele frequencies themselves, but in what they reveal about the species' biology. Information from a single locus, however accurately measured, will be subject to the inherent variation illustrated in Figure 22-3. Indeed, for the estimation of F statistics, Beaumont and Nichols (1996) found that 15 individuals conveyed most of the information available from a locus in a single subpopulation. Thereafter, it can be more efficient to direct effort into typing extra loci from the same sample rather than extra individuals.

Because of the much faster coalescence times of lineages within subpopulations, mutations are less likely to have occurred. Two alleles drawn from the same subpopulation are therefore correlated: they are more likely to match than two drawn from the population at large. As a consequence, there is variation in allele frequency between subpopulations, which may also be viewed as reduced hetero-zygosity within populations. Weir and Cockerham (1984) use a hierarchical analysis of variance to calculate a θ which is essentially Wright's F_{ST} (Wright, 1951), an estimate of the correlation between alleles in a subpopulation; in contrast, Nei (1973) uses the reduced heterozygosity to measure a closely related parameter G_{ST}. Cockerham and Weir (1993) discuss the relative merits of the G_{ST} and F_{ST} approaches to quantifying genetic differentiation. G_{ST} can also be calculated quite efficiently from multilocus VNTR profiles (Lynch, 1990; Jin and Chakraborty, 1994).

These statistics can be related to the proportion of lineages that coalesce within a subpopulation, which is represented by the area of the left-hand peak in Figure 22-3. A similar logic to that used in the derivation of equation (1) can be used to find the proportion. Within a subpopulation, coalescences occur at rate $1/N_s$ where N_s is the subpopulation size. If the proportion of individuals migrating each generation is given by m, so that migration occurs in one or the other lineage with rate $\approx 2m$, then the probability of a coalescence first is

$$\frac{1/N_s}{(1/N_s) + 2m} = \frac{1}{1 + 2N_s m} \tag{2}$$

This has long been known (see Wright, 1969) as the expected value of F_{ST} when the mutation rate is small ($2N_s m$ becomes $4Nm$ if N represents the number of diploid *individuals*). Slatkin (1991) has shown that with low mutation, G_{ST} for a pair of populations is also an estimate of the ratio of mean coalescence times, namely

$$\frac{t_1 - t_0}{t_1 + t_0} \tag{3}$$

where (t_0) and (t_1) are the means for the whole distribution and the right-hand peak (Figure 22-3), respectively.

Some mini- and microsatellite loci have mutation rates of the order of $1/N_s$ and m, so these approximations may not be quite appropriate. A simple modification of equation (2) is

$$\frac{1}{1 + 2N_s(m + \mu)} \tag{4}$$

which holds for realistic mutation processes if the distribution of new alleles is the same for mutants and migrants. Slatkin (1995) found an elegant way of avoiding the dependence on mutation rate in measuring population substructure. He shows that the ratio of coalescence times equation (3) can be estimated by R_{ST}, the proportion of the variance in allele lengths distributed between populations.

POPULATION FOUNDATION

Populations of interest to conservation biologists are often ephemeral, or have been through bottlenecks. Figure 22-4 represents a genealogy characteristic of such a history. The population was founded by $N_0/2$ diploid individuals (N_0 alleles). After foundation, its numbers increased at rate r per generation until it was large enough for genetic drift to be negligible. Migration introduced M new alleles each generation.

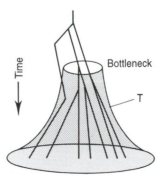

Figure 22-4 A schematic diagram of coalescence in a single growing population. The circle represents the founding population, and the lineages that trace back to it had ancestors in the founding population. Population growth is indicated by expanding boundaries of the population. The lineages of immigrants cross the boundaries of the population. Two coalescent events are shown which occurred in the ancestral population before the subpopulation was founded (drawn above the founding population).

The genealogy shows that the majority of coalescent events occurred in the early generations after foundation, when the population size was small. Migration also had the most impact at this time because the new immigrants made up a larger proportion of the population. Previously the coalescences occurred at a much slower rate. During that period the ancestors were in a much larger population that may have been subdividied.

At one time after foundation, the structure of the genealogy was essentially the same as that in the central population of Figure 22-2. The timings and the branching pattern are indistinguishable. The genealogy is typical of a population with high levels of continuing migration. Histories of this type could explain the phylogeny reconstructed from Hawaiian spider DNA described by Palumbi and Baker (chapter 2).

In such populations, the proportion of coalescences occurring within a subpopulation depends on more than the frequency of migration alone (the only parameter in equation (2)). Nichols and Beaumont (1996) have shown that the proportion is approximately

$$\frac{1 + r}{1 + N_0 r + 2M} \tag{5}$$

As the genealogy determines the pattern of genetic variation within a population, the genetic evidence will not be sufficient to distinguish the very different evolutionary histories in Figures 22-2 and 22-4. Furthermore, Beaumont and Nichols (1996) show that the *distribution* of expected genealogy shapes is essentially identical, so that scoring extra loci from the same location will not be helpful.

A more fruitful alternative is to investigate the geographical distribution of branches of the genealogy. Mitochondrial DNA or closely linked nuclear sequences can be used to attempt a reconstruction of the genealogy. Often major branches are restricted to a particular region, which speaks of a major colonization at some time in the past, or a long period of reproductive isolation (Avise, 1994). Alternatively, the area covered by alleles can be used to estimate the migration rate (O'Connell and Slatkin, 1993).

GENETIC EVIDENCE OF RELATEDNESS WITHIN POPULATIONS

General Patterns

So far we have considered two broad categories of process that might lead to genetic differentiation between subpopulations: population subdivision and founder events. In addition, there is differentiation *within* subpopulations due to different degrees of relatedness. We might hope to use some index of the genetic similarity between individuals to discover how related they were. Lynch (1988) dampened such hopes by showing that even within an outbred population there is substantial inherent imprecision in estimates of relatedness. He considered the band sharing of VNTR profiles (multilocus DNA fingerprints) for different categories of relatives, and found that, even if each allele was vanishingly rare, estimates of relatedness would be biased and have impracticably large coefficients of variation. The use of single-locus VNTR data does not dramatically increase the precision; it would require six loci to reliably distinguish half-sibs from unrelated individuals and more than nine to distinguish cousins (Brookfield and Parkin, 1993). This variability is a consequence of the genealogical history of the population: some pairs of relatives share alleles that are identical by descent, whereas other pairs, just as closely related, do not.

Although individual estimates of relatedness may be imprecise, it is possible to describe the general properties of a population such as the distribution of family sizes (Capy and Brookfield, 1991). Further, there is one category of relationship for which the genealogical variability can be overcome: parent–offspring pairs (Lynch, 1988). If each allele is vanishingly rare, then parents share precisely half their alleles with their offspring. In practice, the proportion of shared alleles is seldom used on its own to establish parentage; genetic information has been more important in *excluding* putative parents. Barring mutation, a putative pair of parents must contain all the alleles found in the offspring. The detection of exclusions accounts for the dramatic effect of multilocus VNTR techniques in revealing the extent of extra-pair copulations and intraspecific brood parasitism in birds (Birkhead et al., 1990) and similar patterns of parentage in other species.

Uncertainty about Parentage

The practicalities of conservation biology can be far removed from a well-designed experiment. It is not always possible to type all the putative parents. Even so, both theoretical and empirical studies suggest that DNA evidence is highly reliable in assigning parentage. By way of illustration, in a broad survey of centers carring out multilocus VNTR paternity testing on humans, Krawczak et al. (1993) found that when paternity was excluded it was generally excluded convincingly with multiple inconsistent bands. Indeed, there appears to be sufficient information to use grandparents' DNA profiles to establish paternity (Beckel et al., 1992).

The reasons why genetic evidence is so convincing deserve closer inspection to see how far the methods can be extended. Typically, only one individual of either sex will be found with a DNA profile consistent with parentage. A pertinent question then is the relative likelihood of one of the untested individuals also having a consistent profile. For unrelated individuals the likelihood ratio calculations may generate impressively large values (e.g., Krawczak et al., 1993); untested individuals may be over 10^4 times less likely.

Such likelihood ratios (LRs) may seem so large that there is no need to report other evidence to implicate the putative parent. The other evidence may appear subjective and imprecise by comparison. In fact, the two lines of evidence *must* be combined (see Balding and Donnelly, 1995, for a discussion in the context of forensic identification). By way of illustration, suppose that, apart from the DNA evidence, a biologist knows little about the parentage of an individual except that there are N_a alternative fathers. The weight of non-DNA evidence can be represented by the odd against a particular individual being the father, say $1:x$.

In the absence of further relevant information, $x = N_a$. Somewhat confusingly, x is known as the *prior* odds. When combined with the DNA evidence, the weight of evidence becomes LR:x (the *posterior* odds). Hence it is the size of the likelihood ratio relative to x that is germane, not its absolute size. In most conservation applications N_a is small, and there may be other evidence to implicate the putative father, so the DNA evidence is dramatically persuasive.

The relative size of likelihood ratios can be much reduced in calculations that take into account population subdivision. We have seen that the parameter F_{ST} can summarize the extent of genetic differentiation. Balding and Nichols (1995) (Tables 22-1 and 22-2) showed that the parameter can be used to adjust likelihood calculations to allow for this differentiation in single locus data. The possibility of making this adjustment provides a distinct advantage to single-locus data. The corrections can alter the likelihood ratio by two orders of magnitude for some four locus profiles when F_{ST} is of the order of 5%. Genetic differentiation of this magnitude and more is found in the relatively small isolated populations of species of conservation interest.

Further dramatic reductions can apply when close relatives are involved. In natural populations, offspring are quite often misallocated to siblings of the natural parent, so the non-DNA evidence implicating siblings is, in general, stronger than 1:N_a. Further, for each locus the likelihood ratio becomes reduced to $1/[r + (1 - r)/LR]$, where r is the coefficient of relatedness ($\frac{1}{2}$ for a sibling, $\frac{1}{4}$ for a half-sib, and so forth). It becomes plausible in such cases that multilocus fingerprints will not exclude the father's relatives. The evidence can then be equivocal. Krawczak et al. (1993) found one instance where only two of the

Table 22-1 Single-locus Likelihood Ratios for Paternity When the Mother's Genotype is *AB*. Blank entries indicate that the alleged father is excluded. *F* is essentially F_{ST}

Alleged Father	Child		
	AA	*AB*	*AC*
AA	$\dfrac{3F + (1 - F)p_A}{1 + 3F}$	$\dfrac{4F + (1 - F)(p_A + p_B)}{1 + 3F}$	
AB	$2\left(\dfrac{2F + (1 - F)p_A}{1 + 3F}\right)$	$\dfrac{4F + (1 - F)(p_A + p_B)}{1 + 3F}$	
AC	$2\left(\dfrac{2F + (1 - F)p_A}{1 + 3F}\right)$	$2\left(\dfrac{3F + (1 - F)(p_A + p_B)}{1 + 3F}\right)$	$2\left(\dfrac{F + (1 - F)p_C}{1 + 3F}\right)$
CC			$\dfrac{2F + (1 - F)p_C}{1 + 3F}$
CD			$2\left(\dfrac{F + (1 - F)p_C}{1 + 3F}\right)$

Source: Balding and Nichols (1995).

Table 22-2 Single-locus Likelihood Ratios for Paternity When the Mother's Genotype is AA. Blank entries indicate that the alleged father is excluded

Alleged Father	Child	
	AA	AB
AA	$\dfrac{4F + (1 - F)p_A}{1 + 3F}$	
AB	$2\left(\dfrac{3F + (1 - F)p_A}{1 + 3F}\right)$	$2\left(\dfrac{2F + (1 - F)p_B}{1 + 3F}\right)$
BB		$\dfrac{2F + (1 - F)p_B}{1 + 3F}$
BC		$2\left(\dfrac{F + (1 - F)p_B}{1 + 3F}\right)$

Source: Balding and Nichols (1995).

paternal bands did not match the presumed father; his brother may be the true father instead, or alternatively the offspring could carry two new mutations. In general, close relatives of the putative father need to be excluded from consideration because of these effects. Often the DNA evidence will not be available, so conservation biologists will have to draw upon their knowledge of the natural history of the species, and the behavior of the individuals in question.

Similar principles can be applied to cases where the genotypes of putative parents are not known. One relatively important case concerns the estimation of the frequency of multiple fertilization of females. The basic methodology can be illustrated in a simple case in which the genotype of the mother and offspring are known in a series of families. It is assumed that with probability m she has been fertilized by one male (she is a monogamous female), and with $(1 - m)$ by two (a bigamous female). We can calculate a likelihood for each possible value of m (L_m) as follows

$$L_m = \prod_f \left(m \sum_{c \in p1} P(G_c)P(O_f|G_c) + (1 - m) \sum_{c \in p2} P(G_c)P(O_f|G_c) \right) \quad (6)$$

where f indicates the product over families; $c \in p1$ indicates the sum over all genotypes of single fathers consistent with the offspring; and $c \in p2$ indicates the sum over all combinations with two fathers. As an example, consider a population with three alleles: A, B, and C. If an AA mother produced four AB offspring, then the possible genotypes of single fathers are BB, AB, and BC. $P(O_f|G_c)$ is the probability of the observed offspring in family f given the genotypes of the father(s) and mother. This can be calculated from classical Mendelian genetics. In the example, $P(O_f|BB) = 1$, whereas $P(O_f|AB) = P(O_f|BC) = 0.5^4$. In the case where a family has two fathers, some assumption must be made about the proportion of offspring that each fertilizes.

$P(G_c)$ is the probability of the genotypes of the parents. It is in this part of the equation that the genetic differentiation between and within subpopulations has a considerable effect. In general it requires nine parameters just to specify the probability of drawing two diploid genotypes from a population (e.g., Cockerham, 1971). However, Balding and Nichols (1994) show that, with some moderate simplifying assumptions, the probability of drawing an arbitrary number of dipolid individuals from a population with $r + 1$ A alleles, s B alleles, t C alleles, and so forth, can be specified by the recursive formula

$$E(\tilde{p}_A^{r+1}\tilde{p}_B^s\tilde{p}_C^t\tilde{p}_D^u\cdots) = E(\tilde{p}_A^r\tilde{p}_B^s\tilde{p}_C^t\tilde{p}_D^u\cdots)\left(\frac{rF + p_A(1 - F)}{1 + (r + s + t + u - 1)F}\right) \qquad (7)$$

where p_A is the frequency of allele A. The parameter used to compensate for the effect of coancestry (F) is essentially F_{ST}, which quantifies the probability of *identity* of alleles, rather than R_{ST} or N_{ST}, which take into account the differences in allele length and sequence, respectively. $\tilde{p}_A^x$ specifies the probability of drawing x A alleles from the subpopulation. Take the example of an AA mother and an AB father. The probability of this combination can be written

$$E(\tilde{p}_A^3\tilde{p}_B)$$

$$= E(\tilde{p}_A^2\tilde{p}_B)\left(\frac{2F + p_A(1 - F)}{1 + 2F}\right)$$

$$= E(\tilde{p}_A\tilde{p}_B)\left(\frac{F + p_A(1 - F)}{1 + F}\right)\left(\frac{2F + p_A(1 - F)}{1 + 2F}\right) \qquad (8)$$

$$= E(\tilde{p}_B)(p_A(1 - F))\left(\frac{F + p_A(1 - F)}{1 + F}\right)\left(\frac{2F + p_A(1 - F)}{1 + 2F}\right)$$

$$= p_B(p_A(1 - F))\left(\frac{F + p_A(1 - F)}{1 + F}\right)\left(\frac{2F + p_A(1 - F)}{1 + 2F}\right)$$

As an indication of the impact of including F, consider our single-locus case in which an AA female produces all AB offspring. One of the most likely combinations $(c \in p1)$ is that there was one BB father. A likely two-father combination $(c \in p2)$ is that there were two BB fathers. Without the correction for F, the probability of drawing two BB males from the subpopulation is p_B^4. If an F of the same magnitude as p_B is included, then the probability (from equation (7)) is around 24 times higher. Multiple paternity is therefore many times more plausible than indicated by the simple calculation. With multiple loci and many families, the modification due to F will be multipled many times over.

CONCLUSION

The general theme of this chapter has been that some aspects of genetic data are inherently variable. Its interpretation requires conservation geneticists to make full use of the range of their biological knowledge, rather than rely on the genetic data alone.

The appropriate choice of data can help overcome this variability. Making use of several loci can characterize the population subdivision accurately; in particular, use of loci with different mutation rates can probe back to different times in a species' evolutionary history. In some cases, the evidence is particularly clearcut. Among the most impressive successes of molecular genetics has been the accurate determination of parentage in natural and seminatural populations. I have argued that this success can be understood in terms of the large likelihood ratios relative to the prior odds in typical applications. Even here a sound knowledge of the species' biology is necessary to underpin the use of the technique.

The increased resolution of molecular data is particularly useful where there is incomplete information about putative parents, including those cases where relatives may be involved, or where the purpose is to estimate the degree of multiple paternity. In the latter case, we have seen how the interpretation of the results requires the biologist to make judgments about issues such as the likely alternatives to monogamy and the pattern of fertilizations when there is multiple mating. It is, perhaps, a cause for satisfaction that molecular technology cannot displace the requirement for biological insight.

ACKNOWLEDGMENTS

This work was supported by SERC grants G11101 and H 08846 to R.A.N.

REFERENCES

Avise, J.C. 1994. Molecular markers, natural history and evolution. New York: Chapman and Hall.

Balding, D.J. and P. Donnelly. 1995. Inference in forensic identification. J. R. Statistic. Soc. A 158: 21–53.

Balding, D.J. and R.A. Nichols. 1994. DNA profile match probability calculation: how to allow for population stratification, relatedness, database selection and single bands. Forensic Sci. Int. 64: 125–140.

Balding, D.J. and R.A. Nichols. 1995. A method for characterising differentiation between populations at multi-allelic loci and its implications for establishing identity and paternity. Genetica 96: 3–12.

Beaumont, M.A. and R.A. Nichols. 1996. How to interpret inter-locus variability in /fst/ Proc. Roy. Soc. B [Submitted].

Bockel, B., P. Nurnberg, and M. Krawczak. 1992. Likelihoods of multilocus DNA fingerprints in extended familes. Am. J. Hum. Genet. 51: 554–561.

Birkhead, T.R., T. Burke, R. Zann, F.M. Hunter, and A.P. Krupa. 1990. Extra-pair paternity and intraspecific brood parasitism in wild zebra finches *Taeniopygia guttata*, revealed by DNA fingerprinting. Behav. Ecol. Sociobiol 27: 315–324.

Brookfield, J.F.Y. and D.T. Parkin. 1993. Use of single locus probes in the establishment of relatedness in wild populations. Heredity 70: 660–663.

Capy, P. and J.F.Y. Brookfield. 1991. Estimation of relatedness in natural populations using highly polymorphic genetic markers. Genet. Select. Evol. 23: 391–406.

Cockerham, C.C. 1971. Higher order probability functions of identity of alleles by descent. Genetics 69: 235–246.

Cockerham, C.C. and B.S. Weir. 1993. Estimation of gene flow from F-statistics. Evolution 47: 855–863.

Di Rienzo, A., A.C. Peterson, J.C. Garza, A.M. Valdes, M. Slatkin, and N.B. Freimer. 1994. Mutational processes of simple-sequence repeat loci in human populations. Proc. Natl. Acad. Sci. U.S.A. 91: 3166–3170.

Ewens, W.J. 1990. Population genetics theory—the past and future. In: S. Lessard, ed. Mathematical and statistical developments of evolutionary theory, pp. 177–227. Amsterdam: Kluwer.

Hewitt, G.M. 1989. The subdivision of species by hybrid zones. In: D. Otte and J.A. Endler, eds. Speciation and its consequences, pp. 85–110. Sunderland, Mass.: Sinauer.

Hudson, R.R. 1990. Gene genealogies and the coalescent process. In: D.J. Futuyma and J. Antonovics, eds. Oxford surveys in evolutionary biology 7, pp. 1–44. Oxford, U.K.: Oxford University Press.

Jeffreys, A.J., A. MacCleod, K. Tamaki, D.L. Neil, and D.G. Monckton. 1991. Minisatellite repeat coding as a digital approach to DNA typing. Nature 354: 204–209.

Jeffreys, A.J., K. Tamaki, A. Macleod, D.G. Monkton, D.L. Neil, and J.A.L. Armour. 1994. Complex gene conversion events in germline mutation at human minisatellites. Nature Genetics 6: 136–145.

Jin, L. and R. Chakraborty. 1994. Estimation of genetic distance and coefficient of gene diversity from single-probe multilocus fingerprint data. Mol. Biol. Evol. 11: 120–127.

Kaplan, N. and R.R. Hudson. 1985. The use of sample genealogies for studying a selectively neutral m-loci model with recombination. Theor. Pop. Biol. 28: 382–396.

Kimura, M. and J.F. Crow. 1964. The number of alleles that can be maintained in a finite population. Genetics 49: 725–738.

Krawczak, M., I. Bohm, P. Nurnberg, J. Hampe, J. Hundrieser, H. Poche, C. Peters, R. Slomski, J. Kwiatkowska, M. Nagy, A. Popper, J.T. Epplen, and J. Schmidtke. 1993. Paternity testing with oligonucleotide multilocus probe (CAC)5/(GTG)5: a multicenter study. Forensic Sci. Int. 59: 101–117.

Lynch, M. 1988. Estimation of relatedness by DNA fingerprinting. Mol. Biol. Evol. 5: 584–599.

Lynch, M. 1990. The similarity index and DNA fingerprinting. Mol. Biol. Evol. 7: 478–484.

Malecot, G. 1948. Les mathematiques de l'hérédité. Paris: Masson.

Menotti-Raymond, M. and S.J. O'Brien. 1993. Dating the genetic bottleneck of the African cheetah. Proc. Natl. Acad. Sci. U.S.A. 90: 3172–3176.

Nei, M. 1973. Analysis of gene diversity in subdivided populations. Proc. Natl. Acad. Sci. U.S.A. 70: 3321–3323.

Nichols, R.A. and M.A. Beaumont. 1996. Is it ancient or modern history we are reading in the genes? In: M. Hochenberg, ed. The origin and maintenance of biological diversity. Oxford, U.K.: Oxford University Press [In press].

O'Connell, N. and M. Slatkin. 1993. High mutation rate loci in a subdivided population. Theor. Pop. Biol. 44: 110–127.

Slatkin, M. 1991. Inbreeding coefficients and coalescence times. Genet. Res. Camb. 58: 167–175.

Slatkin, M. 1993. Isolation by distance in equilibrium and non-equilibrium populations. Evolution 47: 264–279.

Slatkin, M. 1995. A measure of population subdivision based on microsatellite allele frequencies. Evolution 139: 457–462.

Slatkin, M. and R.R. Hudson. 1991. Pairwise comparisons of mitochondrial DNA sequences in stable and exponentially growing populations. Genetics 129: 555–562.

Strobeck, C. 1987. The average number of nucleotide differences in a sample from a single subpopulation: a test for population subdivision. Genetics 117: 149–153.

Weir, B. and C.C. Cockerham. 1984. Estimating hierarchical F-statistics. Evolution 38: 1358–1370.

Wright, S. 1951. The genetical structure of populations. Ann. Eugen. 15: 323–354.

Wright, S. 1969. Evolution and the genetics of populations II: the theory of gene frequencies. Chicago: Chicago University Press.

III
CASE STUDIES

23

Molecular Genetics and the Conservation of Salmonid Biodiversity: *Oncorhynchus* at the Edge of Their Range

JENNIFER L. NIELSEN

As with many extant vertebrates, the evolutionary history of Pacific salmon (*Oncorhynchus* spp.) is complex and includes many periods of extinctions and radiations. This is especially true for the *Oncorhynchus* complex, where fossil records are scarce and current taxonomy is confused by extensive morphological, life history, and ecological variation (Behnke, 1992). The ancestral tetraploidy state of much of the salmonid genome, with the presence of many duplicate genes (Allendorf and Waples, 1996), further complicates the resolution of genetic relatedness, population structure, and local adaptation.

Protein electrophoresis has been used in salmonid population genetics since the late 1960s (Utter, 1991), and remains a significant tool in this field today. With a suite of over 80 allozyme loci available for population differentiation, electrophoresis is a rigorous and necessary tool in salmonid genetic research. Most of the salmonid allozyme research has, in the past, been dedicated to genetic stock identification based on large-scale hatchery production and commercial harvest. In intraspecific studies, however, many of these loci remain conserved and variation using far fewer polymorphic loci is often the result.

A second conflict with allozyme analysis lies in the fact that to access the total suite of allozymes available for salmon (and many of the highly polymorphic loci) one needs to sacrifice the animal. This was not considered a problem in past stock assessment, where the sacrifice of 40–100 individual fish per population seemed irrelevant to population viability. This issue becomes a serious problem when minimum viable population size falls below 50 fish, as is the case in many California salmon and trout populations. Therefore, the intrusive sampling criteria for allozyme investigations preclude certain studies needed to answer significant questions about local stock structure and the preservation of genetic diversity.

In particular, the allozyme approach runs counter to conservation efforts in populations that are considered threatened or endangered, and where local laws restrict the "taking" of individuals. Noninvasive molecular techniques have been widely used in such cases: the endangered black rhinoceros (Ashley et al., 1990);

the desert iguana (Lamb et al., 1992); the European black bear (Paetkau and Strobeck, 1994); the Ethiopian wolf (Gottelli et al., 1994). The new procedures of thermal cycling and DNA cloning (PCR) allow genetic resolution at the molecular level from small fragments of tissue that can easily be collected noninvasively in the field. Fish DNA can be extracted quite simply from fin-clips or scales (see Nielsen et al., 1994b, for protocol). DNA protocols also allow investigation of genetic relationships without removing animals from their natural population (see McDonald and Potts 1994; Morin et al., 1994). Thus, studies of individual behavior and life history can be made over time without the negative effects of destructive sampling impacting population structure.

Finally, DNA analyses open up the possibility of historical reconstruction of genetic phylogeny using archive materials, where proteins will have long since degraded—museum samples, bones, otoliths, and taxidermic collections (Shiozawa et al., 1992; Höss and Pääbo, 1993). This approach to using archived DNA for the development of historic phylogeny adds support to findings of contemporary population structure in trout and salmon, despite the significant (and often undocumented) anthropomorphic manipulations that have been made over the last 100 years. For all these reasons, molecular analyses using DNA will play an important role in the future of salmonid population genetic studies.

Pacific salmon abundance has been coupled with large-scale atmospheric events that affect ocean currents and coastal upwelling intensities (El Niño events) in the north-eastern Pacific Ocean, which in turn affect local salmonid production (Pearcy, 1992). In the western United States, population fragmentation due to habitat destruction resulting from urbanization, forest practices, and overharvesting of ocean stocks have all contributed to recent exponential declines in salmonid abundance (Nehlsen et al., 1991). California's anadromous salmon and trout (steelhead trout, *O. mykiss*; chinook salmon, *O. tshawytscha*; and coho salmon, *O. kisutch*) have diminished significantly since the European colonization of the shores of the eastern Pacific Ocean in the mid nineteenth century (Swift et al., 1993; Titus et al., 1996).

Efforts at salmonid augmentation and stewardship dating from the turn of the century translocated populations of California salmon and trout all over the world. More recently, in the last half of the twentieth century, the combined application of industry and technical husbandry in natural resource management have significantly changed the genetic fabric of salmonids throughout the world (Ryman and Ståhl, 1980; Skaala et al., 1990; Billington and Hebert, 1991; Hindar et al., 1991). Reversing the effects of genetic manipulation, and addressing the continued decline in salmonid abundance, will be nowhere as easy as the slide toward extinction precipitated in these species by the cumulative influences of natural and human-induced activity throughout the twentieth century.

In recent studies of salmon populations at the southern extent of their range, our laboratory has used mitochondrial DNA and nuclear microsatellite analyses to begin to understand the elements of molecular biodiversity that remain in salmon populations today and to trace the evolutionary history available in the DNA of this genus. Salmon and trout from California served as the basis of our studies for two reasons: (1) This group of fish represents the southernmost populations of Pacific *Oncorhynchus* in existence today, and evolution gains momentum at the edge of the range (Avise et al., 1987), where individuals face a continuous environmental challenge to their existence; (2) these populations represent the parental source material for the translocation of salmon and trout all over the world at the turn of the century, and therefore may give insight into the effects of recent (in the last 100 years) manipulations and local adaptations when geographically disparate populations are compared to translocated progeny populations.

This overview of ongoing research includes the three extant California species of *Oncorhynchus*, with investigations covering three general life history scales: broad geographic species distributions or biogeography of steelhead; genetic structure in hatchery versus wild populations of steelhead, coho, and chinook salmon; and within-basin behavioral diversity based on temporal spawning runs in steelhead and chinook. The objective of this chapter is to present a preliminary synthesis of several DNA studies ongoing in our laboratory concerning Pacific salmon and trout in California, and to give the reader an overview of the degree of resolution gained in these analyses using the new tools of molecular conservation genetics.

CASE STUDIES

Biogeography in Coastal Steelhead (*Oncorhynchus mykiss*)

Anadromous steelhead historically ranged in large numbers throughout the entire coastline of California down to the Rio del Presidio in Baja California (Needham and Gard, 1959; MacCrimmon, 1971). Their present limit at the southern edge of their range is in Malibu Creek, which flows into Santa Monica Bay just north of Los Angeles. The remaining populations of southern anadromous steelhead in California are at high to moderate risk of extinction, with extremely low effective population sizes (< 50 individuals) in many streams. Despite the depauperate state of these populations, anadromous steelhead have been reported on and off in numerous streams south of Monterey Bay on the central and southern coasts of California over the last 10 years (Swift et al., 1993).

Extensive allozyme studies divide the *O. mykiss* population in California into two major genetic subdivisions: coastal steelhead, *O. m. irideus* (after Behnke, 1992); and interior redband trout, *O. m. gairdneri* (reviewed in Hershberger, 1992). Both subspecies are thought to have flexible life history criteria allowing both anadromous and exclusively freshwater maturation. Today the coastal stocks are popularly considered as primarily anadromous, although with extensive flexibility (1–4 years) in their timing of ocean migration. The interior redband are thought to be exclusively freshwater resident fish (i.e., rainbow trout). Extensive overlap, however, in morphology, life history, and allelic markers often makes it difficult to differentiate these subspecies, particularly in southern California (Behnke, 1992).

In our salmonid population studies, we have used specific primers (S-phe and P2) to amplify a highly variable segment of mitochondrial DNA (mtDNA), including 188 base pairs (bp) of the control region and 5 bp of the adjacent phenylalanine tRNA gene. Methods, protocols, and complete sequence data for mtDNA amplification and visualization are given in Nielsen et al. (1994b). This segment of mtDNA has shown 14 unique haplotypes in *O. mykiss*, 5 unique haplotypes in *O. kisutch*, and 12 unique haplotypes in *O. tshawytscha*, taken from populations throughout the western United States. The average mtDNA nucleotide diversity for the 12 anadromous steelhead haplotypes (two *O. mykiss* haplotypes were found only in freshwater trout populations) was 1.3%. The two most divergent trout mtDNA haplotypes differed by 3.1%.

A study of 547 coastal steelhead from 37 streams throughout California showed significant geographic structure (Figure 23-1) for four dominant

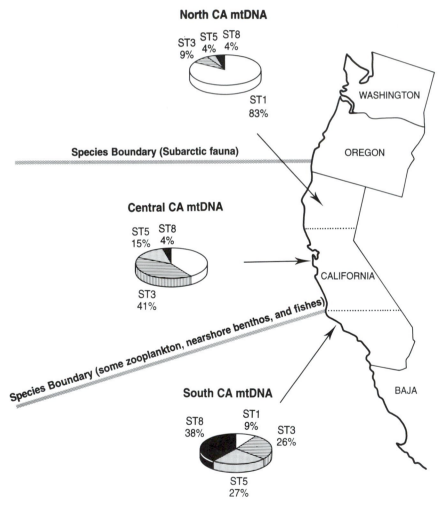

Figure 23-1 Mitochondrial DNA haplotype frequency distributions for three general geographic bioregions in California: North CA from the Eel River to the Gualala River; Central CA from the Russian River to Point Sur; South CA from San Simeon to Santa Monica Bay. Published species boundaries (Cape Blanco and Point Conception) for other oceanic fauna are indicated as dotted lines.

haplotypes (total population frequencies: ST1 32%; ST3 20%; ST5 13%; ST8 13%) based on sequence (188 bp) for the highly variable segment of the mtDNA control region (haplotype base changes given in Nielsen et al., 1994a). Fisher's exact analysis of the mtDNA haplotype frequency data pooled for three bioregions (northern, central, and southern California; see Nielsen et al. (1994a) for exact areas sampled) showed significant differentiation among all pairs of geographic localities (in all cases Fisher's $p < 0.001$). Subsequent statistical analyses of mtDNA distribution in coastal California streams using principal-components analysis showed that 70% of the variation in mtDNA haplotype frequencies could be explained from the first two components (42% and 28%, respectively), describing the greatest portion of the total variance over locations for all haplotypes (Cramer et al., 1995).

Laboratory analyses of preserved steelhead tissue from museum samples have corroborated the occurrence of specific mtDNA haplotypes in geographic areas over time. In a random set of samples taken from the California Academy of Sciences' collection of preserved fishes, we have amplified the following mtDNA haplotypes: ST1 (dominant haplotype in northern California) in 8 of 12 fish collected in the Eel River and the Sacramento River from 1938 to 1945; ST3 (dominant haplotype in central California) in 12 of 14 fish collected in San Francisquito Creek, a tributary to San Francisco Bay in 1941; ST5 (dominant haplotype in southern California) in 6 southern steelhead samples collected from Los Angeles harbor, Mono Lake, and Baja California in the early 1940s. Single-point collections from one location and small sample sizes in most museum samples make statistical analyses of these results difficult. However, these data do suggest that the dominant mtDNA haplotypes found in contemporary populations of steelhead from three general bioregions of California were also present in local streams during the first half of the twentieth century.

Additional resolution for this biogeographic cline in California steelhead genotypes was gained from a concurrent study of nuclear microsatellites in the same populations. In initial tests of 26 dinucleotide microsatellite markers originally used for mouse genome mapping (Hearne et al., 1991), we found only one microsatellite locus that expressed in salmon (MS105; *Ckmm*), and this locus was monomorphic among our steelhead populations. Recent research has shown that organizational difference in microsatellite loci between mammals and teleost fishes (Brooker et. al., 1994) may have contributed to our early results using mammal-derived microsatellite primers.

We have since turned to microsatellite primer pairs developed specifically for salmonids at Dalhousie University in the Marine Gene Probe Laboratory (J.M. Wright, personal communication). Three salmonid-specific, highly polymorphic microsatellite primer pairs are available through this laboratory (Omy77, Omy2, and Omy27), all of which have been tested and shown to be polymorphic to varying degrees in our Pacific salmonid population studies. All three microsatellite loci contain dimeric repeat units: Omy77, $d(GA)_n$; Omy2, $d(CA)_n$; Omy27, $d(GT)_n$. These loci were shown to contain nonrandom alleles inherited in a Mendelian manner in *O. mykiss* through paternity trials on hatchery fish (Morris, 1993). Omy77 allele size in our initial California steelhead samples ranged from 80 to 141 bp, with 20 unique alleles (Nielsen et al., 1994a). Subsequent studies of freshwater trout populations found a total of 43 alleles (62–210 bp) for Omy77 (including 16 of the original 20 found in steelhead), and 41 alleles (64–190 bp) for Omy2. The Omy27 locus has 11 alleles ranging in size from 76–79 bases.

Fisher's exact tests for population independence by geographic area, based on Omy77 allelic frequencies for anadromous steelhead from California's three bioregions described above, were significant (Fisher's $p < 0.02$) in all paired comparisons of geographic bioregions. Chi-square analyses of the distribution of specific Omy77 alleles demonstrated significant differences in geographic abundance at five alleles (Nielsen et al., 1994e). Subsequent preliminary investigations with two other microsatellite loci (Omy2 and Omy27) show similar results in geographic population structure in California steelhead (Nielsen, unpublished data). These findings suggest that with multiple, polymorphic microsatellites we will be able to measure the degree of population subdivision based on microsatellite allele frequencies (Slatkin, 1995) and develop rigorous screening protocols for geographic stock identification in *O. mykiss* from California.

Estimates of migration events among the steelhead genotypes using DNA (Nm; Slatkin 1985) ranged between 1.2 and 1.8 fish per generation for mtDNA and microsatellite alleles, respectively. This estimate is nearly 3–4 times the threshold suggested by Slatkin to maintain genetic panmixia based on drift alone. Despite these high estimates of gene flow among the steelhead populations, both molecular methodologies (mtDNA and microsatellite DNA) showed significant congruence for allelic and haplotype frequency distributions among broad coastal bioregions in California, with unique haplotypes (ST6 and ST14) and microsatellite alleles isolated by a documented southern, biogeographic species boundary near Point Conception (Figure 23-1). Both genetic markers also showed a significant increase in genetic diversity at the southern extent of the species range.

Hatchery Versus Wild *Oncorhynchus* Populations in California

Our search for endemic genotypes of California salmon was complicated by significant stock transfers and hatchery production of *Oncorhynchus* throughout the western United States over the last 100 years. It is interesting that the genetic roots of many of the world's *O. mykiss* hatchery populations were originally derived from wild California stocks. Both steelhead and chinook eggs were transferred from the first U.S. salmon egg-taking station, Baird Station on the McCloud River, to ports of call throughout the United States and all over the world as early as 1897 (U.S. Fisheries Commission Reports 1874–1901). Freshwater populations of *O. mykiss* from these transplantations have been established globally, with the notable exception of Antarctica (MacCrimmon, 1971).

Anadromous populations, however, have been more difficult to establish outside of the species' natural range. The exception to this rule is the transfer of anadromous chinook from the Sacramento River that were introduced in the early 1900s into several New Zealand Rivers (McDowall, 1994). Recent molecular data comparing wild New Zealand chinook populations with the progeny of their putative ancestors indicate that mtDNA frequency distributions remain the same among the New Zealand fish and the fall-run hatchery chinook in the Sacramento River (Quinn et al., 1996), suggesting little divergence for this locus between the two populations, which shared a common ancestor as recently as 1910. Six unique alleles found for the microsatellite locus Omy77 in the New Zealand chinook populations indicate slight but significant divergence among translocated fish when compared with the contemporary Sacramento fall-run fish which are fixed at this locus.

Comparisons between contemporary wild and hatchery populations of California salmon also gave mixed results for mtDNA and microsatellite diversity. In a general survey of anadromous *Oncorhynchus* mtDNA diversity, we found that hatcheries had a significantly higher number of mtDNA haplotypes (see table 5 in Nielsen et al., 1994b). Comparing geographically proximate pairs of hatchery and wild populations for steelhead ($n = 4$), coho ($n = 1$), and chinook ($n = 1$) in California, only three mtDNA haplotypes were found to be exclusive to wild populations, while a total of 15 haplotypes were found only in the hatchery populations.

Since most reported mtDNA diversity is the product of geography (Avise, 1994), these findings were interpreted to be a result of the artificial mixing of geographically divergent gene pools within the hatchery populations by the transfer of eggs from state to state and over broad geographic areas within the state. In fact, hatchery records available from the California Department of Fish and Game (W. Jones, Region 3, Ukiah, CA, personal communication) document large numbers (hundreds of thousands) of introduced eggs and fry from sister hatcheries in Washington and Oregon and numerous transfers among salmonid hatcheries throughout California. Many of these documented transfers, made from 1950 through the 1980s, were used to supplement natural production in years when local stocks were depressed.

The lack of genetic introgression into the wild California stocks by introduced mtDNA haplotypes is surprising. We speculate that these "introduced" mtDNA genotypes are perpetuated in the hatchery by early interception and integration of the geographically divergent females at the hatchery collecting weirs. Their isolation within the hatchery may be the result of unique maturation cycles for the geographically divergent lineages as a result of inherited migration timing and development rates that evolved over long periods of time in their native habitats. Therefore, females with an evolutionary history of a long spawning migration of over 30 miles, such as Alsea River coho, would not mature in their introduced habitats on the Noyo River in Mendocino before hatchery interception at a weir only 3.5 miles up river from the ocean. Because mature females are often the limiting factor in hatchery production in California, most immature females will be held until maturity in pens for use in egg taking, effectively isolating their mtDNA in the hatchery progeny.

Gender-biased dispersal (Avise, 1994), or kin-structured migration patterns (Wade et al., 1994) with low rates of female straying, could contribute to qualitatively different patterns in population structure at maternally (mtDNA) and biparentally (microsatellites) transmitted genes (Moritz et al., 1987; Avise 1991). Analysis of the same anadromous hatchery ($n = 69$) and wild ($n = 56$) steelhead from the Eel River used in our mtDNA study found an opposite trend from our mtDNA results in the frequency distribution of microsatellite alleles at the Omy77 locus (Figure 23-2). One allele (#4) dominated the hatchery

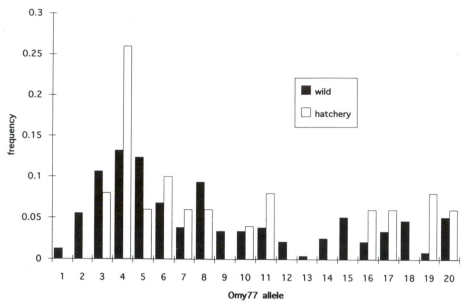

Figure 23-2 Microsatellite (Omy77) allelic frequency distribution for hatchery and wild steelhead from the Eel River basin.

population, while Eel River wild steelhead carried eight alleles not found at all in our hatchery sample. These data (while preliminary because they are based on a single, but highly polymorphic, microsatellite locus) demonstrate an interesting difference in maternally and biparentally inherited genes for steelhead under differing selection regimes represented by artificial propagation and wild spawning.

Nine freshwater rainbow trout stocks (*O. mykiss*) are currently used in California hatcheries. All nine are thought to represent crosses and back-crosses of trout populations derived from the original Baird Station hatchery on the McCloud River at the turn of the century (Busack and Gall, 1980). Unlike the results from hatchery and wild anadromous populations, our analyses of fish from trout hatcheries ($N = 142$) representing all nine brood stocks, showed a significant reduction in mtDNA haplotypes when compared to wild coastal populations found in freshwater (5 mtDNA haplotypes in the hatchery versus 12 wild haplotypes). This diminished mtDNA genetic diversity within rainbow trout hatchery populations suggests that they have undergone significant bottlenecks in the history of their brood-stock programs.

It is interesting to note that in an analysis of mtDNA haplotypes in rainbow trout hatchery stocks from around the globe (New Zealand, Canada, Russia, and Chile) and across the United States (New York, Virginia, Washington, Michigan, and Wyoming), we have found no new nucleotide divergence in the mtDNA control region that we amplify. All rainbow trout hatchery populations that we have looked at were dominated by mtDNA haplotypes that are common to northern California trout and steelhead populations (ST1 and ST3).

These results show the increased degree of resolution available in salmonid studies of hatchery and wild fish using different molecular markers to depict population structure. Differences in the information provided by DNA markers can result from husbandry activities which lead to directionality in artificial selection for brood stocks, divergent genetic origins of transferred stocks, and/or gender-biased gene flow in hatchery or wild fish. These data caution against making conservation decisions from results derived from any single DNA marker (Moritz, 1994) when selection mechanisms (natural or artificial) may affect populations differentially.

Genetic Diversity in Unique Temporal Spawning Runs of *Oncorhynchus*

Chinook populations in the Sacramento–San Joaquin River basin have historically been separated by migration and spawning times into the fall-run (Oct.–Dec.), late fall-run (Jan.–Apr.), winter-run (Apr.–Aug.), and spring-run (Aug.–Oct.) The use of molecular methods (principally allozymes and mtDNA) in stock identification in fisheries has been widely reviewed (Ryman and Utter, 1987; Ovenden, 1990). Recent allozyme studies, however, were unable to identify individual stocks among these four Sacramento River spawning groups (Bartley and Gall., 1990; Bartley et al. 1992). On broad geographic scales (North America versus Europe), mtDNA has been shown to provide stock identification for ocean-caught Atlantic salmon (Bermingham et al., 1991). We were able to show in our laboratory that mtDNA polymorphisms in the four unique runs of chinook salmon differed significantly in frequency distribution (Nielsen et al., 1994c).

Adult migration timing and mtDNA frequency criteria are, however, no help in identifying the parental origins of juvenile fish within the same system the following spring. The recent federal listing of winter-run chinook in the Sacramento River as endangered under the Endangered Species Act has led to a significant demand for stock-specific markers that could be used to identify these populations during their downstream migration as smolts. Economically valuable water withdrawals at federal diversion pumps impede the passage of chinook fry in the Sacramento River and cause significant mortality in juveniles on their way to the San Francisco Bay delta and finally the Pacific Ocean. Stock identification markers would facilitate adaptive control over the withdrawal schedule during "critical" migrations of endangered fish and leave open the possibility of alternate decisions affecting the fate of the hatchery fall-run chinook and the other fish stocks in the river.

Our mtDNA data did not, however, provide individual stock markers for the Sacramento River chinook, where differentiation was possible only through the use of frequency data. In this study, the endangered winter-run chinook carried one unique haplotype (CH6) based on an unusual 81-base repeat found in the mtDNA control region. But this repeat, while not found in any of the other three chinook spawning populations, was rare even in the winter-run fish.

Microsatellite analyses have been difficult in these fish, with many loci tested in our laboratory proving to be monomorphic for all four stocks. Further evidence drawn from a limited number of polymorphic microsatellite markers used in our laboratory and at the University of California's Bodega Bay Marine Laboratory

(Dr. M. Banks, personal communication) show unique frequency distributions of microsatellite markers among the stocks, but have yet to disclose the stock-specific markers required by resource managers.

In a second stock-specific analysis undertaken in our laboratory we examined genetic diversity in summer- and winter-run steelhead from the Eel River. Summer-run steelhead enter the Eel River as reproductively immature adults in late spring. This stock matures in freshwater while holding in deep, thermally-stratified pools in the upper reaches of the river (Nielsen et al., 1994d). The more common winter-run fish enter the river in fall as mature adults and immediately migrate to selected spawning locations lower in the watershed. Therefore, temporal and spatial effects contribute to the reproductive isolation of these two stocks.

Mitochondrial DNA in summer- and winter-run steelhead from the Eel River did not show unique population structure such as that found in the Sacramento River chinook spawning stocks (Figure 23-3). Each run carried a unique rare

M.F. Eel River - summer steelhead

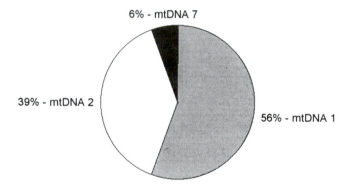

M.F. Eel River - winter steelhead

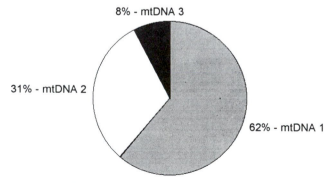

Figure 23-3 Mitochondrial DNA haplotype frequency distributions for summer- and winter-run steelhead in the Middle Fork Eel River captured as spawning adults in 1990–1992.

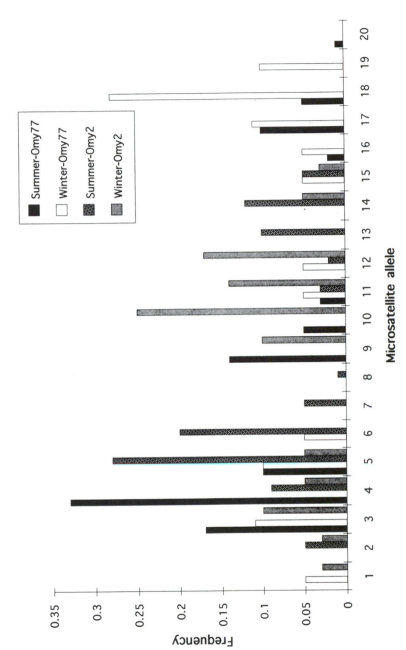

Figure 23-4 Microsatellite (Omy77 and Omy2) allelic frequency distributions for the same summer- and winter-run steelhead in the Middle Fork Eel River used in mtDNA analyses given in Figure 23-3.

haplotype, which suggests some degree of reproductive isolation, but general frequency distributions for the two dominant haplotypes were nearly identical.

Microsatellite analyses of these same Eel River steelhead stocks did add significant resolution to this issue. Both microsatellite loci Omy77 and Omy2 show distinct bimodal distributions of their dominant alleles in the two run types (Figure 23-4. Fisher's exact analyses of the frequency distributions depicted in this graph gave significant resolution ($p < 0.01$) for these loci between the two steelhead spawning populations.

It is interesting to point out that these alleles are numbered sequentially according to the size of the repeat unit. At both microsatellite loci the summer-run steelhead have significantly more smaller repeats and the winter-run more larger repeat units. While these data are preliminary and based on only two loci, studies of human families show that the size of a new mutant microsatellite allele depends on the size of the allele mutated (Weber and Wong, 1993), suggesting that predictable biochemical mechanisms may have contributed to the differential size gradient of microsatellite alleles between the two steelhead stocks.

It has also been reported that there are two separate biochemical mechanisms driving the mutational process of microsatellite repeats in the human genome (Di Rienzo et al., 1994; see also Slatkin, 1995): one mechanism theoretically responsible for simple sequence repeats of the base nucleotide unit (± 1 repeat unit), and a second mechanism for larger changes in repeat number. The bimodal distribution of alleles for both Omy77 and Omy2 in summer- and winter-run steelhead from the Eel River suggests that the dual mechanisms of repeat sequence mutations proposed by Di Rienzo et al. (1994) may have significance for our understanding of the biochemical separation between these two stocks.

CONCLUSION

It is clear from these case studies that the level of information needed in conservation genetics and the molecular tools applied must be carefully matched to reach conclusions about specific hypotheses. The unidirectional inheritance and sensitivity to bottlenecks found in the mtDNA locus make it an excellent tool for biogeographic questions, but without congruence in findings from the recombinant genome it is dangerous to make specific management decisions from mtDNA data alone. In our steelhead study, two molecular markers—mtDNA and micro-satellites—were congruent in their demonstration of geographic structure in California populations. This was not the case in our analyses of hatchery versus wild salmonid populations, where differences in genetic diversity were shown to be highly variable among populations.

Microsatellites add a new suite of molecular loci which can be used to investigate questions at a finer scale of resolution within and among fish populations. Allozymes and mtDNA were not always adequate to provide the degree of resolution needed for conservation questions at the basin ecosystem level, as was shown in our case studies of chinook and steelhead temporal spawning runs. The development of salmonid-specific microsatellites has, however, been slow. The tetraploid nature of the salmonid genome and differences in the organization of microsatellite repeat classes for teleost fishes have prevented direct application of previously described mammalian microsatellite markers to this genus. Until very recently, sufficiently polymorphic markers with strictly diploid allelic expression and demonstrated Mendelian inheritance were rare in the published literature for salmon.

It is especially important in future studies of molecular conservation genetics for salmon to determine the balance of rigor available from the well-described and well-documented systems of protein analysis and mtDNA haplotypes, and the newly developed microsatellite loci. All three genetic systems can be used to depict change in the genome at the population level, but their rates of evolution, mutational mechanisms, and degree of conservation may not be uniform, even within a single species. It is especially important to understand that the mutational processes applicable to microsatellite loci, while still not clearly defined, do not conform to the K-allele model based on low mutation rates that is commonly used for allozyme and mtDNA data (Slatkin, 1995). Therefore, it may not be appropriate to apply the same analytical methods that have been developed for allozymes or mtDNA directly to the analysis of microsatellite data.

Finally, it is important to reiterate the two conditions that remain unique to DNA analyses (be they mtDNA, microsatellites, intron variation, nuclear sequence, or any other PCR-based approach): noninvasive sampling and access to archived material. Working with sampling protocols appropriate to live animals and the ability to trace population trends throughout time seem critical to genetic investigations in conservation biology. As our ability to use these techniques increases, our approach to understand natural systems will change in both temporal and spatial scales.

We must also strive to understand the biochemical systems we work with at many levels. We should realize that today's DNA analyses shed light on biological systems from the metapopulation or evolutionarily significant unit, to the individual living within a population, and on to the cellular mechanisms driving the genes themselves. Only with an appreciation of this interdisciplinary, multiple-scaled approach will the appropriate tools providing the correct rigor of analyses be applied to the most important hypotheses in conservation biology today.

REFERENCES

Allendorf, F.W. and R.S. Waples. 1996. Conservation and genetics of salmonid fishes. In: J.C. Avise and J.L. Hamrick, eds. Conservation genetics: case histories from nature, pp. 238–280. New York: Chapman & Hall.

Ashley, M.V., D.J. Melnick, and D. Western. 1990. Conservation genetics of the black rhinoceros (*Diceros bicornis*), I: Evidence from the mitochondrial DNA of three populations. Conserv. Biol. 4: 71–77.

Avise, J.C. 1991. Ten unorthodox perspectives on evolution prompted by comparative population genetic findings on mitochondrial DNA. Annu. Rev. Genet. 25: 45–69.

Avise, J.C. 1994. Molecular markers, natural history and evolution. New York: Chapman and Hall.

Avise, J.C., J. Arnold, R.M. Ball, E. Bermingham, T. Lamb, J.E. Neigel, C.A. Reeb, and N.C. Saunders. 1987. Intraspecific phylogeography: the mitochondrial DNA bridge between population genetics and systematics. Annu. Rev. Ecol. Syst. 18: 489–522.

Bartley, D.M. and G.A.E. Gall. 1990. Genetic structure and gene flow in chinook salmon populations in California. Trans. Am. Fish. Soc. 119: 55–71.

Bartley, D.M., B. Bentley, J. Brodziak, R. Gomulkiewicz, M. Mangel, and G.A.E. Gall. 1992. Geographic variation in population genetic structure of chinook salmon from California and Oregon. Fish. Bull. 90: 77–100.

Behnke, R.J. 1992. Native trout of western North America. Bethesda, Md.: American Fisheries Society, Monograph 6.

Bermingham, E., S.H. Forbes, K. Friedland, and C. Pla. 1991. Discrimination between Atlantic salmon (*Salmo salar*) of North American and European origin using restriction analyses of the mitochondrial DNA. Can. J. Fish. Aquat. Sci. 48: 884–893.

Billington, N. and P.D.N. Hebert. 1991. Mitochondrial DNA diversity in fishes and its implication for introductions. Can. J. Fish Aquat. Sci. 48(Suppl. 1): 80–94.

Brooker, A.L., D. Cook, P. Bentzen, J.M. Wright, and R.W. Doyle. 1994. Organization of microsatellites differs between mammals and cold-water teleost fishes. Can. J. Fish. Aquat. Sci. 51: 1959–1966.

Busack, C.A. and G.A.E. Gall. 1980. Ancestry of artificially propagated California rainbow trout strains. California Fish and Game Reports 66: 17–24.

Cramer, S.P., D.W. Alley, J.E. Baldridge et al. 1995. The status of steelhead populations in California in regards to the Endangered Species Act. Submitted to National Marine Fisheries, Status Review for Coastal Steelhead. Gresham, Oreg. S.P. Cramer & Associates.

Di Rienzo, A., A.C. Peterson, J.C. Garza, A.M. Valdes, M. Slatkin, and N.B. Freimer. 1994. Mutational processes of simple-sequence repeat loci in human populations. Proc. Natl. Acad. Sci. U.S.A. 91: 3166–3170.

Gottelli, D., C. Sillero-Zubiri, G.D. Applebaum, M.S. Roy, D.J. Girman, J. Garcia-Moreno, E.A. Ostranders, and R.K. Wayne. 1994. Molecular genetics of the most endangered canid: the Ethiopian wolf *Canis simensis*. Mol. Ecol. 3: 301–312.

Hearne, C.M., M.A. McAleer, J.M. Love, T.J. Aitman, R.J. Cornall, S. Ghosh, A.M. Knight, J. Prins, and J.A. Todd. 1991. Additional microsatellite markers for mouse genome mapping. Mammal. Genome 1: 273–282.

Hershberger, W.K. 1992. Genetic variability in rainbow trout populations. Aquaculture 100: 51–71.

Hindar K., N. Ryman, and F. Utter. 1991. Genetic effects of cultured fish on natural fish populations. Can. J. Fish. Aquat. Sci. 48: 945–957.

Höss, M. and S. Pääbo. 1993. DNA extraction from Pleistocene bones by a silica-based purification method. Nucleic Acids Res. 21(16): 3913–3914.

Lamb, T., T.R. Jones, and J.C. Avise. 1992. Phylogeographic histories of representative herperto-fauna of the desert southwest: mitochondrial DNA variation in the chuckwalla (*Sauromalus obesus*) and desert iguana (*Dipsosaurus dorsalis*). J. Evol. Biol. 5: 465–480.

MacCrimmon, H.R. 1971. World distribution of rainbow trout (*Salmo gairdneri*). J. Fish. Res. Boar. Can. 29: 663–704.

McDonald, D.B. and W.K. Potts. 1994. Cooperative display and relatedness among males in a lek-mating bird. Science 226: 1030–1031.

McDowall, R.M. 1994. The origins of New Zealand's chinook salmon, *Oncorhynchus tshawytscha*. Marine Fish. Rev. 56: 1–7.

Morin, P.A., J.J. Moore, R. Chakraborty, L. Ji, J. Goodall, and D.S. Woodruff. 1994. Kin selection, social structure, gene flow, and the evolution of chimpanzees. Science 265: 1193–1201.

Moritz, C. 1994. Applications of mitochondrial DNA analysis in conservation: a critical review. Mol. Ecol. 3: 401–411.

Moritz, C., T.E. Dowling, and W.M. Brown. 1987. Evolution of animal mitochondrial DNA: relevance for population biology and systematics. Annu. Rev. Ecol. Syst. 18: 269–292.

Morris, D.B. 1993. The isolation and characterization of microsatellites from rainbow trout (*Oncorhynchus mykiss*). MSc thesis, Dalhousie University, Halifax, N.S.

Needham, P.R. and R. Gard, 1959. Rainbow trout in Mexico and California with notes on the cutthroat series. Univ. Calif. Publ. Zool. 67(1).

Nehlsen, W., J.E. Williams, J.A. Lichatowich. 1991. Pacific salmon at the crossroads: stocks at risk from California, Oregon, Idaho, and Washington. Fisheries 16: 4–21.

Nielsen, J.L., C.A. Gan, J.M. Wright, D.B. Morris, and W.K. Thomas. 1994a. Biogeographic distributions of mitochondrial and nuclear markers for southern steelhead. Mol. Marine Biol. Biotechnol. 3(5): 281–293.

Nielsen, J.L., C.A. Gan, and W.K. Thomas. 1994b. Differences in genetic diversity for mtDNA between hatchery and wild populations of *Oncorhynchus*. Can. J. Fish. Aquat. Sci. 51 (Suppl. 1): 290–297.

Nielsen, J.L., D. Tupper, and W.K. Thomas. 1994c. Mitochondrial DNA polymorphisms in unique runs of chinook salmon (*Oncorhynchus tshawytscha*) from the Sacramento–San Joaquin River Basin. Conserv. Biol. 8(3): 882–884.

Nielsen, J.L., T.E. Lisle, and V. Ozaki. 1994d. Thermally stratified pools and their use by steelhead in northern California streams. Trans. Am. Fish. Soc. 123: 613–626.

Nielsen, J.L., C.A. Gan, J.M. Wright, and W.K. Thomas. 1994e. Phylogeographic patterns in California steelhead as determined by mtDNA and microsatellite analyses. CalCOFl rep. 35: 90–92.

Ovenden, J.R. 1990. Mitochondrial DNA and marine stock assessment: a review. Aust. J. Mar. Freshwater Res. 41: 835–853.

Paetkau, D. and C. Strobeck. 1994. Microsatellite analysis of the genetic variation in black bear populations. Mol. Ecol. 3: 489–495.

Pearcy, W.G. 1992. Ocean ecology of North Pacific salmonids. Seatle, Wash.: University of Washington Press.

Quinn, T.P., J.L. Nielsen, C.A. Gan, M.J. Unwin, R.Wilmot, C. Guthrie, and F.M. Utter. 1996. Origin and genetic structure of chinook salmon (*Oncorhynchus tshawytscha*) transplanted from California to New Zealand: Allozyme and mtDNA evidence. Fish. Bull. [In press].

Ryman, N. and G. Ståhl. 1980. Genetic changes in hatchery stocks of brown trout. Can. J. Fish Aquat. Sci. 37: 82–87.

Ryman, N. and F. Utter. 1987. Population genetics and fisheries management. Seattle, Wash.: University of Washington Press.

Shiozawa, D.K., J. Kudo, R.P. Evans, S.R. Woodward, and R.N. Williams. 1992. DNA extraction from preserved trout tissues. Gr. Basin Nat. 52(1): 29–34.

Skaala, O., G. Dahle, K.E. Jorstad, and G. Naevdal. 1990. Interactions between natural and farmed fish populations: information from genetic markers. J. Fish Biol. 36: 449–460.

Slatkin, M. 1985. Rare alleles as indicators of gene flow. Evolution 39: 53–65.

Slatkin, M. 1995. A measure of population subdivision based on microsatellite allele frequencies. Genetics 139: 457–462.

Swift, C.C., Haglund, T.R., Ruiz, M., and Fisher, R.N. 1993. The status and distribution of the freshwater fishes of southern California. Bull. S. Calif. Acad. Sci. 92(3): 101–167.

Titus, R.G., Erman, D.C., and Snider, W.M. 1996. History and status of steelhead in California coastal drainages south of San Francisco Bay. Hilgardia [In press].

U.S. Fisheries Commission. 1874–1901. Reports of the Commissioner for the United States Commission of Fish and Fisheries with supplementary papers. B—The Propagation of Food-fishes in the Waters of the United States, Parts II–XXVII. Washington D.C.: U.S. Government Printing Office.

Utter, F.M. 1991. Biochemical genetics and fishery management—an historical perspective. J. Fish Biol. 39(suppl. A): 1–20.

Wade, M.J., M.L. McKnight, and H.B. Shaffer. 1994. The effects of kin-structured colonization on nuclear and cytoplasmic genetic diversity. Evolution 48: 1114–1120.

Weber, J.L. and C. Wong. 1993. Mutation of human short tandem repeats. Hum. Mol. Genet. 2: 1123–1128.

24

Population Genetics of Kenyan Impalas—Consequences for Conservation

PETER ARCTANDER, PIETER W. KAT, BO T. SIMONSEN, AND HANS R. SIEGISMUND

Genetic information is beginning to gain acceptance in the formulation of conservation strategies, especially for species threatened by extinction, and/or those being bred in captivity. However, it has become abundantly clear that our knowledge of the genetic structure of natural populations is less than optimal given the necessity to manage species increasingly represented by small populations relegated to scattered reserves. In Africa, burgeoning human populations and their associated agricultural activities have generally confined wildlife to isolated protected areas. The designation of such reserves was based largely on their unsuitability for cultivation, and therefore they only accidentally represent areas of unique biological diversity. In addition, in eastern as well as southern Africa, wildlife is now being confined by fences erected as veterinary and/or agricultural cordons.

Given this situation, it is equally important to examine the genetics of common species as endangered taxa. Wildlife is increasingly viewed in Africa as an exploitable natural resource. We therefore designed a program to examine the population genetics of a variety of African bovid species, with the ultimate aim of guiding conservation and sustainable utilization programs. In addition, these studies would provide insight into the population genetic structure of large terrestrial mammals, which in Africa have not undergone the repeated population bottlenecks, translocations, and other manipulations characteristic of large mammals on other continents. In this chapter we present data on variation in the mitochondrial control region of impalas from six populations in Kenya. We compare the population structure of this species with that of three other sampled at the same locations.

The impala is a 40–76 kg antelope found in eastern and southern Africa south of the Sahara (Kingdon, 1982). It is an ecotone species, preferring light woodland with little undergrowth and grassland of low to medium height. These preferences, besides its dependence on water, produce an irregular and clumped distribution of populations (Estes, 1991). Impalas predominantly graze on young grasses and browse on foliage, forbs, shoots, and seedpods. This flexibility in feeding on both monocots and dicots is unusual among bovids, and allows establishment of dense local populations (Estes, 1991). The species is seasonally or permanently territorial, gregarious, and relatively sedentary. Females form clans with discrete home territories, and few females leave their clans to join other groups (Murray, 1981, 1982). Males generally leave their natal clans, but only disperse a short distance to enter neighboring clans (Murray, 1982).

Taxonomically, impalas have been classified with alcelaphines (wildebeest, hartebeest, topi) (Gentry and Gentry, 1978), gazelles (Kingdon, 1982), reduncines (waterbuck, reedbuck, kob) (Murray, 1984), and their own subfamily, the *Aepycerotini* (Ansell, 1971; Vrba, 1979). A recent analysis based on enzyme electrophoresis confirmed the genetic distinctiveness of impalas (Geordiadis et al., 1990), but did not satisfactorily resolve their taxonomic relationship to other members of the Bovidae. Since appearing in the fossil record about 4–5 million years ago (Vrba, 1979), impala seems to have been a long-lasting lineage, and there is no evidence that more than a single species existed in the past (Vrba, 1984). This contrasts with the high incidence of speciation among the alcelaphines, for example, and Vrba (1980) postulates a relationship between impala eurytopy and lack of phylogenetic diversification. Currently, only two subspecies are recognized, *A. melampus melampus* and the black-faced impala, *A. m. petersi.*

MATERIALS AND METHODS

Skin biopsies were obtained from free-ranging impalas by a system of remote sample collection described by Karesh et al. (1987). Samples representing a total of 61 individuals were collected from six different localities in Kenya (see Figure 24-1). They were collected from Tsavo East National Park (TS, $n = 10$),

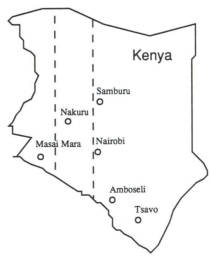

Figure 24-1 Sampling localities in Kenya. Abbreviations: AM, Amboseli; MM, Masai Mara; NA, NB, Nairobi; NK, Nakuru; TS, Tsavo; SA, Samburu.

Samburu National Park (SA, $n = 10$), Nakuru National Park (NK, $n = 11$), Nairobi National Park (NA/NB, $n = 10$), Masai Mara National Reserve (MM, $n = 10$) and Amboseli National Park (AM, $n = 10$). In the field, skin biopsies were immediately transferred to liquid nitrogen cr 25% DMSO saturated with NaCl (Amos and Hoelzel, 1991). In the laboratory, samples were stored at $-80°C$.

Genomic DNA was prepared from skin biopsies using standard protocols involving treatment with SDS and proteinase K, and subsequent phenol–chloroform extraction (e.g., Sambrook et al., 1989). A double-stranded PCR product (Saiki et al., 1988) of the entire control region was obtained with the primers HH00651 and HL15926 (Kocher et al., 1989); primers with H as their first letter are numbered according to the human sequence published by Anderson et al. (1981); B as first letter refers to the cow (*Bos*) sequence (Anderson et al., 1982). Single-stranded DNA was obtained by unbalanced PCR of the double-stranded PCR product (Gyllensten and Erlich, 1988), followed by dideoxy termination sequencing. The nucleotide sequence of a 427 bp fragment of the control region, starting at $tRNA^{PRO}$, was determined for each individual. The primers HL15926, HH16397 (5'-TTTCACGCGGCATGGTGATTAA-3') and BH 16168 (5'-GGTTGCTGGTTTCACGCGGCA-3') were used for single-stranded PCR as well as sequencing; both strands were sequenced. PCR and sequencing followed standard procedures (e.g., Innis et al. 1990).

The nucleotide diversity within and between populations was estimated according to Nei (1987). These measures represent the average probability of observing different bases at a site in the comparison of sequences. Between-population divergence was corrected for within-population diversity to obtain the net nucleotide substitution according to Nei (1987).

Population structure was analyzed according to the method of Hudson et al. (1992a). Nucleotide diversity for the total sample assuming a single population, K_T, was partitioned into the average within-population nucleotide diversity, K_S, and the fraction of the total diversity that was due to population subdivision, $K_{ST} = 1 - K_S/K_T$. The test for subdivision was based on Monte Carlo simulations; individuals were sampled randomly and assigned to populations 1000 times. Each time the parameter K_{ST} was estimated and tabulated. The observed K_{ST} parameter was compared with the distribution of the simulated values. The significance P-value was based on the probability of observing a value as large as or larger than the observed value. The same procedure was used to examine pairwise homogeneity of the populations.

The phylogenetic relationship of the sequences was estimated with the method of Saitou and Nei (1987) using the program NEIGHBOR from the package PHYLIP 3.5 (Felsenstein, 1993). The distance matrix was estimated with the program DNADIST from the sequences using maximum-likelihood distances. Distances were based on the maximum-likelihood method of Felsenstein (1981) as amended by Kishino and Hasegawa (1989). The distances were estimated with an empirical transition/transversion rate of 13.5. Bootstrap values were calculated by generating 100 samples with the program SEQBOOT, which were used as input for the NEIGHBOR program. Resulting trees were analyzed with the program CONSENSE of the same package.

Table 24-1 Segregating Sites in the 46 Haplotypes Found in 427 bp D-loop Sequences of 61 Impalas from Six Locations in Kenya (see Figure 24-1 for abbreviations). A dot indicates equality with the sequence TS2.

Segregating Site

```
                              1111 1111111122 2222222222 2333333333 3333333333 333444
                          23562245 5588889923 3333366667 9002222222 3334555677 799111
                        2701781351 5602390892 3468914573 3032345678 5895234014 818589
                  10        20         30         40         50

 1 TS2  CGTTGAGACC CCTAGTCACT CTTTAAAGA- TTCCTAGACA AGGATCTTAG GATTAC
 2 TS3  .......... .......... ......N... .......... ....C..... ......
 3 TS4  .......... .G...T.TC. ....T..G.. .....G.T.. CT.GA..... A.T...
 4 TS6  .......... .......... .......... .T..AG.... ...C...... ......
 5 TS7  .......... .G...T.TT. ....T..G.. ....GAGT.. ...T..A... .G..T.
 6 TS10 ......G... .G...T.TC. TC.G.T..G. .....G.T.. .CT.A..... ...T..
 7 TS11 .......... .G...T.TT. ....T..G.. .A..GAGT.. ...T..A... .G..T.
 8 TS12 .......... .G...T.TT. ....T..G.. .AGAGT.... ...T..A... .G..T.
 9 SA1  .......... .......... ....A..... ......A.G. ...C...... ......
10 SA3  .......... ........C. .......... .......G.. ...C...... ......
11 SA7  .......... .G........ .......... .....A.G.. ...C.A.... ......
12 SA11 .......... .A....C... .......... .......G.. ...C...... ......
13 NK2  T......... .G...T.TC. .......C.. ......G... .NCT..A... ...GT.
14 NK4  .......... .......... ...A.C.... ......A... ...C...... ......
15 NK5  .......... .......... .......... T..A.G.... ...C...... ...G..
16 NK7  .......T.. TG.A..T... T..G...... ..C.G.G... .TC.A..... ...C..
17 NK8  .......... .GC...TT.. T..G...... ..C..G.T.. .CT..A.... ...T..
18 NK9  .......... .......... .......... T..A.G.... ..TC...... ...G..
19 NK11 .......... ........C. .......... ......G... ...C.A.... ...C..
20 NK12 .......... .......... ....A..... ......A... ...C...... ......
21 NK15 .......... .G...T.TC. T..G...... ......G... .GCT...A.. ..C.GT
```

Numbers of Haplotypes in Populations

Sequence	TS	SA	NK	NA	MM	AM
1 TS2	1					
2 TS3	3					
3 TS4	1					
4 TS6	1					
5 TS7	1					
6 TS10	1					
7 TS11	1					
8 TS12	1					
9 SA1		6				
10 SA3		2				
11 SA7		1				
12 SA11		1				
13 NK2			1			
14 NK4			1			
15 NK5			1			
16 NK7			1			
17 NK8			1			
18 NK9			2			
19 NK11			1			
20 NK12			1			
21 NK15			2			

```
                                                                                                                                                                                    3  1
22 NA1    ........T. TG.A..T. T..G.G... ....C.G.G G...TC..A ...C..
23 NA2    .......... ..A..... .......... ...T..A... ......C... A...     1
24 NA3    .......... ..GA.... .......... ....A.... ......C...          1
25 NB6    .......... ........ ...N...... ...T.GA... .AA..C...           1  1
26 NB9    .......T. TG.A..T. T..G.G... ....C.G. G...TC..A ...C..        1  1
27 NB10   .A..A.... .G...T.. T..-..A. ......G... .......C...            1  1
28 NB11   ......A.T. TG.A..T. T..G.G... ....C.G.G G...TC..A ...C..       1
29 NB12   .......... ..A..... ...C..... ...T..A... ......C...           1
30 MM2    ......... .G...T.. T..G..... ...T.GAGT. .....C.A .G...        1
31 MM3    .......T. TG.A..T. T..G..G. ....C.G.G G...TC..A ...C..         1
32 MM5    .......T. TG.A..T. T..G..... ...C...... G...CTG..A ....        1
33 MM7    .CA...T. .G...CT. T..G..... ...N.GA.T. ....T.C.A G....        1
34 MM10   .A...... .G...CT. T..G..... ...T.GAGT. ....T.C.A G....         1
35 MM11   ......... .GC..T.T. T..G..... C.T..G.T ....CT..A ....T          1
36 MM13   .......T. TG.A..T. T..G..... ...C..G... G...CTC..A ....         1
37 MM14   .CA...... .G...CT. T..G..... ...T.GA.T. ....T.C.A .....         1
38 AM1    .......T. TG.A..T. T..G..... ...C..G... G...CTC...             1
39 AM2    ...A..... .G...T.. T.....A.. ......G... ......A...             1
40 AM5    ......... .G...T.. T.....A.. ......T... ...TC..G...             1
41 AM6    .A....... .G..T.TC T..GG.... ....G.T ...CT..A ....T             1
42 AM8    ......T.. T.G.A..- T..G..... ......T G......A ...T              1
43 AM9    ...A..... ......... ....T..A.. ....C..A.....                    1
44 AM11   ....G... .G...T.TC T..GG.... ...G.T ...CT..A ...T               1
45 AM12   ....G... .G...T.T. T..G..... ...GAGT. ...T..A.G..T              1
46 AM13   .......... .......... ...T..A.. ....C..A......                  2
Number of bases 2222222222 2222222222 2222222221 2222322232 2222223222 222322  10 10 11 10 10 10
```

RESULTS

Each sampled individual was sequenced for 427 bp of the mitochondrial control region. The 61 sequences could be partitioned into 46 different haplotypes (Table 24-1), suggesting a high level of variation. The most common sequence, SAl, was observed six times, and other sequences were represented in at most three individuals. Of the 51 segregating sites with two different base pairs, three were transversions, and the remaining 48 sites were transitions. Only four sites segregated with three different base pairs. Sequences were submitted to GenBank under the accession numbers U12341–U12385.

Owing to the diversity of haplotypes, only two were found in more than one population. These haplotypes, NA1 and NB9, were encountered in individuals in both Nairobi and Masai Mara populations. The high level of haplotype diversity was mirrored by high nucleotide diversity. Except for Samburu, which had a diversity of 0.007, the other populations were characterized by homogeneous values ranging from 0.0228 to 0.0286 (Table 24-2). The average nucleotide diversity over all populations (K_S) was 0.0233. The largest fraction of the total nucleotide diversity ($K_T = 0.0280$) was contained within populations. The contribution due to the differentiation between impala populations (K_{ST}) was 0.167, which was significant at the 0.1% level.

The populations sampled did not form a homogeneous group. In pairwise comparisons of the populations (Table 24-2), 10 comparisons were significant at the 5% level, and one comparison (Masai Mara versus Nairobi) was marginally significant. Three population comparisons were not significant (Tsavo, Nakuru, and Amboseli). A test for heterogeneity among these three populations resulted in a K_{ST} value of -0.004, which was nonsignificant ($P = 0.427$). The level of net nucleotide difference among these populations was also indicative of a low level of divergence (Table 24-2). Two comparisons were negative (Amboseli–Tsavo and Amboseli–Nakuru), and the third (Tsavo–Nakuru) was the smallest positive value of all distances noted. Thus, among the impalas sampled, three populations had diverged significantly from each other; the other three populations formed a homogeneous group. In this homogeneous group, two of the populations were geographically close to each other (Tsavo and Amboseli; ~ 150 km), whereas the

Table 24-2 Sequence Statistics[a]

	Tsavo	Samburu	Nakuru	Nairobi	Masai Mara	Amboseli
Tsavo	**0.0228**	0.0080	0.0003	0.0060	0.0056	−0.0008
Samburu	0.000	**0.0070**	0.0058	0.0091	0.0195	0.0085
Nakuru	0.295	0.003	**0.0272**	0.0040	0.0064	−0.0000
Nairobi	0.015	0.001	0.041	**0.0269**	0.0040	0.0039
Masai Mara	0.013	0.000	0.001	0.052	**0.0286**	0.0036
Amboseli	0.596	0.000	0.349	0.045	0.041	**0.0269**

[a] Diagonal (**bold**): within-population nucleotide diversity. Above diagonal: net population nucleotide distance. Below diagonal: significance level based on Hudson et al. (1992a) K_{ST} statistic.

population from Nakuru was located in the Rift Valley, at a considerable distance from both Tsavo and Amboseli (Figure 24-1).

Based on the net nucleotide differences between populations, the impalas from Samburu diverged furthest from the other five populations (Table 24-2). This divergence was not due to sequence differentiation but rather to a skewed distribution of sequences at this locality.

The phylogenetic relationships among the 46 sequences also did not suggest major branches differentiating populations (Figure 24-2). The branch with the highest bootstrap value (95%) contained sequences from four different populations, suggestive of homogenizing gene flow. There were few branches that contained sequences exclusive to a single population. The Masai Mara sequences MM2, MM7, MM10, and MM14 formed such a group. A similar case applied to four sequences (TS1, TS7, TS12, and AM12) from the two easternmost populations sampled. As most bootstrap values were relatively low, it was not possible to reach conclusions about the evolutionary relationships among the populations from the phylogenetic tree.

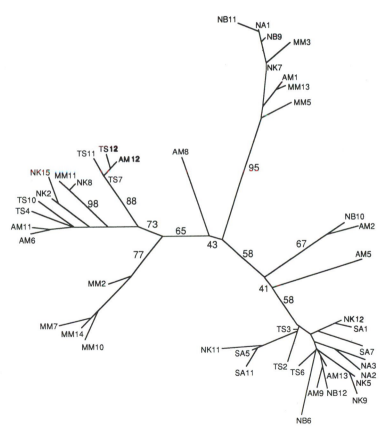

Figure 24-2 Neighbor joining tree of the 46 sequences. The values at branches indicate how often a branch has been observed in 100 bootstrap samples of the sequences. See Figure 24-1 for abbreviations.

Table 24-3 Haplotype and Nucleotide Variation in Four Bovid Species in Kenya.

Species	Sample Size	Sequence Length	Number of Populations	Haplotypes/ Sample Size	Fraction of Segregating Sites	H_T	H_S	H_{ST}	K_T	K_S	K_{ST}
Impala	61	427	6	0.754	0.129	0.984	0.910	0.076	0.0280	0.0233	0.167
Wildebeest	48	420	3	0.792	0.107	0.983	0.927	0.058	0.0181	0.0140	0.225
Waterbuck	105	472	5	0.524	0.197	0.979	0.956	0.035	0.0418	0.0370	0.115
Grant's gazelle	44	371	5	0.886	0.344	0.994	0.970	0.024	0.1085	0.0368	0.661

H_T is the total haplotype diversity for all populations. It is a measure for the probability of drawing two different haplotypes from a population. H_S is the average haplotype diversity within populations. H_{ST} is the part of the total diversity which is due to differentiation between populations. K_T, K_S, and K_{ST} are similar measures for nucleotide diversity.

As a comparative analysis, we examined partitioning of genetic variation within and between populations for several bovid species (wildebeest, Grant's gazelle, and waterbuck) analyzed earlier (Arctander et al., 1996). The control region was highly variable in all species, resulting in a large number of haplotypes. The lowest within-population diversity was found among impalas ($H_S = 0.91$). The highest haplotype diversity (almost 1.00) was found among Grant's gazelles (Table 24-3). In general, the magnitude of haplotype diversity made it difficult to use haplotypes to analyze population structure. Consequently, we based our analyses of population structure on sequence statistics. These differ substantially among the four species. The average within population nucleotide diversity did not differ much among species, ranging from 0.014 to 0.037 (Tables 24-3). A larger difference was observed for the total diversity, where the value found among Grant's gazelles was six times larger than the smallest value observed for wildebeest. These differences were also reflected in the fraction of segregating sites; Grant's gazelles had the largest value (0.344). Finally, the fraction of total diversity due to population differentiation was also highest among Grant's gazelles; about two-thirds of the total nucleotide diversity for this species could be attributed to population subdivision. This value was about 3–6 times larger than that observed for the other species.

DISCUSSION

Biological attributes of impalas, including territoriality, dependence on water, and short distance of observed dispersal, have contributed to an apparently clumped distribution of populations (Murray, 1982; Estes, 1991). Based on these characteristics, we would have predicted a subdivided, geographically structured genetic variation among impala populations. Our observations, however, indicate that the ratio of haplotypes to individuals sampled was high (0.75), and that some populations have diverged to a variable extent, whereas three populations form a homogeneous group. This homogeneity might be spurious, since Hudson et al.'s method has a lower power in resolving population differentiations with non-recombining sequences, like the mtDNA used in our study. On the other hand, the mutation rate in the mitochondrial genome of mammals is much higher than in the nuclear genome—in particular at the control region. In addition, the effective population size of mitochondria is usually smaller than the effective size of nuclear loci (one-half the effective number of females). This increases the power of sequence-based statistics. The power of Hudson et al.'s methods was evaluated under the assumption of equilibrium between mutation, genetic drift, and migration, which is difficult to test for. The test as such does not require the populations to be at that equilibrium. Therefore, we believe that the observed homogeneity is a real phenomenon, indicative of homogenizing gene flow among impala populations. We hypothesize that impala eurytopy and feeding flexibility contributes to such gene flow, and that dispersal distances might exceed those measured by Murray (1982) at least over a longer time span. In addition, there is abundant evidence that climatic fluctuations occurred frequently in eastern Africa during and following Pleistocene glacial episodes (Finney and Johnson, 1991;

Hamilton and Taylor, 1991; Casanova and Hillairemarcel, 1992); such fluctuations contributed to changing local faunal assemblages over time (Marean and Gifford-Gonzalez, 1991). The present genetic structure of impala populations could therefore be influenced by past distributions.

Two populations, Nakuru and Samburu, exhibit genetic characteristics that differentiate them from other impala populations examined in this study. Samburu represents one of the northernmost populations in the geographic range of the species (Estes, 1991), and is located in a corridor of suitable habitat along a seasonal river in an otherwise arid environment. The impalas at Samburu exhibit both the lowest within-population nucleotide diversity observed (0.007) and the smallest fraction of segregating sites (0.19) when compared to other populations in this study. We postulate this reduced diversity is due to a recent bottleneck, associated with relatively recent recolonization of the area. The population at Nakuru, located in the Rift Valley, surprisingly groups with populations from Tsavo East and Amboseli, located a considerable distance to the east. Based on geographic location, we would have predicted a closer relationship to either the Masai Mara or Nairobi populations. We propose that the anomalous genetic affinity of the Nakuru population resulted from a translocation of impalas into the area; before the formation of the reserve, Nakuru had ranchland and farmland. However, no record of such a translocation exists with wildlife authorities.

To date we have examined the population genetics of four bovid species in Kenya: wildebeest, Grant's gazelle (Arctander et al., 1996), waterbuck, and the impalas described in this chapter. We limit the use of sequence statistics to analysis of population subdivision. Natural extensions would be measures of effective population size (Felsenstein 1992a,b) and estimation of gene flow between populations (Hudson et al., 1992b). Some of the methods involve use of phylogenies of sampled sequences, whereas others are based on simple functions of nucleotide diversities within and between populations (Table 24-3). Each of the four species examined to date exhibits a different pattern of population genetic differentiation.

Wildebeest are large, highly mobile bovids that can migrate over vast geographic areas. This mobility would predict extensive gene flow between populations (Avise, 1991), but the wildebeest populations examined in Kenya exhibit considerable control region sequence divergence across the eastern, or Gregory, Rift Valley. This subdivision confirms earlier results (Georgiadis and Kat, 1989) from the nuclear genome (based on enzyme electrophoresis), and provides genetic recognition that the species is divided into separate lineages across this geologic feature: *C. t. mearnsi* to the west, and *C. t. albojubatus* to the east (Kingdon, 1982). The Mara/Serengeti population on the western side of the Gregory Rift also exhibits considerably greater genetic diversity, which is likely explained by the historically and presently large number of animals in this population: the Serengeti migrant population today numbers close to 1.4 million animals. In contrast, the Nairobi and Amboseli populations are much less variable. Such lack of diversity might be explained by a historical bottleneck: as recently as 12 000 BP the large-mammal community structure of the savannas east of the Rift Valley was very different and wildebeest were not present (Marean and

Clifford-Gonzalez, 1991). Apparently, dry grasslands and arid-adapted ungulates expanded at least as far south as northern Tanzania during the last glacial maximum and numerical domination by large herds of wildebeest could be a recent phenomenon. If colonization of the Amboseli/Nairobi savannas occurred through a series of founder events from a source population, in Tanzania, for example, a reduction of genetic variability would be expected. The low level of eastern wildebeest nucleotide diversity compares with that noted among impalas at Samburu. Natural occurrence of a reduced level of genetic diversity among populations of abundant species might therefore be a relatively common phenomenon.

An analysis of control region sequences from Grant's gazelles sampled from six different localities in Kenya has revealed marked differentiation among haplotypes despite lack of dispersal barriers (Arctander et al., 1996). Parapatric populations from Amboseli and Tsavo East exhibit a sequence divergence of about 12%, indicative of long-term reproductive isolation. The amount of divergence among populations of Grant's gazelles is of equal magnitude to the divergence between these gazelles and a closely related species, the Soemmerings gazelle. Despite this high level of population differentiation, Grant's gazelles are more closely related to each other than to related species as measured by sequence divergence at the mtDNA protein-coding cytochrome b gene and the nuclear α-lactalbumin gene. We hypothesize that the present pattern of differentiation among populations is a result of recently established contacts between formerly allopatric populations.

The waterbuck is an antelope species with a widespread distribution in sub-Saharan Africa. Two forms, common and defassa waterbuck, have been described based on coat color and pattern. These forms are largely allopatrically distributed except for a zone of range overlap and reported hybridization in eastern Africa. Statistically significant geographic structuring of mtDNA variation was apparent among populations. Two major mtDNA lineages representing common and defassa waterbucks were observed; both mtDNA lineages were found among individuals of morphologically defined defassa but not among common waterbuck populations. It is hypothesized that common and defassa waterbucks evolved in allopatry with a secondary eastward range expansion of the latter form (e.g., facilitated by climatic changes associated with the most recent post-glacial humid period). Extensive hybridization with common waterbucks has apparently accompanied this range expansion of defassa waterbucks.

Analysis of the population genetic variability of four bovid species sampled from the same localities in Kenya thus reveals four patterns of genetic differentiation, ranging from differentiation by distance (impalas), to differentiation across geographic barriers to migration (wildebeest), to introgression (waterbuck), to differentiation in the absence of apparent geographic or habitat barriers (Grant's gazelle). Given the large body size and mobility of these species, we would not have predicted this extreme population genetic differentiation. These data underline the importance of population genetic studies that are comprehensive, as they illustrate the hazards of extrapolation across species from limited data sets. In addition, these data indicate that populations of large mammals can show surprising amounts of genetic differentiation across small distances. We postulate that climatic and habitat factors no longer apparent today might have been responsible for the generation of such patterns, and that present patterns of distribution of large African mammals might have been established relatively recently.

Programs to conserve genetic diversity among large mammal populations will need to acknowledge that very different patterns of genetic differentiation can exist among populations of different species sampled at the same locality. The establishment of migration corridors to link reserves might therefore not be equally important in all cases, as they might unite populations that diverged long ago. Provenance of breeding stock for game ranches or captive breeding programs also needs to be carefully established, especially if such breeding programs seek to mirror natural patterns of genetic diversity. Our perception of the genetic structure of terrestrial mammalian populations has been heavily influenced by studies of small mammals and endangered species, and the continued analysis of large mammals will undoubtedly add considerably to the design of informed conservation programs.

ACKNOWLEDGMENTS

We thank the Kenya Wildlife Service and the Wardens of Tsavo East, Amboseli, Nairobi, Samburu, and Nakuru National Parks, as well as the Senior Warden, Masai Mara National Reserve, the Senior Warden, Narok District, and the Narok County Council for sampling permission and assistance with logistics. Per Palsbøl kindly provided the sequence for primer HH16397. We also acknowledge the assistance of Ben Kiawa, Joseph Owino, Mohamed Jama, and Paul Mbugwa. This research was carried out as a collaborative venture between the National Museums of Kenya and the Department of Population Biology, University of Copenhagen, and was funded by the Danish International Development Agency (DANIDA) and also supported by the Danish Research Council's Centre for Tropical Biodiversity.

REFERENCES

Amos, B. and A.R. Hoelzel. 1991. Long-term preservation of whale skin for DNA analysis. In: A.R. Hoelzel, ed. Genetic ecology of whales and dolphins. Report of the International Whaling Commission, Special Issue 13. Cambridge, U.K. International Whaling Commission.

Anderson, S., A.T. Bankier, B.G. Barrel, M.H.L. de Bruijn, A.R. Coulson, J. Drouin, I.C. Eperon, D.P. Nierlich, B.A., Roe, F. Sanger, P.H. Schreier, A.J.H. Smith, R. Staden, and I.G. Young. 1981. Sequence and organization of the human mitochondrial genome. Nature 290: 457–465.

Anderson, S., M.H.L. De Bruijn, A.R. Coulson, I.C. Eperon, F. Sanger, and I.G. Young. 1982. Complete sequence of bovine mitochondrial DNA, conserved features of the mammalian mitochondrial genome. J. Mol. Biol. 156: 683–717.

Ansell, W.F.H. 1971. Order Artiodactyla. In: J. Meester and H.W. Setzer, eds. The mammals of Africa, an identification manual. Washington, D.C.: Smithsonian Institution.

Arctander, P., P.W. Kat, R.A. Aman, and H.R. Siegismund. 1996. Extreme genetic differences among populations of Grant's gazelle (Gazella granti) in Kenya. Heredity 76: 465–475.

Avise, J.C. 1991. Ten unorthodox perspectives on evolution prompted by comparative population genetic findings on mitochondrial DNA. Annu. Rev. Genet. 25: 45–69.

Casanova, J. and C. Hillairemarcel. 1992. Late holocene hydrological history of Lake Tanganyika, East Africa, from isotopic data on fossil stromatolites. Palaeogeogr. Palaeoclimatol. Palaeocol. 91: 35–48.

Estes, R.D. 1991. The behavior guide to African mammals. Berkeley, Calif.: University of California Press.

Felsenstein, J. 1981. Evolutionary trees from DNA sequences: a maximum-likelihood approach. J. Mol. Evol. 17: 368–376.

Felsenstein, J. 1992a. Estimating effective population size from samples of sequences: Genet. Res. 59: 139–147.

Felsenstein, J. 1992b. Estimating effective population size from samples of sequences: a bootstrap Monte Carlo integration approach. Genet. Res. 60: 209–220.

Felsenstein, J. 1993. PHYLIP (Phylogeny Inference Package) version 3.5. Distributed by the author. Department of Genetics, University of Washington.

Finney, B.P. and T. Johnson, 1991. Sedimentation in Lake Malawi (East Africa) during the past 10,000 years—a continuous paleoclimatic record from the southern tropics. Palaeogeor. Palaeoclimatol. Palaeocol. 85: 351–366.

Gentry, A.W. and A. Gentry. 1978. The Bovidae (Mammalia) of Olduvai Gorge, Tanzania. Parts 1 and 2. Bull. Br. Mus. Nat. Hist. (Geol.) 29: 289–446; 30; 1–83.

Georgiadis, N.J. and P.W. Kat. 1989. Wildebeest races across the Rift Valley. Swara 12: 21–23.

Georgiadis, N.J., P.W. Kat, J. Patton, and H.A.T. Oktech. 1990. Allozyme divergence within the Bovidae. Evolution. 44: 2135–2149.

Gyllensten, U.B. and H.A. Erlich. 1988. Generation of single-stranded DNA by the polymerase chain reaction and its application to direct sequencing of the HLA-DQA locus Proc. Natl. Acad. Sci. U.S.A. 85: 7652–7656.

Hamilton, A.C. and D. Taylor. 1991. History of climate and forests in tropical Africa during the last 8 million years. Clim. Change 19: 65–78.

Hudson, R.R., D.D. Boos, and N. Kaplan. 1992a. A statistical test for detecting geographic subdivision. Mol. Biol. Evol. 9: 138–151.

Hudson, R.R., M. Slatkin, and W.P. Maddison. 1992b. Estimation of levels of gene flow from DNA sequence data. Genetics 132: 583–589.

Innis, M.A., D.H. Gelfand, J.J. Sninsky, and T.J. White, eds. 1990. PCR protocols. San Diego: Academic Press.

Karesh, W.B., F. Smith, and H. Frazier-Taylor. 1987. A remote method for obtaining skin biopsy samples. Conserv. Biol. 1: 261–262.

Kingdon, J.S. 1982. East African mammals, vol. 3C (Bovids). New York: Academic Press.

Kishino, H. and M. Hasegawa. 1989. Evaluation of the maximum-likelihood estimate of the evolutionary tree topologies from DNA sequence data, and the branching order in Hominoidea. J. Mol. Evol. 29: 170–179.

Kocher, T.D., W.K. Thomas, A. Edwards, S.V. Pääbo, F.X. Villablanca, and A.C. Wilson. 1989. Dynamics of mitochondrial DNA evolution in animals: Amplification and sequencing with conserved primers. Proc. Natl. Acad. Sci. U.S.A. 86: 6196–6200.

Marean, C.W. and D. Gifford-Gonzalez. 1991. Late Quaternary extinct ungulates of East Africa and palaeoenvironmental implications. Nature 350: 418–420.

Murray, M.G. 1981. Structure of association in impala, Aepyceros melampus. Behav. Ecol. Sociobiol. 9: 23–33.

Murray, M.G. 1982. Home range, dispersal, and clan system of the impala. Afr. J. Ecol. 20: 253–269.

Murray, M.G. 1984. Grazing antelopes. In: D.W. MacDonald, ed. The encyclopedia of mammals, pp. 560–569. New York: Facts on File.

Nei, M. 1987. Molecular evolutionary genetics. New York: Columbia University Press.

Saiki, R.K., D.H. Gelfand, S. Stoffel, et al. 1988. Primer-directed enzymatic amplification of DNA with a thermostable DNA polymerase. Science, 239: 487–491.

Saitou, N. and M. Nei. 1987. The neighbor-joining method: a new method for reconstructing phylogenetic trees. Mol. Biol. Evol. 4: 406–425.

Sambrook, J., E.F. Fritsch, and T. Maniatis. 1989. Molecular cloning: a laboratory manual, 2d. ed. New York: Cold Spring Harbor Laboratory Press.

Vrba, E.S. 1979. Phylogenetic analysis and classification of fossil and recent Alcelaphini Mammalia: Bovidae. Biol. J. Linn. Soc. 11: 207–228.

Vrba, E.S. 1980. The significance of bovid remains as indicators of environment and predation patterns. In: A.K. Behrensmeyer and A.P. eds. Hill, Fossils in the making, pp. 242–271. Chicago: University of Chicago Press.

Vrba, E.S. 1984. Evolutionary pattern and process in the sister-group Alcelaphini-Aepycerotini. In: N. Eldredge and S.M. Stanley, eds, Living fossils, pp. 62–74. New York: Springer Verlag.

25

Paternity Studies in Animal Populations

PATRICIA G. PARKER, T. A. WAITE, AND T. PEARE

MATING SYSTEMS AND THE PROBLEM OF PATERNITY

The number of mates attained by members of each sex defines the mating system of animal populations. Under social (apparent) monogamy, one male and one female form an exclusive mating partnership for one or more reproductive attempts. In promiscuous systems, members of each sex typically mate with multiple partners. In polygynous systems, competition among males over mating opportunities with females leads to higher variance in apparent reproductive success (RS) among males than among females; the converse is true for polyandry.

The social mating system can be described through observations of the behavioral associations of individual males and females within populations; actual patterns of gametic exchange, the genetic mating system underlying the social mating system, can differ markedly from the apparent mating system if individuals have copulatory partners besides their social mate (Westneat et al., 1990). This possibility complicates the calculation of individual reproductive success.

For animals with internal fertilization (and for plants), reproductive success of females can be estimated with confidence because they usually remain in physical contact with the zygote (but see Birkhead et al., 1990), and they often provide postzygotic care. However, estimation of RS for males is often problematic because DNA may be the only connection between fathers and offspring. Even when males provide parental care, observational estimates of male RS can be in error. For example, an observational estimate of RS for an apparently monogamous male bird can be an overestimate of actual reproductive success if some of the young in the nest were sired through extra-pair copulation (EPC) by the female, or an underestimate if the male sired young elsewhere by EPC. Recent studies of avian mating systems employing molecular markers to determine parentage have found such errors to be common (e.g., Gibbs et al., 1990; Kempenaers et al., 1992; and references in Birkhead and Møller, 1992). In general, the frequent exchange of gametes across boundaries of social units will mean that some socially monogamous populations are in fact genetically polyandrous, polygynous, or promiscuous.

IMPLICATIONS FOR CONSERVATION

Effective population size (N_e) reflects the vulnerability of small populations to random loss of genetic variation. N_e is strongly influenced by mating system (Chesser, 1991; Nunney, 1993). Mating systems in which some individuals are excluded from reproduction altogether (Nunney, 1991), or in which variance in reproductive success by one or both sexes is high (Nunney, 1993), can produce very small values of N_e. In populations of conservation concern, which may be characterized by elevated rates of reproductive failure (i.e., when a greater than ordinary proportion of those attempting to breed fail to produce any offspring), the response of N_e to failure rate depends on mating system. For example, a nest failure by a genetically monogamous bird would remove both male and female from the effective population, while a similar failure in a promiscuous bird might remove only the female, if the male has mated with another female that succeeded. Similarly, a nest failure in a polygynous system would remove only the female from the effective population, while a similar failure in a polyandrous system in which multiple males are simultaneously and exclusively mated to a single female (e.g., Galapagos hawks; see case studies below) would remove the female and all males from the effective population. Likewise, increases (or decreases) in variance of individual RS due to cryptic reproduction (extra-pair fertilizations, intraspecific brood parasitism, or rapid mate-switching), can have significant effects on estimates of N_e, even when the numbers of reproductive males and females do not change.

Therefore, estimates of effective population size require accurate estimates of both the numbers of adults of each sex that breed and the variance in reproductive success for each sex (Nunney and Elam, 1994). These parameters can now be estimated accurately by molecular determination of parentage in natural populations. Here, we report our efforts to describe the genetic mating systems of four natural populations. We compare apparent (social) with actual (genetic) mating systems to illustrate the extent to which molecular determination of parentage improves our understanding of the breeding structure of natural populations. In future work, our goal is to compare estimates of effective population size based on apparent RS (from observational evidence) with those based on actual RS (from molecular evidence).

CASE STUDIES

Black Vultures (*Coragyps atratus*): Monogamy in a Highly Gregarious Bird?

Black vultures are highly gregarious scavengers that sleep and feed in large groups. Their nests, however, are overdispersed and nest sites are used by mated pairs for many consecutive years (Rabenold, 1986). Mates maintain close association throughout the year, co-occurring more often than expected by chance at communal roosts and feeding sites. Coalitions consisting of many adult black vultures and their immediate nuclear families move about together from roost to roost and interact competitively at carcasses against other coalitions. Despite the wide spacing of nests, females have many opportunities for extra-pair copulations, because they interact daily during the egg-laying period with dozens of males, both at nocturnal roosts and at carcasses. However, among 36 offspring produced in 16 families, no extra-pair parentage was detected (Decker et al., 1993) following determination of parentage by multilocus minisatellite DNA finger-printing (Jeffreys et al., 1985). The actual mating system is identical to the apparent mating system in black vultures.

The ability to use molecular markers to circumvent the intractable demo-graphy and life history of species like black vultures allows quantification of patterns of reproductive success that would not be possible through standard observational techniques. (It is not possible to monitor an individual's activities through an entire day.) Recent work has shown, for example, that the oldest members of coalitions are usually close relatives, suggesting that the competitive coalitions are genetic clans (Parker et al., 1995).

Stripe-backed Wrens (*Campylorhynchus nuchalis*): Do Subordinates Ever Reproduce?

Stripe-backed wrens are cooperatively breeding passerines that inhabit scrub woodlands in the llanos of Venezuela (Rabenold, 1990). Social groups consist of a clearly dominant pair, along with adult "helpers" and dependent juveniles. Until recently (Rabenold et al., 1990), members of the dominant pair were presumed to be the only reproductive individuals. The helpers are physiologically mature and are usually nondispersed offspring of one or both of the dominant group members. They assist the dominant pair in defending the territory, building and maintaining the nest, and feeding of nestlings and fledglings. Many helpers die before attaining dominant (presumed breeding) status. Such individuals, if physiologically capable of reproduction, should be counted as potential breeding adults in the population for the purpose of estimating N_e (Nunney and Elam, 1994). Studies were undertaken to test the assumption that reproduction is the exclusive domain of the dominant group members. The primary motivation for this work was to quantify the costs and benefits of cooperation for the subordinate group members by determining the extent to which they reproduce as subordinates, and to verify the relatedness of subordinates to dominants suggested by pedigrees.

Twenty-two social groups were monitored for up to 3 years each, for a total of 34 group-years. Parentage was examined for 69 young resulting from these 34 breeding attempts by application of multilocus minisatellite DNA fingerprinting. Sex was determined for young birds by means of sex-specific fragments that appeared in minisatellite fingerprints (Rabenold et al., 1991a,b). Parentage was determined for 68 (of 69) young. Dominants (presumed parents) were assigned as actual parents for 62 (91.1%) young. The remaining six (8.8%) young were the offspring of dominant females and subordinate males. Reproduction by sub-ordinate males occurred only in groups in which they were unrelated to the dominant female. This group configuration usually occurred when the mother of the subordinates had died and been replaced by an unrelated immigrant female. No reproduction by subordinate females was detected, despite ample opportunity. Although subordinate females occurred during 19 (55.9%) group-years monitored, no auxiliary female was assigned as mother for any of the young. Auxiliary males occurred in 28 (82.4%) group-years, and sired young in 4 (11.8%) cases (Rabenold et al., 1990). Thus, in 22 presumed monogamous breeding units, there were actually 22 reproductive females and 26 reproductive males.

Galapogas Hawks (*Buteo galapagoensis*): How Many Males Reproduce?

Breeding groups of Galapagos hawks consist of one adult female and from one to eight adult males (modal number of males is two). In polyandrous groups, all males copulate with the female and participate in the provisioning of the young (Faaborg and Bednarz, 1990). Application of multilocus minisatellite DNA fingerprinting to samples from 66 individual hawks from 10 breeding groups on the island of Santiago revealed mixed paternity in most groups (Faaborg et al., 1995). Parentage was determined by matching bands on multilocus DNA finger-prints between offspring and all possible male–female combinations within the group (Figure 25-1) and corroborating the best-fit parental dyad by band-sharing analyses (Wetton et al., 1987; Figure 25-2). Multiple paternity was detected in five of six groups that produced two chicks in one nesting attempt. In addition, different males sired young in consecutive years in five of six groups in which male group membership was constant. Patterns of paternity were consistent with the hypothesis that reproductive success is randomly distributed among males within groups, with males apparently having equivalent probabilities of siring each young.

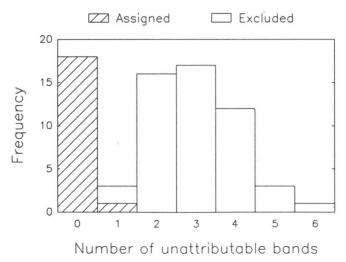

Figure 25-1 Frequency distribution of the number of bands, in fingerprints of 19 Galapagos hawk offspring, that could not be accounted for by the mother's fingerprint combined with that of either the male that was assigned as the father (hatched bars) or with that of a male that was not assigned (i.e., was excluded; open bars) as the father. Unattributable bands were summed across probes (Jeffreys' 33.15 and 33.6).

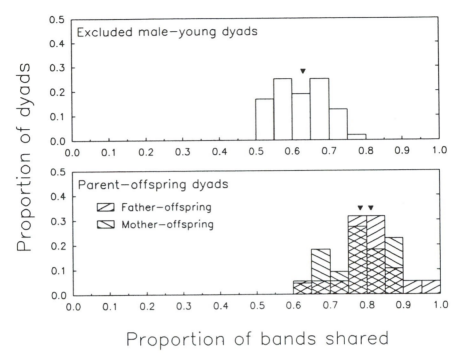

Proportion of bands shared

Figure 25-2 Distributions of band-sharing scores (proportions of bands shared) for dyads of Galapagos hawks taken from HaeIII digests hybridized with Jeffreys' probe 33.15. Each dyad consists of a nestling and an adult male excluded as that individual's father (top panel; N = 48), and either an assigned father and his offspring (N = 19) or a mother and one of her offspring (N = 22; bottom panel). Arrowheads indicate means (bottom panel: father–offspring mean = 0.81; mother–offspring mean = 0.78).

Overall, in 10 social groups, there were 10 reproductive females and 17 reproductive males.

Analysis of genetic similarity indices (band-sharing scores) suggests that males within groups were typically not close relatives (Figure 25-3). These results demonstrate that the Galapagos hawk is polyandrous, with relatively egalitarian relations among unrelated males belonging to the same breeding group (Faaborg et al., 1995). The sharing of paternity among unrelated males should retard the rate of loss of those males' lineages from the population. Simultaneous polyandry, such as that practiced by Galapagos hawks, may have a strong negative effect on N_e with increasing levels of reproductive failure. This decline in N_e results when a nesting failure removes the single female and all of her mates from the effective population. However, this effect would be offset to some extent by the fact that, over time, all males within a social group have opportunities to reproduce.

Green Turtles (*Chelonia mydas*): How Many Fathers Are Represented in Each Clutch?

Adult green turtles migrate hundreds or even thousands of kilometers between nesting and feeding grounds (Meylan, 1982), and immature individuals move

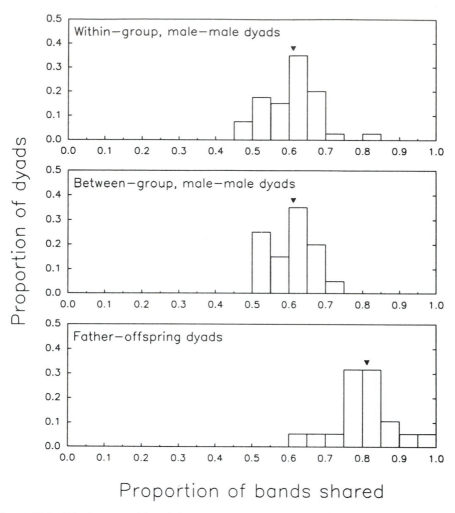

Figure 25-3 Distributions of band-sharing scores (proportions of bands shared) for dyads of Galapagos hawks. Each dyad consists of two males belonging to the same breeding group (top panel; N = 40), two males belonging to different breeding groups (middle panel; N = 20), or an assigned father and one of his offspring (sex unknown; bottom panel; N = 19). Arrowheads indicate means.

widely during their 25- to 30-year development to sexual maturity (Carr, 1980). Because individuals from different nesting populations converge on feeding grounds (Pritchard, 1976; Meylan, 1982) and because females may store sperm (Solomon and Baird, 1979), multiple paternity within clutches seems likely (Harry and Briscoe, 1988). Yet, the mating system of this endangered species remains a conspicuous gap in our knowledge of its life history. If multiple paternity does occur, the effective size of nesting populations may be substantially larger than would be estimated on the basis of censuses of females at nesting beaches and the assumption that each female mates with a single male.

We used multilocus minisatellite DNA fingerprinting to detect multiple paternity. Studies using DNA fingerprinting to analyze parentage generally have had access to the DNA of both putative parents, as in the three studies already discussed. Because male green turtles remain at sea while females go ashore to lay their eggs, only nesting females and emerging hatchlings are easily accessible. Here we present an approach for detecting multiple paternity when samples from candidate fathers are unavailable.

If all hatchling–hatchling dyads within a family comprise full siblings, then the band-sharing scores based on strictly paternally derived bands are expected to be distributed symmetrically around a mean of 0.5. Therefore, evidence for mixed paternity would be provided by an asymmetry where substantially more band-sharing scores fell below than above 0.5. Figure 25-4 shows such an asymmetry in each of three families. Sign tests revealed a significant asymmetry in both family 1 ($p = 0.0004$, one-tailed) and family 2 ($p = 0.036$) but not in family 3 ($p = 0.25$). However, Wilcoxon signed-ranks tests, which take into account the magnitude of deviations, provided suggestive evidence for multiple paternity in

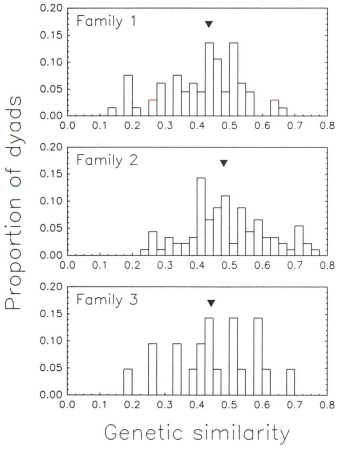

Figure 25-4 Frequency distributions of the genetic similarity values (proportion of strictly paternal bands shared) for hatchling–hatchling dyads within three green turtle families sampled (families 1–3: $N = 55$, 91, and 21 dyads). Arrowheads indicate medians.

family 3 ($p = 0.063$, one-tailed) and a consistent result for family 1 ($p < 0.001$), but a contradictory result for family 2 ($p = 0.16$).

Given this ambiguity, we computed the binomial probability that the two hatchlings of each dyad would share the observed number of bands if the hatchlings were full siblings. In the two families of adequate sample size, this analysis revealed 16 (of 55) dyads in family 1 and 17 (of 91) dyads in family 2 for which the probability of sharing so few bands was <0.1. By contrast, in only two and nine dyads, respectively, was the probability of sharing so *many* bands <0.1. These asymmetries provide further convincing evidence for mixed paternity in family 1 (16 vs. 2: binomial $p = 0.001$, one-tailed) and suggestive evidence in family 2 (17 vs. 9: $p = 0.084$). Taken together, the above analyses provide strong evidence for mixed paternity in family 1 and equivocal evidence in families 2 and 3.

If partial paternal genotypes are to be reconstructed and accurate estimation of effective population size is to be accomplished, then it will be necessary to assign each offspring to a full sibship, corresponding to a common ("cryptic") father. An approach that may prove useful is cluster analysis (Everitt, 1993; Levitan and Grosberg, 1993), where a female's offspring are grouped on the basis of their genetic similarity. Ideally, each distinct cluster would represent a set of full siblings. To illustrate the approach, we present the solution provided by Ward's minimum-variance method (SAS Institute, 1988) using the band-sharing scores of family 1 (Figure 25-5). The pseudo-t^2 procedure for determining the number of "significant" groups (SAS Institute, 1988; Everitt, 1993) suggests two groups, {A,E,C,B,D,I,K} and {G,J,F,H}. However, it is unclear whether these groups represent distinct sets of full siblings; solutions may vary depending on the clustering algorithm adopted. For example, the unweighted pairs group means analysis (UPGMA) method

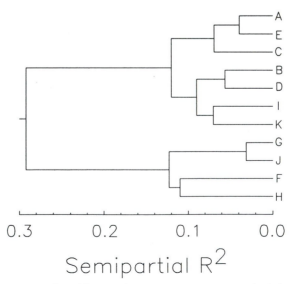

Figure 25-5 Phenogram produced by Ward's minimum-variance method, based on 55 dyadic band-sharing scores among the 11 green turtle hatchlings samples from family 1.

(Rohlf, 1990) suggests two main clusters that strongly resemble those produced by Ward's method, with the important exception that individuals G and J are placed in the other group.

Because of the limitations of multilocus minisatellite DNA fingerprinting in this application, we are developing microsatellite approaches. Availability of single-locus information from hypervariable loci should enhance the resolution of such paternity analyses. Further, reconstruction of paternal genotypes across microsatellite loci will allow us to calculate the probability that fathers come from the same population or a different population.

DISCUSSION

This collection of paternity studies includes three comparisons of apparent and actual mating systems for well-studied avian populations. The comparison revealed one study in which the actual mating system is identical to the apparent mating system (monogamy in black vultures) and two others in which the number of reproductive males exceeded the number of reproductive females (stripe-backed wrens and Galapagos hawks). The fourth study involves a system in which the mating system is simply unknown, as members of one sex (in this case, males) are difficult to monitor and mating events take place in the ocean.

Mating systems in which some individuals are excluded from reproduction altogether (Nunney, 1991), or in which variance in reproductive success by one or both sexes is high (Nunney, 1993), can produce very small values of N_e, particularly in populations characterized by short adult lifespans (Nunney, 1993). The extent to which deviation of genetic mating system from apparent mating system influences N_e is presently unknown. In general, if the number of reproductive individuals of either sex is greater than thought, estimates of N_e will increase. However, if "cryptic" patterns of reproduction suggest that variance of RS is greater than apparent RS would suggest, estimates of N_e will decrease. We are presently working on models that explore this interaction over the range of values reported in recent studies of mating systems employing molecular markers to determine parentage (Waite and Parker, unpublished data).

Until this problem is better understood, whether applying molecular determination of parentage is crucial for accurate estimation of N_e will remain unclear. However, managers of critically small populations who attempt to preserve a minimum viable population (MVP) are advised to consider the true (genetic) mating system of their target species. Even relatively small errors in the estimation of N_e due to failure to account for cryptic reproduction could translate into substantial errors in estimates of MVP (Nunney and Campbell, 1993).

ACKNOWLEDGMENTS

Genetic analyses were supported by the College of Biological Sciences, Ohio State University and the National Science Foundation. The turtle work represents part of a larger project supported by grants from the National Science Foundation (DEB-9322544), Sigma

Xi, the American Museum of Natural History, Wildlife Conservation International, and the Caribbean Conservation Corporation. We thank Monica Guevara, Allison Leslie, Elizabeth Giuliano, Laine Gonzales, Janet Winborne, and Tracy Lenihan for help in the field; Lynn Kramer and Doug Warmolts at the Columbus Zoo for assistance; and Tom Jones for help with data analysis. Other parts of this work were supported by the National Science Foundation (stripe-backed wrens and Galapagos hawks), the American Museum of Natural History, Sigma Xi, and the North Carolina Wildlife Resources Commission (black vultures).

REFERENCES

Birkhead, T.R. and A.P. Møller. 1992. Sperm competition in birds: Evolutionary causes and consequences. London: Academic Press.

Birkhead, T.R., T. Burke, R. Zann, F.M. Hunter, and A.P. Krupa. 1990. Extra-pair paternity and intraspecific brood parasitism in wild zebra finches *Taeniopygia guttata*, revealed by DNA fingerprinting. Behav. Ecol. Sociobiol. 27: 315–324.

Carr, A. 1980. Some problems of sea turtle ecology. Am. Zool. 20: 489–498.

Chesser, R.K. 1991. Influence of gene flow and breeding tactics on gene diversity within populations. Genetics 129: 573–584.

Decker, M.D., P.G. Parker, D.J. Minchella, and K.N. Rabenold. 1993. Monogamy in black vultures: genetic evidence from DNA fingerprinting. Behav. Ecol. 4: 29–35.

Everitt, B.S. 1993. Cluster analysis. New York: Wiley.

Faaborg, J. and J.C. Bednarz. 1990. Galapagos and Harris' Hawks: divergent causes of sociality in two raptors. In: P. Stacey and W. Koenig, eds. Cooperative breeding in birds: Long-term studies of ecology and behavior, pp. 489–526. Cambridge, U.K.: Cambridge University Press.

Faaborg, J., P.G. Parker, L. DeLay, Tj. de Vries, J.C. Bednarz, S. Maria Paz, J. Naranjo, and T.A. Waite. 1995. Confirmation of cooperative polyandry in the Galapagos hawk (*Buteo galapagoensis*). Behav. Ecol. Sociobiol. 34: 83–90.

Gibbs, H.L., P.J. Weatherhead, P.T. Boag, B.N. White, L.M. Tabak, and D.J. Hoysak. 1990. Realized reproductive success of polygynous red-winged blackbirds revealed by DNA markers. Science 250: 1394–1397.

Harry, J.L. and D.A. Briscoe. 1988. Multiple paternity in the loggerhead turtle (*Caretta*). J. Hered. 79: 96–99.

Jeffreys, A.J., W. Wilson, and S.L. Thein. 1985. Individual-specific fingerprints of human DNA. Nature 316: 75–79.

Kempenaers, B., G.R. Verheyen, M. Van Den Broeck, T. Burke, C. Van Broeckhoven, and A.A. Dhondt. 1992. Extra-pair paternity results from female preference for high quality males in the blue tit. Nature 357: 494–496.

Levitan, D.R. and R.K. Grosberg. 1993. The analysis of paternity and maternity in the marine hydrozoan *Hydractinia symbiolongicarpus* using randomly amplified polymorphic DNA (RAPD) markers. Mol. Ecol 2: 315–326.

Meylan, A. B. 1982. Sea turtle migration—evidence from tag returns. In: A. Bjorndal, ed. Biology and conservation of sea turtles, pp. 91–100. Washington, D.C.: Smithsonian Institution Press.

Nunney, L. 1991. The influence of age structure and fecundity on effective population size. Proc. R. Soc. Lond. B. 246: 71–76.

Nunney, L. 1993. The influence of mating system and overlapping generations on effective population size. Evolution 47: 1329–1341.

Nunney, L. and K.A. Campbell. 1993. Assessing minimum viable population size: demography meets population genetics. Trends Ecol. Evol. 8: 234–239.

Nunney, L. and D.R. Elam. 1994. Estimating the effective size of conserved populations. Conserv. Biol 8: 175–184.

Parker, P.G., T.A. Waite, and M.D. Decker. 1995. Behavioral association and kinship in communally roosting black vultures. Anim. Behav. 49: 395–401.

Pritchard, P.C.H. 1976. Post-nesting movements of marine turtles (*Cheloniidae* and *Dermochelyidae*) tagged in the Guianas. Copeia 1976: 749–754.

Rabenold, P.P. 1986. Family associations in communally roosting black vultures. Auk 103: 32–41.

Rabenold, P.P., K.N. Rabenold, W.H. Piper, J. Haydock, and S.W. Zack. 1990. Shared paternity revealed by genetic analysis in cooperatively breeding tropical wrens. Nature 348: 538–540.

Rabenold, K.N. 1990. *Campylorhynchus* wrens: the ecology of delayed dispersal and cooperation in the Venezuelan savanna. In: P.B. Stacey and W.D. Koenig, eds. Cooperative breeding in birds, pp. 159–196. Cambridge, U.K.: Cambridge University Press.

Rabenold, P.P., W.H. Piper, K.N. Rabenold, and D.J. Minchella. 1991a. Polymorphic minisatellite amplified on avian W chromosome. Genome 34: 489–493.

Rabenold, P.P., K.N. Rabenold, W.H. Piper, and D.J. Minchella. 1991b. Density-dependent dispersal in social wrens: genetic analysis using novel matriline markers. Anim. Behav. 42: 144–146.

Rohlf, F.J. 1990. NTSYS-pc: numerical taxonomy and multivariate analysis system. Version 1.60. Setauket, N.Y.: Applied Biostatistics.

SAS Institute. 1988. SAS/STAT User's Guide, Version 6.03. Cary, N.C.: SAS Institute.

Solomon, S.W. and T. Baird. 1979. Aspects of the biology of *Chelonia mydas* L. Oceanogr. Marine Biol. Annu. Rev. 17: 347–361.

Westneat, D.F., P.W. Sherman, and M.L. Morton. 1990. The ecology and evolution of extra-pair copulations in birds. Curr. Ornithol. 7: 331–369.

Wetton, J.H., R.E. Carter, D.T. Parkin, and D. Walters. 1987. Demographic study of a wild house sparrow population by DNA fingerprinting. Nature 327: 147–149.

26

Genetic Structure of Natural *Taxus* Populations in Western North America

STANLEY SCHER

CONSERVATION STATUS OF YEW (*TAXUS*)

Recent field surveys and other supporting data provide evidence that most if not all natural populations of yew (*Taxus*) are in decline (Farjon et al., 1993). A compilation of threatened conifer taxa lists the status of Pacific yew (*T. brevifolia*) as rare, Florida yew (*Taxus floridana*) as endangered, and the Mexican yew (*Taxus globosa*) and Chinese yew (*Taxus yunnanensis*) as vulnerable. Although the Pacific yew is identified as rare, it occurs over a wide geographic range and a variety of habitats, but is sparsely distributed and never abundant.

With few exceptions, systematic efforts to monitor changes in yew population decline are lacking. Studies conducted at approximately 10 year intervals, from 1954 to 1986, provide evidence for recent declining abundance of *T. baccata* in Uttar Pradesh, India (Govil, 1993). Data assembled by Prioton (1979) and Paule et al. (1993) reveal further evidence for yew decline in the forests of western Europe over the last century.

Prior to 1991, no inventory of Pacific yew existed for public lands managed by the USDA Forest Service and USDI Bureau of Land Management, or Forestry Canada. Franklin and Debell (1988) studied mortality and replacement of Pacific yew and other conifers in an old-growth Douglas-fir forest in the Cascade range of southern Washington over a 36-year period. They reported virtually no change in species composition and diameter distribution over that period. Current estimates of Pacific yew abundance on state and private lands in California, Oregon, and Washington are based on forest inventory and analyses conducted by the Pacific Northwest Research Station, USDA Forest Service (USDA Forest Service, 1993).

Genetic structure can be defined as the nonrandom distribution of alleles in space or time (Loveless and Hamrick, 1984). In this chapter, we trace the patterns of spatial distribution and abundance of *Taxus* species, examine the processes that threaten natural yew populations, identify the historical and recent anthropogenic impacts on yew distribution and abundance, document recent efforts to understand the genetic structure of *Taxus* populations in western North America, and briefly discuss the importance of molecular genetic data in conserving remaining yew populations.

PATTERNS OF SPATIAL DISTRIBUTION AND ABUNDANCE

Yews are slow-growing, shade-tolerant, long-lived, clonal evergreen trees or shrubs of temperate and subtropical forests. Four geographically separated species are native to North America (Price, 1990). The common or English yew (*T. baccata*) was once widespread over much of Europe (Bolsinger and Lloyd, 1993; Paule et al., 1993), parts of North Africa (Sauvage, 1941), the Caspian region of southwest Asia (Mossadegh, 1971) and submontane regions of south-central Asia. Four or more *Taxus* species also occur in eastern Asia (Shemluck and Nicholson, 1993; Spjut, 1993). There are no *Taxus* species indigenous to the southern hemisphere.

Within North America, the Florida yew (*T. floridana*) is largely restricted to hardwood forests along the minor tributaries of the Apalachicola River in the panhandle of northwestern Florida on the gulf coastal plain (Ward, 1989; Price, 1990; Redmond and Platt, 1993). In contrast, Canada yew (*Taxus canadensis*) is a widespread but sparsely distributed species from Manitoba to Newfoundland, and south to Virginia. The Mexican yew (*T. globosa*) ranges from northeastern Mexico to Guatamala and El Salvador; but a recent survey reveals few scattered sites (Nicholson and Shemluck, 1993).

Pacific yew occurs from the Coast Range and Sierra Nevada of central California to southeast Alaska (Bolsinger and Jaramillo, 1990). At the southern terminus of its range, it grows in the understory of mixed-conifer forests, largely restricted to mesic sites along small streams and ravines of the Redwood belt in northern and central California (Howell, 1949) and mid-elevation foothills of the Sierra Nevada (Jepson, 1910; Rundel, 1968; Fites, 1993).

In the central part of its range—the Klamath Mountains of northwestern California, the Cascade Range of the Pacific Northwest from northern California to western Canada, the interior forests of central and northern Idaho, and the Rocky mountains of western Montana and British Columbia—the Pacific yew occurs with greatest frequency and abundance in old-growth stands (Spies, 1991; Jimerson and Scher, 1993). At their northern limit, yew populations occupy boggy sites on the coastal islands of southern Alaska (Alaback and Juday, 1989).

Although native peoples in the Pacific Northwest and elsewhere recognized and valued the yew for its physical (hard and resiliant) and chemical (poisonous and medicinal) properties (Jepson, 1910; Hartzell, 1993; Beckstrom-Stemberg et al., 1993), *Taxus* in western North America remained an understudied genus until recently. The discovery of taxol and related taxane anticancer agents in the tissues of the Pacific yew generated interest in this species as a resource for novel therapeutic pharmaceuticals, and called attention to *Taxus* as an object of serious investigation as a threatened genus.

PROCESSES THAT THREATEN NATURAL *TAXUS* POPULATIONS

We identify two classes of anthropogenic threats to yew populations:

1. Human activities directly affecting their habitat, abundance, or diversity.
2. Resource-management policies that indirectly affect yew abundance and distribution.

The physical properties of yew wood—durability and elasticity—contribute to survival of *Taxus* as a forest understory species (Bolsinger and Jaramillo, 1990; Crawford, 1983; Scher, 1992), but also confer vulnerability to human exploitation. In Europe, trade in English yew (*T. baccata*) between Poland and the Netherlands was documented as early as in 1287 (Moewes, 1926/1927). During the Hundred Years War (1337–1453) between England and France, the yew became a strategic resource for constructing long bows. Centuries of conflict in the late Middle Ages contributed to the near extinction of the yew on the European continent (Appendino, 1993). In 1423, the first law to protect "trees of great value. such as the yew," was issued in Cracow, Poland (Gottwald, 1933), but export of yew logs continued until a shortage of yew trees prevented further trade (Appendino, 1993).

In the Pacific Northwest region of North America, road-building, clear-cutting, slash-burning, and conversion of old-growth forests to even-age plantations led to the reduction and fragmentation of yew habitat. Based on estimated reduction in old-growth forests on public lands in Oregon and Washington over the last century, yew habitat in western North America may have declined by an order of magnitude.

With the recognition that *Taxus* represents the only near-term, practical source of taxol—an FDA-approved anticancer agent for treatment of human ovarian and breast cancer—a major threat to yew populations emerged. To provide taxol for clinical trials and to meet the growing demand for therapeutic use, several hundred thousand yew trees were harvested in less than a decade, mainly from National Forests and Bureau of Land Management ownerships in Oregon and Washington, other western states, and British Columbia (USDA Forest Service, 1993; Mitchell, undated). More recently, dysgenic selection—highgading—on private lands suggests that current harvesting practices may contribute to reducing the yew gene pool by selecting against superior individuals.

During the last century, National Forest policies designed to eliminate predators of large game species and to provide winter range for deer, elk, and moose populations resulted in unprecedented browsing pressure on *Taxus* and other forage species (Foster, 1993; Hartzell, 1993; Parks et al., 1993). Browsing negatively affects sexual reproduction in Canada yew by reducing pollen, ovule, and seed production (Allison, 1990a,b,c, 1992). Current wildlife management guidelines maintain densities of deer at several times those present before European settlement (Waller and Alverson, 1993). Ruminant browsing has also contributed to the decline in yew regeneration (Tittensor, 1980; Willits, 1993).

PATTERNS OF GENETIC VARIATION IN *TAXUS*

Yew species vary in gross morphology, microscopic anatomy, chemical composition, and other characteristics. How much of this variation can be attributed to genetic differences as opposed to phenotypic plasticity is not yet understood.

With the exception of Canada yew, which is typically a low shrub, Pacific yew and other *Taxus* species exhibit at least two phenotypes—a tree or arboreal form, and a shrubby growth form (Arno and Hammerly, 1977; Bolsinger and Jaramillo, 1990; Campbell and Russell, 1993; Jimerson and Scher, 1993; Spjut, 1993). Quantitative studies of leaf anatomy, taxane content, and spatial distribution of the tree and shrub forms of Pacific yew reveal significant differences (Jimerson and Scher, 1993; Spjut and Chang, cited in Spjut, 1993).

Studies of molecular markers can contribute to our understanding of the genetic structure of *Taxus* populations. We describe here recent efforts to assess levels and distribution of genetic variation in natural populations of Pacific yew, other related taxa, and their hybrids.

ALLOZYME VARIATION IN PACIFIC YEW POPULATIONS

Allozymes are defined as soluble enzymes that differ in electrophoretic mobility as a result of allelic differences at a single locus. They serve as a convenient measure of genetic variation (Hamrick and Godt, 1990), and contribute to our understanding of the genetic structure and conservation biology of threatened species (Hamrick et al., 1992). We compare three recent studies of allozyme variation in Pacific yew populations in western North America.

Study 1. From over 700 forest-grown trees representing 30 populations throughout the natural range of Pacific yew, Wheeler et al. (1992) characterized a subset of 15 populations for gene frequencies at 24 enzymatic loci (see also Wheeler et al., 1995). In an extended abstract, they reported allozyme variation in four regions: western Oregon, western Washington, eastern Oregon and Washington, and interior British Columbia. Allozyme variation at the species level, summarized in Table 26-1 show that alleles per locus (A_s) average 1.5 (range 1.36–1.62), and the proportion of polymorphic loci (P_s) average 42.6% (range 36–54%). Wheeler et al. (1992) conclude that the Pacific yew is not in danger of becoming genetically depauperate. In commenting on the distribution of allozyme variation in this study, Schepartz (1993) noted that approximately 90% is within populations; variation among regions approaches 10%.

Study 2. In a USDA Forest Service survey of allozyme variation in Pacific yew, Doede et al. (1993) collected tissue samples from 170 locations representing 54 populations from Alaska, California, Idaho, Montana, Oregon, and Washington. In this study, 11 loci were resolved. Ten of the 11 loci examined were polymorphic in at least one population. On average, A_s was 1.5 (range 1.2–1.8) and P_s was 43.8% (range 18.2–63.6%) (Table 26-1). Average expected heterozygosity,

Table 26-1 Average Allozyme Variation in *Taxus brevifolia* Compared to Other Long-lived Woody Species

Species and Authors	Number of Populations	Number of Loci	Alleles per Locus	Percentage Polymorphic Loci
Taxus brevifolia				
Wheeler et al. (1992)	15	24	1.5	42.6
Doede et al. (1993)	54	11	1.5	43.8
El-Kassaby and Yanchuk (1994)	9	21	1.7	42.3
Other long-lived woody species				
Hamrick et al. (1992)				
Geographic range				
Endemic			1.82	42.5
Narrow			2.08	61.5
Regional			1.87	55.7
Widespread			2.11	67.8
Regional distribution				
Boreal–temperate			2.58	82.5
Temperate			2.27	63.5
Temperate–tropical			1.89	62.2

$H_e = 0.171$, ranged from 0.048 to 0.218. Doede et al. (1993) reported the average proportion of genetic diversity among populations (range 0–0.156). Multivariate analysis of allozyme data revealed that much of the genotypic variation was among stands, among individuals within populations, and among populations. When populations were grouped into geographic regions, significant differences in allele frequencies were detected among populations from the Sierra Nevada in California, and the Idaho, Montana, and northeastern Oregon regions, but not for populations sampled in the Cascades and Siskiyou mountains of northwestern California. Genetic distances (the degree of divergence among loci) between populations were small. Canonical trend-surface analysis revealed significant geographic patterns by elevation and latitude.

Study 3. Approximately one-third of the Pacific yew resource in North America resides within Canada. El-Kassaby and Yanchuk (1994) collected needle samples and extracted soluble proteins from yew populations in nine disjunct regions on coastal islands and interior British Columbia. They analyzed allozyme variation at 21 loci, estimated the amount and pattern of genetic diversity, and determined how genetic variation is partitioned within and among regions. They also calculated Wright's gene fixation index (F_{is}) for each locus, estimated gene flow (Nm) for these populations and genetic distances among regions.

El-Kassaby and Yanchuk (1994) detected polymorphisms in 13 of 21 loci; the remaining 8 loci were monomorphic in all regions. A_s averaged 1.7 (range 1.5–1.9) and $P_s = 42.3\%$ (range 33.3–52.4) (Table 26-1). Mean expected heterozygosity (H_e) of 0.166 was higher than the observed 0.085. The proportion of total genetic diversity among regions (G_{st}) was 0.077 (range 0–0.167); thus, approximately 8% of detected variation was due to interregional differences. The remainder (92%)

represents within-population genetic diversity—a pattern typical of widespread, temperate forest tree species (Hamrick and Godt, 1990). Comparison of genetic distances among the nine Pacific yew regions in British Columbia revealed that each was genetically distinct.

El-Kassaby and Yanchuk (1994) reported that, on average, expected hetero-zygosity was higher than that observed. This provided evidence for reduced heterozygosity. Accordingly, they calculated the deviation from the Hardy–Weinberg equilibrium for each locus in each region; several loci were not in agreement with expectations. Significant positive values for Wright's gene fixation index (F_{is}) at six loci averaged 0.472 (range 0.119–1.00), indicating an excess of homozygosity. Based on the proportion of total diversity partitioned among regions (G_{st}), the gene flow (*Nm*) was estimated as 2.99.

To set these values in context, Wheeler et al. (1992) and El Kassaby and Yanchuk (1994) compared measures of genetic variability obtained with Pacific yew populations against results assembled from studies of other plant species with similar life-history characteristics (Hamrick and Godt, 1990); Doede et al. (1993) compared Pacific yew allozyme values against a more recent and larger data set for long-lived woody plants complied by Hamrick et al. (1992) (Table 26-1). An alternative approach is to identify life-history characteristics associated with the Pacific yew, predict expected levels of allozyme variation, and ask whether these expectations are met by the results of these analyses.

The ranges of values for the average A_s (1.5–1.7) and P_s (42.3–43.8) reported in allozyme studies of Pacific yew are lower than the levels expected from long-lived woody species in the temperate zone ($A_s = 2.27$ and $P_s = 63.5$) (Hamrick et al., 1992). In gymnosperm species, A_s average 2.38 (for 89 genetic studies); P_s averages in excess of 70% (Hamrick et al., 1992). For comparison with other western conifer species, A_s range from 1.0 in monomorphic western red cedar (*Thuja plicata*) to 3.9 in Douglas-fir (*Pseudotsuga menziesii*) (Conkle, 1981; Copes, 1981). In western pines, P_s ranges from 3% in Torrey pine (*Pinus torreyana*) to 89% in Lodgepole (*Pinus contorta*) and Ponderosa pine (*Pinus ponderosa*) (Conkle, 1981; Conkle and Westfall, 1983; Ledig and Conkle, 1983).

Within Pacific yew populations, measures of allozyme variation ($P_p = 43.8\%$ and $A_p = 1.5$) are markedly lower than expected from long-lived woody species with a regional or widespread geographic range ($P_p = 69.2–74.3\%$ and $A_p = 2.31–2.56$), and approach values observed in species with a narrow or restricted distribution ($P_p = 44.3\%$ and $A_p = 1.61$). At the population level, allozyme values for Pacific yew are somewhat lower than expected from allozyme studies on long-lived woody species with a temperate-zone distribution ($P_p = 49.2\%$ and $A_p = 1.81$) and similar to temperate-tropical species ($P_p = 43.6\%$ and $A_p = 1.62$) (Hamrick et al., 1992; Doede et al., 1993). It would be instructive to compare measures of allozyme variation in the Florida and Mexican yew with the Pacific yew (N.C. Vance, 1994, personal communication).

Estimates of average expected heterozygosity for Pacific yew ($H_e = 0.166–0.171$) are comparable to those for long-lived woody species with a narrow to regional geographic distribution ($H_e = 0.165–0.169$) (Hamrick et al., 1992; Doede et al., 1993). In western conifers, H_e ranges from 0 in *Pinus torreyana* to 0.33 in *P. menziesii* (Conkle, 1981; Copes, 1981; Millar and Marshall, 1991). Average H_e

values for Pacific yew populations suggest a breeding system observed in other outcrossing, wind-pollinated conifers (H_e = 0.173) restricted to well-defined habitats (Conkle, 1992: Doede et al., 1993) and a mode of reproduction that is sexual (H_e = 0.170) rather than sexual and asexual (H_e = 0.251) (Hamrick et al., 1992).

Among Pacific yew populations, average genetic variation (G_{st}) reported by El Kassaby and Yanchuk (1994) and Doede et al. (1993) ranges from 0.077 to 0.107. These values are slightly higher than expected for long-lived woody species with a widespread or regional geographic range (G_{st} = 0.033–0.065), but similar to other temperate and temperate-tropical species (G_{st} = 0.092–0.109) (Hamrick et al., 1992). For outcrossing, wind-pollinated confiers, G_{st} ranges from 0.015 to 0.068 (Hamrick, 1987). G_{st} values in Pacific yew are similar to *Pseudotsuga* (0.074) and somewhat higher than those for *Picea*, *Abies*, and *Pinus* (0.055–0.065) (Hamrick et al., 1992).

El-Kassaby and Yanchuk (1994) estimated gene flow, Nm (where N is the population size and m the fraction of N replaced by migrants) for Pacific yew populations sampled in British Columbia. They reported Nm = 2.99. Gene flow estimates in natural plant populations vary over a wide range (Hamrick, 1987; Hamrick et al., 1992). For outcrossing, wind-pollinated gymnosperms, Nm ranges from 5.3 in *P. menziesii* (El-Kassaby and Sziklai, 1982) to 37.9 in *P. ponderosa* (Hamrick, 1987). Gene flow varies with pollen and seed disperal characteristics (Hamrick et al., 1991). In addition, widely spaced populations are expected to exhibit less gene flow than closely spaced ones (Hamrick, 1987; Ellstrand, 1992b). Restricted dispersal range in subcanopy species may also limit pollen gene flow (Scher, 1992). Nm estimates for Pacific yew populations are likely to be influenced by avian or other vertebrate-mediated seed dispersal mechanisms (Furnier et al., 1987; Govindaraju, 1990).

The ability of Pacific yew to propagate by both sexual and asexual modes is reflected in H_e values (0.166–0.171) similar to those of other long-lived woody species that reproduce by these methods (H_e = 0.170–0.251) (Hamrick et al., 1992). Dispersal of Pacific yew seeds through transport by birds and rodents would be expected to contribute to levels of genetic diversity higher than through gravity alone (H_e = 0.144) (Hamrick et al., 1992).

EVIDENCE FOR NON-RANDOM MATING IN PACIFIC YEW

The fixation index (F_{is}) defines the average correlation, over all subpopulations, between uniting gametes relative to those of their own subpopulation (Wright, 1933). It serves as a measure of the reduction in heterozygosity of a subpopulation due to random genetic drift. F_{is} can also be viewed as a coefficient of deviation from a reference population (Templeton and Read, 1994), or the probability that two alleles at a locus in an individual are identical by descent (Hartl, 1981).

For Pacific yew populations in British Columbia, F_{is} values average 0.472 (El-Kassaby and Yanchuk, 1994). Positive values indicate an excess of homozygotes, and negative values an excess of heterozygotes. El-Kassaby and Yanchuk (1994) interpret the high, positive, significant value as evidence of high inbreeding levels. Significant heterozygote deficiencies were noted in 6 of 13 polymorphic loci detected in the British Columbia study. However, this high average value is generated primarily by two loci—malate dehydrogenase (MDH) and uridine diphosphoglucose pyrophosphorylase (UGP)—with F_{is} values of 1.0 and 0.8, respectively.

El-Kassaby and Yanchuk (1994) offer several possible explanations for the high F_{is} levels detected in their study. These include positive assortative mating, in which mating occurs among similar phenotypes at frequencies greater than would be expected by chance in a randomly mating population (Crow and Felsenstein, 1968; Lewontin et al., 1968); a neighborhood structure resulting in mating among relatives (Levin and Kerster, 1971, 1974); selection for homozygotes over heterozygotes; and fusion of formerly isolated subpopulations (Wahlund, 1928), resulting in an increase of homozygotes at the expense of heterozygotes (Wallace, 1981).

Assortative mating affects only those genes on which mate selection is based (Hartl, 1988). Unlike assortative mating, inbreeding affects all genes. In Pacific yew populations from British Columbia, significant excess of homozygosity was detected at only 6 of 13 loci. Another potential source of positive assortative mating may arise from variation of allele frequencies in the pollen pool and in the pool of receptive females through or among seasons (Mitton, 1992).

Clonal reproductive strategies—stump sprouting and/or layering (Daoust, 1992; Scher, 1992, 1993; El-Kassaby and Yanchuk, 1994) may contribute to mating among closely related individuals. Limited dispersal of pollen and/or seeds can give rise to a neighborhood structure resulting in mating of close relatives (Shea, 1990; Bacilier et al., 1994; Premoli et al., 1994). El-Kassaby and Yanchuk (1994) called attention to the patchy spatial distribution of Pacific yew in forest communities and their understory status. Accordingly, gene flow via pollen is expected to be very low.

Clustering of genetically related yew individuals may arise from seed deposited in rodent caches (Crawford, 1983) and/or seed hoarding by nuthatches, tits, and other bird species (Bartkoviac, 1975; Crawford, 1983; Sakakibara, 1989). Germination of clustered seeds in caches often results in tree clumps, tree clusters with multiple trunks (Tomback et al., 1993), or fused and contiguous trunks (Schuster and Mitton, 1991), growth forms frequently observed in *T. baccata* (Mossadegh, 1994, personal communication).

MOLECULAR GENETIC EVIDENCE OF PHYLOGENETIC RELATIONSHIPS

Molecular data can help to define phylogenetic relationships and offer insights into the evolutionary position of controversial taxa. We summarize recent studies bearing on the relationship of the yew family (Taxaceae) to other conifers, the distinction between *Taxus* species and hybrids, and the potential for genomic mapping of conifers.

Based on morphological criteria—absence of compound ovulate seed cones in an axillary position—the Taxaceae should be excluded from the conifers. However, polymerase chain reaction (PCR) amplification and analysis of 18S rRNA sequence data argues against this view and indicates that the Taxaceae are clearly related to the Pinaceae (Chaw et al., 1993). Accordingly, the yew family should be subsumed under the Coniferales.

Further support for a close relationship between Taxaceae and other conifers comes from structural studies of chloroplast DNA (cpDNA). Physical mapping of cpDNA from Taxaceae suggests an inverted repeat structure similar to that of the Pinaceae (Raubeson and Jansen, 1994). The inverted repeat includes the coding region for rRNA as well as other genes (Raubeson and Jansen, 1990). Two genera of Taxaceae—*Taxus* and *Torreya*—share the same structural mutation (Raubeson and Jansen, 1992).

Additional evidence for a close affinity of Taxaceae with other conifer families was obtained from analysis of the gene coding for chloroplast ribulose-1,5-bisphosphate carboxylase (*rbc*L) (Price et al., 1993). Sequence data show that Taxaceae are nested within a larger clade of conifers and bear a sister-group relationship with the Cephalotaxaceae.

Immunological comparison of seed proteins of conifer families have been useful in assessing relationships among genera of Cupressaceae, Taxodiaceae, and Sciadopityaceae (Price and Lowenstein, 1989). This approach can also help to distinguish generic differences among the Taxaceae. Guo et al., 1994 developed immunological methods to detect differences between *Taxus* and *Torreya* based on genes that confer ability to synthesize taxanes.

RESTRICTION FRAGMENT LENGTH POLYMORPHISMS (RFLPs)

Measuring genetic variation by direct analysis of DNA fragments avoids limitations associated with allozyme studies. Only a portion of the genome is expressed in the phenotype; accordingly, only a small number of loci are detectable. In contrast, a much larger fraction of genomic DNA is accessible to restriction endonucleases. DNA fragments are separated by gel electrophoresis, denatured, and transferred to a membrane. The resulting single-stranded DNA fragments will hybridize only with homologous sequences on a complementary probe. Genetic variation can be detected by analysis of the DNA fragment pattern (Schaal et al., 1991).

To search for polymorphisms in DNA extracted from needle tissue of *T. brevifolia*, other *Taxus* species, and their hybrids, Vance and Krupkin (1993) used enzymes recognizing six-base restriction sites to cleave the DNA. After electrophoretic separation, denaturation, and blotting, the resulting fragments were probed with cpDNA clones from *Pinus contorta*, or a DNA fragment containing 18S and 5.8S ribosomal genes amplified by PCR. Not surprisingly, polymorphisms detected by cpDNA probes distinguished several *Taxus* species and two hybrid cultivars from one another. Analysis of DNA restriction fragment binding to ribosomal DNA probes provides further evidence for genetic variation within and among *Taxus* species and hybrids (Vance and Krupkin, 1993; Vance, 1994, personal communication).

In a recent report on DNA probes from loblolly pine (*Pinus taeda*), Ahuja et al., 1994) demonstrated the ability of some of these probes to detect

RFLPs in 12 conifer species. This study showed that a small proportion of DNA probes from *Pinus* will hybridize, with California nutmeg (*Torreya californica*), and called attention to the opportunity for comparative genome mapping in conifers.

RANDOM AMPLIFIED POLYMORPHIC DNA (RAPDs)

In a recent study to develop RAPDs as molecular markers for Pacific yew, Gocmen and Neale (1993) used 10-base oligonucleotide primers of arbitrary sequence (random 10-mers) to amplify genomic DNA extracted from haploid mega-gametophyte tissue collected from a single mother tree. Polymerase chain reaction (PCR) products were separated by gel electrophoresis. Preliminary results appear promising: Of 286 primers, 72 revealed at least one polymorphic locus. By identifying segregating loci, linkage relationships were established as a basis for constructing a genetic map of Pacific yew.

RELEVANCE OF MOLECULAR MARKERS IN *TAXUS* CONSERVATION BIOLOGY

Assessing and maintaining genetic diversity are major issues in conservation biology (Schaal et al., 1991). Genetic variation is essential for long-term survival, adaptive change, and evolution (Frankel and Bennett, 1970). According to the neutrality hypothesis, most polymorphisms are selectively neutral (Kimura, 1968) and reveal little or nothing about adaptive changes in evolution. Although evidence in support of the neutral theory has accumulated over the last two decades, recent studies have questioned the neutrality of some allozyme alleles (Mopper et al., 1991; Bush and Smouse, 1992; see Bruford and Wayne, 1994, for other citations). Allozymes and other molecular markers now available are near neutral—their relationships to specific adaptations are not sufficiently understood to be used confidently (Libby, 1992). However, they are clearly useful and informative for analysis of heterozygosity and population structure.

Studies of allozymes have helped to define characteristics such as geographic range and breeding systems that strongly influence genetic diversity in long-lived woody species (Hamrick and Godt, 1990). Seed dispersal also has additional predictive value: (Hamrick et al., 1992).

Assessment of baseline levels of allozyme variability represents a first step in understanding the conservation status of *T. brevifolia* populations in western North America, to set priorities for which populations are in need of monitoring; to identify, analyze, and interpret patterns of spatial and genetic structure that affect levels of inbreeding; to maximize diversity (allelic richness), and to protect geographically localized alleles most vulnerable to loss.

The excess of homozygotes in Pacific yew sampled from British Columbia suggests that isolated patches with low population densities may share breeding characteristics with demes of related individuals (El-Kassaby and Yanchuk, 1994).

Clonal reproductive strategies, restricted disbursal of *Taxus* pollen in the under-story environment, and clustering of yew seed by birds and rodents may contribute to mating between closely related individuals and lead to marked deviation from Hardy–Weinberg expectations. For further discussion of pollen gene flow patterns in small populations and their implications for plant conservation genetics, see Ellstrand (1992a) and Ellstrand and Elam (1993).

Global demand for taxol and related anticancer agents is expected to exceed the limited supply. Currently, the only FDA-approved sources of these therapeutic taxanes are the Himalayan and Pacific yew. In 1994, over 5000 metric tons of Himalayan yew foliage were exported from India and Nepal (CITES, 1994). Continued harvesting and highgrading of the tree phenotype will further reduce abundance of yew and diminish the genetic quality of the resource.

To relieve mounting selection pressure on remaining natural yew populations and assure a sustainable supply of taxol for future clinical trials and cancer treatment, other sources of taxanes need to be developed. Feasible alternatives include semisynthesis of taxol or derivatives (Denis et al., 1988) from precursors isolated from foliage of yew plants grown in plantations using clonal techniques (Libby and Ahuja, 1993; Piesch and Wyant, 1993) or harvested from natural populations as a renewable resource; plant and microbial biosynthesis from abundant, available, and inexpensive starting materials (Lewis and Croteau, 1992; Stierle et al., 1993); *Agrobacterium*-mediated transformation of *Taxus* to generate transgenic cells (Han et al., 1993, 1994; Plaut-Carcasson et al., 1993); and large-scale cell or tissue culture (Shuler et al., 1992). Recent efforts to synthesize taxol do not appear to be commercially feasible (Holton, 1994a,b; Nicolaou, 1994; Wessjohann, 1994).

SUMMARY

Recent studies of allozyme variation inform our current understanding of *T. brevifolia* population structure in western North America. Compared to other long-lived, widespread, temperate-zone plants, Pacific yew reveals levels of genetic diversity consistent with a reproductive mode that is both sexual and asexual. Clonal reproductive strategies and restricted pollen gene flow, combined with patchy avian and rodent-mediated seed dispersal patterns, may create a local neighborhood structure that contributes to significant deviation from random mating and high levels of homozygosity in these populations. Further molecular genetic studies, are needed to test these tentative hypotheses and provide information on the conservation status of other threatened and endangered *Taxus* populations.

ACKNOWLEDGMENTS

I thank M.T. Conkle, N.H. Kashani, W.J. Libby, D.L. Rogers, B.S. Schwarzschild, and P.T. Spieth for thoughtful discussion and constructive comments on previous drafts of this manuscript.

REFERENCES

Ahuja, M.R., M.E. Devey, A.T. Groover, K.D. Jermstad, and D.B. Neale. 1994. Mapped DNA probes from loblolly pine can be used for restriction fragment length polymorphism mapping in other conifers. Theor. Appl. Genet. 88(3–4): 279–282.

Alaback, P.B. and G.P. Juday. 1989. Structure and composition of low-elevation old-growth forests in research natural areas of southeast Alaska. Nat. Areas J. 8: 27–39.

Allison, T.D. 1990a. Pollen production and plant density affect pollination and seed production in *Taxus canadensis*. Ecology 71: 516–522.

Allison, T.D. 1990b. The influence of deer browsing on the reproductive biology of Canada yew (*Taxus canadensis* Marsh). 1. Direct effect on pollen, ovule, and seed production. Oecologia 83: 523–529.

Allison, T.D. 1990c. The influence of deer browsing on the reproductive biology of Canada yew (*Taxus canadensis* Marsh). 2. Pollen limitation: an indirect effect. Oecologia 83: 530–534.

Allison, T.D. 1992. The influence of deer browsing on the reproductive biology of Canada yew (*Taxus canadensis* Marsh). 3. Sex expression. Oecologia 89(2): 223–228.

Appendino, G. 1993. Taxol (paclitaxel): historical and ecological aspects. Fitoterapia 64 (Suppl. 1): 5–25.

Arno, S.F. and R.P. Hammerly. 1977. Northwest trees. Seattle, Wash.: The Mountaineers.

Bacilier, R., T. Labbe, and H. Kremer. 1994. Intraspecific genetic structure in mixed polulations of *Quercus petraea* (Matt.) Leibl. and *Quercus rubra* L. Heredity 73(2): 130–141.

Bartkoviac, S. 1975. Seed dispersal by birds. In: S. Bialobock, ed. The yew—*Taxus baccata* L., pp. 139–146. Polish Academy of Sciences, Institute of Dendrology, and Kornik Arboretum. (Available from the U.S. Department of Commerce, National Technical Information Service, Springfield, Va.).

Beckstrom-Sternberg, S.M., J.A. Duke, S. Scher, and B. Compton. 1993. Ethnomedical uses of yew. In: S. Scher and B. Shimon Schwarzschild, eds. Intern. Yew Resources Conference: Yew (*Taxus*) Conservation Biology and Interactions. Berkeley, Calif. Unpublished proceedings.

Bolsinger, C.L. and A.E. Jaramillo. 1990. *Taxus brevifolia* Nutt. Pacific Yew. In: R.M. Burns and H.B. Honkala, technical coordinators. Silvics of North America. Volume 1. Conifers, pp. 573–579. Agricultural Handbook 654. Washington, D.C.: U.S. Department of Agriculture.

Bolsinger, C.L. and J.D. Lloyd. 1993. Global yew assessment: status and some early results. In: S. Scher and B. Shimon Schwarzschild, eds. Intern. Yew Resources Conference: Yew (*Taxus*) Conservation Biology and Interactions. Berkeley, Calif. Unpublished proceedings.

Bruford, M.W. and R.K. Wayne. 1994. The use of molecular genetic techniques to address conservation questions. In: S.J. Garte, ed. Molecular environmental biology, pp. 11–28. Boca Raton., Fla.: Lewis.

Bush, R.M. and P.E. Smouse. 1992. Evidence for the adaptive significance of allozymes in forest trees. New Forests 6(1–4): 179–196.

Campbell, E. and J. Russell. 1993. Ecology and gene conservation of Pacific yew in British Columbia. In: S. Scher and B. Shimon Schwarzschild, eds. Intern. Yew Resources Conference: Yew (*Taxus*) Conservation Biology and Interactions. Berkeley, Calif. Unpublished proceedings.

Chaw, S.-M., H. Long, B.-S. Wang, A. Zharkikh, and W.-H. Li. 1993. The phylogenetic position of Taxaceae based on 18S rRNA sequences. J. Mol. Evol. 37(6): 642–630.

Cites. 1994 [cited by Sheldon, J.W. Role of plants in the pharmaceutical industry, raw materials from temperate rain forests: *Taxus brevifolia* Nutt. In: The Periwinkle Project. New York Botamical Garden (in press).]

Conkle, M.T. 1981. Isozyme variation and linkage in six conifer species. In: M.T. Conkle et al. eds. Proc. Symp. Isozymes of North American Forest Trees and Forest Insects, pp. 11–17. Berkeley, Calif.: USDA Forest Service General Technical Report PSW-48.

Conkle, M.T. 1992. Genetic diversity—seeing the forest for the trees. New Forests 6(1–4): 5–22.

Conkle, M.T. and R.D. Westfall. 1983. Evaluating breeding zones for ponderosa pine in California. In: Progeny testing. Proc. Servicewide Genetics Workshop, pp. 89–98. Washington, D.C.: USDA Forest Service Timber Management.

Copes, D.L. 1981. Isozyme uniformity in western red cedar seedlings from Oregon and Washington. Can. J. For. Res. 11: 451–453.

Crawford, R.C. 1983. Pacific yew community ecology in north-central Idaho with implications to forest land management. 109 pp. Ph.D. thesis, University of Idaho, Moscow, Idaho.

Crow, J.F. and J. Felsenstein. 1968. The effect of assortative mating on the genetic composition of a population. Eugen. Q. 15: 85–97.

Daoust, D.K. 1992. An interim guide to the conservation and management of Pacific yew. USDA Forest Service, Pacific Northwest Region. Portland, Oregon. 77 pp.

Denis, J.-N., A.E. Greene, D. Guenard, F. Gueritte-Voegelein, L. Mangatal, and P. Potier. 1988. A highly efficient, practical approach to natural taxol. J. Am. Chem. Soc. 110: 5917.

Doede, D.L., E. Carroll, R. Westfall, R. Miller, H.J. Switzer, and K.M. Snader. 1993. Variation in allozymes, taxol, and propagation by rooted cuttings in Pacific Yew. In: S. Scher and B. Shimon Schwarzschild, eds. Intern. Yew Resources Conference: Yew (*Taxus*) Conservation Biology and Interactions. Berkeley, Calif. Unpublished proceedings.

El-Kassaby, Y.A. and O. Sziklai. 1982. Genetic variation in allozyme and quantitative traits in a selected Douglas-fir (*Pseudotsuga menziesii* var, menziesii (Mirb.) Franco) population. For. Ecol. Management 4: 115–126.

El-Kassaby, Y.A. and A.D. Yanchuk. 1994. Genetic diversity, differentiation, and inbreeding in Pacific yew from British Columbia. J. Hered. 85: 112–117.

Ellstrand, N.C. 1992a. Gene flow by pollen: implications for plant conservation genetics. Oikos 63(1): 77–86.

Ellstrand, N.C. 1992b. Gene flow among seed plant populations. New Forests 6(1–4): 241–256.

Ellstrand, N.C. and D.R. Elam. 1993. Population genetic consequences of small population size—implications for plant conservation. Annu. Rev. Ecol. Syst. 24: 217–242.

Farjon, A., C.A. Page, and N. Schellevis. 1993. A preliminary world list of threatened conifer taxa. Biodiversity Conserv. 2: 304–326.

Fites, J.A. 1993. Distribution and habitat of Pacific yew in the northern Sierra Nevada and southern Cascade Mountains of California. In: S. Scher and B. Shimon Schwarzschild, eds. Intern. Yew Resources Conference: Yew (*Taxus*) Conservation Biology and Interactions. Berkeley, Calif. Unpublished proceedings.

Foster, D.K. 1993. *Taxus canadensis* Marsh: its range, ecology and prospects in Wisconsin. In: S. Scher and B. Shimon Schwarzschild, eds. Intern. Yew Resources Conference: Yew (*Taxus*) Conservation Biology and Interactions. Berkeley, Calif. Unpublished proceedings.

Frankel, O.H. and E. Bennett, eds. 1970. Genetic resources in plants: exploration and conservation. Oxford, U.K.: Blackwell Scientific.

Franklin, J.F. and D.S. Debell. 1988. Thirty-six years of tree population change in an old-growth *Pseudotsuga* forest. Can. J. For. Res. 18(5): 633–639.

Furnier, G.R., P. Knowles, M.A. Clydes, and B.P. Dancik. 1987. Effect of avian seed dispersal on the genetic structure of whitebark pine populations. Evolution 41: 607–612.

Gocmen, B. and D.B. Neale. 1993. Development of random amplified polymorphic DNA: Genetic markers in Pacific yew. In: S. Scher and B. Shimon Schwarzschild, eds. Intern. Yew Resources Conference: Yew (*Taxus*) Conservation Biology and Interactions. Berkeley, Calif. Unpublished proceedings.

Gottwald, K.Z. 1933. The oldest protective laws in old Poland. Ochrona Przyrody 3: 16–17.

Govil, K. 1993. Status of *Taxus baccata* in Utter Pradesh, India. In: S. Scher and B. Shimon Schwarzschild, eds. Intern. Yew Resources Conference: Yew (*Taxus*) Conservation Biology and Interactions. Berkeley, Calif. Unpublished proceedings.

Govindaraju, D.R. 1990. Gene flow, spatial patterns, and seed collection zones. For. Ecol. Management 35: 291–302.

Guo, Y., M. Jaziri, B. Dialo, R. Vanhaelen-Fastre, A. Zhiri, M. Canhaelen, J. Homes and E. Bombardelli. 1994. Immunological detection and quantitation of 10-deacetyl-baccatin III in *Taxus* sp. plant and tissue culture. Biol. Chem. Hoppe-Seyler 375(4): 281–287.

Hamrick, J.L. 1987. Gene flow and distribution of genetic variation in plant populations. In: K.M. Urbanski, ed. Differentiation patterns in higher plants, pp. 53–67. London: Academic Press.

Hamrick, J.L. and M.J.W. Godt. 1990. Allozyme diversity in plant species. In: Plant population genetics, breeding, and genetic resources, A.H.D. Brown, M.T. Clegg, A.L. Kahler, and B.S. Weir, eds. pp. 43–63. Sunderland, Mass.: Sinauer.

Hamrick, J.L., M.J.W. Godt, D.A. Murawski, and M.D. Loveless. 1991. Correlations between species traits and allozyme diversity: Implications for conservation biology. In: D.A. Falk and K.E. Holsinger, eds. Genetics and conservation of rare plants, pp. 75–86. New York: Oxford University Press.

Hamrick, J.L., M.J.W. Godt, and S.L. Sherman-Broyles. 1992. Factors influencing levels of genetic diversity in woody plant species. New Forests 6(1–4): 95–124.

Han, K.-H., M.P. Gordon, M. Loper, H.G. Floss, and S. Chilton. 1993. Genetic transformation of *Taxus* as an alternative system for taxol production. In: S. Scher and B. Shimon Schwarzschild, eds. Intern. Yew Resources Conference: Yew (*Taxus*) Conservation Biology and Interactions. Berkeley, Calif. Unpublished proceedings.

Han, K.H., P. Fleming, K. Walker, M. Loper, W.S. Chilton, U. Mocek, M.P. Gordon, and H.G. Floss. 1994. Genetic transformation of mature *Taxus*: an approach to genetically control the *in vitro* production of the anticancer drug, taxol. Plant Sci. 95: 187–196.

Hartl, D.L. 1988. A primer of population genetics, 2d ed. Sunderland, Mass.: Sinauer.

Hartzell, H. Jr. 1993. Anthropogenic impacts on yew resources. In: S. Scher and B. Shimon Schwarzschild, eds. Intern. Yew Resources Conference: Yew (*Taxus*) Conservation Biology and Interactions. Berkeley, Calif. Unpublished proceedings.

Holton, R.A., C. Somoza, H.B. Kim, et al. 1994a. First total synthesis of taxol. 1. Functionalization of the B ring. J. Am. Chem. Soc. 116: 1597–1598.

Holton, R.A., H.B. Kim, C. Somoza, et al., 1994b. First total synthesis of taxol. 2. Completion of the C and D rings. J. Am. Chem. Soc. 116: 1599–1600.

Howell, J.T. 1949. Marin flora. Berkeley, Calif.: University of California Press.

Jepson, W.L. 1910. Silva of California, vol. 2, pp. 165–166.. Berkeley, Calif.: University of California Press.

Jimerson, T.M. and S. Scher. 1993. Analysis of Pacific yew habitat in northwestern California. In: S. Scher and B. Shimon Schwarzschild, eds. Intern. Yew Resources Conference: Yew (*Taxus*) Conservation Biology and Interactions. Berkeley, Calif. Unpublished proceedings.

Kimura, M. 1968. Genetic variability maintained in a finite population due to mutational production of neutral and nearly neutral isoalleles. Genet. Res. 11: 247–269.

Ledig, F.T. and M.T. Conkle. 1983. Gene diveristy and genetic structure in a narrow endemic, Torrey pine (*Pinus torreyana* Parry ex Carr). Evolution 37: 79–85.

Levin, D.A. and H.W. Kerster. 1971. Neighborhood structure in plants under diverse reproductive methods. Am. Nat. 105: 345–354.

Levin, D.A. and H.W. Kerster. 1974. Gene flow in seed plants. Evol. Biol. 22: 130–139.

Lewis, N. and R. Croteau. 1992. Taxol biosynthesis. In: Second National Cancer Institute Workshop on Taxol and *Taxus* (unpaged) Bethesda, Md.: National Cancer Institute, NIH.

Lewontin, R.C., D. Kirk, and J. Crow (1968). Selective mating, assortative mating, and inbreeding: definitions and implications. Eugenics Quart. 15: 141–143.

Libby, W.J. 1992. How knowledge of population structure and mating patterns of forest-tree populations can help improve genetic-conservation procedures. In: Int. Symp. Population Genetics and Gene Conservation of Forest Trees, Carcans Maubuisson, Bordeaux [In press].

Libby, W.J. and M.R. Ahuja. 1993. Clonal forestry. In: M.R. Ahuja and W.J. Libby, eds. Clonal forestry II, pp. 1–8. Berlin: Springer-Verlag.

Loveless, M.D. and J.L. Hamrick. 1984. Ecological determinants of genetic structure in plant populations. Annu. Rev. Ecol. Syst. 15: 65–96.

Millar, C.I. and K.A. Marshall. 1991. Allozyme variation of Port-Orford-Cedar (*Chamaecyparis lawsoniana*): implications for genetic conservation. For. Sci. 37(4): 1060–1077.

Mitchell, A.K. (undated). Pacific yew and taxol. Victoria, B.C.: Pacific Forestry Centre, (unpaged).

Mitton, J.B. 1992. The dynamic mating systems of conifers. New Forests 6(1–4): 197–216.

Moewes, F. 1926–1927. Die deutsche Eibenholzhandel in spateren Mittelalter und im 16 Jahrhundert. Der Naturforscher 3: 245–247.

Mopper, S., J.B. Minton, T.G. Whitham, N.S. Cobb, and K.M. Christensen. 1991. Genetic differentiation and heterozygosity in pinyon pine is associated with resistence to herbivory and environmental stress. Evolution 45: 989–999.

Mossadegh, A. 1971. Stands of *Taxus baccata* in Iran. Revue Forestiere Francaise 23(6): 645–648.

Nicholson, R. and M. Shemluck. 1993. Studies of *Taxus globosa*, the Mexican yew. In: S. Scher and B. Shimon Schwarzschild, eds. Intern. Yew Resources Conference: Yew (*Taxus*) Conservation Biology and Interactions. Berkeley, Calif. Unpublished proceedings.

Nicolaou, K.C., Z. Yang, J.J. Liu, et al. 1994. Total synthesis of taxol. Nature 367: 630–634.

Parks, C.A., A.R. Tiedemann, and L. Bednar. 1993. Browsing ungulates cause decline of Pacific yew. In: S. Scher and B. Shimon Schwarzschild, eds. Intern. Yew Resources Conference: Yew (*Taxus*) Conservation Biology and Interactions. Berkeley, Calif. Unpublished proceedings.

Paule, L., D. Gomory, and R. Longauer. 1993. Present distribution and ecological conditions of the European yew (*Taxus baccata* L.). In: S. Scher and B. Shimon Schwarzschild, eds. Intern. Yew Resources Conference: Yew (*Taxus*) Conservation Biology and Interactions. Berkeley, Calif. Unpublished proceedings.

Piesch, R.F. and V.P. Wyant. 1993. Intensive cultivation of yew species. In: S. Scher and B. Shimon Schwarzschild, eds. Intern. Yew Resources Conference: Yew (*Taxus*) Conservation Biology and Interactions. Berkeley, Calif. Unpublished proceedings.

Plaut-Carcasson, Y.Y., L. Benkrima, L. Dawkins, N. Wheeler, and A. Yanchuk. 1993. Taxol from *Agrobacterium*-transformed cultures. In: S. Scher and B. Shimon Schwarzschild, eds. Intern. Yew Resources Conference: Yew (*Taxus*) Conservation Biology and Interactions. Berkeley, Calif. Unpublished proceedings.

Premoli, A.C., S. Chischilly, and J.B. Mitton. 1994. Levels of genetic variation captured by four descendent populations of pinyon pine (*Pinus edulis* Engelm.). Biodiversity Conserv. 3(4): 331–340.

Price, R.A. 1990. Genera of Taxaceae in southeastern United States. J. Arnold Arboretum 71(l): 69–92.

Price, R.A. and J.H. Lowenstein. 1989. An immunological comparison of the Sciado-pityaceae, Taxodiaceae, and Cupressaceae. System. Bot. 14(2): 141–149.

Price, R.A., J. Thomas, S.H. Strauss, P.A. Gadek, C.J. Quinn, and J.D. Palmer. 1993. Familial relationships of the conifers from *rbc*L sequence data. Am. J. Bot. 80(6 suppl.): 172.

Prioton, J. 1979. Etude biologique et ecologique de l'if (*Taxus baccata* L) en Europe occidentale. Foret Privee 128: 19–34, 129: 19–37.

Raubeson, L.A. and R.K. Jansen. 1990. Variation in the extent of the inverted repeat in the chloroplast genome of vascular plants. Am. J. Bot. 77(6 suppl.): 152.

Raubeson, L.A. and R.K. Jansen. 1992. A rare chloroplast-DNA structural mutation is shared by all conifers. Biochem. Syst. Ecol. 20(l): 17–24.

Raubeson, L.A. and R.K. Jansen. 1994. Conifer chloroplast DNA structure: details (and phylogenetic implication) of the inverted repeat loss. Am. J. Bot. 81(6 suppl.): 181–182.

Redmond, A.M. and W.J. Platt. 1993. Population ecology of *Taxus floridana*. In: S. Scher and B. Shimon Schwarzschild, eds. Intern. Yew Resources Conference: Yew (*Taxus*) Conservation Biology and Interactions. Berkeley, Calif. Unpublished proceedings.

Rundel, P.W. 1968. The southern limits of *Taxus brevifolia* in the Sierra Nevada, California. Madrono 19: 300.

Sakakibara, S. 1989. Role of the varied tit, *Parus varius* T and S in the seed dispersal of Japanese yew. J. Jpn. For. Soc. 71(2): 41–49.

Sauvage, C.H. 1941. L'if dans le Grandatlas. Bull. Soc. Sci. Nat. du Maroc. 21: 82–90.

Schaal, B.A., W.J. Leverich, and S.H. Rogstad. 1991. Comparison of methods for assessing genetic variation in plant conservation biology. In: D.A. Falk and K.E. Holsinger, eds. Genetics and conservation of rare plants, pp. 123–134. New York: Oxford University Press.

Schepartz, S.A. 1993. Summary—Session I: Supply—Harvest of *Taxus brevifolia*. J. Nat. Cancer Ins. Monog. 15: 5–6.

Scher, S. 1992. *Taxus* reproductive stategies: disturbance-initiated clonal regeneration in Pacific yew, a subcanopy species. In: R.R. Harris and D.C. Irwin, eds. Proc. Symp. Biodiversity of Northwestern California, pp. 280–283. University of California Wildland Resources Center Report No. 23. Berkeley, Calif.: University of California.

Scher, S. 1993. Studies on *Taxus* conservation biology: Clonal regeneration—a reactive survival strategy. In: S. Scher and B. Shimon Schwarzschild, eds. Intern. Yew Resources Conference: Yew (*Taxus*) Conservation Biology and Interactions. Berkeley, Calif. Unpublished proceedings.

Schuster, W.S.F. and J.B. Mitton. 1991. Relatedness within clusters of a bird-dispersed pine and the potential for kin interactions. Heredity 67(1): 41–48.

Shea, K.L. 1990. Genetic variation between and within populations of Englemann spruce and subalpine fir. Genome 33(1): 1–8.

Shemluck, M. and R. Nicholson. 1993. Conservation status of *Taxus* in Taiwan and the Philippines. In: S. Scher and B. Shimon Schwarzschild, eds. Intern. Yew Resources Conference: Yew (*Taxus*) Conservation Biology and Interactions. Berkeley, Calif. Unpublished proceedings.

Shuler, M.L., T.J. Hirasuna, and D.M. Willard. 1992. Kinetics of taxol production by tissue culture. In: Second National Cancer Institute Workshop on Taxol and *Taxus*. (unpaged). Bethesda, Md.: National Cancer Institue, NIH.

Spies, T.A. 1991. Plant species diversity and occurrence in young, mature, and old-growth Douglas-fir stands in western Oregon and Washington. In: L.F. Ruggerio, K.B. Aubry, A.B. Carey, and M.H. Huff, Technical Coordinators. Wildlife and vegetation of unmanaged douglas-fir forests, pp. 111–121. Gen. Tech. Rep. PNW-285. Portland, Oreg.: USDA Forest Service, Pacific Northwest Research Station.

Spjut, R. 1993. Reliable morphological characters for distinguishing Pacific yew. In: S. Scher and B. Shimon Schwarzschild, eds. Intern. Yew Resources Conference: Yew (*Taxus*) Conservation Biology and Interactions. Berkeley, Calif. Unpublished proceedings.

Stierle, A., G. Strobel, and D. Stierle. 1993. Taxol and taxane production by *Taxomyces andreanae*, an endophytic fungus on Pacific yew. Science 260: 214–216.

Templeton, A.R. and B. Read. 1994. Inbreeding: One word, several meanings, much confusion. In: L. Loeschcke, J. Tomiuk, and S.K. Jain, eds. Conservation genetics, pp. 91–105. Basel: Birkhauser Verlag.

Tittensor, R.M. 1980. Ecological history of yew *Taxus baccatta* L. in southern England. Biol. Conserv. 17: 243–265.

Tomback, D.F., F.-K. Holtmeier, H. Mattes, K.S. Carsey, and M.L. Powell. 1993. Tree clusters and growth form distribution in *Pinus cembra*, a bird-dispersed pine. Arctic Alpine Res. 25(4): 374–381.

USDA Forest Service. 1993. Pacific yew: Final environmental impact statement. Portland, Oreg.: Pacific Northwest Region.

Vance, N.C. and A.B. Krupkin. 1993. Using restriction fragment length polymorphisms to assess genetic differences among *Taxus* species and taxa. In: S. Scher and B. Shimon Schwarzschild, eds. Intern. Yew Resources Conference: Yew (*Taxus*) Conservation Biology and Interactions. Berkeley, Calif. Unpublished proceedings.

Wahlund, S. 1928. Zusammensetzung von populationen und korrelationserscheinungen vom standpunkt der vererbungslehre ausbetrachtet. Hereditas 11: 65–106.

Wallace, B. 1981. Basic population genetics. New York: Columbia University Press.

Waller, D. and W. Alverson. 1993. Do deer limit yew? Patterns in the occurrence of *Taxus canadensis* in northern Wisconsin and the western Upper Peninsula of Michigan. In: S. Scher and B. Shimon Schwarzschild, eds. Intern. Yew Resources Conference: Yew (*Taxus*) Conservation Biology and Interactions. Berkeley, Calif. Unpublished proceedings.

Ward, D.B. 1989. Atlantic white cedar (*Chamaecyparis thyoides*) in the southern states. Florida Sci. 52(1): 8–47.

Wessjohann, L. 1994. The first total synthesis of taxol. Angewandte Chemie, Int. Ed. 33: 959–961.

Wheeler, N.C., R.W. Stonecypher, K.S. Jech, S.A. Masters, C. O'Brien, and A. Dettmering. 1992. Genetic variation in the Pacific yew: practical application (unpaged). In: Second National Cancer Institute Workshop on Taxol and *Taxus*. Bethesda, Md.: National Cancer Institute, NIH.

Wheeler, N.C., K.S. Jach, S.A. Masters, C.J. O'Brien, D.W. Timmons, R.W. Stonecypher, and A. Lupkes. 1995. Genetic variation parameter estimates in *Taxus brevifloia* (Pacific yew) Can. J. Forest Research 25: 1913–1927.

Willits. M. 1993. Regeneration of *Taxus brevifolia* in Douglas-fir dominated forests in the Klamath Mountains. In: S. Scher and B. Shimon Schwarzschild, eds. Intern. Yew Resources Conference: Yew (*Taxus*) Conservation Biology and Interactions. Berkeley, Calif. Unpublished proceedings.

Wright, S. 1933. Inbreeding and homozygosis. Proc. Natl. Acad. Sci. U.S.A. 19: 411–420.

27

Applications of Genetics to the Conservation and Management of Australian Fauna: Four Case Studies from Queensland

CRAIG MORITZ, JESSICA WORTHINGTON WILMER, LISA POPE, WILLIAM B. SHERWIN, ANDREA C. TAYLOR, AND COLIN J. LIMPUS

THE PROBLEM

The Australian vertebrate fauna presents a profound challenge to conservation biology (Table 27-1). For marsupials alone, 10 species have become extinct over the 200 years since European colonization and a further 27 species are currently regarded as endangered or vulnerable (Kennedy, 1992). The causes of these extinctions and declines are debated, but there seems little doubt that habitat clearing and degradation combined with predation by introduced species (such as cats and foxes) are major threats (Morton, 1990; Recher and Lim, 1990). Particularly notable is the large number of small mammals formerly widespread on the continent, but now restricted to islands (Burbidge, 1994). Substantial problems have also been identified for other groups of terrestrial vertebrates (Table 27-1); for example, there is a suite of rapidly declining frog species in the wet tropical rainforests of north Queensland (Richards et al., 1993). These figures are likely to be underestimates given that new species of Australian vertebrates are still being reported at a high rate (Baverstock et al., 1994): 36% of amphibians and reptiles have been described in the last 20 years (Cogger et al., 1993).

Increasingly, these species are being managed through a formal recovery process involving the formation of expert recovery teams, ideally including research and managment perspectives. Recovery actions might include habitat protection or restoration, control of introduced predators or competitors, or translocation to new or previously occupied sites (although most attempts at translocation have failed; Short et al., 1992), and, in a small number of cases, captive breeding. In many cases there is interest in the use of genetics to complement other approaches, although the range of problems and practical issues that can be addressed has only recently become apparent.

Table 27-1 Conservation Status of the Terrestrial Vertebrate Fauna of Australia According to the Australian and New Zealand Environment and Conservation Council (ANZECC 1991)

	Presumed Extinct	Endangered	Vulnerable
Fish	0/0	0/7	0/6
Amphibia	0/0	0/7	0/2
Reptiles	0/0	0/6	1/14
Birds	11/10	13/11	12/15
Mammals	1/20	2/26	1/17
Total	12/30	15/57	14/54

The numbers refer to named subspecies/species listed in each category. The numbers of amphibian species in each category have subsequently increased as a result of the declining amphibian phenomenon (Richards et al., 1993) and the numbers of vulnerable and endangered reptile species have increased to 41 and 11, respectively (Cogger et al. 1993).

THE GENETIC TOOLS

Although managers typically are aware of genetic issues relating to conservation, particularly the effects of inbreeding depression, the contribution of molecular genetics to the management of natural populations of Australian fauna has been limited (but see Robinson et al., 1993; Broderick et al., 1994; Norman et al., 1994). This, in part, stems from the limited power of techniques available until recently. For example, marsupials typically have low diversity at allozyme loci (Sherwin and Murray, 1990), restricting the inferences that can be made about genetic effects of isolation and the like (e.g., Sherwin et al., 1991; Southgate and Adams, 1993). A notable exception has been the use of allozyme electrophoresis and/or karyotyping to identify cryptic species, some of which are vulnerable or endangered (Baverstock et al., 1994).

As other chapters in this volume demonstrate, the range of molecular genetic techniques that can be applied to conservation problems has expanded enormously in the past few years. Now, in addition to allozyme electophoresis, we can examine sequence or restriction site variation in mitochondrial or nuclear genes and changes in hypervariable nuclear sequences using multilocus or single-locus methods. Particularly notable is the rapid development of single-locus microsatellite analysis, which combines the utility of gene amplification methods, permitting analysis of small amounts of nondestructively obtained or preserved tissue with very high levels of allelic diversity. At the same time, population genetics theory is providing new analytical approaches for dealing with these data, especially for making use of the information on allele phylogeny as well as frequency.

GENETICS AND MANAGEMENT

Clearly, we now have the tools necessary to assay and interpret patterns of molecular variation in threatened species. It now needs to be demonstrated that this information is relevant to managing species for recovery. Some have questioned the need for information on levels of genetic variation when attention should be focused on identifying and ameliorating the ecological causes of population declines (Caughley, 1994) or on demographic processes operating in the remnant populations (Lande, 1988). One response is to suggest that genetic processes such as drift and inbreeding are inextricably linked with, and contribute to, demographic fluctuations in small populations (Gilpin and Soulé, 1987; Mills and Smouse, 1994). Another is to make it clear that molecular genetics has a dual role in conservation (Moritz, 1994a): (1) "gene conservation," in which the extent and distribution of genetic diversity is described in order to recognize and maintain this aspect of biodiversity, and (2) "molecular ecology," where genetic markers are used as a complement to ecological studies to contribute to demographic management. For the latter, genetics can be used to obtain information that is impossible or difficult to derive from ecological methods.

In this broader context, genetics addresses several questions that arise in delineating and managing threatened populations (e.g., Table 27-2). As documented in detail by Avise (1994), molecular techniques enable analysis of population structure from genealogies of individuals to phylogenetic structuring of allelic variation among geographic populations (Figure 27-1). For management purposes, this translates into the ability to examine mating systems and to identify two types of conservation unit (Moritz, 1994b): management units (MUs), identified as sets of populations with distinct allele frequencies; and evolutionarily significant units (ESUs), identified as sets of populations distinguished by strong

Table 27-2 Examples of Issues and Questions Raised by Wildlife Managers That Can Be Addressed Using Molecular Genetic Studies, Together with Suggested Genetic Tools and Australian Examples Discussed Here (*) or Drawn from the Literature

Conservation Issue	Genetic Tools	Examples
Gene conservation:		
Identification of evolutionarily significant units	mtDNA sequencing or RFLPs + allozymes or microsatellites or nuclear RFLPs	Ghost bats*, Prickly skinks[1]
Genetic variation in remnant or captive populations	Hypervariable nuclear sequences: multilocus minisatellites or single locus microsatellites	Eastern barred bandicoot[2], Northern hairy-nosed wombat*, Bridled nailtail wallaby*, Western swamp tortoise[3]
Molecular ecology		
Identification of management units	mtDNA or nuclear RFLPs, allozymes, microsatellite	Yellow-footed rock wallaby*, Marine turtles[4,5]
Analysis of mating systems, inbreeding, etc.	Microsatellites, minisatellites, allozymes	Northern hairy-nosed wombat*, Bridled nailtail wallaby
Genetic tags for individuals or stocks	mtDNA, microsatellites	Northern hairy-nosed wombat*, Marine turtles[4,5]

References: 1, Moritz et al. (1993); 2, Robinson et al. (1993); 3, Hall et al. (1992); 4, Norman et al. (1994); 5, Broderick et al. (1994).

CONSERVATION UNITS

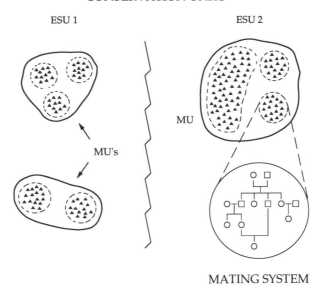

MATING SYSTEM

Figure 27-1 Molecular genetic information can be used at several hierarchical levels relevant to conservation. From the distribution and relationships of alleles, two types of conservation unit can be defined: evolutionarily significant units (ESUs) and management units (MUs) (see Moritz, 1994b). At a finer scale, relationships within populations can be ascertained, allowing detailed analysis of mating system and population structure.

phylogenetic structuring of mtDNA variation and divergence in the frequencies of nuclear alleles. The definition of ESUs primarily relates to gene conservation, whereas the recognition of MUs lies more in the domain of molecular ecology (Moritz, 1994b).

This chapter illustrates some of these applications using examples from our ongoing studies on several Australian vertebrates. It is not intended as an exhaustive review. Rather, the purpose is to illustrate strengths and limitations of different techniques in relation to current management issues. The examples represent a range of management problems; some occur as single remnant populations, others as naturally fragmented systems. Accordingly, the goals of the molecular genetic studies vary from assessment of phylogeographic variation through to analysis of mating systems and levels of genetic diversity remaining in remnant or captive populations. In all cases, the genetic data are most powerful when complemented by information obtained from long-term studies of demography and dispersal.

CASE STUDIES

The Ghost Bat (*Macroderma gigas*): Defining Conservation Units

The ghost bat requires warm and humid caves for reproduction and consequently has a naturally, patchy distribution of breeding populations (Figure 27-2). The

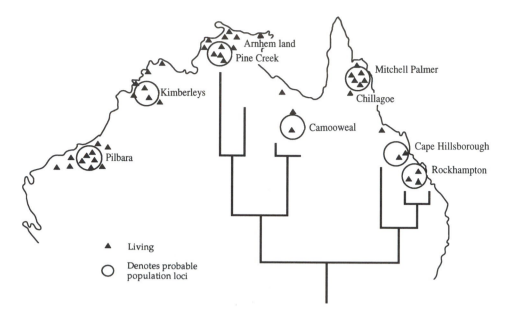

Macroderma gigas

Figure 27-2 Map of Australia showing the distribution of extant populations of the ghost bat, overlain with mtDNA phylogeny derived from control region sequences (Worthington Wilmer et al., 1994). Alleles sampled from each population were monophyletic and the presence of fixed differences was confirmed by screening of additional individuals using restriction enzymes.

species is ecologically important as the only specialist carnivorous bat in Australia; and it has a long history, being present in Miocene fossils (Hand, 1985). The species was formerly widespread, but contracted northward before and after European colonization (Molnar et al., 1984). The remaining populations are found in suitable caves or disused mine shafts in the tropics and the species is regarded as vulnerable by the IUCN and the Australian Nature Conservation Agency. It has also been the subject of considerable controversy because of impacts of mining operations on extant breeding colonies in Queensland (e.g., Mt. Etna near Rockhampton) and the Northern Territory (Pine Creek).

Genetic studies are in progress to determine (1) the significance of individual populations to the overall genetic diversity of the species (are they ESUs?) and (2) the connectedness of populations and potential for recolonization following local disturbance (are they MUs?). Sequencing of a 330 bp segment of the mtDNA control region from 22 bats drawn from four major populations, three in Queensland and one in the Northern Territory, revealed extreme structuring of variation among populations (Worthington Wilmer et al., 1994; Figure 27-2). Each population was monophyletic for mtDNA alleles and analysis of a further 100 individuals for differences at diagnostic restrictions sites confirmed the presence of fixed differences in mtDNA alleles among populations. Overall, 87% of the total nucleotide diversity was distributed among populations. Variation at six micro-satellite loci was substantial both within (Table 27-3) and among populations, and

Table 27-3 Genetic Diversity Revealed in Populations of Endangered or Vulnerable Species by Sequencing of mtDNA (either control region from tRNA^Pro or, for the wombat, cytochrome b) and Microsatellite Analysis

| Population | mtDNA Diversity | | Microsatellites | |
	Haplotype	Nucleotide	Expected Heterozygosity	Number of Alleles
Ghost bat[1]				
within populations	0.45 (5.5, 330 bp)	0.68%	0.67 (133, 6 loci)	6.3
total	0.90 (22, 330 bp)	4.3%	NA	NA
Yellow-footed rock wallaby[2]				
Hill of Knowledge, Idalia N.P.	0.09 (11, 591 bp)	0.05%	0.73 (50, 4 loci)	7.5
Captive population	0.71 (10, 591 bp)	1.8%	0.69 (10, 4 loci)	4.0
Bridled nailtail wallaby[3]				
Taunton N.P.	NA	NA	0.83 (73, 7 loci)	11.6
Captive colony	0.00 (4, 577 bp)	0.00	0.78 (14, 4 loci)	6.25
Hairy-nosed wombats[4]				
Northern HNW,				
Epping Forest	0.00 (2, 383 bp)	0.00	0.27 (28, 16 loci)	1.81
Deniliquin	NA	NA	0.41 (5, 8 loci)	2.10
Southern NHW, Brookfield	0.04 (2, 383 bp)	0.50	0.65 (16, 16 loci)	4.38

NA = not applicable.

Figures in parentheses are sample sizes and numbers of loci or base pairs, as appropriate. The Deniliquin population of the northern hairy-nosed wombat is probably extinct and was analyzed from museum skins.

References: 1, Worthington Wilmer et al. (1994); 2, Pope et al., in press; 3, Moritz et al. (submitted); 4, Taylor et al. (1994).

preliminary analyses indicate significant differences between all pairs of populations tested (Worthington Wilmer, unpubished data).

These results have several implications for understanding of the species' biology and for its management. First, they confirm the site fidelity of females indicated by previous mark–release–recapture studies and indicate that the philopatry of females extends over numerous generations. Second, they indicate that these local breeding concentrations represent effectively closed populations for demographic analysis. Third, it is clear that the extirpation of regional breeding populations will not be reversed by natural immigration within the time frame relevant to management. Fourth, each of the four regional assemblages so far examined represents a substantial fraction of the species' genetic diversity and each appears to be an ESU as well as a MU, as defined by Moritz (1994b).

The Yellow-footed Rock Wallaby (*Petrogale xanthopus*): Defining Conservation Units and Analysis of Genetic Diversity

This species also has a naturally patchy distribution because of specialized habitat requirements, in this case for rock outcrops with large boulders and deep crevices combined with appropriately structured vegetation (Gordon et al., 1993). On a large scale, the species has a disjunct distribution with the nominate subspecies consisting of a relatively large but fragmented population in South Australia and a small isolate in north-western New South Wales, and *P.x. celeris* being recorded only from south-western Queensland. Several populations of the latter appear to have declined, possibly because of habitat degradation and competition with goats, predation by foxes, and so on (Copley, 1983; Gordon et al., 1993), and the Queensland form is currently classified as potentially vulnerable to extinction (Kennedy, 1992).

Previous ecological studies have assumed that local concentrations of individuals represent discrete colonies. The extent to which these colonies are part of a larger metapopulation and the effects of behavioral philopatry or habitat heterogeneity in isolating colonies are not known, although Lim (1987) suggested that individuals disperse to find new water sources in times of drought and that young males disperse from their natal colony. Given the importance of intercolony migration for understanding the dynamics of the larger metapopulation and its susceptibility to extinction, we used a combination of mtDNA sequencing and microsatellite analysis to test for restrictions on gene flow between colonies within and between suitable habitats (Pope et al., in press; Figure 27-3).

Sequencing of ~590 bp of the mtDNA control region reveal marked differentiation (6% net divergence) between the Queensland populations (*P.x. celeris*) and samples of *P.x. xanthopus* from South Australia. Within Queensland, there was limited phylogenetic structuring and significant differences in mtDNA allele frequency between samples from geographically isolated rock outcrops separated by 70 km. These populations also differed in the frequencies of alleles at all three of four microsatellite loci studied. Colonies closer together (10 km) and located on the same rock ridge had statistically indistinguishable mtDNA allele frequencies, but also had very low mtDNA diversity. However, they differed in allele frequency at one of the four microsatellite loci analyzed (Figure 27-3). Within the largest colony there was no evidence for microgeographic variation and genotype frequencies were in Hardy–Weinberg equilibrium. Thus, there was a gradation in genetic differences, ranging from phylogeographic structure among subspecies, to divergence in mtDNA and microsatellite allele frequencies between colonies from separate habitats, to differences at microsatellite loci only between nearby colonies from the same habitat, to panmixia within a colony. Genetic variation in a captive colony maintained by Queensland Department of Environment (QDE) was not noticeably lower than in the field populations (Table 27-3); mtDNA diversity was higher and expected heterozygosity for microsatellites was similar. (The lower number of alleles at microsatellite loci may be an artifact of the relatively small sample size from the captive colony.)

These results support the continued recognition of the Queensland populations as a separate entity for the purposes of conservation management (although the north-west NSW population needs to be examined), and they indicate that although colonies are connected in evolutionary time, they are substantially isolated in the short term and should be considered as separate entities (i.e., MUs) for monitoring and manipulation (Pope et al., in press). The data suggest a greater restriction of gene flow where colonies are separated by unsuitable habitat, but our comparisons were confounded by geographic distance and need to be replicated in a larger-scale study.

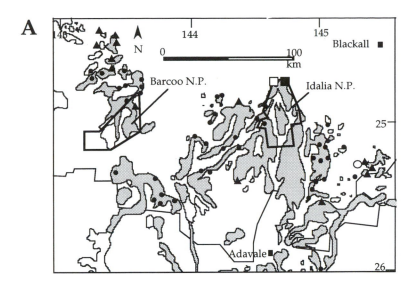

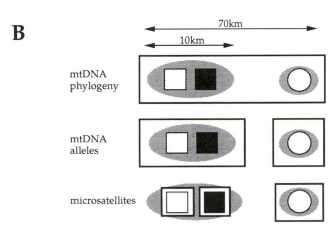

Figure 27-3 (A) Map of a portion of south-western Queensland showing the distribution of Yellow-footed rock wallabies and the localities from which wallabies were sampled for genetic analysis (modified from Gordon et al., 1993). Stippled areas indicate areas of suitable habitat. Symbols: ●, wallabies present: ▲, wallabies absent; □, Emmet Pocket sample; ■, Hill of Knowledge sample; ○, Lisburne sample. The boundaries of Barcoo and Idalia National Parks are also indicated. (B) Diagrammatic summary of distinct population units recognized using various approaches to genetic analysis (Pope et al., in press). The shading indicates the distribution of suitable habitat. Increasing resolution was observed in moving from phylogeographic analysis of mtDNA to analysis of mtDNA allele frequencies to analysis of nuclear microsatellite loci. For the latter, each colony (locations as per symbols defined in (A)) constitutes a management unit and no further subdivision was indicated by finer-scale analysis within the Hill of Kknowledge population on Idalia National Park (■).

The Bridled Nailtail Wallaby (*Onychogalea fraenata*): Analysis of Genetic Diversity and Mating System

The bridled nailtail wallaby is endangered, having contracted to a single population of a few hundred individuals from a former range extending through semi-arid
woodlands from central Queensland to northern Victoria (Gordon and Laurie,
1980). The rapid and extensive decline of the species is probably due to a
combination of habitat destruction and predation from introduced foxes and cats.
The current recovery plan calls for management of the remnant population on
Taunton National Park and for the establishment of new populations within the
former range by translocation from either the remnant or captive populations. In
this case, genetics will be used to examine mating systems in the source versus
translocated populations and to assess levels of variation in the captive versus
source populations.

Sequencing of mtDNA from 577 bp of control region from four individuals
from Taunton did not reveal any variation, indicating that mtDNA analysis is
unlikely to produce useful information for population management. Using a
combination of primers developed for the yellow-footed rock wallaby and for the
bridled nailtail wallaby (Moritz, unpublished data) we have assayed variation
at five microsatellite loci in animals sampled from Taunton and the captive colony
maintained by QDE. The analyses revealed abundant variation (Table 27-3),
with heterozygosity values at the different loci ranging from 0.69 to 0.84 and
observed numbers of alleles from 5 to 10. There was no substantial difference in
heterozygosity between the captive and source populations and the latter appeared
to be panmictic. The diversity in the remnant population of the bridled nailtail
wallaby is similar to that in the yellow-footed rock wallaby.

These preliminary results indicate that there is sufficient diversity at these
microsatellite loci to allow for a high-resolution analysis of mating system and
population subdivision within the remnant population and to detect changes in
genetic diversity or mating system accompanying translocations. This is the focus
of current studies. The captive population, mostly derived from seven individuals
taken from Taunton in 1991 and another four obtained earlier, does not appear
to have lost diversity relative to the source population, perhaps because of the
high reproductive rate of the wallabies.

The Northern Hairy-nosed Wombat (*Lasiorhinus krefftii*): ESUs, Mating Systems, and Remote Identification of Individuals

This is probably the most critically endangered species of mammal in Australia,
being restricted to a single population of fewer than 70 individuals in Epping
Forest National Park located in central Queensland. The species formerly ranged
as far south as Victoria and its decline has been attributed to loss of habitat and
competition from cattle, particularly during droughts (Crossman et al., 1994). The
immediate goals for recovery are to protect and increase the size of the remnant
population through habitat management at Epping Forest National Park and to
carry out the basic studies needed to support future translocations to new sites.

In this context, the primary purposes of the genetic studies (Taylor et al., 1994) were to examine the mating system and to test for inbreeding and loss of genetic diversity within the remnant population. MtDNA sequencing confirmed that the northern hairy-nosed wombat is a distinct ESU compared to its relative, the southern hairy-nosed wombat and that the extinct population from southern New South Wales was part of the former. The cytochrome *b* sequences were uniform within the Epping Forest population and more extensive analysis of control region sequences is in progress. Analysis of 16 microsatellite loci revealed substantially less variation in the remnant population than observed in extant populations of the southern hairy-nosed wombat (Table 27-3), suggesting a relatively low effective population size over an extended period. However, the variation at nine polymorphic loci was sufficient for each of the 28 individuals assayed to have a unique multilocus genotype and to provide a preliminary analysis of the genetic structure of the Epping Forest population. Genotype proportions were in Hardy–Weinberg equilibrium, suggesting that inbreeding within the colony is not a major problem. There was, however, evidence for slight differentiation between animals sampled from the southern and northern extremes of the 4 km distribution. This is somewhat surprising given previous evidence for movements of adult females

1 2 3 4 5 6 7 8 9 10

Figure 27-4 PCR products from locus 67CA, amplified from 1/20 of the extract of each of 10 hair samples from northern hairy-nosed wombats collected from Epping Forest N.P. Genotypes are: lanes 1,4 and 9, 151/165; lanes 2,3,5,7,8 and 10, 151/151; lane 6, 165/165. (See Taylor et al. (1994) for details.)

among burrow complexes (Johnson et al., 1991) and is being examined further with more extensive samples now in hand.

These data indicate that the microsatellites provide sufficient resolution for fine-scale analysis of mating systems in the remnant population. This will provide a guide to the need for managed movement of individuals within the remnant population and will help the design of an appropriate translocation strategy. Another application of immediate relevance to management is the remote identification of individuals via unique microsatellite profiles amplified from hairs (e.g., Figure 27-4). A major problem with managing these burrowing nocturnal marsupials is lack of information on the numbers and distribution of individuals. Wombat numbers typically are monitored indirectly via activity measures based on the presence of fresh droppings and tracks at the mouths of burrows (e.g., Crossman et al., 1994), a measure fraught with difficulty. Every three to five years the population is sampled by extensive trapping at the mouths of burrows, but this is a difficult and stressful exercise for both wombats and researchers. The diversity of microsatellite genotypes combined with the power of PCR means that individuals can now be monitored using samples of hair obtained from hair-traps located at the mouths of burrows (Figure 27-4), reducing the need for trapping and providing a more rigorous method for regular censuses of the population. This technology has now been incorporated in the revised recovery plan (Horsup, 1994).

DISCUSSION

The case studies reviewed above provide an interesting perspective on the relative strengths of mtDNA and microsatellite analysis and on how genetic information can be used effectively for the management of endangered or vulnerable species. For species with multiple geographic populations that are either naturally or anthropogenically fragmented, mtDNA sequencing provides a phylogenetic perspective allowing insights into the connectedness of populations on an evolutionary time-scale and, in combination with information on nuclear allele frequencies, the delineation of ESUs. To analyze allele frequencies in more detail, diagnostic restriction sites identified from the available sequences can be screened in much larger sample sizes, providing a perspective on contemporary as well as historical population processes (that is, for defining MUs; e.g., Norman et al., 1994).

Microsatellite loci complement mtDNA in providing information on allele frequencies at nuclear loci and are clearly very powerful for defining MUs. Although mtDNA is typically more sensitive than nuclear loci for detecting population subdivision (Dowling et al., 1990), microsatellites can detect genetic differences among populations in situations where mtDNA diversity is low because of small effective population size or matrilineal founder events.

For species reduced to a single population, mtDNA diversity may be too low for this locus to be of any use for population management. However, even in critically endangered species such as the northern hairy-nosed wombat, variation at microsatellite loci is sufficient for precise analyses of mating systems and population structure. The main current limitation on the use of microsatellites is the need to develop primers for the species concerned. The methods for doing this are relatively straightforward (see Rassman et al., 1991; Queller et al., 1993) in a laboratory equipped for cloning and sequencing. Also, primers designed against one species may amplify homologous polymorphic segments in related species or genera (e.g., cetaceans, Schlotterer et al. (1991); wombats, Taylor et al. (1994); marine turtles, FitzSimmons et al. (1995); macropods, Moritz et al., submitted), so that microsatellite technology will become more accessible as systems are developed for phylogenetically divergent groups. As an alternative, sequence variation in nuclear gene introns can be assayed using primers designed against conserved portions of exons (e.g., Lessa and Appelbaum, 1993; Slade et al., 1993; Palumbi and Baker, 1994). This approach should prove useful for analysis of population structure in a wide variety of species, but will probably be less appropriate than microsatellites for studies of mating systems and the like in small populations.

The information gained from these studies varies from the esoteric to the immediately relevant. To translate the genetic results into effective management, it is important for the geneticists to be directly involved in the recovery process as member of multidisciplinary teams including researchers and managers. This is essential for managers to direct the efforts of geneticists to immediate management needs and for geneticists to make managers aware of the potential and limitations of the techniques and concepts. In this way, the power of modern molecular genetics can make a meaningful contribution to the management and recovery of threatened species.

ACKNOWLEDGMENTS

Thanks to Anita Heideman for laboratory assistance and production of illustrations, to Peter Johnson, Alan Horsup, John Toop and Andrew Sharp for access to samples and to Hamish McCallum and Scott Edwards for comments on the manuscript. This research was supported by an Australian Research Council grant for collaborative research between the Queensland Department of Environment and the Universities of Queensland and New South Wales.

REFERENCES

ANZECC. 1991. List of endangered vertebrate fauna. Canberra: Australian Nature Conservation Agency.

Avise, J.C. 1994. Molecular markers, natural history and evolution. New York: Chapman and Hall.

Baverstock, P.R., L. Joseph, and S. Degnan. 1994. Units of management in biological conservation. In: C. Moritz and J. Kikkawa, eds. Conservation biology in Australia and Oceania, pp. 287–293. Chipping Norton: Surrey Beatty and Sons.

Broderick, D., C. Moritz, J.D. Miller, M. Guinea, R.I.T. Prince, and C.J. Limpus. 1994. Genetic studies of the Hawsbill Turtle (*Eretmochelys imbricata*): evidence for multiple stocks in Australian waters. Pacific Conserv. Biol. 1: 123–131.

Burbidge, A.A. 1994. Conservation biology in Australia: where should it be heading, will it be applied? In: C. Moritz and J. Kikkawa, eds. Conservation biology in Australia and Oceania. pp. 27–37. Chipping Norton: Surrey Beatty and Sons.

Caughley, G. 1994. Directions in conservation biology. J. Animal Ecol. 63: 215–244.

Cogger, H.G., E.E. Cameron, R.A. Sadlier, and P. Eggler. 1993. The action plan for Australian reptiles. Canberra: Aust. Nature Conservation Agency.

Copley, P.B. 1983. Studies on the Yellow-footed Rock Wallaby, *Petrogale xanthopus* Gray (Marsupialia: Macropodidea). 1. Distribution in South Australia. Aust. Wildlife Res. 10: 47–61.

Crossman, D.G., C.N. Johnson, and A.B. Horsup. 1994. Trends in the population of the Northern Hairy-nosed Wombat, *Lasiorhinus krefftii* in Epping Forest National Park, Central Queensland. Pacific Conserv. Biol. 1: 141–149.

Dowling, T.E., C. Moritz, and J.D. Palmer. 1990. Nucleic acids II: restriction site analysis. In: D.M. Hillis and C. Moritz, eds. Molecular systematics, pp. 250–317. Sunderland, Mass.: Sinauer.

FitzSimmons, N.N., C. Moritz, and S.S. Moore. 1995. Conservation and dynamics of microsatellite loci over 300 million years of marine turtle evolution. Mol. Biol. Evol. 12: 432–440.

Gilpin, M.E. and M.E. Soulé. 1986. Minimum viable populations: processes of species extinction. In: M. Soulé, ed. Conservation biology: the science of scarcity and diversity, pp. 19–34. Sunderland, Mass.: Sinauev.

Gordon, G. and B.C. Laurie. 1980. The rediscovery of the Bridled Nailtail Wallaby, *Onychogalea fraenata* (Gould) (Marsupialia: Macropodoidea) in Queensland. Aust. Wildlife Res. 7: 339–345.

Gordon, G., P. McRae, L. Lim, D. Reimer, and G. Porter. 1993. The conservation status of the yellow-footed rock wallaby in Queensland. Oryx 27: 159–168.

Hall, G., D. Groth, and J. Wetherall. 1992. Application of DNA profiling to the management of endangered species. Int. Zoo Yearbook 31: 103–108.

Hand, S. 1985. New Miocene megadermatids (Chiroptera: Megadermatidae) from Australia with comments on megadermatid phylogeny. Aust. Mammalogist 8: 5–43.

Horsup, A. 1994. Recovery plan for the Northern Hairy-nosed Wombat. Brisbane: Queensland Department of Environment.

Johnson, C.R. and D.G. Crossman. 1991. Dispersal and social organisation of the Northern Hairy-nosed Wombat. J. Zool., London 225: 605–613.

Kennedy, M., ed. 1992. Australasian marsupials and monotremes: an action plan for their conservation. Gland, Switzerland: IUCN.

Lande, R. 1988. Genetics and demography in biological conservation. Science 241: 1455–1460.

Lessa, E.P. and G. Appelbaum. 1993. Screening techniques for detecting allellic variation in DNA sequences. Mol. Ecol. 2: 119–129.

Lim, L. 1987. The ecology and management of the rare Yellow-footed Rock Wallaby, *Petrogale xanthopus* Gray (Macropodoidea). Ph.D thesis, Macquarie University, Sydney.

Mills, L.S. and P.E. Smouse. 1994. Demographic consequences of inbreeding in remnant populations. Am. Nat. 144: 412–431.

Molnar, R.E., L.S. Hall, and J.H. Mahoney. 1984. New fossil localities for *Macroderma* in

New South Wales and its past and present distribution in Australia. Aust. Mammalogist 7: 63–73.

Moritz, C. 1994a. Applications of mitochondrial DNA analysis in conservation: a critical review. Mol. Ecol. 3: 403–413.

Moritz, C. 1994b. Defining evolutionarily significant units for conservation. Trends Ecol. Evol. 9: 373–375.

Moritz, C., L. Joseph, and C. Moritz. 1993. Cryptic diversity in an endemic rainforest skink (*Gnypetoscincus queenslandiae*). Biodiversity Conserv. 2: 412–425.

Morton, S.R. 1990. The impact of European settlement on the vertebrate animals of arid Australia. In: D.A. Saunders, A.J.M. Hopkins, and R.A. How, eds. Australian ecosystems: 200 years of utilization, degradation and reconstruction, pp. 201–213. Chipping Norton: Surrey Beatty and Sons.

Norman, J., C. Moritz, and C.J. Limpus. 1994. Mitochondrial DNA control region polymorphisms: genetic markers for ecological studies of marine turtles. Mol. Ecol. 3: 363–373.

Palumbi, S.R. and C.S. Baker. 1994. Opposing views of population structure from nuclear intron sequences and mtDNA of humpback whales. Mol. Biol. Evol. 11: 426–435.

Pope, L., C. Moritz, and A. Sharp. Genetic structure of populations of the Yellow-footed Rock Wallaby inferred from mitochondrial DNA and microsatellites. Mol. Ecol., in press.

Queller, D.C., J.E. Strassman, and C.R. Hughes. 1993. Microsatellites and kinship. Trends Ecol. Evol. 8: 285–288.

Rassman, K., C. Schlotterer, and D. Tautz. 1991. Isolation of simple-sequence loci for use in polymerase chain reaction based DNA fingerprinting. Electophoresis 12: 113–118.

Recher, H.F. and L. Lim. 1990. A review of current ideas of the extinction, conservation and management of Australia's terrestrial vertebrate fauna. In: D.A. Saunders, A.J.M. Hopkins and R.A. How, eds. Australian ecosystems: 200 years of utilization, degradation and reconstruction pp. 287–301. Chipping Norton: Surrey Beatty and Sons.

Richards, S.R., K.R. McDonald, and R.A. Alford. 1993. Declines in populations of Australia's endemic tropical rainforest frogs. Pacific Conserv. Biol. 1: 66–77.

Robinson, N.A., N.D. Murray, and W.B. Sherwin. 1993. VNTR loci reveal differentiation between and structure within populations of the eastern barred bandicoot *Perameles gunnii*. Mol. Ecol. 2: 195–207.

Schlotterer, C., B. Amos, and D. Tautz. 1991. Conservation of polymorphic simple sequences in cetacean species. Nature 354: 63–65.

Sherwin, W.B. and N.D. Murray. 1990. Population and conservation genetics of marsupials. In: J.A.M. Graves, R.M. Hope, and D.W. Cooper, eds. Mammals from pouches and eggs: genetics, breeding and evolution of marsupials and monotremes pp. 19–38. Melbourne: CSIRO.

Sherwin, W.B., N.D. Murray, J.A.M. Graves, and P.B. Brown. 1991. Measurement of genetic variation in endangered populations: bandicoots (Marsupialia: Peramelidae) as an example. Conserv. Biol. 5: 103–108.

Short, J., S.D. Bradshaw, J. Giles, R.I.T. Prince, and G.R. Wilson. 1992. Reintroduction of macropods (Marsupialia: Macropodoidea) in Australia—a review. Biol. Conserv. 62: 189–204.

Slade, R.W., C. Moritz, A. Heideman, and P.T. Hale. 1993. Rapid assessment of single-copy nuclear DNA variation in diverse species. Mol. Ecol. 2: 359–373.

Southgate, R. and M. Adams. 1993. Genetic variation in the Greater Bilby. Pacific Conserv. Biol. 1: 46–52.

Taylor, A.C., W.B. Sherwin and R.K. Wayne. 1994. Genetic variation of microsatellite loci in bottlenecked species: the northern hairy-nosed wombat *Lasiorhinus krefftii*). Mol. Ecol. 3: 277–290.

Worthington Wilmer, J., C. Moritz, L. Hall, and J. Toop. 1994. Extreme population structuring in the threatened Ghost Bat, *Macroderma gigas*: evidence from mitochondrial DNA. Proc. R. Soc. Lond. B 257: 193–198.

IV

PERSPECTIVE

28

Conservation Genetics and Molecular
Techniques: A Perspective

PHILIP W. HEDRICK

As the human population and human impact on natural areas has increased over the past few decades, there have been heightened threats to rare and endangered species throughout the world. Largely because of this trend, the rate of extinction has increased and there has been great public and scientific interest in trying to reverse the trend toward extinction of many rare and endangered species. Although the factors that appear to be important in ultimately causing extinction are thought to be multifold (e.g., Shaffer, 1981; Soulé, 1986; Lande, 1988), in the scientific efforts in conservation biology, there has often been an emphasis on the importance of genetic factors in influencing extinction.

For example, the simple population genetic prescriptions for endangered species suggested by the 50/500 guidelines (e.g., Soulé, 1980) (50 founders to start a population of endangered organisms and to avoid inbreeding depression, and an effective population size of 500 to maintain its genetic variation) have been beguiling but are now thought to be only very general guidelines (e.g., Lande and Barrowclough, 1987; Hedrick and Miller, 1992). In fact, these guidelines have been the cause of criticism of the role of genetics in conservation (e.g., Simberloff, 1988; Caughley, 1994). Further, Lynch et al. (1995a,b) have suggested that small populations may decline in fitness owing to the accumulation of detrimental mutations, and Lande (1995) has recently suggested that it may take an effective population size of 5000 to maintain adaptive genetic variation in a population.

The past decade has seen the rapid development of a number of molecular techniques, such as those using restriction fragment polymorphisms, DNA finger-printing, microsatellite loci, sequences of particular genes, and so on. The simultaneous development of these techniques and the crisis in extinction of species have already resulted in the widespread application of these techniques to answer questions in conservation. As with the population genetics prescriptions mentioned earlier, the sophisticated techniques from molecular genetics are seductive in their glamour and potential to help make quick decisions in conservation. However, as they begin to be used for conservation decisions, their power (or lack of power) should be fully evaluated before adopting recommendations based on molecular data.

Much of the advancement in molecular approaches in conservation genetics has developed around better ways of identifying genetic variation so that differences between species and populations can be identified, a task that often was not possible with allozymes because rare and endangered species generally are low in allozyme variation. As illustrated in the contributions to this book and other recent publications (e.g., Avise, 1994; Sherwater et al., 1994), there are now many techniques that can be used for this objective. Appropriately, most of these approaches utilize neutral genetic variants to determine phylogenetic relationships between groups, identification of relationships between individuals, and so on. However, it should be kept in mind that it is very likely that loci that have adaptive significance are not sampled by most of these techniques and that important adaptive variants may have different patterns of variation from neutral variants.

In the past several years, a number of concerns have been raised about the relevance of genetics in conservation (e.g., Caro and Laurenson, 1994; Caughley, 1994; Merola, 1994; see, however, Hedrick et al., 1996) It is indisputable that many of the critical factors that appear to cause extinction are not genetic ones. However, in an effort to give recommendations for conservation and management (and perhaps to tell a good story), some conservation genetics research has been oversold by overly enthusiastic researchers and science writers. Now, as some elements of these conservation genetics stories have been questioned, there appears to be a backlash among some conservation biologists against the value of genetics in conservation just when a new generation of molecular approaches are maturing.

Although I do not want to dampen the enthusiasm for the application of new molecular techniques to conservation, I think that we need to keep our potential contributions to conservation in perspective and to not oversell their value. The purpose of the application of molecular approaches to endangered species should be as a tool to prevent their extinction. The fact that sophisticated molecular techniques are available does not mean that the answer is always apparent in the DNA and, in fact, common sense recommendations are in the best interest of both endangered species and the future acceptance of molecular genetics research in conservation. Remember, that if a recommendation based on molecular genetics is not completely predictive or correct in some important way, a wrong recommendation could contribute to the extinction of a species.

WHICH MOLECULAR APPROACH TO USE?

Mace et al. (chapter 1) have discussed the merits of a number of the molecular approaches available to conservation genetics. The molecular toolbox that has become available in the past few years is indeed impressive, particularly if one is interested in questions about the genetic relationship of different species, populations, or individuals. However, now that there are a number of molecular approaches, it is important to know which ones are appropriate for specific tasks (for various opinions, see the general discussion in chapter 1, and discussions of specific techniques in other chapters). Further, as various techniques become more widespread, they need to be constantly evaluated for their efficacy, reliability, and appropriateness for the investigation of various questions. For example, the determination of segregation and independent assortment for molecular markers using related individuals can support the genetic interpretation of molecular data. Obviously, it is also important to thoroughly identify the question that one is investigating before deciding upon the molecular or other technique to use. Such an orientation may be backward for some researchers who have expertise and laboratory facilities for a particular laboratory technique. For example, pedigree analysis and not molecular genetics investigation may be most appropriate for some questions when a reliable pedigree of a captive population is at hand (see the discussion of pedigree analysis in Hedrick and Miller, 1992).

For example, variation in microsatellite loci (see Ashley and Dow, 1994; and chapter 17), the present favorite among many researchers, has been quite successful in identifying differences among populations of the same species but may not be appropriate at the level of the species or above. However, Goldstein et al., (1995) have recently suggested that, even for species that diverged several million years ago, microsatellites may be an appropriate molecular approach if a distance metric based on the number of repeats is used.

The usefulness of microsatellite loci as compared to several other molecular techniques is illustrated by the recent study of Mexican wolves, *Canis lupus baileyi*, by García-Moreno et al. (1996). There are no known Mexican wolves in the wild, but there is a Certified (by the United States Fish and Wildlife Service, USFWS) lineage of Mexican wolves in captivity. This subspecies is thought from morphology to be one of the more divergent of the gray wolf subspecies. There are two other lineages, the Ghost Ranch/Arizona-Sonoran Desert Museum (Ghost Ranch) lineage and the Aragón lineage, that are also thought to be Mexican wolves, but it is not certain whether there might be some dog ancestry in these lineages (see Hedrick, 1995a, for a summary). Several studies using allozymes, mtDNA, and multilocus fingerprinting have not been definitive in demonstrating a difference of Mexican wolves from northern gray wolves and in showing that the captive lineages did not contain dog (or coyote) alleles (Shields et al., 1987; Wayne et al., 1992; Fain, unpublished data).

However, when García Moreno et al. (1996) examined ten polymorphic microsatellite loci, they found that the three captive lineages clearly clustered together and away from northern gray wolves. Further, it is quite unlikely that the Mexican wolf lineages have any microsatellite alleles resulting from introgression from dogs or coyotes.

As an example of these data, Table 28-1 gives the frequencies of five of the most diagnostic alleles in the survey. Notice that all of these alleles are high in frequency in the three Mexican wolf lineages and low or missing in the other three taxa. All of the microsatellite data can be summarized in a neighbor-joining tree (Figure 28-1). Here the three Mexican wolf lineages cluster together and away from the other taxa.

Of course, the ultimate information is the sequence of the various genes of interest. This type of information is now beoming available at the population level because of techniques such as SSCP (single-strand conformation polymorphism; chapter 11) in which particular DNA sequences can be identified as bands on a gel. For such data, association of variation at different sites that are physically closely linked may result in correlation of variation at different sites (linkage or gametic disequilibrium). For endangered species, which are now low in number

Table 28-1 Frequencies of Five Diagnostic Microsatellite Alleles in the Three Captive Mexican Wolf Lineages (Certified, Ghost Ranch, and Aragón) and Three Related Taxa (García-Moreno et al., 1996)

Gene	Allele	Certified	Ghost Ranch	Aragón	Gray Wolf	Coyote	Dog
172	G	0.810	1.000	1.000	0.089	0.015	0.000
204	D	0.500	1.000	0.929	0.364	0.000	0.026
213	L	0.929	1.000	1.000	0.236	0.202	0.100
225	C	0.952	0.800	0.786	0.288	0.366	0.012
250	G	0.548	1.000	1.000	0.264	0.142	0.214

and may have been rare for a number of generations, stochastic factors, such as genetic drift and historical effects may greatly influence molecular variation, and factors normally thought to be more deterministic, such as selection and gene flow, may be of lesser significance for molecular variation.

IS THE MOLECULAR VARIATION ADAPTIVE OR INDICATIVE OF ADAPTATION?

When using molecular variation to identify different individuals, populations, or species, it is appropriate to have genetic markers that are neutral and therefore good indicators of ancestry or relationship. To know that the variation being

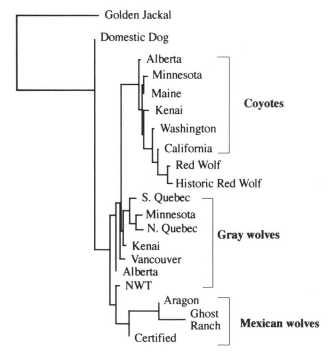

Figure 28-1 A neighbor-joining tree for the three Mexican wolf lineages and three related taxa using data from 10 microsatellite loci (García-Moreno et al., 1996).

assayed is, and has always been, neutral, and that changes in it have not been influenced by selection at the locus being examined or selective loci associated with it, is a critical assumption when using molecular variation for phylogenetic purposes. On the other hand, with new techniques, it may be possible to identify molecular variants that are indicative of adaptive differences. Determining what molecular variation is the "stuff of evolution" is one of the most difficult unsolved questions of evolutionary biology, and of course may be critical to the survival of an endangered species.

Ever since Lewontin and Hubby (1966) introduced allozyme variation in an evolutionary perspective, researchers have tried to determine what molecular variation is adaptive. Some researchers have suggested that populations that have more genetic variation or individuals that are more heterozygous using allozymes or other molecular markers should be used in conservation efforts. If the amount of genetic variation in the population or in particular individuals actually indicates the amount of past inbreeding, then such molecular markers may be of consequence. However, with a small number of independent loci, it is often difficult to adequately estimate the heterozygosity of either individuals or populations and there are wide confidence limits on such estimates (e.g., Chakraborty, 1981; Hedrick et al., 1986). Further, the estimation of either individual or population fitness is quite difficult and is probably environmentally specific in most instances (e.g., Hedrick and Murray, 1983).

As a start in determining adaptive variation, if a variety of molecule information is known, then it is more likely that DNA variation that results in amino acid differences would be adaptive than synonymous variation. However, even some of this variation may not be of adaptive significance in endangered species in which the population size is small. For example, elegant techniques have been developed to detect very small selective differences using DNA sequence data (e.g., Kreitman and Akashi, 1995). Such very small selective differences may be of evolutionary significance in *Drosophila* and other species that have very large population sizes, but since a selective difference smaller than the reciprocal of twice the effective population size, $1/(2N_e)$, is effectively neutral (Kimura, 1979), small selective differences are unlikely to be of adaptive significance in most endangered species.

The genes in the major histocompatibility complex (Hedrick, 1994a; chapter 14) are probably the best candidates for genes that are of adaptive value in vertebrates. These genes play a central role in the immune system of all vertebrates and are thought to be of adaptive significance even by most neutralists (e.g., Nei and Hughes, 1991). For the major histocompatibility complex molecules, the amino acids that interact with the antigen that is being presented to initiate the immune response are known from x-ray crystallography studies. These positions are much more variable than other positions that do not interact with the antigen or the T-cell receptor. An example of this difference is given in Figure 28-2 for two class I human genes, *HLA-A* and *HLA-B* (Hedrick et al., 1991). The average heterozygosity for the 29 amino acid positions over the two genes that interact with the antigen is 0.300, while the average for the 312 or 309 positions (for genes *HLA-A* and *HLA-B*, respectively) that are not thought to interact with either the antigen or the T-cell receptor have a heterozygosity of only 0.034, almost an order

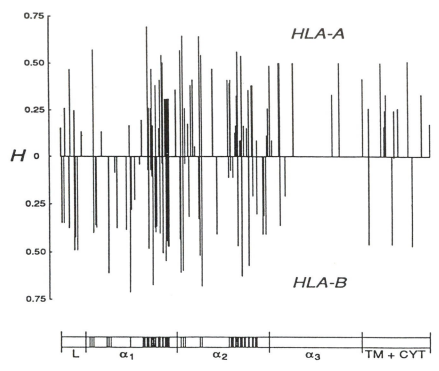

Figure 28-2 The heterozygosity at individual amino acid sites for two human class I MHC loci, *HLA-A* (above) and *HLA-B* (below) (Hedrick et al., 1991). Across the bottom, sites that interact with the antigen and T-cell receptor are indicated and L, α_1, α_2, TM, and CYT refer to different exons.

of magnitude lower than for the antigen binding sites, even though a number of these sites are adjacent to the antigen-binding sites.

Even though there is substantial evidence that variation at the MHC is adaptive in some manner (e.g., Hedrick, 1994a; chapter 14), the actual demonstration of selective differences between MHC variants has been a difficult task. Further, some organisms have very low levels of MHC variation but there are no organisms in which causal relationships between the level of MHC variability and disease resistance or susceptibility have been demonstrated (e.g., chapter 14). For example, cheetahs (see below) have a low level of MHC variation and may have relatively high disease susceptibility, but the demonstration of a direct cause–effect relationship between low MHC variation and disease susceptibility in this species has not been shown.

Another potential class of genes that may be of great adaptive significance are the QTL (quantitative trait loci). Although many quantitative traits over the decades have been shown to have heritable variation, in nearly all instances it was not possible to identify these loci. Because of the current intense gene mapping projects in many organisms using molecular genetic markers, such as RAPDs and microsatellite loci, in the near future it may be possible to identify a number of the major loci that influence quantitative traits. However, even identifying QTLs for a given trait may not tell how the phenotype for that trait is determined because of the complication of intergene interaction or epistasis (see recent potential examples of epistasis; e.g., Lynch and Deng, 1994; Cheverud and Routman, 1995; Lark et al., 1995; Mackay et al., 1995). In addition, the relationship of the phenotype to fitness may be difficult to determine and there may be genotype–environment interaction (e.g., Hedrick, 1985; Falconer, 1989).

ASSOCIATION OF INDIVIDUAL HETEROZYGOSITY AND FITNESS

Although some researchers assume that a positive association of heterozygosity and fitness is a general phenomenon, a number of careful studies do not show clear-cut associations of molecular variation and quantitative traits. For example, some studies have shown positive associations between allozyme heterozygosity among different individuals and various quantitative traits related to fitness, primarily in trees and mollusks, although this finding is far from universal (see citations in Houle, 1989; Pogson and Zouros, 1994, Savolainen and Hedrick, 1995). In trees, the quantitative traits for which a positive association have been found are primarily growth rate and size, although upon close inspection the findings of many of these studies do not appear to be strong and may be caused by statistical association of the allozyme loci with other genes or genotypic association and not by intrinsic heterozygous advantage (Savolainen and Hedrick, 1995).

Savolainen and Hedrick (1995) examined the association of 12 polymorphic allozyme loci and six fitness-related traits in three populations of Scots pine, *Pinus sylvestris*, an organism in which associations with the genetic background are unlikely because of its large population size and extensive gene flow. They found an association of allozyme heterozygosity and these quantitative traits at the 5% level in only 7–8% of the comparisons; the actual significance for the 156 comparisons in the study are plotted in Figure 28-3. However, of the few comparisons that are significant at the 5% level, the heterozygote was superior in half and inferior in the other half. Further, a detailed power analysis showed that if there had been a selective advantage for heterozygotes, it would have been detected in many cases. In addition, analysis of the association of multiple locus heterozygosity and these traits explained 5.8% of the variation in the traits, but only 1.5% of this variation was explained by a positive association between heterozygosity and the quantitative traits.

In a recent, elegant study, Pogson and Zouros (1994) compared the association between multiple locus heterozygosity and height in scallops, *Placopecten magellanicus*, for a group of allozyme loci and a group of DNA markers in the same individuals. They did find a slight positive association for the allozymes (the association explained 3% of the variance in height and was due primarily to one locus) but found no effect for the DNA markers. While this study appears to provide some support the hypothesis that allozyme heterozygotes have higher phenotypic values than homozygotes, because

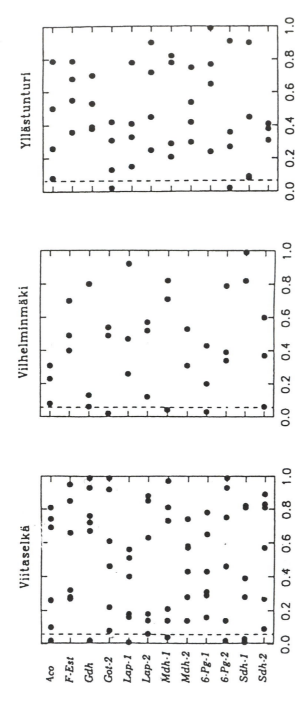

Figure 28-3 The probability and significance level for a difference between heterozygostes and homozygotes for 12 polymorphic allozyme loci in three different populations of Scots pine (Savolainen and Hedrick, 1995). The individual points are the probability levels for a given quantitative trait and the vertical broken lines indicate significance at the 5% level.

only a single locus (or the region around it) had a significant effect, other similar studies need to be carried out to determine the generality of these results.

ASSOCIATION OF POPULATION HETEROZYGOSITY AND FITNESS

Some researchers feel that populations with higher heterozygosity generally have higher fitness. In fact, there is often the concern that populations with low heterozygosity for molecular variants have low fitness and therefore may be much more likely to become extinct (Soulé, 1980, 1986). Although there may be some truth to this dictum, the reality is probably much more complicated for a number of reasons. First, there is a high variation in heterozygosity among populations and much of the range of this variation is probably not strongly related to fitness. Second, different measures of genetic variation may give substantially different values in the same species and it is not generally clear that which measure is more reliable. Perhaps a measure of the variation for fitness-related traits, using either traditional quantitative genetics or molecular analysis of QTLs, may be of more significance than any molecular measure. Third, measurement of fitness is complicated in any species, and in a rare and endangered species it may be doubly difficult. For example, the environment presently inhabited by a endangered species may be a marginal one and the type of genetic variation that gives adaptation to a marginal environment cannot easily be defined.

For many endangered species, avoiding the reduction in fitness when relatives mate, or inbreeding depression, is a high priority (e.g., Hedrick and Miller, 1992). If the level of heterozygosity, here allozyme heterozygosity, could be used to predict inbreeding depression, this would be an important conservation tool. However, scenarios can be constructed in which a population may have any combination of allozyme heterozygosity and inbreeding depression (Table 28-2). These two measures may be positively correlated, as in large populations at equilibrium (both heterozygosity and inbreeding depression high) or a population after a severe bottleneck (both are low).

However, they may also be negatively correlated, as in a population some time after a bottleneck in which there has been time for recovery of quantitative variation that causes inbreeding depression but not enough time for recovery of heterozygosity. The rate of input of additive genetic variance to quantitative traits from mutation is approximately 0.001 per generation (Lande, 1975; however, see

Table 28-2 Situations That May Result in Various Associations of Population Heterozygosity and Inbreeding Depression (Hedrick and Miller, 1992)

Heterozygosity	Inbreeding Depression	Situation
High	High	Large population at equilibrium
Low	Low	Population shortly after severe bottleneck
Low	High	Population some time after severe bottleneck
High	Low	Mixture of populations after severe bottlenecks

Lande, 1995) while mutation rates for allozyme loci are approximately three orders of magnitude lower (e.g., Voelker et al., 1980). Finally, high heterozygosity and low inbreeding depression may occur when two (or more) populations have combined after bottlenecks that reduced inbreeding depression in both but differentially fixed them for allozyme variants. For example, an extensive study in subspecies of the mouse genus *Peromyscus* found little association between levels of allozyme heterozygosity and inbreeding depression (Brewer et al., 1990). Perhaps there may be a tighter correlation between inbreeding depression and heterozygosity as measured by variants with a higher mutation rate, such as microsatellites or minisatellites, but there have been no comprehensive studies to date.

Cheetahs

In the early 1980s, O'Brien and his coworkers published their first studies showing that there was lower genetic variation in the cheetah, *Acinonyx jubatus*, than in most other big cats. They have shown that the cheetah have lower variation in allozymes, proteins, and RFLPs at MHC (O'Brien et al., 1983, 1985, 1987, Yukhi, and O'Brien, 1990) than most other species and populations of big cats. However, for mtDNA, minisatellites, and microsatellites, there is substantial genetic variation in cheetahs (Menotti-Raymond and O'Brien, 1993, 1995). Because the mutation rate for mtDNA and minisatellites is higher than for allozymes, proteins, and MHC loci, Menotti-Raymond and O'Brien (1993) suggested that mtDNA and minisatellites have accumulated variation since a hypothesized bottleneck approximately 10 000–12 000 years ago.

Pimm et al. (1989) and Gilpin (1991) suggested that the low level of genetic variation in cheetahs may not be the result of the one (or two) bottlenecks that the whole cheetah population underwent simultaneously as originally hypothesized by O'Brien but that metapopulation dynamics, extinction and recolonization of patches by a few individuals, could result in low genetic variation (a mechanism acknowledged by Menotti-Raymond and O'Brien, 1993). Hedrick and Gilpin (1996) have shown quantitatively that the effective population size of a metapopulation may be quite small even though the census number is always large.

If it is assumed that the effective population size of cheetahs is indeed lower than that of other big cats owing to metapopulation dynamics, then the equilibrium level of genetic variation for variants that have a high mutation rate, such as mtDNA, minisatellites, and microsatellites, would be expected to be much higher than for other variants with a low mutation rate, such as allozymes, proteins, and RFLPs at MHC (equilibrium heterozygosity is a function of the product of the mutation rate and the effective population size under the neutrality model; e.g., Nei, 1987). In other words, an alternative explanation to the bottleneck scenario is that the higher levels of variation in mtDNA, minisatellites, and microsatellites may not be due to a recovery of variation since the hypothesized bottleneck but may be just the level of standing variation expected for genetic variants with a high mutation rate (Hedrick, 1996).

Cheetahs also have low levels of normal sperm and appear to be susceptible to infectious diseases. Even though it is tempting to suggest that the low genetic variation for some types of genetic variation may cause some of the fitness problems in cheetahs, there is no definitive evidence to support such a cause–effect relationship between these observations (e.g., Caro and Laurenson, 1994; Caughley, 1994; Merola, 1994). Further, because of the conclusions of O'Brien et al., 1985) suggesting that inbred and outbred offspring had similar mortality rates, it was assumed that cheetahs were already highly inbred and unfit because of high inbreeding. However, if the data of O'Brien et al. (1985) are examined with proper statistics, there is evidence that inbred offspring had statistically significantly higher mortality than outbred offspring, a finding not consistent for a population lacking genetic variation for fitness (Hedrick, 1987, see also Caughley, 1994). A later study (Marker and O'Brien, 1989) suggested that there was no evidence for inbreeding depression, but because of differences of husbandry in different zoos and changes in husbandry over time, it is difficult to evaluate these findings (e.g., Caro and Laurenson, 1994).

The elegant molecular studies of O'Brien and his colleagues are a landmark for studies in other endangered species. It does appear that the value of these measures of genetic variation has been oversold, a fact almost admitted by O'Brien (1994). However, as suggested by O'Brien (1994), if we ignore the potential for high disease susceptibility in cheetahs, which may be of genetic origin, then we run the risk of further endangering a species that is already in trouble in many areas. Although it appears that lions may be the chief cause of mortality for cheetahs in the wild, and husbandry problems may be of great significance in captivity (e.g., Caro and Laurenson, 1994), not attempting to incorporate the genetic, reproductive, and disease susceptibility data from O'Brien and his colleagues into a comprehensive picture of the biology of cheetahs is also a gamble.

Cheetahs are one of the most studied of all endangered species from a molecular genetics viewpoint. This detailed information may be of great benefit in making some future conservation decisions and it is not sensible to ignore this information just because some aspects of the cheetah story have been questioned. For example, the relatively high level of minisatellite and microsatellite variation may suggest that cheetahs might have substantial variation for fitness components and other quantitative traits which also have relatively high mutational levels.

Gila Topminnow

Another example of an association of population heterozygosity and fitness in an endangered species that had substantial impact on policy decisions is the research of Vrijenhoek and his colleagues of the Gila topminnow, *Poeciliopsis occidentalis occidentalis*. For example, Quattro and Vrijenhoek (1989) examined four traits related to fitness—growth rate, survival, brood size, and fluctuating asymmetry (the association of the latter trait with fitness is not clear; Hedrick, 1994b; however, see Clarke, 1995)—in two different populations from Arizona. One population, Monkey Spring, was not polymorphic at 25 allozyme loci and the other population, Sharp Spring, was polymorphic at 2 of 25 loci (Vrijenhoek et al., 1985; Meffe and Vrijenhoek, 1988). In their laboratory study, Quattro and Vrijenhoek (1989) found that the more heterozygous population had higher brood size, survival rate, and growth rate and less asymmetry.

Recently, Sheffer et al. (1996) have examined these same two populations and

Table 28-3   Allozyme Heterozygosity and Measures of Fecundity (standard errors in parentheses) in Four Populations of the Endangered Gila Topminnow, *Poecilopsis occidentalis occidentalis*

Trait	Monkey Spring	Sharp Spring	Bylas Spring	Cienega Creek
Heterozygosity[a]	0.000	0.037	0.000	0.000
Egg number[b]	5.8 (1.5)	13.3 (1.1)	–	–
Brood size (wild-caught)[c]	12.1 (1.3)	15.8 (1.2)	12.8 (1.1)	17.8 (1.1)
Brood size (captive-reared)[c]	6.0 (0.7)	5.5 (0.7)	6.1 (0.8)	6.6 (0.6)

[a] Vrijenhoek et al. (1985).
[b] Quattro and Vrijenhoek (1989).
[c] Sheffer et al. (1966)

populations from two other main sites for the species in the United States, both of which similarly had no heterozygosity at the 25 loci (Vrijenhoek et al., 1985; Meffe and Vrijenhoek, 1988). In the laboratory studies of Sheffer et al. (1996), there were no differences in survival and asymmetry among the four populations and the population with the highest growth rate was not Sharp Spring, the polymorphic population. Additionally, the 12-week survival was above 90%, while the survival in Quattro and Vrijenhoek (1989) was only approximately 50%. It appears that the laboratory environment of Quattro and Vrijenhoek (1989) was in some way stressful to the fish and resulted in lower survival overall.

Quattro and Vrijenhoek (1989) also found that egg number was higher in females from the heterozygous population, Sharp Spring (Table 28-3). Similarly, Sheffer et al. (1996) found the brood size of wild-caught females from Sharp Spring higher than that from Monkey Spring. However, the brood size for a Cienega Creek sample, which also had no heterozygous loci in the allozyme sample, was the highest of all the four populations. When the female progeny of the wild-caught females were bred in the laboratory, the brood sizes of the four samples were not significantly different (Table 28-3). In other words, in the study of Sheffer et al. (1996) in an apparently less stressful environment than that of Quattro and Vrijenhoek (1989), there was no association between population heterozygosity, as measured by allozymes, and fitness-related traits.

USFWS in their restocking program for the Gila topminnow initially used a stock from Monkey Spring (e.g., Hendrickson and Brooks, 1991). When the studies suggesting that Monkey Spring had both low genetic variation and low values for the four fitness correlates were published, USFWS abandoned this stock and established one from the Sharp Spring population, the population with 2 of 25 loci polymorphic and higher fitness-correlate values. Thus, even though Quattro and Vrijenhoek (1989) made a number of disclaimers in their paper, their study was taken as support for changing the source population for the stocking program (Hendrickson and Brooks, 1991). As stated by Quattro and Vrijenhoek (1989) and recommended by the Draft Gila Topminnow Recovery Plan (USFWS, 1993) and Williams et al. (1988), it is preferable that source populations from near the site for restocking be used. In this case, the genetic and fitness data, the latter of which now seem to hold only in a specific and apparently stressful environment, were used as a basis for changing the source of the population used for restocking.

Northern Elephant Seals

One of the classic studies of molecular variation in an endangered species is that by Bonnell and Selander (1974), which showed no variation at 20 allozyme loci in the northern elephant seal, *Mirounga angustrirostris*. This observation is consistent with the severe historical bottleneck for this species when it was nearly hunted to extinction at the end of the nineteenth century. Hoezel et al. (1993) (see also chapter 21) examined mtDNA diversity in both northern and southern elephant seals, which are not thought to have gone through a similar bottleneck, and estimated the duration and size of the bottleneck in the northern elephant seals using the southern elephant seal data as a pre-bottleneck value. Both their simulations and the population genetic analysis of Hedrick (1995b) using the mtDNA data were consistent with an approximate bottleneck duration of one or two generations and 10–20 individuals.

However, as pointed out by Hedrick (1995b), a bottleneck of this size and duration cannot, by itself, account for the much lower allozyme heterozygosity in the northern elephant seals than in the southern elephant seals. In other words, the genetic variation for mtDNA and allozymes leads to quite different interpretations of past historical events influencing genetic variation, a difference for which there is no ready explanation. Often in an endangered species there is no comparison with genetic variation in the same species before it became endangered (there are some exceptions in which museum specimens can be used to estimate historical genetic variation). Obviously, to understand the basis of the genetic variation in the northern elephant seal, an examination of museum specimens taken before the bottleneck (and there should be plenty because collecting by museums was one of the major causes of the final crash of the northern elephant seals) could give an explanation.

Northern elephant seals are a species that has rebounded from near extinction to over 100 000 individuals today. Obviously, hunting was the major factor causing the crash and banning of hunting appears to have been just as effective in resulting in a rapid increase in the species. In this case, it would be useful to determine the level of variation for traits related to fitness, for example, MHC loci and some quantitative traits related to fitness, to determine whether northern elephant seals are indeed genetically depauperate for adaptive variants.

CONCLUSIONS

For most endangered species there is some environmental problem, such as lack of adequate habitat or protection, that is a critical issue for the survival of the species. In other words, molecular information may not be as critical for the immediate survival of the species as improving the habitat (e.g., Caughley, 1994; see also Hedrick et al., 1996). However, it behooves us as scientists to provide as up-to-date information as possible and also to realize that what appears to be concrete natural history, ecological, or morphological data should also be carefully scrutinized. For example, behavioral studies suggested that the rate of mating among sibs in the splendid fairy wren may be the highest in any vertebrate

(Ralls et al., 1986). However, molecular markers have now shown that there are few, if any, matings between sibs and that the behavioral observations suggesting sib matings were incorrect (Rowley et al., 1993). We should make recommendations based on the best information available, and molecular data can be an important step forward from strictly observational information.

It is tempting to use molecular information that is state-of-the-art when strong recommendations are urged by the managers involved, but we need to be cautious because the underlying genetics or population genetics of some of the new measures may not be completely understood or they may be telling us about only a small part of the genome, as for mtDNA. Wildlife managers often want a black or white answer and it is tempting to give them one if the molecular results appear conclusive. However, if the value of the results are overstated, then they may come back to haunt us. One lesson that should be apparent from the discussion here is that it is important to understand the limitations of the data. For example, confidence limits should be given for the data and the assumptions that go into these confidence limits should be explained.

A number of researchers over the years have suggested that breeding programs be changed to make sure that new alleles discovered by molecular techniques are preserved. The general response to such proposals has been to use pedigree and other information first and only then to cautiously utilize such single-gene information (e.g., Hedrick et al., 1986). However, Hughes (1991), in a rather simplistic proposal, suggested, because variation at the MHC genes appear to be maintained by heterozygote advantage, that captive breeding programs be reoriented to save any MHC "allele declining in frequency by drift... by selective breeding." This suggestion resulted in several critical replies (Gilpin and Wills, 1991; Miller and Hedrick, 1991; Vrijenhoek and Leberg, 1991) which pointed out, among other things, that other loci may be of adaptive significance and that some MHC alleles may be neutral or even detrimental. Further, selecting for particular alleles in a pedigreed population may influence the overall genetic variation (Hedrick and Miller, 1994), although the impact of such a program is somewhat mitigated if current recommended breeding protocols are employed (Miller, 1995).

It is useful to have more than one independent researcher working on a given organism or organisms even though territoriality in science is often a difficult problem. Although this recommendation may seem like duplication of effort, different approaches or perspectives may give different insights or interpretations. In addition, researchers, like any other individuals, may have vested interests in their research and interpret any new results to be consistent with their previous results. Unlike the case with models organisms, such as *Drosophila melanogaster* or house mice, the results of research on endangered species are unlikely to be thoroughly examined, and if the researchers have a conscious or unconscious bias then this may never come to light.

It is important to attempt to generate guidelines for conservation genetics research. In fact, the 50/500 recommendations that were used in the 1980 are not completely invalid, but they must be evaluated in the context of all the information for each species. Likewise, molecular genetics information should also be utilized, but only in the context of understanding the particular molecular technique or techniques, population genetic effects on them, and the knowledge that adaptive variation may not show the same patterns. It is useful to have information from several molecular techniques if possible, but a real problem occurs if different approaches give different results. If there is detailed information about the techniques and the particular species, then perhaps such disparate information can be combined or evaluated in some logical way.

As discussed above, it is important to try to determine what genetic variants are adaptively important. In addition, it is important to identify loci that have variants that may cause low fitness. For example, it is likely that the low fitness in Florida panther males (high levels of cryptorchidism and low sperm quality) are genetically based (Roelke et al., 1993). There is at present a program to introduce Texas cougars into Florida to restore the fitness of the Florida panther (Seal, 1994; Hedrick, 1995c), but it would be very useful to identify genes that have caused problems in male reproductive fitness, maybe using the mapping studies and segregation in house cats or other mammals.

Perhaps the perspective given by Lewontin (1991) on allozyme variation is important to keep in mind. Allozymes revolutionized evolutionary genetics and allowed researchers to investigate organisms genetically that were previously inaccessible. However, Lewontin (1991) suggests that it is not clear how much progress in understanding the genetic basis of evolutionary change and adaptation was accomplished from the hundreds of studies documenting allozymic variation in various species. Hopefully, the allure of DNA sequences will not divert all of our energy from studies to determine the adaptive significance of genetic variation. My fear is that the refrain, "the answer is in the DNA," may result in conservation and evolutionary geneticists becoming technicians and not creative, ingenious scientists. Only time will tell whether we will keep a balanced enough perspective to gain insight into basic evolutionary processes and successfully translate this information into reasonable guidelines for conservation.

REFERENCES

Ashley, M. and B.D. Dow. 1994. The use of microsatellite analysis in population biology: background, methods and potential applications. In: B. Sherwater, B, Streit, G.P. Wagner, and R. DeSalle, eds. Molecular ecology and evolution: Approaches and applications, pp. 185–201. Basel: Birkhauser.

Avise, J. 1994. Molecular markers, natural history, and evolution. London: Chapman and Hall.

Bonnell, M.L. and R.K. Selander. 1974. Elephant seals: genetic variation and near extinction. Science 184: 908–909.

Brewer, B.A., R.C. Lacy, M.L. Foster, and G. Alaks. 1990. Inbreeding depression in insular and central populations of *Peromyscus* mice. J. Hered. 81: 257–264.

Caro, T.M. and Laurenson. 1994. Ecological and genetic factors in conservation: a cautionary tale. Science 263: 485–486.

Caughley, G. 1994. Directions in conservation biology. J. Animal Ecol. 63: 215–244.

Chakraborty, R. 1981. The distribution of the number of heterozygous loci in an individual in natural populations. Genet. 98: 461–466.

Cheverud, J.M. and E.J. Routman. 1995. Epistasis and its contribution to genetic variance components. Genetics 139: 1455–1461.

Clarke, G.M. 1995. Relationship between developmental stability and fitness: application for conservation biology. Conserv. Biol. 9: 18–24.

Falconer, D.S. 1989. Introduction to quantitative genetics. 3d ed. London: Longman.

García-Moreno, J., M.S. Roy, E. Geffen, and R.K. Wayne. 1996. Relationships and genetic purity of the endangered Mexican wolf based on analysis of microsatellite loci. Conserv. Biol. [In press].

Gilpin, M. 1991. The genetic effective size of a metapopulation. Biol. J. Linn. Soc. 42: 165–175.

Gilpin, M. and C. Wills. 1991. MHC and captive breeding: a rebuttal. Conserv. Biol. 5: 554–555.

Goldstein, D.B., A.R. Linares, L.L. Cavalli-Sforza, and M.W. Feldman. 1995. An evaluation of genetic distances for use with microsatellite loci. Genetics 139: 463–471.

Hedrick, P. W. 1985. Genetics of Populations. Boston: Jones and Bartlett.

Hedrick, P.W. 1987. Genetic bottlenecks. Science 237: 963.

Hedrick, P.W. 1994a. Evolutionary genetics of the major histocompatibility complex. Am. Nat. 143: 945–964.

Hedrick, P.W. 1994b. Summary for population biology and conservation. In: T. Markow, ed. Developmental instability: its origin and evolutionary implications, pp. 434–436. Dordrecht: Kluwer Academic.

Hedrick, P.W. 1995a. Genetic evaluation of the three captive Mexican wolf lineages and consequent recommendations. Report of the Genetics Committee of the Mexican Wolf Recovery Team, United States Fish and Wildlife Service, Albuquerque, N.M.

Hedrick, P.W. 1995b. Elephant seals and the estimation of a population bottleneck. J. Hered. 86: 232–235.

Hedrick, P.W. 1995c. Gene flow and genetic restoration: the Florida panther as a case study. Conserv. Biol. 9: 996–1007.

Hedrick, P.W. 1996. Bottlenecks or metapopulation in cheetahs. Conserv. Biol. [In press].

Hedrick, P.W. and M. Gilpin. 1996. Metapopulation genetics: effective population size. In: I. Hanski and M. Gilpin, eds. Metapopulation dynamics: Ecology, genetic and evolution. New York: Academic Press. [In press].

Hedrick, P.W. and P.S. Miller. 1992. Conservation genetics: techniques and fundamentals. Ecol. Appl. 2: 30–46.

Hedrick, P.W. and P.S. Miller. 1994. Rare alleles, MHC and captive breeding. In: V. Loeschcke, J. Tomiuk, and K. Jain, eds. Conservation genetics, pp. 187–204. Basel: Birkhauser.

Hedrick, P.W. and E. Murray. 1983. Selection and measures of fitness. In: J. Thompson and M. Ashburner, eds. Genetics and biology of drosophila, Volume 3d, pp. 61–104. New York: Academic Press.

Hedrick, P.W., P.F. Brussard, F.W. Allendorf, J.A. Beardmore, and S. Orzack. 1986. Protein variation, fitness and captive propagation. Zoo Biol. 5: 91–99.

Hedrick, P.W., T.S. Whittam, and P. Parham. 1991. Heterozygosity at individual amino acid sites: extremely high levels for *HLA-A* and *-B* genes. Proc. Natl. Acad. Sci. U.S.A. 88: 5897–5901.

Hedrick, P.W., R.C. Lacy, F.W. Allendorf, and M. Soulé. 1996. Directions in conservation biology: comments on Caughley. Conserv. Biol. [In press].

Hendrickson, D.A. and J.E. Brooks, 1991. Transplanting short-lived fishes in North American deserts: review, assessment, and recommendation. In: W. Minckley and J. Deacon, eds. Battle against extinction, pp. 283–301. Tucson: University of Arizona Press.

Hoezel, A.R., J. Halley, S.J. O'Brien, C. Campaga, T. Arborn, B. Le Boeuf, K. Ralls, and G.A. Dover. 1993. Elephant seal genetic variation and the use of simulation models to investigate historical population bottlenecks. J. Hered. 84: 443–449.

Houle, D. 1989. Allozyme-associated heterosis in *Drosophila melanogaster*. Genetics 123: 789–801.

Hughes, A. 1991. MHC polymorphism and the design of captive breeding programs. Conserv. Biol. 5: 249–251.

Kimura, M. 1979. The neutral theory of molecular evolution. Sci. Am. 241(5): 94–104.

Kreitman, M. and H. Askashi. 1995. Molecular evidence for natural selection. Annu. Rev. Ecol. Syst. 26: 403–422.

Lande, R. 1975. The maintenance of genetic variability by mutation in a polygenic character with linked loci. Genet. Res. 26: 221–235.

Lande, R. 1988, Genetics and demography in biological conservation. Science 241: 1455–1460.

Lande, R. 1995. Mutation and conservation. Conserv. Biol. 9: 782–791.

Lande, R. and G.R. Barrowclough. 1987. Effective population size, genetic variation, and their use in population management. In: M. E. Soulé, ed. Viable populations for conservation, pp. 87–123. Cambridge, U.K.: Cambridge University Press.

Lark, K.G., K. Chase, F. Adler, L.M. Mansur, and J.H. Orf. 1995. Interactions between quantitative trait loci in soybean in which trait variation at one locus is conditional upon a specific allele at another. Proc. Natl. Acad. Sci. U.S.A. 92: 4656–4660.

Lewontin, R.C., 1991. Twenty-five years ago in genetics: electrophoresis in the development of evolutionary genetics: milestone or millstone? Genetics 128: 657–662.

Lewontin, R.C. and J.L. Hubby. 1966. A molecular approach to the study of genetic heterozygosity in natural populations. II. Amount of variation and degree of heterozygosity in natural populations of *Drosophila pseudoobscura*. Genetics 54: 595–609.

Lynch, M. and H.-W. Deng. 1994. Genetic slippage in response to sex. Am. Nat. 144: 242–261.

Lynch, M., J. Conery, and R. Burger. 1995a. Mutational meltdowns in sexual populations. Evolution 49: 1067–1080.

Lynch, M., J. Conery, and R. Burger. 1995b. Mutational accumulation and the extinction of small populations. Am. Nat. 146: 489–518.

Mackay, T.F.C., R.F. Lyman, and W.G. Hill. 1995. Polygenic mutation in *Drosophila melanogaster*: non-linear divergence among unselected strains. Genetics 139: 849–859.

Marker, L. and S.J. O'Brien. 1989. Captive breeding of the cheetah (*Acinonyx jubatus*) in North American zoos (1871–1986). Zoo Biol. 8: 3–16.

Meffe, G.K. and R.C. Vrijenhoek. 1988. Conservation genetics in the management of desert fishes. Conserv. Biol. 2: 157–169.

Menotti-Raymond, M. and S.J. O'Brien. 1993. Dating the genetic bottleneck of the African cheetah. Proc. Natl. Acad. Sci. U.S.A. 90: 3172–3176.

Menotti-Raymond, M. and S.J. O'Brien. 1995. Evolutionary conservation of ten microtellite loci in four species of Felidae. J. Hered. 86: 319–322.

Merola, M. 1994. A reassessment of homozygosity and the case for inbreeding depression in the cheetah *Acinonyx jubatus*: implications for conservation. Conserv. Biol. 8: 961–971.

Miller, P.S. 1995. Evaluating selective breeding programs for rare alleles: examples using the Prezwalski's horse and California condor pedigrees. Zoo Biol. 9: 1262–1273.

Miller, P.S. and P.W. Hedrick. 1991. MHC polymorphism and the design of captive breeding programs: simple solutions are not the answer. Conserv. Biol. 5: 556–558.

Nei, M. 1987. Molecular evolutionary genetics. New York: Columbia University Press.

Nei, M. and A. Hughes. 1991. Polymorphism and evolution of the major histocompatibility complex loci in mammals. In: R. Selander, A. Clark, and T. Whittam, eds. Evolution at the molecular level. pp. 222–247. Sunderland, Mass.: Sinauer.

O'Brien, S.J. 1994. The cheetah's conservation controversy. Conserv. Biol. 8: 1153–1155.

O'Brien, S.J., D.E. Wildt, D.Goldman, C.R. Merril, and M. Bush. 1983. The cheetah is depauperate in genetic variation. Science 221: 459–462.

O'Brien, S.J., M.E. Roelke, L. Marker, A. Newman, C.A. Winkler, D. Meltzer, L. Colly, J.F. Evermann, M. Bush, and D.E. Wildt. 1985. Genetic basis for species vulnerability in the cheetah. Science 227: 1428–1434.

O'Brien, S.J., D.E. Wildt, M. Bush, T.M. Caro, C. FitzGibbon, I. Aggundey, and R.E. Leakey. 1987. East African cheetahs: evidence for two population bottlenecks. Proc. Natl. Acad. Sci. U.S.A. 84: 508–511.

Pimm, S.L., J.L. Gittlemam, G.F. McCracken, and M. Gilpin. 1989. Plausible alternatives to bottlenecks to explain reduced genetic diversity. Trends Ecol. Evol. 4: 176–178.

Pogson G.H. and E. Zouros. 1994. Allozyme and RFLP heterozygosities as correlates of growth rate in the scallop *Placopecten magellanicus*: a test of the associative overdominance hypothesis. Genetics 137: 22–231.

Quattro, J.M. and R.C., Vrijenhoek. 1989. Fitness differences among remnant populations of the endangered Sonoran topminnow. Science 245: 976–978.

Ralls, K., P.H. Harvey, and A.M. Lyles. 1986. Inbreding in natural populations of birds and mammals. In: M.E. Soulé, ed. Conservation biology: The science of scarcity and diversity, pp. 35–56. Sunderland, Mass.: Sinauer.

Roelke, M.E., J.S, Martenson, and S.J. O'Brien. 1993. The consequences of demographic reduction and genetic depletion in the endangered Florida panther. Curr. Biol. 3: 340–350.

Rowley, I., E. Russell, and M. Brooker. 1993. Inbreeding in birds. In: N. Thornhill, ed. The natural history of inbreeding and outbreeding, pp. 304–328. Chicago.: University of Chicago Press.

Savolainen, O. and P.W. Hedrick. 1995. Heterozygosity and fitness: no association in Scots pine. Genetics 140: 755–766.

Seal, U.S. 1994. A plan for the genetic restoration and management of the Florida panther (*Felis concolor coryi*). Report to the U.S. Fish and Wildlife Service, Captive Breeding Specialist Group, SSC/IUCN. Apple Valley, Minn.

Shaffer, M.L. 1981. Minimum population sizes for species conservation. BioScience 31: 131–134.

Sheffer, R., P.W. Hedrick, W.L. Minckley, and A.L. Velasko. 1996. Fitness in the endangered Gila topminnow. Conser. Biol. [in press].

Sherwater, B., B. Streit, G.P. Wagner, and R. DeSalle, ed. 1994. Molecular ecology and evolution: approaches and applications. Basel: Birkhauser.

Shields, W.M., A. Templeton, and S. Davis. 1987. Genetic assessment of the current captive breeding program for the Mexican wolf (*Canis lupus baileyi*). Final Contract Report # 516.6-73-13. New Mexico Department of Game and Fish, Santa Fe, N.M.

Simberloff, D. 1988. The contribution of population and community biology to conservation science. Annu. Rev. Ecol. Syst. 19: 473–511.

Soulé, M.E. 1980. Thresholds for survival: maintaining fitness and evolutionary potential. In: M.E. Soulé and B.A. Wilcox, eds. Conservation biology, pp. 151–168. Sunderland, Mass.: Sinauer.

Soulé, M.E., ed. 1986. Conservation biology: The science of scarcity and diversity. Sunderland, Mass.: Sinauer.

USFWS, (U.S. Fish and Wildlife Service). 1993. Draft Gila topminnow recovery plan. Albuquerque, N.M. U.S. Fish and Wildlife Service.

Voelker, R.A., H.E. Schaffer, and T. Mukai. 1980. Spontaneous allozyme variation in *Drosphila melanogaster*: rate of occurrence and nature of the mutants. Genetics 94: 961–968.

Vrijenhoek, R.C. and P.L. Leberg. 1991. Let's not throw the baby out with the bathwater: a comment on management for MHC diversity in captive populations. Conserv. Biol. 5: 252–254.

Vrijenhoek, R.C., M.E. Douglas, and G.K. Meffe. 1985. Conservation genetics of endangered fish populations in Arizona. Science 229: 400–402.

Wayne, R.K., N. Lehman, M.W. Allard, and R.L. Honeycutt. 1992. Mitochondrial DNA variability of the gray wolf: genetic consequences of population decline and habitat fragmentation. Conserv. Biol. 6: 559–569.

Williams, J.E., D.W. Sada, and C.D. Williams. 1988. American Fisheries Society guidelines for introduction of threatened and endangered fishes. Fisheries 13: 5–11.

Yuhki, N. and S.J. O'Brien. 1990. DNA variation of the mammalian major histo-compatibility complex reflects genomic diveristy and population history. Proc. Natl. Acad. Sci. U.S.A. 87: 836–840.

Index